Game Theory

Ana Espinola-Arredondo ·
Felix Muñoz-Garcia

Game Theory

An Introduction with Step-by-Step Examples

Ana Espinola-Arredondo
School of Economic Sciences
Washington State University
Pullman, WA, USA

Felix Muñoz-Garcia
School of Economic Sciences
Washington State University
Pullman, WA, USA

ISBN 978-3-031-37576-7   ISBN 978-3-031-37574-3 (eBook)
https://doi.org/10.1007/978-3-031-37574-3

© The Editor(s) (if applicable) and The Author(s), under exclusive license to Springer Nature Switzerland AG 2023

This work is subject to copyright. All rights are solely and exclusively licensed by the Publisher, whether the whole or part of the material is concerned, specifically the rights of translation, reprinting, reuse of illustrations, recitation, broadcasting, reproduction on microfilms or in any other physical way, and transmission or information storage and retrieval, electronic adaptation, computer software, or by similar or dissimilar methodology now known or hereafter developed.
The use of general descriptive names, registered names, trademarks, service marks, etc. in this publication does not imply, even in the absence of a specific statement, that such names are exempt from the relevant protective laws and regulations and therefore free for general use.
The publisher, the authors, and the editors are safe to assume that the advice and information in this book are believed to be true and accurate at the date of publication. Neither the publisher nor the authors or the editors give a warranty, expressed or implied, with respect to the material contained herein or for any errors or omissions that may have been made. The publisher remains neutral with regard to jurisdictional claims in published maps and institutional affiliations.

Cover credit: © lecosta/Shutterstock

This Palgrave Macmillan imprint is published by the registered company Springer Nature Switzerland AG
The registered company address is: Gewerbestrasse 11, 6330 Cham, Switzerland

Preface

This textbook offers an introduction to Game Theory for undergraduate students, although some of the advanced topics can also be used in graduate-level courses. Our presentation differs from current Game Theory textbooks—such as Gibbons (1992), Osborne (2003), Rasmusen (2006), Harrington (2014), Dixit et al. (2015), or Dutta and Vergote (2022)—along several dimensions:

- *Tools.* We provide step-by-step "tools" or "recipes" on how to solve different classes of games, helping students apply a common approach to solve similar games on their own.
- *Algebra support and step-by-step examples.* We do not require a strong mathematical background in algebra and calculus. Instead, we walk readers through each algebra step and simplification. From our recent experience, providing algebra steps helps students more easily follow the material, ultimately facilitating them reproduce all the results on their own.
- *Worked-out examples.* Every chapter starts providing the main definitions, but then focuses its attention to the application of every solution concept to different settings in economics, business, and other social sciences.
- *Length.* The book is significantly shorter than many current books in this topic, such as Harrington (2014) or Dixit et al. (2015). Providing shorter explanations, we seek to make the material more accessible to students, who can read each chapter in approximately one hour.
- *Writing style.* The book's style is relaxed, avoids mathematical notation when possible, and should be easy to read for undergraduate students. Its style and step-by-step examples should, then, make it more accessible than Osborne and Rubinstein (1994), Osborne (2003), or Rasmusen (2006), which assume readers have a strong math background. Yet, the presentation is sufficiently rigorous and advanced to make it appropriate for courses at the upper-undergraduate and Masters-level. Overall, the topics we cover and our style are closest to those in Watson (2013) and Tadelis (2013), but our step-by-step approach and "tools" should make the presentation and examples easier to follow by undergraduate students.
- *Incomplete information emphasis.* Six chapters cover games of incomplete information (Chapters 8–13), which are often presented in just 2–3 chapters in most

textbooks. Particular attention is given to sequential-move games of incomplete information, signaling games with discrete and continuous actions, equilibrium refinements, and cheap-talk games in Chapters 10–13 which is generally covered in only 1–2 chapters.
- *Guided Exercises with Answers.* An accompanying website provides detailed answer keys to 1–2 exercises per chapter. In addition, it offers step-by-step explanations and emphasizes the economic intuition behind the mathematical results. The combination of textbook and website seeks to help students improve both their theoretical and practical preparation.

Therefore, we expect the book to be especially useful for students in programs in economics, business administration, finance, or related fields in social sciences. Given its step-by-step approach to examples and intuition, it should be appropriate for Game Theory courses, both at the undergraduate and graduate level.

Organization of the Book

Chapter 1 defines Game Theory, presents the main ingredients in a game and its graphical representation, discusses the solution concepts that help us identify equilibrium behavior in a game, and how to assess these solution concepts according to different criteria.

Chapters 2–5 study simultaneous-move games of complete information. Specifically, Chapter 2 presents our first solution concept (dominance), where players rule out strategies that yield an unambiguously lower payoff regardless of what her opponent choose. We apply this solution concept to standard games, and study the notion of weakly dominated strategies and strictly dominant strategies.

While dominance helps us rule out strategies that a rational player would never use, it may provide us with little bit in many games, or no bite at all in other games. To help provide more precise equilibrium predictions, Chapter 3 presents the notion of best response, where a player seeks to maximize her payoff given her opponents' strategies. Using best responses, we discuss what we mean when we say that a player's strategy is "never a best response," its associated concept of "rationalizable" strategies, and the Nash equilibrium (NE) solution concept, where every player chooses a mutual best response to her opponents' strategies). Chapter 4, then, applies the concept of best response and NE to games where players choose from a continuous strategy space, such as oligopoly models of quantity competition, price competition, public good games, or electoral competition. They all help us illustrate how to find best responses and NEs in typical games in economics and social sciences, identifying how equilibrium predictions differ from socially optimal outcomes.

Chapter 5 studies settings where players have incentives to randomize their strategies (mixed strategies), as opposed to choosing one of their strategies with certainty (pure strategies).

Using the mixed strategy Nash equilibrium concept, we explore strictly competitive games, present security strategies, correlated equilibrium, and equilibrium refinements in simultaneousmove games (such as the trembling-hand perfect equilibrium and the proper equilibrium). In Chapters 6 and 7, we switch our attention to sequential-move games of complete information. In particular, Chapter 6 first shows that using the NE solution concept in this class of games may yield outcomes that are based on incredible beliefs. Then, it presents the subgame perfect equilibrium (SPE) solution concept, which identifies equilibria where players are "sequentially rational," meaning that they maximize their payoffs at every section of the game tree when they are called to move. Chapter 7 continues our study of sequential-move games, but now allowing for games to be repeated, either finite or infinite times, and focusing on whether more cooperative outcomes can be sustained as a SPE than when the game is unrepeated.

Chapters 8 and 9 consider simultaneous-move games of complete information. In Chapter 8, we analyze the Bayesian Nash equilibrium (BNE) solution concept, and different tools to find BNEs in games where one or more players are privately informed. Chapter 9 applies this solution concept to auctions—a strategic setting where players are privately informed about their valuations for the object on sale. We examine different auction formats, such as the first-price, second-price, and common-value auctions, how to find equilibrium bids (the BNE of this game), and the seller's expected revenue from each auction format.

Chapters 10–13 focus on sequential-move games of incomplete information. First, Chapter 10 presents the Perfect Bayesian Equilibrium (PBE) solution concept, and why using the BNE from Chapter 8 may lead to outcomes based on incredible beliefs. This solution concept is particularly useful to study signaling games, which refer to settings where the actions of the first mover may convey or conceal her private information to other players. We, then, assess how our results are affected if we allow for two or more responses, messages, and sender types.

While signaling games can be readily analyzed with the PBE solution concept, some of these games often have a large number of PBEs. More importantly, some of these PBEs are based on insensible beliefs "off-the-equilibrium" path (i.e., at sections of the game tree which are not reached in equilibrium). Chapter 11 discusses two "refinement criteria", the Cho and Kreps' (1987) Intuitive Criterion and Banks and Sobel's (1987) D1 Criterion; and a solution concept, sequential equilibrium, which helps us only identify PBEs that are based on sensible off-the-equilibrium beliefs.

Chapters 12 and 13 apply the PBE solution concept to two common classes of games, namely, signaling games with continuous messages and cheap-talk games. In Chapter 12, we consider Spence's (1973) labor market signaling game, where a worker chooses her education level (among a continuum) and the firm, without observing the worker's innate productivity, but observing her education level, responds with a wage offer. For illustration purposes, we apply the Intuitive and D1 Criteria in this setting, and study how the equilibrium outcome is affected when the worker has more than two types. Finally, Chapter 13 examines a special class of signaling games where messages are costless, such as the message that a lobbyist sends to an uninformed politician (cheap talk).

How to Use This Textbook

The writing style of the textbook allows for flexible uses by instructors of the following courses:

- **Game Theory (undergraduate).** This book probably fits best with this type of course, with instructors recommending most chapters in the book as the main reading reference for students (Chapters 1–10). In addition, instructors could assign the reading of specific exercises from the *Guided Exercises* website, which should help students better understand the application of solution concepts to different games, ultimately allowing them to become better prepared for homework assignments and exams.
- **Game Theory (graduate).** The book is also appropriate for instructors teaching upper-undergraduate or Masters-level courses in Game Theory. In this case, instructors could assign the reading of the first chapters in the book (Chapters 1–7, covering games of complete information), perhaps briefly discussing them in class; focusing, instead, on the remaining chapters in the book (Chapters 8–13). These chapters cover games of incomplete information and equilibrium refinements and are often more challenging for undergraduate students, particularly the last few chapters on signaling games with continuous action spaces, equilibrium refinement criteria, and cheap talk games.
- **Economics of Information (undergraduate).** Instructors teaching this type of course often start by covering game of incomplete information to provide the foundations of Bayesian Nash Equilibrium, Perfect Bayesian Equilibrium, and Sequential Equilibrium; as these tools are frequently used to identify equilibrium behavior in contexts where at least one player enjoys an information advantage. As a consequence, instructors in this course could assign the reading of the chapters studying games of incomplete information (such as Chapters 8–12).

The length of this book also can be attractive to instructors of online version of the abovementioned courses, as students can easily read on their own the main theory, step-by-step examples, and the guided exercises.

Ancillary Materials

- *Guided exercises with Answers.* An accompanying website includes step-by-step answer keys with intuitive explanations to 1–2 exercises per chapter. This can be a useful practice for students, as they can see the common approach that we follow to solve similar exercises.

- To access Guided Exercises with Answers for this book, please scan this QR code with a mobile device or visit this link online: https://bit.ly/GameTheoryMaterials.

- *Solutions Manual for Game Theory* (available only to instructors). It includes step-bystep answer keys to all end-of-chapter exercises (207 exercises in total). These exercises are ranked in order of difficulty, with A next to the title for the easiest exercises (often involving no calculus), B for the intermediate-level exercises, and C for the most difficult exercises, typically requiring a formal proof.
- *Microsoft PowerPoint slides* (available only to instructors). They cover the main topics in every chapter of the textbook. The slides include all definitions, equations, explanations, and figures, thus facilitating class preparation. Slides can also be distributed to students as a first set of lecture notes that they can complement with in-class explanations.

Acknowledgments

We would first like to thank several colleagues who encouraged us in the preparation of this manuscript: Ron Mittelhammer, Alan Love, and Jill McCluskey. We are, of course, especially grateful to our teachers and advisors at the University of Pittsburgh (Andreas Blume, Oliver Board, Alexander Matros, and Esther Gal-Or) and at the University of Barcelona (Antonio Manresa and Marina Nuñez) for instilling a passion for research and teaching in microeconomics. We are thankful to several teaching assistants at Washington State University, who helped us with this project over several years; and to a number of professors and students who provided feedback on earlier versions of the manuscript: Alessandro Tampieri, Miguel A. Melendez, Ramakanta Patra, Tommaso Reggiani, Dusan Paredes, Ismail Sağlam, Giuseppe Pignataro, and Xiaoyong Cao, among others; and to several students for their comments, especially Pak-Sing Choi, Eric Dunaway, John Strandholm, Kiriti Kanjilal, and Joseph Navelski. Last, but not least, we would like to thank our family and friends for encouraging us during the preparation of the manuscript.

Pullman, WA, USA
Ana Espinola-Arredondo
Felix Muñoz-Garcia

References

Banks, J. and J. Sobel (1987) "Equilibrium selection in signaling games," Econometrica, 55, pp. 647–661.
Cho, I. and D. Kreps (1987) "Signaling games and stable equilibria," Quarterly Journal of Economics, 102, pp. 179–221.
Dixit, A. K., S. Skeath, and D. H. Reiley Jr. (2015) *Games of Strategy*, Fourth edition, W.W. Norton & Company.
Dutta, P. K., and W. Vergote (2022) *Strategies and Games, Theory and Practice*, Second edition, MIT Press.
Gibbons, R. (1992) *Game Theory for Applied Economists*, Princeton University Press.
Harrington, J. E. (2014) *Games, Strategies, and Decision Making*, Worth Publishers.
Osborne, M. (2003) *An Introduction to Game Theory*, Oxford University Press.
Osborne, M. and A. Rubinstein (1994) *A Course in Game Theory*, The MIT Press.
Rasmusen, E. (2006) *Games and Information: An Introduction to Game Theory*, Blackwell Publishing.
Spence, M. (1973) "Job Market Signaling," The Quarterly Journal of Economics, 87(3), pp. 355–374.
Tadelis, S. (2013) *Game Theory, An Introduction*, Princeton University Press.
Watson, J. (2013) *Strategy: An Introduction to Game Theory*, Third edition, Worth Publishers.

Contents

1 Introduction to Games and Their Representation 1
   1.1   Introduction 1
   1.2   What Is Game Theory? 1
   1.3   Main Elements in a Game 3
        1.3.1   Players 3
        1.3.2   Strategies 3
        1.3.3   Payoffs 5
   1.4   Two Graphical Approaches 5
        1.4.1   Matrices 6
        1.4.2   Game Trees 7
   1.5   Introducing Imperfect Information in Game Trees 9
   1.6   Identifying Equilibrium Behavior 10
        1.6.1   Does an Equilibrium Exist? 11
        1.6.2   Is the Equilibrium Unique? 11
        1.6.3   Is the Equilibrium Robust to Small Payoff Changes? 11
        1.6.4   Is the Equilibrium Pareto Optimal? 12
   References 12

2 Equilibrium Dominance 13
   2.1   Introduction 13
   2.2   Strictly Dominated Strategies 14
   2.3   Iterated Deletion of Strictly Dominated Strategies 15
        2.3.1   Does the Order of Deletion Matter in IDSDS? 18
        2.3.2   Deleting More Than One Strategy at a Time 18
        2.3.3   Multiple Equilibrium Predictions 19
   2.4   Applying IDSDS in Common Games 20
        2.4.1   Prisoner's Dilemma Game 20
        2.4.2   Coordination Games—The Battle of the Sexes Game 23
        2.4.3   Pareto Coordination Game—The Stag Hunt Game 25
        2.4.4   Anticoordination Game—The Game of Chicken 26
        2.4.5   Symmetric and Asymmetric Games 28

	2.5	Allowing for Randomizations to Bring IDSDS Further	30
	2.5.1	What If IDSDS Has No Bite?	33
	2.6	Evaluating IDSDS as a Solution Concept	33
	2.7	Weakly Dominated Strategies	34
	2.7.1	Deletion Order Matters in IDWDS	36
	2.7.2	IDSDS Vs. IDWDS	38
	2.8	Strictly Dominant Strategies	39
	2.8.1	Evaluating SDE as a Solution Concept	40
	Exercises		41
	Reference		46
3	**Nash Equilibrium**		47
	3.1	Introduction	47
	3.2	Best Response	47
	3.2.1	Finding Best Responses with Discrete Strategy Spaces	48
	3.2.2	Finding Best Responses with Continuous Strategy Spaces	50
	3.3	Deleting Strategies That Are Never a Best Response	52
	3.4	Rationalizability	53
	3.4.1	Evaluating Rationalizability as a Solution Concept	55
	3.5	Applications of Rationalizability	56
	3.5.1	Finding NBRs in the Beauty Contest	56
	3.5.2	Finding NBRs in the Cournot Duopoly	57
	3.6	Nash Equilibrium	58
	3.7	Finding Nash Equilibria in Common Games	60
	3.7.1	Prisoner's Dilemma Game	60
	3.7.2	Coordination Game—The Battle of the Sexes Game	61
	3.7.3	Pareto Coordination Game—The Stag Hunt Game	62
	3.7.4	Anticoordination Game—The Game of Chicken	63
	3.7.5	Multiple Nash Equilibria	64
	3.8	Relationship Between NE and IDSDS	65
	3.9	What If We Find No NEs?	65
	3.10	Evaluating NE as a Solution Concept	67
	Appendix: Equilibrium Selection		68
	Exercises		69
	References		75
4	**Nash Equilibria in Games with Continuous Action Spaces**		77
	4.1	Introduction	77
	4.2	Quantity Competition	78
	4.2.1	Quantity Competition with Homogeneous Goods and Two Firms	78

		4.2.2	Extending Quantity Competition to $N \geq 2$ Firms	80

		4.2.2	Extending Quantity Competition to $N \geq 2$ Firms	80

            4.2.2   Extending Quantity Competition to $N \geq 2$ Firms 80
            4.2.3   Quantity Competition with Heterogeneous Goods 82
    4.3   Price Competition 84
            4.3.1   Price Competition with Homogeneous Goods 84
            4.3.2   Price Competition with Heterogeneous Goods 87
    4.4   Public Good Game 90
            4.4.1   Inefficient Equilibrium 93
    4.5   Electoral Competition 94
            4.5.1   Alternative Proof to the Electoral Competition game 96
    Exercises 97
    References 102

5 Mixed Strategy Nash Equilibrium 105
    5.1   Introduction 105
    5.2   Mixed Strategy 107
    5.3   Mixed Strategy Nash Equilibrium 108
    5.4   Finding Mixed Strategy Equilibria 109
            5.4.1   Graphical Representation of Best Responses 113
    5.5   Some Lessons 115
    5.6   Extensions 117
            5.6.1   Mixed Strategy Equilibria in Games with $k \geq 3$ Pure Strategies 117
            5.6.2   Finding Mixed Strategy Equilibria in Games with $N \geq 2$ Players 119
    5.7   Strictly Competitive Games 121
            5.7.1   Strictly Competitive Games 121
            5.7.2   Zero-Sum Games 123
            5.7.3   Security Strategies 124
    5.8   Security Strategies and NE 126
    5.9   Correlated Equilibrium 129
            5.9.1   Public or Private Recommendations? 131
    5.10   Equilibrium Refinements in Strategic-Form Games (Technical) 133
            5.10.1   Trembling-Hand Perfect Equilibrium 133
            5.10.2   Proper Equilibrium 136
    Appendix—NE Existence Theorem (Technical) 138
    Exercises 141
    References 148

6 Subgame Perfect Equilibrium 151
    6.1   Introduction 151
    6.2   Tree rules 152
            6.2.1   Actions vs. Strategies 155

6.3	Why Don't We Just Find the Nash Equilibrium of the Game Tree?	156
6.4	Subgames	158
	6.4.1 What If the Game Tree Has Information Sets?	158
6.5	Subgame Perfect Equilibrium	160
	6.5.1 Finding SPEs in Games Without Information Sets	161
	6.5.2 Finding SPEs in Game Trees with Information Sets	164
6.6	Evaluating SPE as a Solution Concept	168
6.7	Applications	169
	6.7.1 Stackelberg Game of Sequential Quantity Competition	169
	6.7.2 Sequential Public Good Game	172
	6.7.3 Ultimatum Bargaining Game	175
	6.7.4 Two-Period Alternating-Offers Bargaining Game	177
	6.7.5 Some Tricks About Solving Alternating-Offer Bargaining Games	180
	6.7.6 Alternating-Offer Bargaining Game with Infinite Periods	181
Appendix—Mixed and Behavioral Strategies		184
Exercises		186
References		202

7 Repeated Games — 203

7.1	Introduction	203
7.2	Repeating the Game Twice	204
7.3	Repeating the Game $T \geq 2$ Times	208
7.4	Repeating the Game Infinitely Many Times	210
	7.4.1 Uncooperative Outcome	211
	7.4.2 Cooperative Outcome	212
	7.4.3 Cooperative Outcome—Extensions	214
7.5	Folk Theorem	217
	7.5.1 Feasible and Individually Rational Payoffs	218
	7.5.2 Folk Theorem and Cooperation	221
7.6	Application to Collusion in Oligopoly	224
	7.6.1 Minimal Discount Factor Supporting Collusion	228
	7.6.2 Other Collusive GTS	229
7.7	What if the Stage Game has More than One NE?	230
7.8	Modified GTSs	232
	7.8.1 An Eye for an Eye	232
	7.8.2 Short and Nasty Punishments	235
	7.8.3 Imperfect Monitoring	236

	Exercises	240
	References	251
8	**Bayesian Nash Equilibrium**	**253**
	8.1 Introduction	253
	8.2 Background	254
	8.2.1 Players' Types and Their Associated Probability	254
	8.2.2 Strategies Under Incomplete Information	256
	8.2.3 Representing Asymmetric Information as Incomplete Information	256
	8.2.4 Best Response Under Incomplete Information	257
	8.3 Bayesian Nash Equilibrium	259
	8.3.1 Ex-ante and Ex-post Stability	259
	8.4 Finding BNEs—First Approach: Build the Bayesian Normal Form	260
	8.5 Finding BNEs—Second Approach: Focus on the Informed Player First	263
	8.6 Evaluating BNE as a Solution Concept	267
	8.7 What If Both Players Are Privately Informed?	268
	Exercises	270
	Reference	279
9	**Auction Theory**	**281**
	9.1 Introduction	281
	9.2 Auctions as Allocation Mechanisms	282
	9.3 Second-price Auctions	284
	9.3.1 Case 1: Bid Equal To Her Valuation	284
	9.3.2 Case 2: Downward Deviations, Bidding Below Her Valuation	285
	9.3.3 Case 3: Upward Deviations, Bidding Above Her Valuation	286
	9.3.4 Discussion	286
	9.4 First-Price Auctions	287
	9.5 Efficiency in Auctions	292
	9.6 Seller's Expected Revenue	293
	9.6.1 Expected Revenue in the FPA	293
	9.6.2 Expected Revenue in the SPA	297
	9.6.3 Revenue Equivalence Principle	299
	9.7 Common-Value Auctions and the Winner's Curse	300
	9.7.1 Bid Shading Is a Must!	300
	9.7.2 Equilibrium Bidding in Common-Value Auctions	301
	Exercises	303

10	**Perfect Bayesian Equilibrium**	309
	10.1 Introduction	309
	10.2 Sequential-Move Games of Incomplete Information—Notation	310
	10.3 BNE Prescribing Sequentially Irrational Behavior	314
	10.4 Perfect Bayesian Equilibrium—Definition	316
	10.5 A Tool to Find PBEs in Signaling Games	320
	10.6 Finding PBEs in Games with one Information Set	321
	10.6.1 Separating Strategy Profile (O^B, N^F)	322
	10.6.2 Pooling Strategy Profile (O^B, O^F)	323
	10.7 Finding PBEs in Games with Two Information Sets	325
	10.7.1 Separating Strategy Profile (E^H, NE^L)	325
	10.7.2 Pooling Strategy Profile (NE^H, NE^L)	328
	10.7.3 Insensible Off-the-Equilibrium Beliefs	331
	10.8 Evaluating PBE as a Solution Concept	332
	10.9 Semi-Separating PBE	334
	10.10 Extensions	337
	10.10.1 What if the Receiver has More than Two Available Responses?	337
	10.10.2 What if the Sender has More than Two Available Messages?	338
	10.10.3 What if the Sender has More than Two Types?	342
	10.10.4 Other Extensions	346
	Exercises	348
	Reference	358
11	**Equilibrium Refinements**	359
	11.1 Introduction	359
	11.2 Intuitive Criterion	360
	11.2.1 A Six-Step Tool to Apply the Intuitive Criterion	361
	11.2.2 Separating Equilibria Survive the Intuitive Criterion	362
	11.3 D1 Criterion	363
	11.3.1 Applying the D1 Criterion—An Example	364
	11.3.2 Discrete and Continuous Responses	365
	11.3.3 Comparing Intuitive and Divinity Criteria	365
	11.3.4 Other Refinement Criteria	366
	11.4 Sequential Equilibrium	366
	11.4.1 Finding Sequential Equilibria	367
	11.4.2 Separating PBEs that Are Also SEs	368
	11.4.3 A Pooling PBE that Is Not a SE	369
	11.4.4 A Pooling PBE that Is Also A SE	371
	Exercises	372
	References	378

Contents

12 Signaling Games with Continuous Messages 379
- 12.1 Introduction 379
- 12.2 Utility Functions with Continuous Actions 380
- 12.3 Complete Information 381
- 12.4 Separating PBE 382
  - 12.4.1 Separating PBE—Applying the Intuitive Criterion 387
  - 12.4.2 Separating PBE—Applying the D1 Criterion 388
- 12.5 Pooling PBE 390
  - 12.5.1 Other Pooling PBEs 393
  - 12.5.2 Pooling PBE—Applying the Intuitive Criterion 394
- 12.6 Can Signaling Be Welfare Improving? 396
- 12.7 What If the Sender Has Three Types? 397
  - 12.7.1 Separating PBEs 397
  - 12.7.2 Separating PBE—Applying the Intuitive Criterion 398
  - 12.7.3 Separating PBE—Applying the D1 Criterion 401
- Appendix: Equilibrium Refinements 402
- Exercises 405
- References 408

13 Cheap Talk Games 409
- 13.1 Introduction 409
- 13.2 Cheap Talk with Discrete Messages and Responses 410
  - 13.2.1 Separating PBE 411
  - 13.2.2 Pooling PBEs 413
- 13.3 Cheap Talk with Discrete Messages But Continuous Responses 415
  - 13.3.1 Separating PBE 416
  - 13.3.2 Pooling PBEs 418
- 13.4 Cheap Talk with Continuous Messages and Responses 420
  - 13.4.1 Separating PBE 420
  - 13.4.2 Equilibrium Number of Partitions 423
  - 13.4.3 Interval Lengths in Equilibrium 427
- 13.5 Extensions 429
- Exercises 430
- References 434

Mathematical Appendix 435

References 447

Index 451

List of Figures

Fig. 1.1	**a** Example of a game tree. **b** Strategy profile (H,l)	7
Fig. 1.2	Example of a game tree—Larger version	8
Fig. 1.3	Example of a game tree—Larger version with an information set	9
Fig. 1.4	**a** Nodes with different number of actions. **b** Nodes with the same number of actions but different action labels	10
Fig. 3.1	Firm 1's best response function	51
Fig. 3.2	Comparing strategy profiles: **a** Two players. **b** More than two players	65
Fig. 4.1	**a** Firm 2's best response function, $q_2(q_1)$. **b** NE output in the Cournot game	79
Fig. 4.2	Firm i's best response function	83
Fig. 4.3	NE output levels with heterogeneous products	83
Fig. 4.4	Firm i's best response function when competing in prices	86
Fig. 4.5	Firm j's best response function when competing in prices	86
Fig. 4.6	Price equilibrium	87
Fig. 4.7	Firm i's best response function, $p_i(p_j)$, with heterogeneous goods	88
Fig. 4.8	Equilibrium price pair with heterogeneous goods	89
Fig. 4.9	Player i's best response function, $x_i(x_j)$	91
Fig. 4.10	Contribution profiles in equilibrium	93
Fig. 4.11	Candidate i's best response correspondence	96
Fig. 4.12	Votes to candidate i and j when $x_i < x_j$ and $x_i > 1 - x_j$	97
Fig. 5.1	The goalie's best responses	114
Fig. 5.2	Kicker's best responses	114
Fig. 5.3	Both players' best responses	115
Fig. 5.4	Equilibrium probability p^*	121
Fig. 5.5	Lower envelope and security strategies	126
Fig. 5.6	Lower envelope and security strategies—Corner solution	129

Fig. 5.7	Totally mixed strategy converging to $\sigma_i = (1, 0)$	135
Fig. 5.8	(**a**) Function f(x) and the 45-degree line (**b**)	139
Fig. 6.1	Games with one or two initial nodes	152
Fig. 6.2	Game with two predecessors	153
Fig. 6.3	Games with same/different action labels	154
Fig. 6.4	Information sets with one or more players	154
Fig. 6.5	**a**. Different number of immediate successors. **b**. Different immediate successors	155
Fig. 6.6	The entry game	156
Fig. 6.7	**a**. Subgames in the entry game. **b**. Four subgames in a game tree	159
Fig. 6.8	Finding subgames in a game with an information set	159
Fig. 6.9	Applying backward induction in the entry game—last mover	161
Fig. 6.10	Applying backward induction in the entry game—first mover	162
Fig. 6.11	SPEs as a subset of NEs	163
Fig. 6.12	Modified entry game	163
Fig. 6.13	Modified entry game—subgames	164
Fig. 6.14	Equilibrium path vs. SPE	165
Fig. 6.15	A more involved game tree	166
Fig. 6.16	**a**. Finding subgames; **b**. Not subgames	166
Fig. 6.17	The reduced game tree from Fig. 6.16a	167
Fig. 6.18	Equilibrium quantities in Cournot vs. Stackelberg games	172
Fig. 6.19	**a**. Ultimatum bargaining game. **b**. Ultimatum bargaining game—Smallest subgame	176
Fig. 6.20	Two-period alternating-offers bargaining game	178
Fig. 6.21	Mixed and behavioral strategies in a game tree	185
Fig. 6.22	More general version of the entry game	187
Fig. 6.23	Backward induction with three moves	190
Fig. 6.24	Backward Induction with information sets	190
Fig. 6.25	Entry deterring investment	201
Fig. 7.1	Twice-repeated prisoner's dilemma game	205
Fig. 7.2	Twice-repeated prisoner's dilemma game—First stage	207
Fig. 7.3	Incentives from cheating in the infinitely repeated prisoner's dilemma game	215
Fig. 7.4	Temporary punishments	215
Fig. 7.5	More attractive cheating	216
Fig. 7.6	More severe punishment	216
Fig. 7.7	Lag at detecting cheating	217
Fig. 7.8	FP set in the Prisoner's Dilemma game	219
Fig. 7.9	FIR set in the prisoner's dilemma game	220
Fig. 7.10	Minimal discount factor $\underline{\delta}(p)$	239
Fig. 8.1	**a** Simultaneous-move game where $x = 20$. **b** Simultaneous-move game where $x = 12$	257

Fig. 8.2	Combining both games of Fig. 8.1 to allow for incomplete information	258
Fig. 8.3	**a** Firm 2—high or low costs. **b** Equilibrium output under incomplete information	265
Fig. 8.4	Cournot competition with asymmetrically informed firms—equilibrium output	267
Fig. 9.1	Equilibrium bids in the first-price auction with uniformly distributed valuations	292
Fig. 10.1	Sequential-move game with incomplete information—One information set	313
Fig. 10.2	Labor market signaling game	314
Fig. 10.3	Separating strategy profile (O^B, N^F)	322
Fig. 10.3a	Separating strategy profile (O^B, N^F)—Responses	323
Fig. 10.4a	Pooling strategy profile (O^B, O^F)—Responses	324
Fig. 10.4	Pooling strategy profile (O^B, O^F)	325
Fig. 10.5	Separating strategy profile (E^H, NE^L)	326
Fig. 10.5a	Separating strategy profile E^H, NE^L—Responses	327
Fig. 10.6	Pooling strategy profile (NE^H, NE^L)	328
Fig. 10.6a	Pooling strategy profile (NE^H, NE^L)—Responses (C', M)	331
Fig. 10.6b	Pooling strategy profile (NE^H, NE^L)—Responses (C', C)	332
Fig. 10.7	Poker game	334
Fig. 10.7a	Poker game—Semi-separating strategy profile	336
Fig. 10.8	Labor market signaling game with three responses	338
Fig. 10.9	Labor market signaling game with three messages	339
Fig. 10.9a	Labor market signaling game with three messages—Alternative representation	340
Fig. 10.9b	Labor market signaling game with three messages—Strategy profile (A^H, E^L)	340
Fig. 10.9c	Labor market signaling game with three messages—Optimal messages	342
Fig. 10.10	Labor market signaling game with three worker types	343
Fig. 10.10a	Labor market signaling game with three worker types—Alternative representation	344
Fig. 10.10b	Labor market signaling game with three worker types—Strategy profile (E^H, E^M, NE^L)	345
Fig. 10.10c	Labor market signaling game with three worker types—Optimal responses	347
Fig. 10.11	Beer-Quiche game	348
Fig. 10.12	- Labor market signaling when education is productivity enhancing	348
Fig. 10.13	- Labor market signaling game with cost differentials	350
Fig. 10.14	- Labor market signaling game with profit differentials	351
Fig. 10.15	- Brinkmanship game	352

Fig. 10.16a	- Reputation without commitment, Complete information	353
Fig. 10.16b	- Reputation without commitment, Incomplete information	354
Fig. 10.17	Selten's horse	355
Fig. 10.18	- Players holding asymmetric beliefs	356
Fig. 11.1	Pooling strategy profile (NE^H, NE^L)—Responses (C', C)	360
Fig. 11.2	Modified labor market signaling game	368
Fig. 11.3	Separating strategy profile $(E^H, NE^L; M, C')$	369
Fig. 11.4	Updated beliefs μ^k and γ^k as a function of k	370
Fig. 11.5	Pooling strategy profile $(NE^H, NE^L; M, M')$	370
Fig. 11.6	Pooling strategy profile $(NE^H, NE^L; M, M')$	372
Fig. 11.7	PBEs when all players are uninformed-I	375
Fig. 11.8	PBEs when three players are uninformed	376
Fig. 12.1	Indifference curves for a representative worker	380
Fig. 12.2	Separating PBE—Two wage schedules	383
Fig. 12.3	Separating PBE—Low-productivity worker	384
Fig. 12.4	Separating PBE—Low- and high-productivity worker	385
Fig. 12.5	Separating PBE: **a** Case I, **b** Case II	385
Fig. 12.6	Applying the Intuitive Criterion to the separating PBEs	387
Fig. 12.7	Applying the D1 Criterion to the separating PBEs	388
Fig. 12.8	Pooling PBE—Example of wage schedule	390
Fig. 12.9	Pooling PBE—Low-productivity worker	391
Fig. 12.10	Pooling PBE: **a** High-productivity worker, **b** Both worker types	392
Fig. 12.11	Pooling PBE—Other equilibria	393
Fig. 12.12	Applying the Intuitive Criterion to Pooling PBEs	395
Fig. 12.13	Utility comparison across information settings	396
Fig. 12.14	Separating PBEs with three worker types	398
Fig. 12.15	Applying the Intuitive Criterion with three worker types	399
Fig. 12.16	Applying the D1 Criterion with three worker types	401
Fig. 12.17	Unique separating PBE surviving the D1 Criterion with three types	402
Fig. 13.1	A cheap talk game with two sender types and two messages	410
Fig. 13.2	Separating strategy profile (m_1^H, m_2^L) in the cheap talk game	411
Fig. 13.2a	Separating strategy profile (m_1^H, m_2^L) in the cheap talk game—Optimal responses	411
Fig. 13.3	Pooling strategy profile (m_1^H, m_1^L) in the cheap talk game	412
Fig. 13.3a	Pooling strategy profile (m_1^H, m_1^L) in the cheap talk game—Optimal responses	413
Fig. 13.4	Cheap talk with two messages but continuous responses	415

Fig. 13.5	Quadratic loss function for each player 416
Fig. 13.6	Cheap talk with two messages and continuous responses—Separating profile (θ_H, θ_L) 417
Fig. 13.7	Cheap talk with two messages and continuous responses—Pooling profile (θ_H, θ_H) 418
Fig. 13.8	Partially informative strategy profile 423
Fig. 13.9	Cutoff $\overline{N}(\delta)$ as a function of the bias parameter δ 426

List of Tables

Table 2.1	The Prisoner's Dilemma in movies	22
Table 2.2	Anticoordination games in different movies	28
Table 6.1	An approach to solve alternating-offers bargaining games	181
Table 7.1	Modified GTS inducing partial cooperation	221
Table 7.2	Timing of the an-eye-for-an-eye GTS	233
Table 7.3	Timing of the short-and-nasty-punishments GTS	236
Table 7.4	Minimal discount factors sustaining cooperation in the game in Matrix 7.7 when using different GTS	236
Table 11.1	Separating PBEs and the number of sender types	363

Introduction to Games and Their Representation

1.1 Introduction

In this chapter, we discuss the main topics that Game Theory studies. We also list the key elements in strategic settings (which we denote as "games"), such as the set of players interacting, their available strategies, and payoffs they earn. We then describe the two main graphical approaches to illustrate games, namely, matrices and game trees. We focus our attention on how these illustrations can help us clarify the order of play and the information that each player observes at each point of the game.

1.2 What Is Game Theory?

Game theory studies settings where the actions of an individual (such as a firm, a consumer, a government, or a non-governmental organization) affect the payoffs of other individuals. A common example is a firm deciding to lower its price, which affects the sales and profits of other firms selling similar products. Another example is an individual contributing to a public good (or paying taxes to fund a public project, such as a highway) as her contributions potentially improve the project's quality. In this case, other individuals (even those who did not contribute) can benefit from the project. Finally, in the context of the Covid-19 pandemic, examples of strategic settings abound: should I wear a face mask before leaving home? Should I get vaccinated?

Day-to-day life is packed with scenarios in which your actions affect the well-being of other agents (individuals, firms, or governments) and, as a consequence, many contexts can be modeled as games.

We next provide a more formal definition and then highlight its main ingredients.

Definition. **Game Theory** studies the interaction between a group of rational agents who behave strategically.

Let us separately consider the key elements of this definition.

1. **Group of individuals**. Game theory studies *interaction between a group of agents*. Scenarios with a single individual or firm can be analyzed as an individual decision-making problem, but they are not a game between two or more players.
2. **Rationality**. Game theory assumes that individuals are *rational*, meaning that they seek to maximize their payoff function. In addition, every player knows this information, which is referred to as "common knowledge."

    In a two-player game, common knowledge entails that player 1 tries to maximize her payoff, that player 2 knows that player 1 seeks to maximize her payoff, that player 1 knows that player 2 knows that she seeks to maximize her payoff... and, so on, ad infinitum. Intuitively, common knowledge helps each player put herself in the shoes of her rival, anticipating the actions that her rival chooses (either now if players choose their actions simultaneously or in subsequent stages if they do so sequentially). This assumption is known as "common knowledge of rationality."[1]
3. **Strategic behavior**. Finally, the definition tells us that every individual behaves *strategically*, meaning that she seeks to maximize her payoff function. If a player chooses her actions without taking into account a well-defined payoff function (as an automaton), she does not behave strategically.

    The payoff function that a player seeks to maximize can be, for instance, a profit function in the case of firms, a utility function when we consider consumers, or a social welfare function of a whole country. The payoff function of an individual (e.g., consumer) may be a function of her monetary benefits and costs alone, but can also include payoffs from other individuals. This occurs, for instance, when individuals exhibit a preference for equity or envy, as shown in Fehr and Schmidt (1999) and Bolton and Ockenfels (2000) among others. Therefore, players could be altruistic and still be strategic.

---

[1] A scene in movie *The Princess Bride* (1987) provides a humorous example of common knowledge of rationality, where Vizzini, one of the characters, must drink from one of two wine goblets, not knowing which one has been poisoned by his rival, the Man in Black. In this context, Vizzini says: "A clever man would put the poison into his own goblet, because he would know that only a great fool would reach for what he was given. I am not a great fool, so I can clearly not choose the wine in front of you. But you must have known I was not a great fool [...] so I can clearly not choose the wine in front of me." You can watch the complete scene at: www.youtube.com/watch?v=9s0UURBihH8.

1.3 Main Elements in a Game

We next describe the main ingredients of a game and their notation.

1.3.1 Players

The set of agents, such as individuals, firms, or countries interact in the game. For instance, when only two firms compete in an industry, we say that the number of players, N, is $N = 2$, but when ten firms interact we have $N = 10$ players, and similarly for games with more players. Formally, since we need two or more players to interact in a game, we write that $N \geq 2$.

1.3.2 Strategies

Definition 1.1. **Strategy**. A strategy is a complete contingent plan describing which actions a player chooses in each possible situation (contingency) that she faces along the game.

In a two-player game with firms A and B, we can denote firm A's strategy as s_A where $s_A \in S_A$, meaning that firm A selects s_A from a set of available strategies S_A (known as the "strategy set"). A strategy can then be informally understood as an instruction manual: a player opens the manual, looks for the page describing the actions other players chose, and the manual indicates how to respond, recommending that she chooses a specific action.

We can face different strategy sets depending on the problem we analyze:

- **Discrete strategies**. If firm A can only select an output level in $S_A = \{6, 10\}$, we say that its strategy is discrete. (In this example, where firm A can only select an output of either 6 or 10, we say that its strategy is "binary.") In this setting, we may have that firm A chooses, for instance, an output level of $s_A = 6$ units. More generally, firm A's strategy set can be $S_A = \{1, 2, ..., 10\}$, which is still discrete but allows firm A to choose among more output levels. Similarly, if firm A can choose any positive integer output levels, $S_A = \{1, 2, ...\}$ or alternatively $S_A = \mathbb{N}$, implying that firm A's strategy space coincides with the set of natural numbers.
- **Continuous strategies**. If, instead, firm A can choose its output level s_A from a continuous strategy space, we can have that $S_A = [0, 10]$, meaning that the firm can now choose non-integer amounts from 0 to 10 (e.g., $s_A = 5.7$ units). Still in a context with continuous strategy spaces, firm A can select its output from $S_A = [0, +\infty)$. Alternatively, we say that its strategy space coincides with the positive real numbers $S_A = \mathbb{R}_+$.

A similar argument applies to firm B's strategy, s_B, and its strategy set S_B. More generally, we denote player i's strategy as s_i, where $s_i \in S_i$. Similarly, let s_j represent player j's strategy, where $j \neq i$ denotes a player different than i (which, more compactly, we often refer to as "player i's rival"), where $s_j \in S_j$.

Symmetric and asymmetric strategy sets. In some settings, players face the same strategy space, $S_i = S_j = S$ for every player $j \neq i$, such as when firms competing in the same industry and have access to the same technology, so they both choose an output level in, for instance, $S = [0, 10]$. In other games, however, players may face different strategy sets, such as when firms have access to different technologies (or regulations affecting each firm differently), entailing that $S_i \neq S_j$. In the context of two firms choosing their output levels, this asymmetry in their strategy sets can mean that firm i may produce a larger output level than firm j, for instance, $S_i = [0, 100]$ while $S_j = [0, 75]$.

Definition 1.2. **Strategy profile.** A list describing the strategies that each player selects, $s = (s_1, s_2, ..., s_N)$.

For example, in a two-player game between two firms (A and B), a strategy profile could be $s = (12, 8)$ where $s_A = 12$ denotes firm A's output level while $s_B = 8$ represents firm B's output. In the context of more than two players, strategy profiles can be relatively long vectors. For instance, in a setting with N players, a strategy profile is

$$s \equiv (s_1, s_2, ..., s_{i-1}, s_i, s_{i+1}, ..., s_N).$$

For compactness, however, we use $s = (s_i, s_{-i})$ to denote a strategy profile where player i chooses strategy s_i while its rivals select s_{-i}, defined as

$$s_{-i} \equiv (s_1, s_2, ..., s_{i-1}, s_{i+1}, ..., s_N),$$

which includes the strategies selected by all players but i (note that the list of strategies jumps from s_{i-1} to s_{i+1}, skipping s_i). In a context with four firms, each choosing its output level, we could write

$$s = (12, 8, 10, 13)$$

or, more compactly, $s = (12, s_{-1})$, where the first term, $s_1 = 12$, represents the units of output that firm 1 chose to produce, and the second term, $s_{-1} = (8, 10, 13)$, denotes the output levels of firm 1's rivals.[2]

Listing all strategy profiles. A strategy profile $s = (s_i, s_{-i})$ is an element in the Cartesian product $S_i \times S_{-i}$. To understand this point, consider a two-player game

---

[2] A similar argument applies if we focus on firm 2, writing this strategy profile as $s = (8, s_{-2})$ where $s_2 = 8$ units for firm 2 and $s_{-2} = (12, 10, 13)$ for its rivals. As a practice, you can write this strategy profile focusing on firm 3's strategy, $s_3 = 10$, and then when considering firm 4's strategy, $s_4 = 13$.

between firms A and B. In this context, a strategy profile simplifies to the vector $s = (s_A, s_B)$, where $s \in S_A \times S_B$, since firm A chooses a strategy from its strategy space, $s_A \in S_A$ and, similarly, firm B chooses a strategy from its strategy set, $s_B \in S_B$. For instance, if S_A and S_B are both binary (i.e., each firm can only choose between two output levels $S_A = \{s_A^1, s_A^2\}$ for firm A, and $S_B = \{s_B^1, s_B^2\}$ for firm B), the Cartesian product $S_A \times S_B$ has a total of four elements, that is, the combinations of strategies from firms A and B, namely,

$$(s_A^1, s_B^1), (s_A^1, s_B^2), (s_A^2, s_B^1), \text{ and } (s_A^2, s_B^2).$$

Since each firm has two available strategies and two firms interact in this game, there are a total of $2 \times 2 = 4$ different strategy profiles. More generally, in a two-player game where player i has k available strategies[3] and player j has m available strategies, there are $k \times m$ different strategy profiles. (This argument holds for games where players choose their strategies simultaneously, but may not be satisfied when players act sequentially, as we discuss in Chapter 6).

If, instead, every firm i can choose any positive output level, $S_i = \mathbb{R}_+$, the Cartesian product becomes $S_A \times S_B = \mathbb{R}_+^2$ in a two-player game, that is, every output pair (s_A, s_B) lies in the positive quadrant. Similarly, in an N-player game, the Cartesian product is $S_1 \times S_2 \times \cdots \times S_N = \mathbb{R}_+^N$.

1.3.3 Payoffs

A game must also list the payoff that every player obtains under each possible strategy.

Definition 1.3. **Payoff function**. Player i's payoff function is a mapping from a strategy profile $s = (s_i, s_{-i})$, where $s \in S_i \times S_{-i}$, into a real number, $u_i : S_i \times S_{-i} \to \mathbb{R}$, yielding $u_i(s_i, s_{-i})$.

If in our ongoing two-firm industry example, firm A chooses to produce 5 units and firm B chooses 6 units, the strategy profile would be $(s_A, s_B) = (5, 6)$, yielding a payoff, for instance, of $u_A(s_A, s_B) = 8$ for firm A, and a payoff of $u_B(s_A, s_B) = 10$ for firm B.

1.4 Two Graphical Approaches

We generally use two approaches to graphically represent games: matrices and trees, as we describe next. Matrices usually depict "simultaneous-move games," where every player chooses her strategy at the same time or, more precisely, without observing the strategy that her opponents select. In contrast, game trees

---

[3] Technically, we say that the "cardinality" of her strategy space is k, expressing it as $\#S_i = k$.

commonly illustrate "sequential-move games," where a first mover chooses her strategy first, the second mover responds by selecting her strategy, the third mover then chooses her strategy, and so on until the game ends. In sequential-move games, we consider settings where, for instance, the second mover observes the specific strategy that the first mover chose before her; but also contexts where, instead, the second mover does not observe which strategy the first mover chose in the previous stage.

1.4.1 Matrices

In the case of matrices, player 1 typically chooses rows and is referred to as the "row player" (see Matrix 1.1). Player 2, however, chooses columns and is known as the "column player". For easier reference, we use uppercase letters for player 1 (H or L, indicating high or low prices, respectively) and lowercase letters for player 2 (h and l to indicate high and low prices too for this player). In Matrix 1.1, for instance, if player 1 chooses high prices, H, while player 2 picks low prices, l, their payoff becomes $(0, -7)$, indicating that player 1's payoff is zero, while player 2's is -7. (Negative payoffs may represent, for instance, that the player's revenue is lower than her costs, thus losing money.)

	Player 2	
	h	l
H	4, 4	0, −7
L	8, 1	2, 2

Player 1 (rows)

Matrix 1.1. Example of a two-player game.

For presentation purposes, Matrix 1.1 assumed that every player had only two available strategies, implying that

$$S_1 = \{H, L\} \text{ and } S_2 = \{h, l\},$$

so each player faces a binary decision. In other games, however, a player could have more available strategies than her rival, a rich strategy spaces with hundreds of possible strategies to choose from, or continuous strategy spaces (such as when firms compete in prices). When players interact choosing their strategies simultaneously but the number of available strategies in S_i, S_j, or both, makes the matrix too large to represent graphically, we will analyze the game keeping in mind that players interact in a context which is, essentially, equivalent to that of a matrix form.

1.4 Two Graphical Approaches

Fig. 1.1 a Example of a game tree. b Strategy profile (H, l)

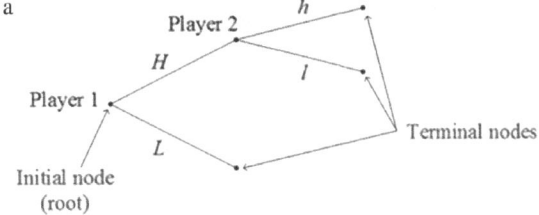

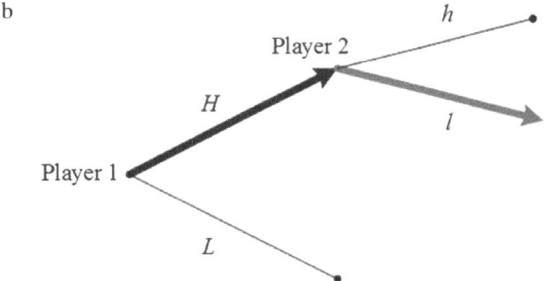

1.4.2 Game Trees

When players act sequentially (one after the other), we frequently use game trees, as trees can help us illustrate who acts when, representing which information a player observes in each part of the game. Figure 1.1a depicts a game tree, with player 1 acting first at the initial node, which we can interpret as the "root" of the tree. The first mover is often referred to as the leader and player 2, who observes player 1's choice and responds with her own action, is known as the follower.

In Fig. 1.1a, player 1 can choose between H and L. If player 1 chooses L, the game is over and payoffs are distributed, whereas if she chooses H, player 2 must respond with either h or l. The nodes at the end of the game tree are denoted as "terminal nodes," which list the payoffs players earn at each strategy profile. For instance, if the game proceeds through H and then l, we say that the strategy profile is (H, l) which, in the context of game trees, can also be understood as a path of play. Figure 1.1b illustrates strategy profile (H, l) by shading the branch corresponding to H for player 1 and, then, the branch of l for player 2, thus following a distinctive path from the initial node to one of the terminal nodes.[4]

We could, of course, consider more involved game trees with additional stages and more available strategies to each player, like that in Fig. 1.2, where player 1 can choose between H, M, and L. If player 1 chooses L, the game is over, as in Fig. 1.1a. However, if the first mover chooses H or M, it is player 2's turn to move.

---

[4] When we discuss equilibrium solution concepts in next chapters, this path of play will often be denoted as "equilibrium path" and it will be graphically interpreted as the shaded branches in the game tree, from the initial node to one of the terminal nodes.

Fig. 1.2 Example of a game tree—Larger version

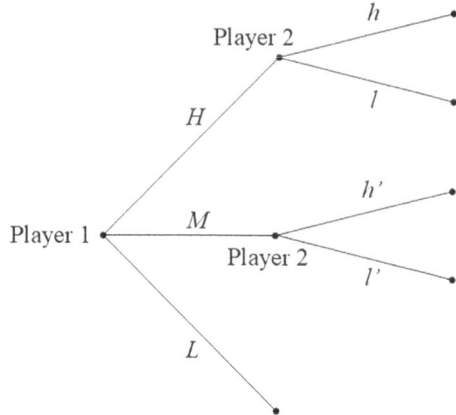

As a practice, we can find the strategy space of each player in this context. Player 1, being called on to move at a single node (the initial node) has as many strategies as action branches stemming from the initial node, $S_1 = \{H, M, L\}$. Player 2, in contrast, can respond in two different nodes (responding to H or to M), entailing four different strategies for this player (two actions in each node times two nodes) which implies a strategy space

$$S_2 = \{hh', hl', lh', ll'\}.$$

For instance, strategy hl' prescribes that player 2 responds with h after observing that player 1 chooses H but with l' when player 1 chooses M. We use different labels to describe the available actions (branches) of player 2 upon observing that player 1 chose H than after observing she chose M, helping us precisely denote player 2's responses.[5]

The above example illustrates our definition of strategies as "complete contingent plans" in a sequential context. Specifically, each strategy in S_2 serves as an instruction manual for player 2, telling her what to do after player 1 chooses H and what to do after L.

A common question at this point is, if player 1 chooses, for instance, L, why does player 2's strategy describe how would she respond to H? Strategies must be so detailed because player 2's responses at nodes that may not be reached can make player 1 behave in one way or another in the first stage. We return to this point when studying equilibrium behavior in sequential-move games, where we will show that every player, anticipating her rival's response at each of the nodes where she is called on to move, chooses the action that maximizes her payoff.

---

[5] Otherwise, if we used h and l to label player 2's actions after observing both H and M, and we wrote that she responds with, for instance, h, we would have to specify if that is after observing H or M. Instead, by using h (h'), we know that we refer to player 2's response to player 1 choosing H (M, respectively).

1.5 Introducing Imperfect Information in Game Trees

Figure 1.3 is identical to Fig. 1.2, except for the dashed line connecting the two nodes where player 2 is called to move. This dashed line indicates that player 2 cannot distinguish whether player 1 chose H or L in the first stage of the game, and it is known as player 2's "information set," as we define next.

Definition 1.4. **Information set**. An information set connects all nodes that player i cannot distinguish. In addition, these nodes must have the same number of action branches, which must have the same labels.

In the context of Fig. 1.3, player 2 is imperfectly informed about player 1's previous moves. In that setting, player 2 must respond without being able to condition on player 1's choice. This explained why, unlike in Fig. 1.2, player 2's available actions in Fig. 1.3 have the same labels (h and l). A common example is that of a firm which observes that its rival either built a new plant or that it did not. However, upon observing a new plant, the follower cannot distinguish whether the new plant allows for a high (H) or medium (M) capacity.

Needless to say, all nodes connected by an information set must have the same number of actions stemming from them. Figure 1.4a illustrates what would happen otherwise: player 2 chooses between two actions when player 1 selects H (h and l, at the top node) but a different number of actions when player 1 chooses M (e.g., h, m, and l at the bottom node). Then, she would be able to infer the node she is in just by looking at the number of available actions on that node. In that scenario, player 2 would know the action that player 1 chose, so no information set would need to be depicted.

Importantly, not only the number of available branches must be the same across all nodes connected by an information set, but these must have the same action labels stemming from them. Figure 1.4b considers that, instead, player 2 chooses between h and l after player 1 selects H, but between h and x after player 1 chooses

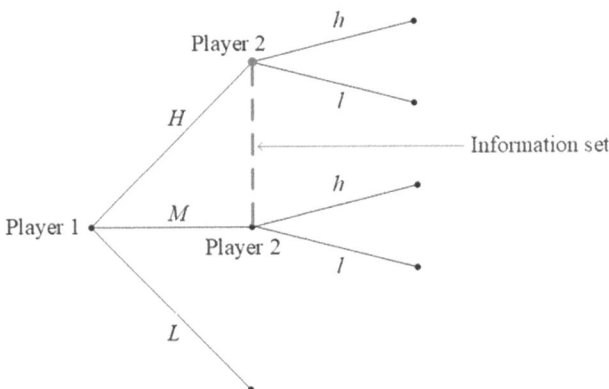

Fig. 1.3 Example of a game tree—Larger version with an information set

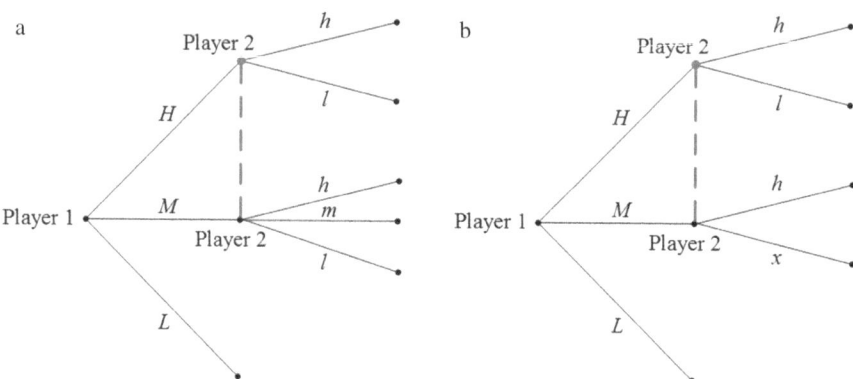

Fig. 1.4 **a** Nodes with different number of actions. **b** Nodes with the same number of actions but different action labels

M. In this setting, player 2 would be able to infer the node she is at by looking at the specific list of actions she can choose from (either h and l, or h and x), implying that nodes must not be connected with an information set either.

Alternative interpretation. Another interesting point of information sets is that they represent a game which, despite being sequential in nature, has every player choosing her actions "in the dark." In this context, a player cannot observe her rival's previous choices. As a consequence, the sequential-move game in Fig. 1.3 becomes strategically equivalent to a simultaneous-move game. This applies regardless of whether player 1's and 2's choices happen simultaneously (in the same second) or sequentially (days or years apart) as long as player 2 cannot observe whether player 1 chose H or M when called to respond.

1.6 Identifying Equilibrium Behavior

After describing the main elements in simultaneous and sequential games in this chapter, the remaining chapters study how to predict equilibrium behavior in a strategic setting. In other words, we seek to understand how players behave when facing different types of games. Specifically, we will identify scenarios where no player has incentives to change her choice, given the strategy of her opponents, ultimately making the strategy profile stable, which we refer to as "equilibria."

In our search for solution concepts that help us find equilibria, we start with a solution concept that, rather than identifying strategies that maximize a player's payoffs, just tries to discard all strategies that a rational player would never use (deletion of strictly dominated strategies). This is because these strategies yield an unambiguously lower payoff than other strategies, regardless of the actions taken by her rivals.

While the application of this solution concept is straightforward, we will show that it sets a relatively "high bar" because there are relatively few games where players can find strategies they would never use *regardless* of the actions chosen by their opponents. As a consequence, this solution concept may have little or no bite in several games, not allowing us to offer precise equilibrium predictions. We then discuss other solution concepts, such as the Nash equilibrium in simultaneous-move games and the Subgame Perfect equilibrium in sequential-move games, along with their extensions to games of incomplete information (Bayesian Nash equilibrium and Perfect Bayesian equilibrium, respectively).

We finish this chapter with a short list of criteria that we can use to evaluate solution concepts, based on Tadelis (2013).

1.6.1 Does an Equilibrium Exist?

When applying a solution concept, we would like to find that *at least one* strategy profile is an equilibrium outcome. Intuitively, this means that, if we face a payoff matrix, the solution concept finds one or more cells to be equilibria. Or, in the context of sequential-move games, the solution concept identifies a path of play to be an equilibrium of our game tree. If, instead, the solution concept did not find at least one equilibrium in all types of games, it does not help us prescribe how players behave in equilibrium, making the solution concept not applicable to different settings.

1.6.2 Is the Equilibrium Unique?

If a solution concept finds a single equilibrium outcome in all games, we can claim that it provides a precise prediction about how players behave. Graphically, uniqueness means that, when we apply a solution concept to simultaneous-move games, we find a single cell being the equilibrium of the game in all matrices. Similarly, when we apply the solution concept to sequential-move games, we find a single path of play in all trees. This property is, of course, rather demanding and, if we seek it, we increase the chances of facing no equilibria at all, thus conflicting with the above criterion (existence).

1.6.3 Is the Equilibrium Robust to Small Payoff Changes?

Another desirable property of a solution concept is that it yields the same equilibrium predictions if we make small changes in players' payoffs. For instance, in a payoff matrix like 1.2, if we alter player 1's payoff in one of the cells, without affecting the ranking of payoffs across outcomes (so player 1 would still rank her payoffs in each cell in the same way), we would like that the solution concept is unaffected. That is, if (H, l) was the equilibrium before we changed player 1's payoff, this strategy profile should remain the equilibrium after we change this player's payoff.

Intuitively, this property seeks to detect whether the solution concept would suddenly change its equilibrium prediction—jumping, for instance, from (H, l) to (L, h) in Matrix 1.1—if we change player 1's payoff by a relatively small amount.

1.6.4 Is the Equilibrium Pareto Optimal?

Finally, the equilibrium outcome that a solution concept provides can be interpreted as how players behave if they independently choose their strategies, as in a decentralized (or unregulated) market. In this context, we can evaluate whether other strategy profiles, such as $s = (s_i, s_{-i})$, can be Pareto improving relative to the equilibrium strategy profile $s^* = (s_i^*, s_{-i}^*)$. That is, we test whether strategy profile s makes every individual i weakly better off than in equilibrium, $u_i(s_i, s_{-i}) \geq u_i(s_i^*, s_{-i}^*)$ for every i; and at least one individual i is strictly better off, $u_i(s_i, s_{-i}) > u_i(s_i^*, s_{-i}^*)$.

This property relates to a common tension between equilibrium and Pareto optimal results in economics and other social sciences. Players, when independently choosing their strategies (such as output levels or contribution to a public project), converge toward equilibrium strategy profiles which often yield lower payoffs than alternative strategy profiles. However, players do not have incentives to unilaterally deviate from their equilibrium strategies and could only achieve a Pareto improvement if they coordinate their decisions, or if the rules of the game were changed. We return to this tension in future chapters.

References

Bolton, G., and A. Ockenfels (2000) "ERC: A Theory of Equity, Reciprocity, and Competition," American Economic Review, 90(1), pp. 166–193.

Fehr, E., and K. M. Schmidt (1999) "A Theory of Fairness, Competition, and Cooperation," The Quarterly Journal of Economics, 114(3), pp. 817–868.

Tadelis, S. (2013) *Game Theory, An Introduction*, Princeton University Press.

Equilibrium Dominance   2

2.1 Introduction

In this chapter we present the first solution concept, strict dominance. Unlike the solution concepts that we explore in all subsequent chapters, which directly search for the strategy (or strategies) that yield the highest possible payoff to a player, strict dominance seeks to just rule out those strategies that a rational player "would never choose." This occurs when a strategy provides a player with a strictly lower payoff than other available strategies *regardless* of the strategy chosen by her opponents. Intuitively, a strictly dominated strategy is "always a bad idea" and can, therefore, be eliminated from the set of strategies a player considers. This deletion process can, then, continue with other players, until we cannot rule out any other strategies for any player.

The result of this iterative process (iterated deletion of strictly dominated strategies, IDSDS) is straightforward and, importantly, helps us "clean the game" from strategies that a rational player would never choose. In some games, IDSDS provides us with a unique equilibrium (i.e., only one cell in the matrix surviving the iterative process), thus giving us a precise prediction. In other games, however, IDSDS only eliminates some strategies, still leaving several strategy profiles (cells) that survive the iterative process and, therefore, provide a relatively imprecise prediction of how players behave. Despite that, applying IDSDS can be quite helpful as a first tool before we consider other solution concepts, as it cleans up the matrix, removing those strategies that a rational player would never choose, yielding a reduced matrix easier to analyze. Nonetheless, IDSDS has no bite in some games, meaning that it does not help us eliminate strategies for any player. In these games, we will need to rely on other solution concepts to provide a meaningful equilibrium prediction.

Section 2.4, then, applies IDSDS to typical games, which serves as a first introduction to these games, which are used in next chapters to illustrate other results. We finish the chapter evaluating IDSDS as a solution concept, following the four criteria introduced in Chapter 1, and presenting two related solution concepts, weak dominance and strictly dominant equilibrium.

2.2 Strictly Dominated Strategies

In this first solution concept, we seek to delete those strategies that a rational player would never select. Recall, from Chapter 1, that player i's strategies satisfy $s_i, s_i' \in S_i$, for all $s_i' \neq s_i$, and that the strategy profile of player i's rivals, s_{-i}, satisfies $s_{-i} \in S_{-i}$.

Definition 2.1. **Strictly dominated strategy**. Player i finds that strategy s_i strictly dominates strategy $s_i' \neq s_i$ if

$$u_i(s_i, s_{-i}) > u_i(s_i', s_{-i})$$

for every strategy profile s_{-i} of player i's rivals.

Intuitively, strategy s_i yields a higher payoff than s_i' regardless of the strategy that her rivals choose. In other words, a strictly dominated strategy s_i' yields a strictly lower payoff than s_i regardless of her rivals' choice and, hence, we should expect a rational player to never choose such a strategy. We can, essentially, delete strictly dominated strategies from player i's strategy set, reducing the list of potential strategies this player considers.

Tool 2.1 provides a step-by-step road map (or "recipe") on finding strictly dominated strategies in two-player games, while Example 2.1 puts the tool to work.

Tool 2.1. How to find strictly dominated strategies:

1. Fix your attention on one strategy of the column player, s_2 (i.e., one specific column). Find a strategy s_1' that yields a strictly lower payoff than some other strategy s_1 for the row player, that is, $u_1(s_1, s_2) > u_1(s_1', s_2)$ for a given strategy of player 2.
2. Repeat step 1, but now fix your attention on a different column. That is, find if the above payoff inequality also holds, $u_1(s_1, s_2') > u_1(s_1', s_2')$, which is now evaluated at a different strategy of the column player, s_2'.
3. If, after repeating step 1 for all possible strategies of player 2 (all columns), you find that strategy s_1' yields a strictly lower payoff for player 1 than strategy s_1, you can claim that strategy s_1' is strictly dominated by s_1.

    Otherwise, player 1 does not have a strictly dominated strategy (or we may need to rely on randomizations, as described below).

The method is analogous when we search for strictly dominated strategies for the column player, but fix the strategy of the row player (looking at one specific row). We now compare the payoffs of the column player (second component of every pair) across columns. Specifically, we seek to find a strategy s_2' that yields a

2.3 Iterated Deletion of Strictly Dominated Strategies

strictly lower payoff than some other strategy s_2 for player 2, that is, $u_2(s_1, s_2) > u_2(s_1, s_2')$ for a given strategy of player 1, s_1 (that is, a given row). We then repeat this step across all possible strategies of player 1 to check if the payoff inequality remains.

Example 2.1 Finding strictly dominated strategies. Matrix 2.1 considers a setting in which two firms simultaneously and independently choose their output levels, High, Medium, or Low, denoted as H, M or L for firm 1 (in rows), and h, m, or l for firm 2 (in columns).

		Firm 2		
		h	m	l
	H	2, 2	3, 1	5, 0
Firm 1	M	1, 3	2, 2	2.5, 2.5
	L	0, 5	2.5, 2.5	3, 3

Matrix 2.1. Strictly dominated strategies in the output game.

We can easily show that L is strictly dominated for firm 1 because it yields a strictly lower payoff than H when its rival, firm 2, chooses:

- h in the left column (since $0 < 2$),
- m in the middle column (given that $2.5 < 3$), and
- l in the right-hand column (given that $3 < 5$).

A similar argument applies to row M, because it yields a strictly lower payoff than H when its rival, firm 2, chooses:

- h in the left column (since $1 < 2$),
- m in the middle column (given that $2 < 3$), and
- l in the right-hand column (given that $2.5 < 5$).

Therefore, strategies M and L are both strictly dominated by H. As payoffs are symmetric across firms, a similar argument applies to Firm 2, which finds strategies m and l to be strictly dominated by h. Try to confirm this result as a practice!
□

2.3 Iterated Deletion of Strictly Dominated Strategies

We can now use the above definition of strictly dominated strategies to obtain our first solution concept: Iterative Deletion of Strictly Dominated Strategies (IDSDS). Intuitively, we seek to delete all strictly dominated strategies for each player until we are left with those strategies that cannot be "ruled out" as being attractive for some players in at least some contingencies.

Tool 2.2. Applying IDSDS:

Step 1   From the definition of rationality, we know that a player would never use strictly dominated strategies, so we can delete them from her original strategy set, S_i, obtaining S'_i (which is technically a subset of S_i).

Step 2   We can then proceed by also using common knowledge of rationality, which in this context entails that every player j can put herself in her opponent's shoes, identify all strictly dominated strategies for her opponent, and delete them from j's strategy set, S_j, obtaining S'_j.

Step 3   We continue this process for one more step: player i considers now her rival's reduced strategy set S'_j and finds whether some of her own strategies in $S'_i \times S'_j$ become strictly dominated.[1] At the end of this step, we obtain a further reduced strategy set S''_i. In some games, this set may coincide with the original set in the third step, S'_i, if we could not identify more strictly dominated strategies for player i; while in other games strategy set S''_i is a subset of S'_i.

Step 4   Starting with strategy sets S''_i and S''_j, we repeat the above process, seeking to find more strictly dominated strategies for either player.

Step k The process continues until no player can identify more strictly dominated strategies to delete. The remaining strategies are referred to as the strategy profile/s surviving IDSDS.

In summary, rationality helps a player delete her strictly dominated strategies, but common knowledge of rationality helps her anticipate the strictly dominated strategies of her rivals. This step lets player i assess which of her own strategies can become strictly dominated in successive rounds of iteration.

Example 2.2 Applying IDSDS. Consider Matrix 2.2a describing a simultaneous-move game between firms 1 and 2.

		Firm 2	
		h	l
Firm 1	H	4, 4	0, 2
	M	1, 4	2, 0
	L	0, 2	0, 0

Matrix 2.2a. IDSDS yields a unique equilibrium.

---

[1] These strategies were not deleted in the first round of IDSDS, where player i only seeks to delete strictly dominated strategies in S_i, without putting herself in player j's shoes, but can be deleted now thanks to common knowledge of rationality.

2.3 Iterated Deletion of Strictly Dominated Strategies

Starting with firm 1, row L yields a strictly lower payoff (zero) than M regardless of the column that firm 2 chooses (both in h and in l). Therefore, we can claim that L is strictly dominated by M, deleting row L (from Matrix 2.2a), as depicted in Matrix 2.2b.

$$
\begin{array}{c c}
 & \text{Firm 2} \\
 & \begin{array}{c c} h & l \end{array} \\
\text{Firm 1} \quad \begin{array}{c} H \\ M \end{array} & \begin{array}{|c|c|} \hline 4,4 & 0,2 \\ \hline 1,4 & 2,0 \\ \hline \end{array}
\end{array}
$$

Matrix 2.2b. IDSDS yields a unique equilibrium-After one round of IDSDS.

We can now move to firm 2, the column player, finding that l yields a strictly lower payoff than h, regardless of the row that firm 1 chooses. Indeed, if firm 1 chooses H (in the top row), firm 2's payoff from l is only 2 while that from h is 4. Similarly, when firm 1 selects M (bottom row), firm 2's payoff from l is zero while that of h is 4. In short, l is strictly dominated by h, implying that we can delete the column corresponding to l from Matrix 2.2b, leaving us with Matrix 2.2c.

$$
\begin{array}{c c}
 & \text{Firm 2} \\
 & h \\
\text{Firm 1} \quad \begin{array}{c} H \\ M \end{array} & \begin{array}{|c|} \hline 4,4 \\ \hline 1,4 \\ \hline \end{array}
\end{array}
$$

Matrix 2.2c. IDSDS yields a unique equilibrium-After two rounds of IDSDS

We now turn to firm 1 again. At this point of the analysis, firm 1 anticipates that firm 2 has deleted l because of being strictly dominated, which leaves firm 2 with only one undominated strategy (h), as shown in Matrix 2.2c. In this third round of IDSDS, firm 1 finds that strategy M is strictly dominated by H since $4 > 1$. After deleting row M from Matrix 2.2c, we are left with only one cell, as shown in Matrix 2.2d.

$$
\begin{array}{c c}
 & \text{Firm 2} \\
 & h \\
\text{Firm 1} \quad H & \begin{array}{|c|} \hline 4,4 \\ \hline \end{array}
\end{array}
$$

Matrix 2.2d. IDSDS yields a unique equilibrium-After three rounds of IDSDS

In summary, we iteratively deleted one strictly dominated strategy for firm 1, then one for firm 2, and then one more for firm 1, leaving us with a single strategy profile surviving IDSDS, (H, h), with corresponding payoffs $(4, 4)$, meaning that every firm earns a payoff of 4. □

2.3.1 Does the Order of Deletion Matter in IDSDS?

The order of deletion *does not matter* when applying IDSDS. This is an important property, as otherwise we would predict different equilibrium results if we started deleting strictly dominated strategies for player 1 or 2. The next example illustrates this property, considering again Example 2.2, but starting to delete strictly dominated strategies for firm 2.

Example 2.3 Applying IDSDS but starting with firm 2. In the first step of IDSDS, firm 2 finds that the column corresponding to l is strictly dominated by h in the original Matrix 2.2a. Matrix 2.2e deletes column l.

	Firm 2
	h
H	4, 4
Firm 1 M	1, 4
L	0, 2

Matrix 2.2e. Applying IDSDS - Starting with firm 2.

At this point, it is easy to find that, for firm 1, row M is strictly dominated by H and so is L. After deleting the rows corresponding to M and L, we are left with a single strategy profile surviving IDSDS: (H, h). Therefore, starting IDSDS with firm 2, we obtained the *same* equilibrium result as when we started with firm 1. □

2.3.2 Deleting More Than One Strategy at a Time

An interesting point from Example 2.3 is that it illustrates that, at any step of our application of IDSDS, we can delete all strictly dominated strategies we found for that player—so we do not need to delete one strictly dominated strategy at a time, saving us valuable time. This was the case when we started with firm 2 and then moved on to firm 1, where we deleted the rows corresponding to M and L (as both were strictly dominated by H) in the *same step* of IDSDS. In addition, Example 2.3 highlights that, when one player finds one of its strategies to strictly dominate all its other strategies, we can delete all of them, leaving us with a single surviving strategy for this player (only one row in this example).

2.3.3 Multiple Equilibrium Predictions

In Examples 2.2 and 2.3, the use of IDSDS yields a unique equilibrium prediction: strategy profile (H, h). When IDSDS provides a unique equilibrium, we say that the game is "dominance solvable." In other games, however, the use of IDSDS does not necessarily provide such a precise prediction, but rather two or more strategy profiles, as Example 2.4 illustrates.

Example 2.4 When IDSDS does not provide a unique equilibrium. Consider Matrix 2.3a. Starting IDSDS with firm 1, we find that H is strictly dominated by L, since it yields a lower payoff than L regardless of the strategy that firm 2 chooses (i.e., independently of the column that firm 2 selects). To see this point, note that, when firm 2 chooses h, firm 1's payoffs from H and L satisfy $2 < 3$; when firm 2 selects m, firm 1's payoffs satisfy $1 < 4$; and when firm 2 chooses l, they satisfy $3 < 5$.

		Firm 2		
		h	m	l
	H	2, 3	1, 4	3, 2
Firm 1	M	5, 3	2, 1	1, 2
	L	3, 6	4, 7	5, 4

Matrix 2.3a. IDSDS yields more than one equilibrium-I.

After deleting the strictly dominated strategy H from firm 1's rows in Matrix 2.3a, we are left with the reduced Matrix 2.3b, which only has two rows. We can put ourselves in the shoes of firm 2, to see if we can find a strictly dominated strategy for this firm. Specifically, l is strictly dominated by h, since l yields a strictly lower payoff than h regardless of the row that firm 1 selects. Indeed, when firm 1 chooses M, firm 2's payoffs satisfy $2 < 3$; and when firm 1 selects L, firm 2's payoffs satisfy $4 < 6$.

		Firm 2		
		h	m	l
Firm 1	M	5, 3	2, 1	1, 2
	L	3, 6	4, 7	5, 4

Matrix 2.3b. IDSDS yields more than one equilibrium-II.

After deleting column l from firm 2's strategies, we are left with a further reduced matrix (see Matrix 2.3c). We can now move again to analyze firm 1. At this point, however, we cannot identify any more strictly dominated strategies for this firm since there is no strategy (no row in Matrix 2.3c) yielding a lower payoff regardless of the column firm 2 plays.[2]

$$
\begin{array}{c}
\textit{Firm 2} \\
\begin{array}{cc}
h & m
\end{array} \\
\textit{Firm 1} \quad
\begin{array}{c|c|c|}
\cline{2-3}
M & 5,3 & 2,1 \\
\cline{2-3}
L & 3,6 & 4,7 \\
\cline{2-3}
\end{array}
\end{array}
$$

Matrix 2.3c. IDSDS yields more than one equilibrium-III.

Therefore, the remaining four cells in Matrix 2.3c are our equilibrium prediction after applying IDSDS:

$$\text{IDSDS} = \{(M, h), (M, m), (L, h), (L, m)\},$$

entailing four possible outcomes. Importantly, the order of deletion does not affect the strategy profiles surviving IDSDS. As a practice, you can redo this example but starting with firm 2, trying to delete strictly dominated strategies, and then moving to firm 1. □

2.4   Applying IDSDS in Common Games

2.4.1   Prisoner's Dilemma Game

Consider the following scenario. Two people have been arrested by the police and held in different cells so they cannot communicate with each other. The prosecutor has only minor evidence against them, which would lead to only two years in jail. However, the prosecutor suspects that these individuals committed a felony so she separately offers the following deal to each prisoner:

> If you confess the crime and your partner doesn't, we will let you go home, while your partner will serve 8 years in jail. If, instead, you don't confess but your partner does, he will go home and you will serve 8 years in jail. If both of you confess, you will both serve 4 years in jail. Finally, if none of you confess, both of you will only serve 2 years in jail.

---

[2] Indeed, firm 1 prefers M to L if firm 2 chooses h (in the left column) since $5 > 3$; but prefers L to M if firm 2 chooses m (in the right-hand column) given that $4 > 2$. A similar argument applies to firm 2, since there is no strategy (column) yielding a lower payoff regardless of the row that firm 1 selects: if firm 1 chooses M, at the top row, firm 2 prefers h to m since $3 > 1$; but when firm 1 selects L, at the bottom row, firm 2 prefers m to h because $7 > 6$.

2.4 Applying IDSDS in Common Games

Matrix 2.4a describes this game, often known as the "Prisoner's Dilemma" game, where all amounts are negative to represent years in jail. We start finding player 1's strictly dominated strategies.

		Player 2	
		Confess	Not confess
Player 1	Confess	−4, −4	0, −8
	Not confess	−8, 0	−2, −2

Matrix 2.4a. The Prisoner's Dilemma game.

Player 1. For player 1, we find that Not confess is strictly dominated by Confess because he earns a higher payoff, both when player 2 confesses (in the left column) given that $-4 > -8$ and when player 2 does not confess (in the right column) since $0 > -2$. We can, then, delete the bottom row—corresponding to Not confess for player 1—from the above matrix, yielding Matrix 2.4b.

		Player 2	
		Confess	Not confess
Player 1	Confess	−4, −4	0, −8

Matrix 2.4b. The Prisoner's Dilemma game after one round of IDSDS.

Player 2. We can now turn to player 2 who anticipates player 1's deleting Not confess as being strictly dominated. In Matrix 2.4b, player 2 anticipates player 1 choosing Confess. In this context, player 2 only compares his payoff from Confess, −4, against that of Not confess, −8 (recall that player 2's payoffs are the second component in every cell). Therefore, player 2 finds Not confess as being strictly dominated in our second round of applying IDSDS, leaving us with only one cell surviving IDSDS, as shown in Matrix 2.4c.

		Player 2
		Confess
Player 1	Confess	−4, −4

Matrix 2.4c. The Prisoner's Dilemma game after two rounds of IDSDS.

In summary, both players find Not confess as strictly dominated, leaving them with a single strategy, Confess.

$$\text{IDSDS} = \{\text{Confess, Confess}\}.$$

This is a rather unfortunate result, because both prisoners end up serving 4 years in jail (as they both confess) rather than only two years (if none of them confessed). However, no player has incentives to unilaterally deviate to Not confess regardless of what his opponent does, driving them to this bad outcome.

In terms of the Pareto optimality criterion described in Chapter 1, the equilibrium of the Prisoner's Dilemma game is not Pareto optimal, as we can find another outcome (Not confess, Not confess) where both players strictly improve their payoffs.

Other examples of Prisoner's Dilemma games. Similar conflicts between individual and social incentives are common in economics, such as price wars between firms (both firms would be better-off by setting high prices, but each firm has individual incentives to lower its own price to capture a larger market share); tariff wars between countries (where both countries would be better-off by setting low tariffs, but each country has individual incentives to raise its own tariff to protect domestic firms); or the use of negative campaigning in politics (where all candidates would be better-off by not spending money on negative campaigning, but each candidate has incentives to spend some money on it to win the election). Finally, note that the Prisoner's Dilemma game is also part of the plot in several movies, such as those listed in Table 2.1.

General form. In Matrix 2.4d, players exhibit the same incentives as in the Prisoner's Dilemma of Matrix 2.4a, but allowing for more general payoffs. For players to face the same incentives as in Matrix 2.4a, payoffs in Matrix 2.4d must satisfy $a > c$ and $b > d$. (In Matrix 2.4a, for instance, $a = -4$, $c = -8$, $b = 0$, and $d = -2$, entailing that $b > d > a > c$.) This payoff ranking entails that Confess strictly dominates Not confess for every player (i.e., player 1 finds that $a > c$

Table 2.1 The Prisoner's Dilemma in movies

Movie	Scene
The Dark Knight: https://www.youtube.com/watch?v=K4GAQtGtd_0	
Murder by Numbers: https://www.youtube.com/watch?v=UAR7WDrL3Ec	
The Hunger Games: https://www.youtube.com/watch?v=rCWL-pRX_hE	

2.4 Applying IDSDS in Common Games

when player 2 confesses, and $b > d$ when player 2 does not confess). As a result, (Confess, Confess) is the unique strategy profile surviving IDSDS. In addition, if payoffs satisfy $d > a$, players would be better-off if they could coordinate their actions, moving from (Confess, Confess) to (Not confess, Not confess).

		Player 2	
		Confess	Not confess
Player 1	Confess	a, a	b, c
	Not confess	c, b	d, d

Matrix 2.4d. The Prisoner's Dilemma game - General form.

2.4.2 Coordination Games—The Battle of the Sexes Game

Consider a couple without cell phone signal in separate areas of the city. In the morning, they talked about where to go after work, the football game or the opera, but they did not agree on an event. Every individual must simultaneously and independently choose whether to attend a football game or the opera. As illustrated in Matrix 2.5a, the husband prefers to attend the football game if his wife is also there (earning a payoff of 10), followed by attending the opera with her (earning a payoff of 8), followed by being at the football game alone (payoff of 6), and attending the opera without her (payoff of 4). The wife's payoffs are symmetric: her most preferred event is being at the opera with him (payoff of 10), followed by being at the football game with him (payoff of 8), being at the opera without him (payoff of 6), and followed by being at the football game without him (payoff of 4).

		Wife	
		Football, F	Opera, O
Husband	Football, F	10, 8	6, 6
	Opera, O	4, 4	8, 10

Matrix 2.5a. The Battle of the Sexes game.

Husband. Starting with the row player (husband), we cannot find any strictly dominated strategy. Indeed, when his wife chooses Football (in the left column) his payoff from Football is higher than from Opera ($10 > 4$), but when she chooses Opera (in the right column) his payoff from Opera is actually higher than from Football ($8 > 6$). In summary, we did not find a strategy that yields strictly lower payoffs for the husband regardless of the strategy that his wife chooses.

Wife. We can move to the column player (wife) who still considers Matrix 2.5a because she can anticipate that her husband has no strictly dominated strategies (i.e., no rows that can be deleted from the matrix). She does not have any strictly dominated strategy either. In particular, when her husband chooses Football (in the top row) her payoff from Football is higher than from Opera (8 > 6), but when he chooses Opera (in the bottom row) her payoff is higher from Opera than from Football (10 > 4).

Overall, we could not delete any strictly dominated strategy for either player, leaving us with all four strategy profiles in the original matrix (all cells), that is,

$$IDSDS = \{(F, F), (F, O), (O, F), (O, O)\}.$$

In future chapters, we introduce solution concepts different from IDSDS, which produce more precise equilibrium predictions for this game. As you may expect, players in the Battle of the Sexes game have incentives to attend the same event as their spouse, explaining why this class of games is known as "coordination games." We next describe applications of this game to other settings.

Other examples of coordination games. We can represent the interaction between two or more firms investing in different technologies as a coordination game (e.g., firms can share files only if they both invest in the same technology). Similarly, friends purchasing game consoles that run on different operating systems face similar incentives as those in the Battle of the Sexes game. Other examples include choosing the sides of the road to drive (driving conventions) and, more generally, games where players enjoy a positive network externality when choosing the same strategy as her rivals, such as social media platforms.

General form. Matrix 2.5b presents a coordination game allowing for more general payoffs. For players to face coordination incentives, payoffs must satisfy $a_i > c_i$ and $d_i > b_i$ for every player $i = \{H, W\}$ where H denotes husband and W represents wife. This payoff structure entails that players do not have strictly dominated strategies, as in the Battle of the Sexes in Matrix 2.5a. Indeed, if his wife goes to the Football game, the husband would go to the Football game as well since $a_H > c_H$; and if his wife goes to the Opera, he also goes to the Opera because $d_H > b_H$. (A similar argument applies to the wife's payoffs: $a_W > c_W$ when her husband is at the Football game, and $d_W > b_W$ when he is at the Opera.) Therefore, all four strategy profiles in the matrix survive IDSDS.

		Wife	
		Football, F	Opera, O
Husband	Football, F	a_H, a_W	b_H, c_W
	Opera, O	c_H, b_W	d_H, d_W

Matrix 2.5b. Coordination games - General form.

2.4 Applying IDSDS in Common Games

When players attend the same event, Matrix 2.5b allows, for generality, that they earn the same payoffs, which happens if $a_H = a_W$ and $d_H = d_W$; or that they earn different payoffs, which occurs if $a_H \neq a_W$, $d_H \neq d_W$, or both. This was the case in Matrix 2.5a, where $a_H = 10$ and $a_W = 8$ when both players go the Football game, but their payoffs switch to $d_H = 8$ and $d_W = 10$ when they both go to the Opera. Similarly, when players miscoordinate (attending different events) Matrix 2.5b allows that players earn the same payoff, which occurs if $c_H = b_W$ and $b_H = c_W$ (as in Matrix 2.5a, where $c_H = b_W = 4$ and $b_H = c_W = 6$); or that they earn different payoffs otherwise.

2.4.3 Pareto Coordination Game—The Stag Hunt Game

Matrix 2.6a represents a special class of coordination games, the so-called "Stag Hunt game," originally proposed by philosopher Jean-Jacques Rousseau. Two hunters simultaneously and independently choose whether to hunt a stag or a hare. Each hunter can catch a hare alone, as it is a small animal, but needs her rival's help to hunt a stag. The stag is, however, a larger animal, providing more meat. As a consequence, both players would be better-off if they coordinated hunting the stag.

		Player 2	
		Stag, S	Hare, H
Player 1	Stag, S	6, 6	1, 4
	Hare, H	4, 1	2, 2

Matrix 2.6a. The stag hunt game.

We next show that, as the Battle of the Sexes game, the stag hunt game has no strictly dominated strategies for either player.

Player 1. Starting with the row player, we cannot find strictly dominated strategies. Indeed, when player 2 chooses Stag (in the left column) her payoff from also choosing Stag is higher than from Hare ($6 > 4$), but when player 2 selects Hare (in the right column) her payoff from Hare is actually higher than from Stag ($2 > 1$). In summary, we did not find a strategy that yields strictly lower payoffs for player 1 regardless of how player 2 behaves.

Player 2. Since payoffs are symmetric across players, a similar argument applies to player 2, who does not have strictly dominated strategies either. In particular, when player 1 chooses Stag (in the top row) her payoff from also choosing Stag is higher than from Hare ($6 > 4$) but when she selects Hare (in the bottom row) her payoff is higher from Hare than from Stag ($2 > 1$).

Overall, we cannot delete any strictly dominated strategy for either player, leaving us with all four strategy profiles in the original matrix (all cells), that is,

$$\text{IDSDS} = \{(S, S), (S, H), (S, H), (H, H)\}.$$

Generic form. Matrix 2.6b presents a generic Pareto coordination game, where payoffs satisfy $a > b \geq d > c$. As in the stag hunt game, no player has strictly dominated strategies. When player 2 selects Stag, player 1 is better-off choosing Stag as well since $a > b$; but when player 2 selects Hare, player 1 is better-off choosing Hare because $d > c$. (As a reference, note that in Matrix 2.6a, payoffs satisfy this ranking given that $a = 6$, $b = 4$, $c = 1$, and $d = 2$.)

		Player 2	
		Stag	Hare
Player 1	Stag	a, a	c, b
	Hare	b, c	d, d

Matrix 2.6b. Pareto coordination games - General form.

2.4.4 Anticoordination Game—The Game of Chicken

Matrix 2.7a presents a game with the opposite strategic incentives as the Battle of the Sexes game. The matrix illustrates the Game of Chicken, as seen in movies like *Rebel without a Cause*, where two teenagers in cars drive toward each other (or toward a cliff). If both swerve, they avoid the accident, but both are regarded as "chicken" by their friends, yielding a negative payoff of -1 to each player. If only one player swerves, he is declared the chicken, earning a payoff of -8, while his friend (who stayed) is the top dog, getting a payoff of 10. Finally, if both players stay, they crash in a serious car accident, yielding a payoff of -30 for both of them (they almost die!).

		Player 2	
		Swerve	Stay
Player 1	Swerve	$-1, -1$	$-8, 10$
	Stay	$10, -8$	$-30, -30$

Matrix 2.7a. Anticoordination game.

2.4 Applying IDSDS in Common Games

Player 1. As usual, we start with the row player.[3] Player 1 has no strictly dominated strategy: when player 2 chooses Swerve (in the left column) player 1's payoff from Stay is higher than from Swerve ($10 > -1$), but when player 2 selects Stay (in the right column) player 1's payoff from Swerve is higher than from Stay ($-8 > -30$). In other words, when player 2 swerves, player 1 earns a higher payoff by staying and becoming the top dog of the gang. However, when player 2 stays, player 1 prefers to be named a chicken than risking his life in a serious car accident. Ultimately, we cannot delete any row from player 1 as being strictly dominated.

Player 2. Anticipating that player 1 has no strictly dominated strategies, player 2 still considers the whole Matrix 2.7a when finding strictly dominated strategies. As player 2's payoffs are symmetric, we find that he does not have strictly dominated strategies either. When player 1 chooses Swerve (in the top row), player 2's payoff from Stay is higher than from Swerve ($10 > -1$) while when player 1 chooses Stay (in the bottom row) player 2's payoff from Swerve is higher than from Stay ($-8 > -30$), exhibiting a similar intuition as to player 1 above.

In summary, we could not find any strictly dominated strategies for either player (as in coordination games such as the Battle of the Sexes), leaving us with the four strategy profiles in the original matrix as our equilibrium prediction, that is,

IDSDS $= \{(Swerve, Swerve), (Swerve, Stay), (Stay, Swerve), (Stay, Stay)\}$.

Because every player earns a higher payoff by choosing a different strategy than his opponent, the Game of Chicken is an example of "anticoordination games." In Chapter 3, we revisit this type of game, seeking to find more precise equilibrium predictions with the use of another solution concept (Nash equilibria).

Other examples of anticoordination games. We can also use anticoordination games to model two firms simultaneously deciding whether to enter into an industry with low demand, which would yield positive profits for only one firm; or ex-couples seeking to avoid each other by not attending the same social events. More generally, games where players suffer from a negative network externality when selecting the same strategy as her rivals (also known as "congestion games") exhibit similar incentives as the anticoordination game described above, as players seek to choose opposite strategies. To finish this section on a humorous note, Table 2.2 lists some movies where characters interact in the Game of Chicken or anticoordination games.

Generic form. Matrix 2.7b presents an anticoordination game with generic payoffs. For players to exhibit incentives to anticoordinate, payoffs must satisfy $c > d > b > a$. As in Matrix 2.7a, no player finds strictly dominated strategies. When player 2 selects Swerve, player 1 is better-off choosing to Stay (since $b > a$), but when player 2 selects Stay, player 1 is better-off choosing Swerve (because $c > d$).

---

[3] As a practice, however, you can start with the column player, showing that you would obtain the same equilibrium prediction.

Table 2.2 Anticoordination games in different movies

Movie	Scene
Rebel without a Cause: https://www.youtube.com/watch?v=BGtEp7zFdrc	
Footloose: https://www.youtube.com/watch?v=ZL57muBck0w	
Stand by Me: https://www.youtube.com/watch?v=6L-IbkWHsOE	
The Hunt for Red October: https://www.youtube.com/watch?v=mh599VtMB1c	
Thirteen days: https://www.youtube.com/watch?v=1DqzjoXz22c	
A Beautiful Mind: https://www.youtube.com/watch?v=LJS7Igvk6ZM	

		Player 2	
		Swerve	Stay
Player 1	Swerve	a, a	c, b
	Stay	b, c	d, d

Matrix 2.7b. Anticoordination game - General form.

2.4.5 Symmetric and Asymmetric Games

Definition 2.2. **Symmetric game.** A two-player game is symmetric if both players' strategy sets coincide, $S_A = S_B$, and payoffs are unaffected by the identity of the player choosing each strategy, that is,

$$u_A(s_A, s_B) = u_B(s_A, s_B)$$

for every strategy profile (s_A, s_B).

2.4 Applying IDSDS in Common Games

This property is also known as the payoff function satisfying *anonymity*: starting from a setting where player A chooses s_A and player B chooses s_B, if we were to switch the identities of players A and B, so that player A becomes B and player B becomes A, every player's payoff would be unaffected. Indeed, player A's payoff in the original setting was $u_A(s_A, s_B)$ where she chooses s_A and her rival selects s_B; while after the "identity switch" player A becomes player B and her payoff is now $u_B(s_A, s_B)$, taking into account that she chooses s_B while her rival picks s_A.[4]

Therefore, the Prisoner's Dilemma game and the Game of Chicken are symmetric, and so is the game in Matrix 2.1, but the Battle of the Sexes game is not. As a practice, consider the Prisoner's Dilemma game and strategy profile $(s_1, s_2) = (C, NC)$, where player 1 confesses but player 2 does not. In this setting, player 1's payoff is $u_1(C, NC) = 0$ while player 2's is $u_2(NC, C) = -8$. If we switch their identities, so that player 1 becomes 2 (and does not confess) and player 2 becomes 1 (and confesses), their payoffs are $u_2(NC, C) = 0$ for player 2 (who originally was player 1) and $u_1(NC, C) = -8$ for player 1 (who originally was player 2), thus being unaffected by the identity switch. A similar argument applies to all other strategy profiles in this game, (C, C), (NC, NC), and (NC, C), which you can confirm as a practice.

Visually, the matrix of a symmetric game must satisfy the following properties:

1. Same number of rows and columns, so both players have the same number of available strategies.
2. Same action labels in rows and columns, so players A and B face the same strategy sets.
3. Payoffs along the main diagonal must coincide for every player (i.e., both players earn the same payoff when they choose the same strategy). Formally, this property entails that, if both players choose the same strategy, $s_A = s_B = s$, their payoffs must also coincide, $u_A(s, s) = u_B(s, s)$.
4. Cells above the main diagonal must be mirror images of those below the main diagonal, so that when $s_A \neq s_B$, we can satisfy condition $u_A(s_A, s_B) = u_B(s_A, s_B)$.

As a practice, you can confirm that the Game of Chicken satisfies these four properties.

Definition 2.3. **Asymmetric game.** A two-player game is asymmetric if players face different strategy sets, $S_A \neq S_B$; or if, despite facing the same strategy sets, $S_A = S_B$, their payoffs satisfy

$$u_A(s_A, s_B) \neq u_B(s_A, s_B)$$

---

[4] The extension of this definition to a setting with more than two players is a bit involved, as it requires the use of a perturbation $\pi(i)$, which maps the identity of individual i into another individual j (potentially still i). We then check that the payoff of every individual i remains unaffected before and after the perturbation.

for at least one strategy profile (s_A, s_B).

In other words, a game is asymmetric if it is not symmetric, i.e., it violates at least one of the above four properties. For instance, switching the identity of the two players yields a different payoff for at least one of the players and at least one of the strategy profiles. The Battle of the Sexes game satisfies this definition. To see this point, take for example strategy profile $(s_H, s_W) = (F, F)$, where both husband and wife go to the football game. In this context, the husband's payoff is $u_H(F, F) = 10$ while his wife's is $u_W(F, F) = 8$. If we were to switch their identities, the wife's payoff becomes $u_W(F, F) = 8$ while the husband's is $u_H(F, F) = 10$, implying that their payoffs are affected.[5]

Graphically, this property could be checked in Matrix 2.5a, as players do not earn the same payoff in symmetric strategy profiles (when they both choose the same strategy) along the main diagonal, as required by property 3 above, implying that the game is not symmetric.

2.5 Allowing for Randomizations to Bring IDSDS Further

The application of IDSDS may have little bite in some games, such as Example 2.2 where we could only delete one strategy for each player. In other games, however, IDSDS has no bite, such as the Battle of the Sexes game (as an example of coordination games) and the Game of Chicken (representing anticoordination games). This "no bite" result happened, nonetheless, because we assumed that players use one strategy with 100% probability (which we refer as "pure strategies"). When we allow players to randomize between two or more strategies, IDSDS may still have some bite, as Example 2.5 illustrates.

Example 2.5 **Allowing for randomizations in IDSDS.** Consider Matrix 2.8 describing a game between firms 1 and 2, simultaneously choosing a high, medium, or low price. We can immediately notice that firm 1 has no strictly dominated strategies we can delete. While row H at the top of the matrix yields strictly lower payoffs than L when firm 2 chooses h or m ($0 < 10$ and $4 < 6$, respectively), it yields the same payoff when firm 2 chooses column l (4). A similar argument applies for firm 2, which does not have a strictly dominated strategy either.

---

[5] In particular, the identity switch decreased the husband's payoff by two units but increased his wife's by two units.

2.5 Allowing for Randomizations to Bring IDSDS Further

			Firm 2		
			h	m	l
		H	0, 10	4, 6	4, 6
Firm 1	Prob. $(1-q) \to$	M	4, 6	0, 10	6, 4
	Prob. $q \to$	L	10, 0	6, 4	4, 6

Matrix 2.8. IDSDS allowing for randomizations-I.

However, strategy H for firm 1 seems to produce a lower payoff than a combination of rows M and L. Specifically, M and L yield weakly higher payoffs than H does when firm 2 picks columns h or l, but yields extreme payoffs when firm 2 chooses column m (which are clearly above or below that of H). This argument indicates that, if we could create a linear combination between the payoffs in rows M and L, player 1 would receive an expected utility that lies strictly above that of strategy H, ultimately helping us to delete this top row from the matrix. Let us first compute the expected utility from assigning a probability weight $q \in (0, 1)$ to strategy L and $1-q$ to M, entailing

$$EU_1(\sigma|h) = q10 + 4(1-q) = 4 + 6q$$
$$EU_1(\sigma|m) = q6 + 0(1-q) = 6q$$
$$EU_1(\sigma|l) = q4 + 6(1-q) = 6 - 2q$$

where $\sigma = qL + (1-q)M$ denotes the randomization between rows M and L, so $EU_1(\sigma|h)$ represents firm 1's expected payoff from this randomization given that firm 2 chooses h in the left column of the matrix. A similar argument applies to its expected payoff $EU_1(\sigma|m)$ and to $EU_1(\sigma|l)$. Intuitively, a randomization could be firm 1 flipping a coin and choosing L when heads comes up and M otherwise, implying that $q = \frac{1}{2}$. However, at this point, we do not know the exact probability weight q that firm 1 must assign to row L that would produce an expected utility higher than the certain utility of choosing H. We next seek to find the values of q that yield this result.

We can now insert the expected payoffs in Matrix 2.9 to more visually compare them against H, as follows. As we only seek to compare firm 1's payoffs, we only report its payoffs.

		Firm 2		
		h	m	l
Firm 1	H	0	4	4
	$qL + (1-q)M$	$4+6q$	$6q$	$6-2q$

Matrix 2.9. IDSDS allowing for randomization-II.

We can then say that, for the randomization $\sigma = qL+(1-q)M$ to strictly dominate strategy H, we need that firm 1's expected payoff to satisfy:

$$4 + 6q > 0, \text{ or } q > -\frac{2}{3} \text{ when firm 2 chooses } h,$$

$$6q > 4, \text{ or } q > \frac{2}{3} \text{ when firm 2 chooses } m, \text{ and}$$

$$6 - 2q > 4, \text{ or } q < 1 \text{ when firm 2 chooses } l.$$

The first and last condition on probability q holds because q must satisfy $q \in (0, 1)$ by assumption. However, the second condition, $q > \frac{2}{3}$, restricts the range of probability weights we can use in randomization $qL + (1 - q)M$. Therefore, any randomization with $q > \frac{2}{3}$, helps us show that strategy H is strictly dominated, so we can delete it from Matrix 2.8 to obtain Matrix 2.10.

		Firm 2		
		h	m	l
Firm 1	M	4, 6	0, 10	6, 4
	L	10, 0	6, 4	4, 6

Matrix 2.10. IDSDS allowing for randomizations-III.

At this point, we can move to firm 2 to note that h is strictly dominated by m since h yields a strictly lower payoff than m does, both when firm 1 chooses M in the top row (where $10 > 6$) and when firm 1 chooses L in the bottom row (where $4 > 0$). After deleting this strictly dominated strategy for firm 2, we are left with Matrix 2.11.

		Firm 2	
		m	l
Firm 1	M	0, 10	6, 4
	L	6, 4	4, 6

Matrix 2.11. IDSDS allowing for randomizations-IV.

We can now move again to firm 1, noticing that it no longer has strictly dominated strategies. In particular, M yields a strictly lower payoff than L when firm 2 chooses m, but a strictly higher payoff than L when firm 2 chooses l. If we move to firm 2, we cannot find strictly dominated strategies for this player either. When firm 1 chooses row M, firm 2 is better-off selecting m; but when firm 1 chooses row L, firm 2 is better-off selecting l.

In summary, the application of IDSDS did not have a bite when we assumed players could not randomize. In that context, our equilibrium prediction would have been the game as a whole (3 × 3 = 9 different outcomes!). Considering the possibility of randomizations, however, helped us discard one strategy for each player as being strictly dominated, leaving us with a 2 × 2 matrix (4 different equilibrium outcomes), that is,

$$\text{IDSDS} = \{(M, m), (M, l), (L, m), (L, l)\}.$$

2.5.1 What If IDSDS Has No Bite?

In some games, players may have no strictly dominated strategies that we can delete using IDSDS (not even after allowing players to randomize!), which is informally described as that IDSDS "has no bite." In this class of games, and generally in those where IDSDS provides a relatively imprecise equilibrium prediction like in Example 2.3, we can apply the Nash equilibrium solution concept that we define in Chapter 3, which offers the same (or more precise) equilibrium predictions as IDSDS.

2.6 Evaluating IDSDS as a Solution Concept

Following our discussion in Sect. 1.6, we next evaluate IDSDS as a solution concept according to the following criteria:

1. **Existence? Yes.** When we apply IDSDS to any game, we find that at least one equilibrium exists. Graphically, this means that at least one cell must survive IDSDS. While we showed that in many games IDSDS may have no bite, or little bite (in the sense of only deleting some rows or columns), it still points to one or more strategy profiles as equilibrium outcomes of the game, thus satisfying existence.
2. **Uniqueness? No.** As shown in previous sections, IDSDS does not necessarily provide a unique equilibrium outcome for all games since more than one cell may survive IDSDS, giving us more than one equilibrium prediction. Recall that the "uniqueness" criterion means that a solution concept, IDSDS in this case, must yield a unique equilibrium prediction in *all* games. This criterion is, of course, rather demanding, and does not hold for IDSDS.
3. **Robust to small payoff perturbations? Yes.** If we change the payoff of one of the players by a small amount (e.g., 0.001, but generally, for any ε that approaches zero), IDSDS still yields the same equilibrium outcome/s, implying that IDSDS is robust to small payoff perturbations. This is due to the fact that, if a strategy s_i strictly dominates another strategy s'_i in our original game, it

must strictly dominate s'_i after we apply a small payoff perturbation, meaning that we can delete strategy s'_i before and after the payoff perturbation.[6]

4. **Socially optimal? No**. The application of IDSDS does not necessarily yield socially optimal outcomes. To see this point, recall the Prisoner's Dilemma game, where IDSDS provides us with a unique equilibrium prediction, (Confess, Confess), which does not coincide with the strategy profile that maximizes the sum of players' payoffs, (Not Confess, Not Confess).[7] A similar argument applies to other examples presented in this chapter, and you should confirm this point as a practice.

2.7 Weakly Dominated Strategies

In some games, player i may find that her payoff from two strategies, s_i and s'_i, may coincide, which yields the following weaker version of the definition of strictly dominated strategies.

Definition 2.4. **Weakly dominated strategies**. Player i finds that strategy s_i weakly dominates another strategy s'_i if:

1. $u_i(s_i, s_{-i}) \geq u_i(s'_i, s_{-i})$ for every strategy profile s_{-i} of player i's rivals, and
2. $u_i(s_i, s_{-i}) > u_i(s'_i, s_{-i})$ for at least one strategy profile s_{-i}.

Therefore, we say that player i finds that strategy s_i weakly dominates another strategy s'_i if choosing s_i provides her with the same or higher payoff than selecting s'_i against all her rivals' strategies, but provides her with a strictly higher payoff than s'_i for at least one of her rivals' strategies.

Technically, we need requirement #2 in the definition of weakly dominated strategies since, otherwise, we could have a tie in the payoffs of strategies s_i and s'_i, across all strategy profiles of player i's rivals (that is, for all s_{-i}). Requirement #2 says that strategy s_i yields a strictly higher payoff than s'_i gainst, at least,

---

[6] You may consider a special case where this argument does not seem to work. For instance, strategy s_i provides player i with a payoff of $3, and strategy s'_i provides her with $2.99. Applying a small payoff perturbation of 0.01 to the payoff that she earns with strategy s_i, so she now earns $2.99, would make her indifferent between s_i and s'_i. This argument is, however, incorrect, as the "robustness to small payoff perturbations" criterion needs to hold for all payoff perturbations, ε, regardless of how small they are (that is, $\varepsilon \to 0$). This means that, in the above example, applying a small payoff perturbation on the payoff that player i earns with strategy s_i yields a payoff infinitely close to $3.

[7] Recall from Sect. 1.6.4 that an equilibrium outcome is socially optimal if there is no other strategy profile that could increase the payoff of one individual without decreasing that of her rivals. At (Not Confess, Not Confess), we cannot increase the payoff of one player without hurting her rival. At (Confess, Confess), however, both players can improve their payoffs if they move to (Not Confess, Not Confess), implying that (Confess, Confess) is not socially optimal.

2.7 Weakly Dominated Strategies

one strategy of i's rivals. Graphically, this means that in a two-player game, for instance, player 1 finds that two of her rows yield the same payoff against every column of player 2, *except* for one column of player 2, where player 1 finds strategy s_i to yield a strictly higher payoff than s'_i. A similar argument applies for player 2. Example 2.6 illustrates this point.

Example 2.6 Finding weakly dominated strategies. Matrix 2.12 presents a version of the game in Example 2.3 where we changed firm 1's payoffs from choosing M (see middle row).

		Firm 2		
		h	m	l
	H	2, 2	3, 2	5, 0
Firm 1	M	2, 3	3, 2	2.5, 2.5
	L	0, 5	2.5, 2.5	3, 3

Matrix 2.12. Finding weakly dominated strategies.

Given these changes, we can claim that strategy H weakly dominates M for firm 1 because H yields the same payoff as M when firm 2 chooses h in the left column (2 in both cases), and when it chooses m in the middle column (3 in both settings), but H still yields a strictly higher payoff than M when firm 2 selects l in the right-hand column (5 > 2.5). If firm 1's payoffs coincide with H and M for every strategy of firm 2 (i.e., for every column), we would not be able to claim that H weakly dominates M or the other way around. □

Example 2.6 also illustrates that, if strategy s_i is strictly dominated, then s_i must also be weakly dominated; but the converse is not necessarily true. More compactly,

$$s_i \text{ is strictly dominated} \underset{\nLeftarrow}{\Longrightarrow} s_i \text{ is weakly dominated}$$

In Matrix 2.12, strategy L is strictly dominated by H, since L yields an unambiguously lower payoff for firm 1 than H does, regardless of the column that firm 1 chooses. This means that requirement #2 of the weak dominance definition holds for all firm 2's strategies, meaning that L is also weakly dominated by H. However, strategy M is weakly dominated by H, as described in Example 2.6, but M is not strictly dominated by H. For that to occur, firm 1's payoff when choosing M would have to be strictly lower than that with H regardless of the strategy selected by firm 2 (for all columns). This is not the case in our example because firm 1 earns the same payoff with H and M when firm 2 chooses h in the left column.

2.7.1 Deletion Order Matters in IDWDS

Unlike IDSDS, where the order of deletion does not matter, the iterative deletion of weakly dominated strategies (IDWDS) may yield different equilibrium predictions depending on which player we start with. This is, of course, a serious drawback on this solution concept, making it too dependent on the deletion order, ultimately reducing its applicability to many types of games.

Starting with firm 1. To illustrate this point, we can start with firm 1 in Matrix 2.12, deleting row M because it is weakly dominated by H, yields Matrix 2.13a.

		Firm 2		
		h	m	l
Firm 1	H	2,2	3,2	5,0
	L	0,5	2.5,2.5	3,3

Matrix 2.13a. IDWDS starting with firm 1 - First step.

It is straightforward to see that firm 1 finds that H strictly dominates L, which helps us delete row L and obtain Matrix 2.13b.

		Firm 2		
		h	m	l
Firm 1	H	2,2	3,2	5,0

Matrix 2.13b. IDWDS starting with firm 1 - Second step.

Turning now to firm 2, it finds that h strictly dominates l, so we can delete the right-hand column and obtain Matrix 2.13c.

		Firm 2	
		h	m
Firm 1	H	2,2	3,2

Matrix 2.13c. IDWDS starting with firm 1 - Third step.

At this point, firm 2 does not have any more weakly dominated strategies: for h to weakly dominate m, for instance, we would need that at least one strategy of firm 1 yields a strictly higher payoff when firm 2 chooses h than when firm 2 selects m (see requirement #2 in the definition of weakly dominated strategies). Therefore,

2.7 Weakly Dominated Strategies

we cannot delete any further strategies as being weakly or strictly dominated for either firm, yielding an equilibrium outcome

$$\text{IDWDS} = \{(H, h), (H, m)\}.$$

Starting with firm 2. If, however, we start deleting weakly dominated strategies for firm 2 in Matrix 2.12, we find that column h weakly dominates m since it yields the same payoff as m does when firm 1 chooses H, but generates a strictly higher payoff when firm 1 selects M and L. After deleting column m from Matrix 2.12, we obtain Matrix 2.14a.

		Firm 2	
		h	l
	H	2,2	5,0
Firm 1	M	2,3	2.5, 2.5
	L	0,5	3,3

Matrix 2.14a. IDWDS starting with firm 2 - First step.

Still analyzing firm 2, we find that h strictly dominates l, which leaves us with the reduced Matrix 2.14b.

		Firm 2
		h
	H	2,2
Firm 1	M	2,3
	L	0,5

Matrix 2.14b. IDWDS starting with firm 2 - Second step.

We can now turn to firm 1, which finds that H strictly dominates L, yielding Matrix 2.14c.

		Firm 2
		h
Firm 1	H	2,2
	M	2,3

Matrix 2.14c. IDWDS starting with firm 2 - Third step.

At this point, however, we cannot delete any other strategies for firm 1, as none are strictly or weakly dominated.[8] Therefore, our equilibrium prediction when we apply IDWDS starting with firm 2 is

$$\text{IDWDS} = \{(H, h), (M, h)\},$$

which does not coincide with our equilibrium prediction when we applied IDWDS starting with firm 1.

2.7.2 IDSDS Vs. IDWDS

The previous example also illustrates that set of strategy profiles surviving IDWDS is a subset of those surviving IDSDS. In other words, if a strategy profile $s = (s_i, s_{-i})$ survives IDWDS, it must also survive IDSDS, that is,

$$s \text{ survives IDWDS} \Longrightarrow s \text{ survives IDSDS}$$
$$\nLeftarrow$$

To see this point, let us apply IDSDS on Matrix 2.12, finding that row L is strictly dominated by H for firm 1, and that column l is strictly dominated by h for firm 2. After deleting row L and column l, we obtain Matrix 2.15.

		Firm 2	
		h	m
Firm 1	H	2, 2	3, 2
	M	2, 3	3, 2

Matrix 2.15. Applying IDSDS in matrix 2.11.

Firm 1 cannot delete any strictly dominated strategies from Matrix 2.15, and neither can firm 2, yielding the following four equilibrium outcomes

$$\text{IDSDS} = \{(H, h), (H, m), (M, h), (M, m)\}.$$

Therefore, the application of IDWDS has more bite than IDSDS, producing more precise equilibrium predictions. However, IDWDS depends on the order of deletion (e.g., starting with player 1 or 2), thus limiting its applicability.

---

[8] If you think that H could be weakly dominated by M, see requirement #2 in the definition of weakly dominated strategies.

2.8 Strictly Dominant Strategies

In our discussion of strictly dominated strategies and IDSDS, we could find games where, in the first round of IDSDS, every player can delete all but one of her strategies as being strictly dominated, leaving us with only one surviving strategy for each player. This was the case, for instance, in the Prisoner's Dilemma game, where every player deletes Not confess in the first round of IDSDS leaving each prisoner with a unique surviving strategy, Confess.

A similar argument applies in larger games where every player has three or more strategies to choose from. In these settings, if player i finds that strategy s_i strictly dominates all her other available strategies we say that s_i is a *strictly dominant* strategy, and we would expect her choosing that strategy, as defined below.

Definition 2.5. **Strictly dominant strategy.** Player i finds that strategy s_i is strictly dominant if

$$u_i(s_i, s_{-i}) > u_i(s'_i, s_{-i})$$

for every strategy $s'_i \neq s_i$ and every strategy profile s_{-i} of player i's rivals.

Intuitively, all of her available strategies yield a strictly lower payoff than s_i regardless of the strategy profile that her rivals select. This means that we can delete all strictly dominated strategies $s'_i \neq s_i$, leaving her with only one surviving strategy, s_i. This strategy, then, provides player i with an unambiguously higher payoff than any of her available strategies and regardless of the strategy that her opponents select.

In Matrix 2.1, strategy H is strictly dominant for firm 1 as it yields a strictly higher payoff than all other strategies for this firm (i.e., a strictly higher payoff than either M or L). Similarly, strategy h is strictly dominant for firm 2.

We finally use the above discussion to define a "strictly dominant equilibrium," as follows.

Definition 2.6. **Strictly dominant equilibrium**. A strategy profile $s^{SD} = (s_i^{SD}, s_{-i}^{SD})$ is a strictly dominant equilibrium (SDE) if every player i finds her strategy, s_i^{SD}, to be strictly dominant.

It is straightforward to show that an SDE is the only strategy profile surviving IDSDS after just one round of deleting strictly dominated strategies for each player. Consider, for instance, Matrix 2.1, where player 1 (2) finds that strategy H (h) is strictly dominant, implying that strategy profile (H, h) is the unique SDE, and it is also the only strategy profile surviving IDSDS.

The next example illustrates how to find SDEs in some of the common games described in previous sections of this chapter.

Example 2.7 Finding SDEs. Consider the Prisoner's Dilemma game again. Confess is an SDE since Confess is a strictly dominant strategy for every player. Intuitively, every player's payoff from Confess is higher than from Not confess regardless of what her opponent does. In contrast, in the coordination and anticoordination games we discussed in previous sections, players do not have strictly dominant strategies and, hence, an SDE does not exist. □

In the game of Example 2.2, where we found that a unique strategy profile survives IDSDS, it is easy to see that players have no strictly dominant strategies and, as a consequence, there is no SDE. In other words, while we found a strictly dominated strategy for at least one player, and then we could delete strictly dominated strategies for other players, there is no strictly dominant strategy for either player in the original payoff Matrix 2.2a. As a consequence, that game does not have a SDE. A similar argument applies to games in Examples 2.3 and 2.4, where we found that several strategy profiles survived IDSDS.

As a result, if strategy profile $s = (s_i, s_{-i})$ is a SDE, it must also survive IDSDS, such as (*Confess, Confess*) in the Prisoner's Dilemma game. In contrast, a strategy profile surviving IDSDS does not need to be an SDE too, such as (*Football, Football*) in the Battle of the Sexes game. More compactly,

$$s \text{ is an SDE} \underset{\nLeftarrow}{\Longrightarrow} s \text{ survives IDSDS.}$$

2.8.1 Evaluating SDE as a Solution Concept

Following our discussion in Sect. 1.6, we next evaluate IDSDS as a solution concept according to the following criteria:

1. **Existence? No.** When we seek to find strictly dominant strategies for each player, in our search of an SDE, we may find games where one or more players do not have a strictly dominant strategy, implying that an SDE does not exist. While in other games we can find strictly dominant strategies for each player, and thus point to clear SDE, this is not guaranteed for all games, as required by the "existence" criterion.
2. **Uniqueness? Yes.** While an SDE may not exist in some games, when we find one, it must be unique. This statement can be easily proven by contradiction. To do that, consider a game where we found that strategy profile s^{SD} is an SDE, but assume that we find *another* strategy profile s_1, different from s^{SD}, that is also an SDE (so there are multiple SDEs). Then, by the definition of strict dominance, every player i must strictly prefer strategy profile s_1 to s^{SD}, contradicting that s^{SD} is an SDE. We can conclude that, if we find an SDE in a game, it must be unique.
3. **Robust to small payoff perturbations? Yes.** If we change the payoff of one of the players by a small amount ($\varepsilon \to 0$), SDE still yields the same equilibrium

prediction. To understand this property, note that if a strategy s_i strictly dominates all other strategies of player i in our original game, it must still strictly dominate all her other strategies after we apply a small payoff perturbation.

4. **Socially optimal? No.** The SDE of a game does not need to be socially optimal. We can see this with an example: in Prisoner's Dilemma game, the SDE is (Confess, Confess), which does not coincide with the strategy profile that maximizes the sum of players' payoffs, (Not Confess, Not Confess).

Exercises

2.1. **General Prisoner's Dilemma game.**[A] Consider the following Prisoner's Dilemma game where every player simultaneously chooses between confess (C) and not confess (NC). Payoffs satisfy $a_i > b_i$ for every player $i = \{1, 2\}$, implying that everyone earns a higher payoff at (NC, NC) than at (C, C).

		Player 2	
		C	NC
Player 1	C	b_1, b_2	c_1, d_2
	NC	d_1, c_2	a_1, a_2

(a) Find conditions on payoffs a_i through d_i that make C a strictly dominant strategy for every player i.
(b) If the conditions that you found in part (a) are satisfied, which strategy profile/s survive the application of IDSDS?

2.2. **Finding dominated strategies in four categories of simultaneous-move games.**[A] Consider the following simultaneous-move game with payoff matrix.

		Player 2	
		A	B
Player 1	A	a, a	b, c
	B	c, b	d, d

Find players' strictly dominated strategies in the following settings. Interpret and relate your results to common games.
(a) Payoffs satisfy $a > c$ and $b > d$.
(b) Payoffs satisfy $a > c$ and $b < d$.
(c) Payoffs satisfy $a < c$ and $b < d$.

(d) Payoffs satisfy $a < c$ and $b > d$.

2.3. **Equilibrium predictions from IDSDS vs. IDWDS.**[A] While the order of deletion of dominated strategies does not affect the equilibrium outcome when applying IDSDS, it can affect the set of equilibrium outcomes when we delete weakly (rather than strictly) dominated strategies. Use the payoff matrix below to show that the order in which weakly dominated strategies are eliminated can affect equilibrium outcomes.

		Player 2		
		L	M	R
Player 1	U	2, 1	1, 1	0, 0
	C	1, 2	3, 1	2, 1
	D	2, -2	1, -1	$-1, -1$

2.4. **IDSDS and IDWDS.**[A] Consider the following simultaneous-move game. Player 1 can choose *Up* or *Down*, while player 2 can choose *Left* or *Right*. If player 2 chooses *Left*, then every player receives a payoff of 1 regardless of whether player 1 chooses *Up* or *Down*. If player 1 chooses *Up* and player 2 chooses *Right*, player 1 (2) receives a payoff of -1 (1). Otherwise, if player 1 chooses *Down* and player 2 chooses *Right*, player 1 (2) receives a payoff of 1 (-1).

(a) Represent this game in extensive form.
(b) Represent this game in normal form.
(c) Identify the strategy profiles that survive IDSDS.
(d) Identify the strategy profiles that survive IDWDS.

2.5. **Strict dominance, IDSDS, and IDWDS.**[B] Consider the following two-player game, adapted from Tadelis (2013).

		Player 2		
		L	C	R
Player 1	U	3, 3	5, 1	6, 2
	M	4, 1	8, 4	3, 6
	D	4, 0	9, 6	6, 8

(a) Find the strict dominant equilibrium of this game.
(b) Which strategy profile/s survive IDSDS?
(c) Which strategy profile/s survive IDWDS?

Exercises

2.6. **Applying IDSDS in a large matrix.**[B] Consider the game in the matrix below. Player A in rows and player B in columns, simultaneously choose to swerve, go straight, dive, reverse, jump, or stop. Use IDSDS to find the equilibrium outcome/s of this game.

		Player B					
		Swerve	Straight	Dive	Reverse	Jump	Stop
Player A	Swerve	6 , 6	5 , 3	5 , 7	8 , 2	5 , 4	4 , 6
	Straight	3 , 5	9 , 9	7 , 6	9 , 1	4 , 2	6 , 4
	Dive	7 , 5	6 , 7	8 , 8	4 , 5	6 , 3	5 , 7
	Reverse	2 , 8	1 , 9	5 , 4	3 , 3	5 , 6	2 , 5
	Jump	4 , 5	2 , 4	3 , 6	6 , 5	5 , 5	3 , 7
	Stop	6 , 4	4 , 6	7 , 5	5 , 2	7 , 3	4 , 4

2.7. **Existence of dominant and dominated strategies.**[C] Consider a simultaneous-move game with N players, and denote player i's strategy space as S_i.
   (a) Show that, if we find that strategy s'_i is strictly dominant for player i, then there must be at least another strategy $s''_i \neq s'_i$ which is strictly dominated, where $s'_i, s''_i \in S_i$.
   (b) Show that, if we find that strategy s'_i is strictly dominated for player i, we may not find any strictly dominant strategy for this player. [*Hint*: An example suffices.]

2.8. **Unique prediction in IDWDS and IDSDS.**[C] Consider a game with N players, where player i's strategy space is denoted as S_i. In addition, assume that in strategy profile $s = (s_1, ..., s_N)$ strategy s_i is strictly dominant for every player i, that is,

$$u_i(s_i, s_{-i}) > u_i(s'_i, s_{-i}) \text{ for every } s'_i \neq s_i \text{ and for all } s_{-i} \in S_{-i}.$$

   (a) Show that s_i must be the only strategy profile surviving IDSDS.
   (b) Show that s_i must be the only strategy profile surviving IDWDS.
   (c) Show that, when applying IDSDS, the order of deletion does not produce different equilibrium outcomes.

2.9. **IDWDS and Sophisticated equilibrium.**[B] Consider a game with N players, where player i's strategy space is denoted as S_i. Assume that the game is solvable by IDWDS, yielding the surviving strategy set S_i^{Sur} for each player i, that is, $S_i^{Sur} \subset S_i$. Therefore, the strategy profile surviving IDWDS is denoted as the Cartesian product of surviving strategy sets $S^{Sur} \in S_1^{Sur} \times ... \times S_N^{Sur}$. We say that strategy profile surviving IDWDS, S^{Sur}, is a "sophisticated equilibrium" if every player i is indifferent between any two of his surviving strategies s_i and s_i', that is,

$$u_i(s_i, s_{-i}) = u_i(s_i', s_{-i}) \text{ for every } s_i, s_i' \in S_i^{Sur} \text{ and all } s_{-i} \in S_{-i}.$$

Consider the following normal form game.

		Player 2		
		L	M	R
Player 1	U	5,4	5,4	9,0
	C	1,7	2,5	8,6
	D	2,3	1,4	8,3

Find the set of strategy profiles surviving IDWDS. Can you identify any sophisticated equilibria?

2.10. **Applying IDSDS in a three-player games.**[B] Consider the following three-player game between Annie (choosing rows), Bonnie (choosing columns), and Carol (choosing matrices).

Carol choosing *Front*

		Bonnie		
		Top	*Middle*	*Down*
Annie	*Left*	4,5,9	6,3,6	5,1,8
	Center	1,9,4	2,5,3	7,5,6
	Right	2,1,3	6,8,9	9,6,7

Carol choosing *Back*

		Bonnie		
		Top	*Middle*	*Down*
Annie	*Left*	3,7,7	4,2,5	6,5,4
	Center	1,9,3	0,7,1	4,8,5
	Right	5,2,2	7,9,8	2,1,6

(a) Which strategy profile(s) survive iterated deletion of strictly dominated strategies (IDSDS)?

2.11. **Applying IDWDS in three-player games.**[B] Consider the game in Exercise 2.10. We will now apply iterated deletion of weakly dominated strategies (IDWDS).

(a) Find the strategy profile(s) that survive IDWDS.

(b) Compare your results from applying IDSDS and IDWDS. Interpret.

2.12. **IDWDS is affected by the order of deletion.**[A] Consider the following simultaneous-move game between Austin, in rows, and Charlene, in columns. In this exercise, we apply iterated deletion of weakly dominated strategies (IDWDS), showing that its equilibrium prediction depends on which strategies we delete first, and which we delete second (i.e., IDWDS is path-dependent).

	Charlene			
	North	South	East	West
Up	4 , 6	3 , 2	2 , 5	3 , 6
Austin Down	4 , 4	4 , 5	2 , 4	3 , 2
Left	2 , 3	3 , 4	4 , 3	1 , 2
Right	4 , 4	5 , 4	3 , 2	4 , 3

(a) Beginning with Austin, find the equilibrium prediction/s according to IDWDS.

(b) Beginning with Charlene, find the equilibrium prediction according to IDWDS.

(c) Compare your results from parts (a) and (b). Interpret.

2.13. **Finding Dominant Strategies in games with $N \geq 2$ players.**[B] Consider a public project between $N \geq 2$ individuals, each of them simultaneously and independently contributing effort x_i (such as a group project in class), incurring cost $c_i x_i^2$ where $c_i > 0$. Individual i's payoff function is

$$u_i(x_i, x_{-i}) = x_i + \alpha \sum_{j \neq i} x_j - c_i x_i^2$$

where $x_{-i} = (x_1, x_2, ..., x_{i-1}, x_{i+1}, ..., x_N)$ represents the effort profile from all i's rivals, and parameter $\alpha > 0$.

(a) Provide a verbal interpretation behind parameter α.

(b) Do players have a strictly dominant strategy? If so, find the SDE.

(c) Is the SDE found in part (b) socially optimal? Interpret.

(d) Are your equilibrium results in part (b) affected if the number of players, N, increases? What if parameter α becomes negative?

(e) Are your equilibrium results in part (c) affected if the number of players, N, increases? What if parameter α is asymmetric across players (i.e., player i's parameter is α_i, where $\alpha_i \neq \alpha_j$ for all $j \neq i$)? What if parameter α is symmetric across players but becomes negative?

2.14. **IDSDS and IDWDS with a more general payoff matrix-I.**[B] Consider the following payoff matrix between players 1 and 2.

	Player 2	
	L	R
Player 1 U	a , 0	0 , 0
D	0 , 0	0 , b

(a) What are the strategy profiles that survive IDSDS?

(b) Suppose that payoffs satisfy $a, b < 0$. Find the strategy profiles that survive IDWDS.

(c) Suppose that payoffs satisfy $a, b > 0$. Find the strategy profiles that survive IDWDS.
(d) Suppose $a > 0 > b$. Find the strategy profiles that survive IDWDS.
(e) Suppose $b > 0 > a$. Find the strategy profiles that survive IDWDS.
(f) Compare your results from parts (b) to (e). Interpret.

2.15. **IDSDS and IDWDS allow for a more general payoff matrix-II.**[B] Consider the following payoff matrix, where payoffs a and b are allowed, for generality, to be positive or negative.

		Player 2	
		L	R
Player 1	U	a, a	$0, 0$
	D	$0, 0$	b, b

(a) For what values of a and b is (D, R) the only strategy profile surviving IDSDS?
(b) For what values of a and b is (U, L) the only strategy profile surviving IDWDS?
(c) Suppose that $a, b > 0$. What are the strategy profiles surviving IDSDS?

Reference

Tadelis, S. (2013). *Game Theory, An Introduction*. Princeton University Press.

Nash Equilibrium

3.1 Introduction

In this chapter, we present a different solution concept which may have "more bite" than IDSDS. Fewer strategy profiles can, then, emerge as equilibria of the game. This solution concept, known as the Nash Equilibrium after Nash (1950), builds upon the notion that every player finds the strategy that maximizes her payoff against each of her rivals' strategies, which we refer as her "best response." In a context where player i has only two available strategies, "maximizing" her payoff means choosing the strategy that yields the highest payoff, taking her rivals' strategies as given. A similar argument applies when player i has three or more available strategies to choose from, where she selects that yields the highest payoff, fixing her rivals' strategies.

In a setting where player i chooses her strategy from a continuum of available strategies (e.g., output or price), her best response is found by maximizing this player's payoff, literally, by solving a utility or profit maximization problem, where we will take the strategies of her rivals as given (as if they were parameters). We start with the definition of best response.

3.2 Best Response

Definition 3.1. **Best response**. Player i regards strategy s_i as a best response to strategy profile s_{-i} if

$$u_i(s_i, s_{-i}) \geq u_i(s'_i, s_{-i})$$

for every available strategy $s'_i \neq s_i$.

Intuitively, in a game with two players, i and j, strategy s_i is player i's best response to player j's strategy, s_j, if s_i yields a weakly higher payoff than any other strategy s_i' against s_j. In other words, when player j chooses s_j, player i maximizes her payoff by responding with s_i than with any other available strategies.

The next tool describes how to find best responses in matrix games.

Tool 3.1 How to find best responses in matrix games:

1. Focus on the row player by fixing your attention on one strategy of the column player (i.e., one specific column).
   a. Cover with your hand (or with a piece of paper) all columns that you are not considering.
   b. Find the highest payoff for the row player by comparing the first component of every pair.
   c. For future reference, underline this payoff. This is the row player's best response payoff to the column that you considered from the column player.
2. Repeat step 1, but now fix your attention on a different column.
3. For the column player, the method is analogous, but now direct your attention on one strategy of the row player (i.e., one specific row). Cover with your hand all the rows that you are not considering, and compare the payoffs of the column player (i.e., second component of every pair).

3.2.1 Finding Best Responses with Discrete Strategy Spaces

Consider Matrix 3.1a describing players 1 and 2's interaction. Player 1 can choose between U and D in rows (denoting up and down, respectively) and player 2 chooses between ll and r in columns (representing left and right, respectively).

	Player 2	
	l	r
U	5, 3	2, 1
D	3, 6	4, 7

Player 1

Matrix 3.1a. Finding best responses in a 2×2 matrix.

Player 1. We can start finding player 1's best responses.

1. BR_1 *to player 2 choosing column l.* We start by fixing player 2's strategy at one of her columns (l), finding that player 1's payoff is higher when responding with U than D because $5 > 3$ (see first payoffs in the left column). We compactly write player 1's best response to l in this context as

$$BR_1(l) = U.$$

3.2 Best Response

For easier reference, Matrix 3.1b underlines the payoff that player 1 earns from choosing this best response, 5, which is known as her "best response payoff."

2. BR_1 *to player 2 choosing column r*. In contrast, when player 2 chooses r (in the right column), player 1's best response is

$$BR_1(r) = D$$

since $4 > 2$ (see first payoff in the right column).

We summarize player 1's best response as follows

$$BR_1(s_2) = \begin{cases} U \text{ if } s_2 = l, \text{ and} \\ D \text{ if } s_2 = r. \end{cases}$$

The best response payoffs for player 1 are underlined in Matrix 3.1b (see first element in every payoff pair).[1] These underlined payoffs will be quite useful when we seek to find Nash equilibria in subsequent sections of this chapter, since they help us keep track of those cell/s where every player chooses best responses to her opponent's strategies.

	Player 2	
	l	r
Player 1 U	$\underline{5}, 3$	2, 1
D	3, 6	$\underline{4}, 7$

Matrix 3.1b. Finding best responses in a 2×2 matrix - Player 1.

Player 2. We can now focus on player 2's best responses.

1. BR_2 *to player 1 choosing row U*. When player 1 chooses U in the top row, player 2's best response is

$$BR_2(U) = l$$

since her payoff from l is higher than that from r, $3 > 1$. (Recall that player 2's payoffs are the second element in every pair implying that, at the top row, she only compares her payoff from responding with l, 3, against that of responding with r, 1.)

---

[1] Other textbooks, such as Harrington (2014), use circles around a player's best response payoff. We use underlined payoffs throughout this book.

2. *BR_2 to player 1 choosing row D*. When player 1 chooses D in the bottom row, player 2's best response is

$$BR_2(D) = r$$

because her payoff from responding with r, 7, is higher than that from responding with l, 6.

Player 2's best responses can, then, be summarized as follows

$$BR_2(s_1) = \begin{cases} l \text{ if } s_1 = U, \text{ and} \\ r \text{ if } s_1 = D. \end{cases}$$

Matrix 3.1c underlines player 2's best response payoffs for easier reference.

	Player 2	
	l	r
Player 1 U	5,<u>3</u>	2, 1
Player 1 D	3, 6	4,<u>7</u>

Matrix 3.1c. Finding best responses in a 2×2 matrix - Player 2.

You may have noticed that at strategy profile (U, l), on the top left of the matrix, player 1 plays a best response to player 2 choosing l and, similarly, player 2 chooses a best response to player 1 selecting U. As we show in Sect. 3.7, (U, l) is a Nash equilibrium of the game since players choose best responses to each other's strategies.

3.2.2 Finding Best Responses with Continuous Strategy Spaces

What if players have a continuous strategy space? In that setting, our search for player i's best responses is also straightforward. If player i faces a payoff function $u_i(s_i, s_{-i})$, we only need to fix her rival's strategies at a generic s_{-i} and differentiate with respect to s_i to find the value of the strategy s_i that maximizes player i's payoff. The following example, based on the Cournot model of simultaneous quantity competition, illustrates this approach.

Example 3.1 Finding best responses under Cournot quantity competition. Consider an industry with two firms, 1 and 2, simultaneously and independently choosing their output level. Every firm i faces a symmetric cost function $C(q_i) = cq_i$, where $c > 0$ denotes the marginal cost of additional units of output, and an inverse demand function $p(Q) = 1 - Q$, where $Q = q_1 + q_2$ represents aggregate output. To find the

3.2 Best Response

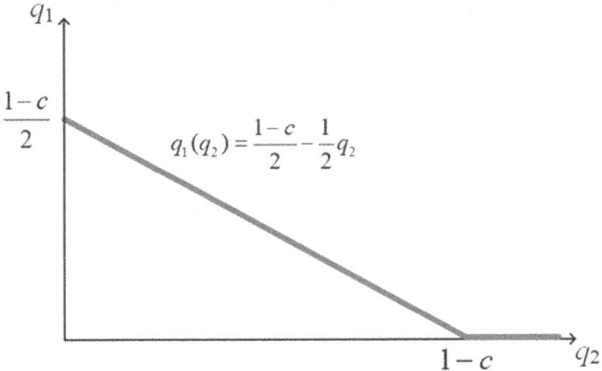

Fig. 3.1 Firm 1's best response function

best responses of firm 1, we need to fix firm 2's output at a generic output level q_2, which just means that firm 1 cannot control firm 2's output decisions, taking them as given. After fixing q_2, firm 1 chooses its output q_1 to maximize its profits, as follows

$$\max_{q_1 \geq 0} \ (1 - q_1 - q_2)q_1 - cq_1$$

where the first term captures firm 1's total revenue while the second term represents its total costs. Differentiating with respect to q_1, we find $1 - 2q_1 - q_2 - c = 0$ or, alternatively,

$$1 - 2q_1 - q_2 = c.$$

Intuitively, the left term indicates the marginal revenue from producing one more unit, while the right term denotes its marginal cost.[2] Rearranging the above equation yields $2q_1 = 1 - c - q_2$. Solving for q_1, we obtain

$$q_1 = \frac{1-c}{2} - \frac{1}{2}q_2.$$

Graphically, this equation is a straight line originating at $q_1 = \frac{1-c}{2}$ when $q_2 = 0$ (if firm 2 is inactive), but decreasing at a rate of $1/2$ for every unit of output that firm 2 produces, as depicted in Fig. 3.1. The line, then, hits the horizontal axis at the point where $0 = \frac{1-c}{2} - \frac{1}{2}q_2$ which, solving for q_2, yields $q_2 = 1 - c$.

---

[2] Note that second-order conditions hold since, differentiating $1 - 2q_1 - q_2 = c$ with respect to q_1 again, we find $-2 < 0$, thus confirming that firm 1's profit function is concave in its output level, q_1.

We can then summarize firm 1's best response function (allowing for both interior solutions, where firm 1 is active; and corner solutions where it is inactive) as follows

$$q_1(q_2) = \begin{cases} \frac{1-c}{2} - \frac{1}{2}q_2 & \text{if } q_2 < 1-c \\ 0 & \text{otherwise.} \end{cases}$$

Intuitively, firm 1 responds by decreasing its output when firm 2 produces more units, but eventually chooses to remain inactive. Otherwise, a positive output would lead to negative profits.[3] Because firm 1 responds with a lower output q_1 when its rival increases its output, q_2, we say that firms' output decisions in this context are "strategic substitutes." A symmetric best response function applies for firm 2. □

This approach is applicable to several games where players face continuous strategy spaces, such as firms choosing prices, advertising levels, investment in research and development, individuals choosing how much effort to exert in a group project, or how much to contribute to a charity. We will apply it to several games in other sections of the book.

3.3 Deleting Strategies That Are Never a Best Response

Before we use best responses to find the Nash equilibrium of a game, we can apply this concept using a similar approach as that of strictly dominated strategies in the last chapter. Specifically, when applying IDSDS, we argued that, if a player finds one of her strategies to be strictly dominated, we can essentially delete it from the strategy set of this player.

We can now apply a similar approach, arguing that, if a player never uses one or more of her available strategies as a best response to her opponents' strategies, we can label these strategies as "never a best response" (NBR) and delete them from her strategy set. We formally define this concept below.

Definition 3.2. **Never a best response (NBR).** Strategy s_i is never a best response if

$$u_i(s_i, s_{-i}) > u_i(s'_i, s_{-i}) \text{ for every } s'_i \neq s_i$$

does not hold for any strategy profile of her rivals, s_{-i}.

Alternatively, if a strategy s_i is NBR, there are no beliefs that player i can sustain about how her opponents behave (no profile s_{-i}) that would lead her to use strategy s_i as a best response. In other words, player i cannot rationalize (explain) why would she ever choose strategy s_i since it is never a best response to her opponents' choices. As a result, we can delete s_i from player i's strategy set S_i.

---

[3] This happens when firm 2 produces at least $q_2 = 1-c$ units, which coincides with the total output in a perfectly competitive industry. To see this, set price equal to marginal cost, $1-q = c$, and solve for q to obtain $q = 1-c$.

3.4 Rationalizability

It is easy to prove that, if player i finds that strategy s_i is strictly dominated by $s_i' \neq s_i$, then s_i yields a strictly lower payoff than s_i' regardless of the specific strategy profile that her opponents select. As a consequence, strategy s_i cannot be a best response against any strategy profile of i's opponents, that is, s_i is NBR. In short,

$$s_i \text{ is strictly dominated} \Longrightarrow s_i \text{ is NBR}.$$

3.4 Rationalizability

We can go through an iterative process—analogous to IDSDS—to identify strategies that are NBR for either player rather than strategies that are strictly dominated. This iterative process is known as "rationalizability" because it finds strategies that player i can rationalize to be best responses to at least one of her opponents' strategies.

Tool 3.2 Applying rationalizability:

Step 1 Starting with player i, delete every strategy that is NBR, obtaining the reduced strategy set S_i', where $S_i' \subset S_i$. (This step only uses the definition of rationality as no player would choose a strategy that is NBR.)

Step 2 Using common knowledge of rationality, we continue the above reasoning, arguing that player $j \neq i$ can anticipate player i's best responses and, as a consequence, the strategies that are NBR for player i, deleting them from S_i. Given this reduced strategy space $S_i' \subset S_i$, player j can now examine her own best responses to player i, seeking to identify if one or more are never used, and further restricting her strategy space to $S_j' \subset S_j$.

Step 3 We obtain the Cartesian product $S_i' \times S_j'$, representing the remaining rationalizable strategies after deleting NBRs for two steps. Player i then finds if some of her strategies in this reduced game are NBR, deleting them from her strategy set S_i', obtaining $S_i'' \subset S_i'$

Step k The process continues until we cannot find more strategies that are NBR for either player.

The strategy profile (or set of strategy profiles) that survives this iterative process are referred as "rationalizable" strategy profiles because every player can sustain beliefs about her rivals' behavior (i.e., which strategies her rivals choose) that would lead her to best respond using one of these surviving strategies.

Example 3.2 applies rationalizability to the same payoff matrix as Example 2.2, showing that both IDSDSand rationalizability produce the same equilibrium outcomes. It is easy to show that, in games with two players, both solution concepts yield identical equilibrium results. However, in games with three or more players, equilibrium outcomes do not necessarily coincide, with rationalizability producing more precise equilibrium outcomes than IDSDS. That is, for a given strategy profile s,

$$s \text{ is rationalizable} \implies s \text{ survives IDSDS}$$
$$\not\Leftarrow$$

For more details, see Pearce (1984) and for a more advanced presentation, see Fudenberg and Tirole (1991, pages 51-53 and 62-63).

Example 3.2 Rationalizability, NBR, and IDSDS. Consider the payoff matrix in Example 2.2, which we reproduce in Matrix 3.2a for easier reference.

$$
\begin{array}{c|c|c|}
 & \multicolumn{2}{c}{Firm\ 2} \\
 & h & l \\
\hline
H & 4,4 & 0,2 \\
\hline
M & 1,4 & 2,0 \\
\hline
L & 0,2 & 0,0 \\
\hline
\end{array}
$$

Matrix 3.2a. Applying rationalizability - First step.

Starting with firm 1, it is easy to see that row L is NBR: when firm 2 chooses h, firm 1's best response is H; while when firm 2 chooses l, firm 1's best response is M. In other words, firm 1 does not have incentives to respond with L regardless of the beliefs that firm 1 sustains on firm 2's behavior. After deleting row L from firm 1's strategy space, S_1, we obtain the reduced strategy space $S'_1 = \{H, M\}$, as depicted in Matrix 3.2b.

$$
\begin{array}{c|c|c|}
 & \multicolumn{2}{c}{Firm\ 2} \\
 & h & l \\
\hline
H & 4,4 & 0,2 \\
\hline
M & 1,4 & 2,0 \\
\hline
\end{array}
$$

Matrix 3.2b. Applying rationalizability - Second step.

We can now examine firm 2's best responses: when firm 1 chooses H (top row), firm 2's best response is h; and, similarly, when firm 1 chooses M (bottom row), firm 2's best response is also h. Therefore, we can claim that strategy l is NBR for firm 2, entailing a reduced strategy space $S'_2 = \{h\}$. As a result, the Cartesian product $S'_1 \times S'_2$ has only two remaining strategy profiles, as illustrated in Matrix 3.2c.

	Firm 2
	h
Firm 1 H	4, 4
M	1, 4

Matrix 3.2c. Applying rationalizability - Third step.

Firm 2 only has one available strategy at this point, $s_2 = h$, implying that firm 1's best response is H and, therefore, strategy M is NBR at this stage of our analysis. Hence, firm 1's strategy set reduces to $S_1'' = \{H\}$, yielding a unique rationalizable strategy profile, (H, h). As expected, this strategy profile coincides with that surviving IDSDS in Example 2.2.□

3.4.1 Evaluating Rationalizability as a Solution Concept

From our discussion in Sect. 3.4, we know that, if a strategy s is rationalizable, it survives IDSDS, but the converse is not necessarily true. As a consequence, rationalizability satisfies the same properties as IDSDS:

1. **Existence? Yes**. Rationalizability satisfies existence, meaning that at least one strategy profile in every game must be rationalizable.
2. **Uniqueness? No**. Rationalizability does not satisfy uniqueness, since more than one strategy profiles may survive rationalizability. Although rationalizability may provide more precise equilibrium outcomes than IDSDS in games with three or more players, it does not yield a unique equilibrium prediction in all games; entailing that neither rationalizability nor IDSDS satisfy uniqueness.
3. **Robust to small payoff perturbations? Yes**. Rationalizability is robust to small perturbations because it does not change equilibrium outcomes if we alter the payoff of one of the players by a small amount, ε, where ε is close to zero.
4. **Socially optimal? No**. Finally, rationalizability does not necessarily yield socially optimal outcomes, as illustrated by the Prisoner's Dilemma game, where the only strategy profile surviving rationalizability is (Confess, Confess), which is not socially optimal.

3.5 Applications of Rationalizability

3.5.1 Finding NBRs in the Beauty Contest

Consider that your instructor of the Game Theory course shows up in the classroom proposing to play the following "Guess the Average" game (also known as "Beauty Contest"). She distributes small pieces of paper and asks each student to write a number (an integer) between 0 and 100. There are $N > 2$ students and each of them simultaneously and independently writes her number, taking into account the following rules of the game: the instructor will collect all numbers, write them on the board, compute the average, and then multiply that average by 1/2 (half of the average). The instructor, then, declares as a winner the student who submitted a number closest to half the average. (A similar argument applies if the instructor multiplies the average by a generic ratio p, where $0 < p < 1$, instead of $p = 1/2$.)

Which strategies survive rationalizability?

- In the first step, we can try to eliminate strategies that are NBR for any student. Starting with the highest number, $s_i = 100$, we can see that, even if all other students submit $s_j = 100$ for all $j \neq i$, and the number of students is large enough to yield an average of $\bar{s} = 100$, we would obtain that half of the average approaching 50.[4] Therefore, every student i's best response would be 50, i.e., $BR_i(s_j) = 50$ where $s_j = 100$ for all $j \neq i$, entailing that i cannot rationalize $s_i = 100$ being a best response, *regardless* of which numbers she believes her rivals submit. In summary, $s_i = 100$ is NBR, helping us restrict her strategy space to $S'_i = \{0, 99\}$ in the first step of rationalizability.
- We operate similarly in the second step of rationalizability. Even if all other students submit the highest number (in the reduced strategy space S'_i), $s_j = 99$ for all $j \neq i$, the average would be $\bar{s} = 99$, entailing that half of the average is 49.5. As a result, every student i finds that submitting $s_i = 99$ is NBR either, so we can further restrict her strategy space to $S''_i = \{0, 98\}$ in the second step of rationalizability.
- ...
- After 97 more rounds, we are left with a restricted strategy space of $\{0, 1\}$ which cannot be further reduced. Indeed, submitting $s_i = 0$ ($s_i = 1$) is a best response if student i believes that her classmates will all submit $s_j = 0$ ($s_j = 1$), so we cannot rule out strategies as being NBR. Therefore, all symmetric strategy profiles, where every student i submits the same number (all submit $s_i = 0$ or all submit $s_i = 1$) survive rationalizability; and so do all asymmetric strategy profiles, where at least one student submits $s_i = 0$ and all other $j \neq i$ students submit $s_j = 1$.

---

[4] Technically, we do not need a large number of students for the argument to be correct, but we assume it here to keep the analysis easier to follow. Indeed, half of the average is $\frac{(N-1)100}{2}$, which satisfies $\frac{(N-1)100}{2} > 50$ given that $N > \frac{3}{2}$ holds because $N > 2$ by assumption.

3.5 Applications of Rationalizability

Experimental tests. This game has been tested in experimental labs quite often; see, for instance, Nagel (1995) and Stahl and Wilson (1995), which initiated these studies, and, for a literature review, see Camerer (2003) and Crawford et al. (2013). If you participate in one of these experiments and believe that all other participants randomly submit a number uniformly distributed between 0 and 100, often referred by the literature as "Level-0" players, then you can anticipate that the average will be around 50 (assuming a sufficiently large number of participants), entailing that half of the average would be 25. Therefore, if you think that your rivals are "Level-0" players, you would submit $s_i = 25$. However, if you think that your rivals must have gone through that same thought process, and they are submitting $s_j = 25$, so they are "Level-1" players, you could outsmart them anticipating that the average will be 25, and half of it will be 12.5, inducing you to submit $s_i = 12.5$. If you think that your rivals went through that thought process too, and are submitting $s_j = 12.5$, thus being "Level-2" players, then half the average will be 6.25, implying that, for you to outsmart them, you must submit $s_i = 3.125$. The argument, of course, extends to further steps, so that if a player thinks that her rivals are "Level-k" players, she must submit $s_i = \frac{50}{2^{k+1}}$.

Interestingly, this argument implies that, as a participant, it is not necessarily optimal for you to submit some of the strategies that survive rationalizability ($s_i = 0$ or $s_i = 1$, as shown above). Instead, you submit a different number depending on the type of Level-k rivals that you deal with, submitting lower numbers as your rivals become more sophisticated (higher k, so they can go through more iterations). Nagel (1995), for instance, shows that many subjects submit $s_i = 25$, suggesting they are Level-1 players; but undergraduate students who took Game Theory courses tend to submit $s_i = 12.5$ or $s_i = 6.25$ (thus being Level-2 or -3 players), and so do Caltech students in the Economics major (who generally have a very strong Math background) and usual readers of financial newspapers (*Financial Times* in the UK and *Expansion* in Spain).

3.5.2 Finding NBRs in the Cournot Duopoly

Consider again the Cournot duopoly from Example 3.1, where we found that firm 1's best response function is

$$q_1(q_2) = \begin{cases} \frac{1-c}{2} - \frac{1}{2}q_2 & \text{if } q_2 < 1-c \\ 0 & \text{otherwise.} \end{cases}$$

and similarly for firm 2. Therefore, firm 1's output ranges between $q_1 = \frac{1-c}{2}$ (which occurs when firm 2 is inactive, $q_2 = 0$) and $q_1 = 0$ (which happens when firm 2 produces the aggregate output under perfect competition, $q_2 = 1 - c$), implying that we can delete all output levels $q_1 > \frac{1-c}{2}$ as NBR for firm 1. In other words, there are no beliefs about firm 2's output for which firm 1's best response lies above $\frac{1-c}{2}$. Then, we can delete $q_1 > \frac{1-c}{2}$ from firm 1's strategy space, S_1, which restricts to $S_1' = \left[0, \frac{1-c}{2}\right]$. Following a similar argument with firm 2's best response function (which is symmetric to firm 1's), we claim that $q_2 > \frac{1-c}{2}$ is NBR for firm 2, yielding a reduced strategy space $S_2' = \left[0, \frac{1-c}{2}\right]$. Therefore, the first step of rationalizability entails that $S_i' = \left[0, \frac{1-c}{2}\right]$ for every firm i.

In the second step of rationalizability, we consider firm 1's best response function again, but taking into account that firm 2's rationalizable output levels are those in $S_2' = \left[0, \frac{1-c}{2}\right]$, that is,

$$q_1(q_2) = \begin{cases} \frac{1-c}{2} - \frac{1}{2}q_2 & \text{if } q_2 < \frac{1-c}{2} \\ 0 & \text{otherwise.} \end{cases}$$

As a consequence, firm 1's output now ranges between $q_1 = \frac{1-c}{2}$ (which occurs when $q_2 = 0$) and $q_1 = \frac{1-c}{4}$ (which happens when firm 2 produces $q_2 = \frac{1-c}{2}$ units since $q_1 = \frac{1-c}{2} - \frac{1}{2}\frac{1-c}{2} = \frac{1-c}{4}$), helping us further reduce firm 1's strategy space to $S_1'' = \left[0, \frac{1-c}{4}\right]$. Again, a similar argument applies to firm 2, yielding $S_2'' = \left[0, \frac{1-c}{4}\right]$.

In the third step of rationalizability, firm 1's best response function becomes

$$q_1(q_2) = \begin{cases} \frac{1-c}{2} - \frac{1}{2}q_2 & \text{if } q_2 < \frac{1-c}{4} \\ 0 & \text{otherwise.} \end{cases}$$

yielding that firm 1's output now ranges between $q_1 = \frac{1-c}{2}$ (which occurs when $q_2 = 0$) and $q_1 = \frac{3(1-c)}{8}$ (which happens when $q_2 = \frac{1-c}{4}$ because $q_1 = \frac{1-c}{2} - \frac{1}{2}\frac{1-c}{4} = \frac{3(1-c)}{8}$), entailing that its strategy space further shrinks to $S_1''' = \left[0, \frac{3(1-c)}{8}\right]$; and similarly for firm 2's strategy space.

Repeating the process again, we see that firm 2's output levels that induce firm 1 to remain active (upper part of firm 1's best response function) keeps shrinking, from $q_2 = 1 - c$ to $q_2 = \frac{1-c}{2}$, to $q_2 = \frac{1-c}{4}$, and to $q_1 = \frac{3(1-c)}{8}$, until the point where it does not change in further iterations of rationalizability. For this to occur, we need that $q_1 = \frac{1-c}{2} - \frac{1}{2}q_1$, since $q_2 = q_1$, which, solving for q_1, yields $q_1 = \frac{1-c}{3}$. In future sections, we confirm that this output level coincides with the Nash equilibrium of the Cournot game, where both firms produce $q_1 = q_2 = \frac{1-c}{3}$.

3.6 Nash Equilibrium

We next use the concept of best response to define a Nash Equilibrium.

Definition 3.3. Nash Equilibrium (NE). A strategy profile $s^* = \left(s_i^*, s_{-i}^*\right)$ is a Nash Equilibrium if every player chooses a best response given her rivals' strategies.

In a two-player game, the above definition says that strategy s_i^* is player i's best response to s_j^* and, similarly, strategy s_j^* is player j's best response to s_i^*. For compactness, we often write player i's best response to s_j^* as $BR_i(s_j^*) = s_i^*$, and that of player j as $BR_j(s_i^*) = s_j^*$.

3.6 Nash Equilibrium

Therefore, a strategy profile is an NE if it is a *mutual* best response: the strategy that player i chooses is a best response to that selected by player j, and vice versa. As a result, no player has unilateral incentives to deviate since doing so would lower her payoff, or keep it unchanged.[5] The next tool describes how to use best responses to find NEs in a matrix game.

Tool 3.3 How to find NEs in matrix games:

1. Find the best responses to all players (see Tool 3.1 for details).
2. Identify which cell or cells in the matrix has every payoff underlined, meaning that all players have a best response payoff. These cells are the NEs of the game.

Example 3.3 Finding NEs in a two-player game. Consider Sect. 3.2.1 again, where we found the best responses of firms 1 and 2, and underlined their best response payoffs in Matrices 3.1b and 3.1c, respectively. Matrix 3.3 combines our underlined best response payoffs for both firms, illustrating that in two cells both players' payoffs are underlined.

	Firm 2	
	h	m
Firm 1 M	<u>5</u>,<u>3</u>	2, 1
Firm 1 L	3, 6	<u>4</u>,<u>7</u>

Matrix 3.3. Finding best responses and NEs.

This means that players choose mutual best responses to each other's strategies, which is the definition of an NE. In summary, by identifying best response payoffs for each player we found two NEs:

$$\text{NE} = \{(M, h), (L, m)\}.$$

As a practice, show that no player has strictly dominated strategies, implying that IDSDS has no bite.[6] We obtain similar results if we find the set of rationalizable

---

[5] Another common definition of NE focuses on every player i's beliefs about how her rivals behave. Using that approach, an NE is a system of beliefs (that is, a list of beliefs for each player i, assigning probability weights to each of her opponents' strategies) and a list of strategies satisfying two properties: (1) every player i uses a best response to her beliefs about how her rivals behave; and (2) the beliefs that players sustain are, in equilibrium, correct. For simplicity, however, we focus on the above definition.

[6] Intuitively, no player has a strategy that yields unambiguously lower payoffs than another strategy. As a remark, note that the application of IDWDS would produce the same equilibrium predication as IDSDS in this game, since no player has weakly dominated strategies.

strategy profiles. In particular, firm 1 does not have any strategy that is NBR since it uses M against h and L against m. The same argument applies to firm 2, which does not have NBR strategies either. As we cannot delete NBR strategies, the game as a whole (with the four original strategy profiles) is our equilibrium prediction according to rationalizability. □

3.7 Finding Nash Equilibria in Common Games

In this section, we apply the NE solution concept to four common games in economics and other social sciences: the Prisoner's Dilemma game, the Battle of the Sexes game, Coordination games, and Anticoordination games.

3.7.1 Prisoner's Dilemma Game

Consider again the Prisoner's Dilemma game that we described in Chapter 2. Matrix 3.4a reproduces the payoff matrix for easier reference. In Chapter 2, we found that Confess is a strictly dominant strategy and that Not confess is a strictly dominated strategy, leaving a unique strategy profile surviving IDSDS, (*Confess, Confess*). We now show that the NE solution concept produces the same equilibrium prediction.

As in previous examples, we first analyze the best responses of each player and then examine if one (or more) mutual best response exists, which will reveal to us the NE of the Prisoner's Dilemma game.

		Player 2	
		Confess	Not confess
Player 1	Confess	$-4, -4$	$0, -8$
	Not confess	$-8, 0$	$-2, -2$

Matrix 3.4a. The Prisoner's Dilemma game.

Player 1's best responses. For player 1, we first fix player 2's strategy at Confess (left column), obtaining $BR_1(C) = C$ because $-4 > -8$. Similarly, fixing player 2's strategy at Not confess (right column), yields $BR_1(NC) = C$ because $0 > -2$. Therefore, player 1 responds with Confess both when player 2 confesses and when she does not, confirming our result from Chapter 2 where we found that Confess is a strictly dominant strategy.

Player 2's best responses. Player 2's best responses are symmetric to player 1's as her payoffs are symmetric. Fixing player 1's strategy at Confess (top row), we

3.7 Finding Nash Equilibria in Common Games

obtain $BR_2(C) = C$ because $-4 > -8$, and fixing player 1's strategy at Not confess (bottom row) we find that $BR_2(NC) = C$ because $0 > -2$. Therefore, every player responds with Confess regardless of her opponent's strategy. Underlining best response payoffs, the payoff matrix becomes Matrix 3.4b.

		Player 2	
		Confess	Not confess
Player 1	Confess	$\underline{-4}, \underline{-4}$	$\underline{0}, -8$
	Not confess	$-8, \underline{0}$	$-2, -2$

Matrix 3.4b. The Prisoner's Dilemma game – Underlining best response payoffs.

As a result, (*Confess,Confess*) is the unique NE of the game because both players choose mutual best responses. As discussed in Chapter 2, the Prisoner's Dilemma game helps us illustrate scenarios with a tension between individual and collective incentives. In other words, the NE does not coincide with the socially optimal outcome, (*Not confess, Not confess*) in this context. We will return to this tension at several other parts of the book.

3.7.2 Coordination Game—The Battle of the Sexes Game

Consider again the Battle of the Sexes game as an example of a coordination game, reproduced for easier reference in Matrix 3.5a. Recall from Chapter 2 that no player had a strictly dominant strategy and, thus, the application of IDSDS had no bite, leaving us with all four strategy profiles in the original matrix as our equilibrium prediction. We next show that the NE solution concept provides more precise equilibrium outcomes. As usual, we start finding each player's best responses.

		Wife	
		Football	Opera
Husband	Football	10, 8	6, 6
	Opera	4, 4	8, 10

Matrix 3.5a. The Battle of the Sexes game.

Husband's best responses. Fixing the Wife's strategy at Football (left column), we find that her husband's best response is $BR_H(F) = F$ because $10 > 4$. In contrast, when we fix her strategy at Opera (right column), the Husband's best response becomes $BR_H(O) = O$ as $8 > 6$. Intuitively, the Husband seeks to attend to the same event as her wife does.

Wife's best responses. Similarly, when we fix the Husband's strategy at Football (in the top row), we can see that the Wife's best response is $BR_W(F) = F$ because $8 > 6$, while when we fix the Husband's strategy at Opera (in the bottom row) the Wife's best response becomes $BR_W(O) = O$ as $10 > 4$. In summary, they both prefer to be together than separated, but each has a more preferred event.

Matrix 3.5b becomes Matrix 3.5a (with the best response payoffs underlined).

		Wife	
		Football	Opera
Husband	Football	<u>10</u>,<u>8</u>	6, 6
	Opera	4, 4	<u>8</u>,<u>10</u>

Matrix 3.5b. The Battle of the Sexes game–underlining best response payoffs.

Therefore, we found two cells where the payoff of both players are underlined, entailing mutual best responses. These two cells constitute the two NEs in this game:

$$NE = \{(F,F),(O,O).\}$$

3.7.3 Pareto Coordination Game—The Stag Hunt Game

Matrix 3.6a depicts the "Stag Hunt game" from Sect. 2.4.3. Like in the Battle of the Sexes game, no player has strictly dominated strategies, entailing that IDSDS has no bite. Next, we find each player's best response, showing that the NE solution concept has more bite than IDSDS in this game.

		Player 2	
		Stag, S	Hare, H
Player 1	Stag, S	6, 6	1, 4
	Hare, H	4, 1	2, 2

Matrix 3.6a. The stag hunt game.

Player 1's best responses. Fixing player 2's strategy at Stag (left column), player 1's best response is $BR_1(S) = S$ because $6 > 4$. However, when we fix player 2's strategy at Hare (right column), player 1's best response is $BR_1(H) = H$ given that $2 > 1$. Intuitively, player 1 chooses the same strategy as her opponent, thus exhibiting similar incentives as in the Battle of the Sexes game.

3.7 Finding Nash Equilibria in Common Games

Player 2's best responses. Because players' payoffs are symmetric, player 2's best responses are also $BR_2(S) = S$ when player 1 hunts the stag (in the top row); and $BR_2(H) = H$ when player 1 hunts the hare (in the bottom row).

Matrix 3.6b underlines best response payoffs, implying that the two NEs in this game are:

$$NE=\{(S,S), (H,H)\}$$

which associated payoffs (6, 6) and (2, 2), respectively.

		Player 2	
		Stag, S	Hare, H
Player 1	Stag, S	<u>6,6</u>	1, 4
	Hare, H	4, 1	<u>2,2</u>

Matrix 3.6b. The stag hunt game–underlining best response payoffs.

Therefore, while both strategy profiles are NEs, (S, S) Pareto dominates (H, H) since both players are better-off in the former than the latter. In other words, it is in the players' common interest to coordinate into the (S, S) equilibrium, which explains why this type of game is also known as "common interest games." It has been used to model countries' cooperation in international problems (such as climate change), individuals' incentives to row a boat, and orcas' cooperative hunting practices.

3.7.4 Anticoordination Game—The Game of Chicken

Matrix 3.7a reproduces the game of Chicken that we presented in Chapter 2, played between two teenagers who are driving their cars toward each other. Like the Battle of the Sexes game, we could not find strictly dominated strategies for either player, and thus four strategy profiles (all cells in the matrix) survived the application of IDSDS. We next show that the NE solution concept has more bite than IDSDS in this game, thus helping us provide more precise equilibrium predictions. We start by finding each player's best responses.

		Player 2	
		Swerve	Stay
Player 1	Swerve	−1, −1	−8, 10
	Stay	12, −8	−30, −30

Matrix 3.7a. Anticoordination game.

Player 1's best responses. Fixing player 2's strategy at Swerve (left column), player 1's best response is $BR_1(Swerve) = Stay$ because $12 > -1$. However, when we fix player 2's strategy at Stay (right column), player 1's best response is $BR_1(Stay) = Swerve$ given that $-8 > -30$. Intuitively, player 1 chooses the opposite strategy as her opponent: when player 2 Swerves, player 1 responds with Stay, becoming the top dog of the group, whereas when player 2 Stays, player 1 avoids a serious car accident by swerving.

Player 2's best responses. Because players' payoffs are symmetric, player 2's best responses are also $BR_2(Swerve) = Stay$ when player 1 Swerves (in the top row); and $BR_2(Stay) = Swerve$ when player 1 Stays (in the bottom row).

Underlining best response payoffs, our original Matrix 3.7a becomes Matrix 3.7b.

		Player 2	
		Swerve	Stay
Player 1	Swerve	$-1, -1$	$-8, \underline{10}$
	Stay	$\underline{12}, -8$	$-30, -30$

Matrix 3.7b. Anticoordination game–underlining best response payoffs.

We have then found two strategy profiles where players choose mutual best responses, thus becoming the NEs of the game of Chicken:

$$NE = \{(Swerve, Stay), (Stay, Swerve)\}.$$

In summary, every player seeks the opposite strategy as her opponent, explaining why this type of games is often known as "anticoordination games."

3.7.5 Multiple Nash Equilibria

In some of the games, such as the Battle of the Sexes, the Stag Hunt game, and the Game of Chicken, we found more than one NE. The literature has often relied on some equilibrium refinement tools such as Pareto or risk dominance (presented in the Appendix of this chapter) and other, more technical tools, such as Trembling-Hand Perfect Equilibrium or Proper Equilibrium (see Appendix of Chapter 5).

An alternative often used are the so-called "focal points" from Schelling (1960), which players tend to choose in the absence of preplay communication. However, it is unclear how a specific focal point arises (i.e., what makes an NE more salient than another) and it may vary depending on the time or place where players interact, leading many researchers to experimentally test the emergence of focal points in simultaneous-move games with more than one NE. For a more technical presentation, see Stahl and Wilson (1995), and for a literature review, see Camerer et al. (2004).

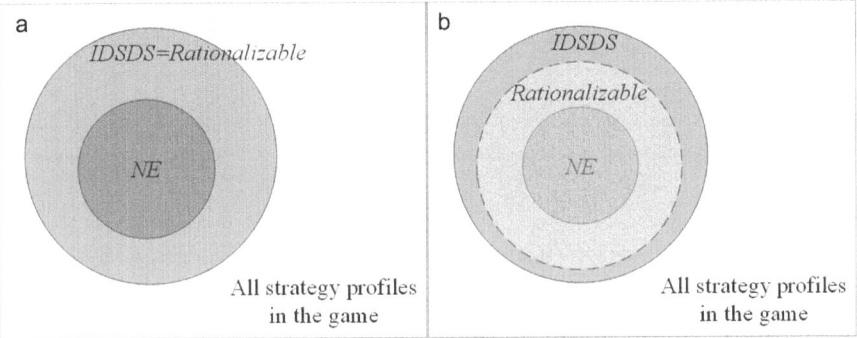

Fig. 3.2 Comparing strategy profiles: **a** Two players. **b** More than two players

3.8 Relationship Between NE and IDSDS

The previous example highlights an important point connecting strategy profiles that survive IDSDS and those that are NEs in a game: if a strategy profile $s = (s_i, s_{-i})$ is an NE, then it must survive IDSDS, but the opposite does not necessarily hold:

$$s \text{ is an NE} \implies s \text{ survives IDSDS.}$$
$$\not\Leftarrow$$

Therefore, if we find an NE, that strategy profile would also survive IDSDS (as in the Prisoner's Dilemma game), but strategy profiles that survive IDSDS are not necessarily an NE of that game (think, for instance, of the Battle of the Sexes game which has four strategy profiles surviving IDSDS but only two of these strategy profiles are NEs). In other words, the NEs in a game are a subset of the strategy profiles surviving IDSDS, as illustrated in Fig. 3.2.

From Sect. 3.4, recall that the set of rationalizable strategies in two-player games coincides with the strategy profiles surviving IDSDS, implying that the former coincides with the latter, as depicted in Fig. 3.2a. In games with three or more players, however, the set of rationalizable strategies is a subset of those surviving IDSDS, but weakly larger than the set of NEs, as shown in Fig. 3.2b.

3.9 What If We Find No NEs?

Consider the following Matching Pennies game, where two players simultaneously and independently flip a coin each. If both coins show the same side (both show Heads or both Tails), player 1 earns player 2's coin, as illustrated in the cells along the main diagonal of Matrix 3.8, (Heads, Heads) and (Tails, Tails). In contrast, if the coins show opposite sides, player 1 must give her coin to player 2, as shown in the cells below and above the main diagonal of Matrix 3.8, (Tails, Heads) and (Heads, Tails).

$$
\begin{array}{c c}
 & \text{Player 2} \\
 & \begin{array}{c c} \text{Heads} & \text{Tails} \end{array} \\
\text{Player 1} \quad \begin{array}{c} \text{Heads} \\ \text{Tails} \end{array} & \begin{array}{|c|c|} \hline 1,-1 & -1,1 \\ \hline -1,1 & 1,-1 \\ \hline \end{array}
\end{array}
$$

Matrix 3.8. Matching Pennies game.

We seek to find the NE of this game. To do that, we start identifying the best responses of player.

Player 1's best responses. Conditional on player 2 choosing Heads (in the left column), player 1's best response is also to select Heads given that $1 > -1$, that is, $BR_1(H) = H$. When player 2 chooses Tails (in the right column), player 1's best response is Tails because $1 > -1$, which we report as $BR_1(T) = T$. We underline player 1's best response payoffs in Matrix 3.9.

$$
\begin{array}{c c}
 & \text{Player 2} \\
 & \begin{array}{c c} \text{Heads} & \text{Tails} \end{array} \\
\text{Player 1} \quad \begin{array}{c} \text{Heads} \\ \text{Tails} \end{array} & \begin{array}{|c|c|} \hline \underline{1},-1 & -1,\underline{1} \\ \hline -1,\underline{1} & \underline{1},-1 \\ \hline \end{array}
\end{array}
$$

Matrix 3.9. Matching Pennies game - Underlining best response payoffs.

Player 2's best responses. We now move to player 2. Conditional on player 1 choosing Heads (in the top row), player 2's best response is Tails since $1 > -1$, or more compactly, $BR_2(H) = T$. However, when player 1 chooses Tails (in the bottom row), player 2's best response is Heads given that $1 > -1$, which we report as $BR_1(T) = H$. Matrix 3.9 also underlines the best response payoffs of player 2 in the second payoff of each cell.

Inspecting Matrix 3.9, we observe that no cell has the payoffs of both players underlined. This means that there is no strategy profile (no cell) where players choose best responses to their opponent's strategy. As a consequence, we can claim that no NE exists when we restrict players to choose one of their available strategies (Heads or Tails) with *certainty* (with probability one). Nonetheless, as we examine in Chapter 5, an NE exists, even in the Matching Pennies game, when we allow players to randomize over their available strategies (i.e., choosing Heads with some probability p and Tails with the remaining probability $1 - p$).

3.10 Evaluating NE as a Solution Concept

We next evaluate NE as a solution concept according to the same criteria as in previous chapters.

1. **Existence? Yes.** This result is often known as "Nash's existence theorem." Its proof requires the use of mixed strategies where players randomize their choices and, thus, we postpone it until Chapter 5 (Appendix). Intuitively, games with $N \geq 2$ players, each facing a discrete strategy space S_i, have at least one NE, either involving pure strategies (if players choose one strategy with certainty) or mixed strategies (if players randomize over two or more of their available strategies). As shown in the previous section, the use of best responses, and the underlining of best response payoffs, will lead to no cell where both players' payoffs are underlined, meaning that no NE exists when players are restricted to use pure strategies. (Recall the Matching Pennies game.) In those games, we will need to rely on mixed strategies (i.e., randomizations) to find at least one NE.

   A similar result applies when players face a continuous strategy space S_i, where for NE to exist we require that, in essence, players' payoff functions not to be extremely discontinuous. All games we consider in the book will be regular enough to yield at least one NE.

2. **Uniqueness? No.** While the above discussion helps us rest assured that an NE arises in most games, we cannot claim that *every* game will have a unique NE. We already encountered games with a unique NE, such as the Prisoner's Dilemma, but also games with two NEs, such as coordination and anticoordination games, confirming that the NE solution concept does not imply uniqueness.[7]

3. **Robust to small payoff perturbations? Yes.** If we alter the payoff of one of the players by a small amount, ε, we would find the same NE than before the payoff change. In other words, if player i finds that strategy s_i is her best response to her opponents playing s_{-i}, then s_i must still be her best response to s_{-i} after we apply a small payoff perturbation. Recall that this must hold for all $\varepsilon \to 0$, so the payoffs of one of the players are almost unchanged, thus keeping her best response unaffected.

4. **Socially optimal? No.** As discussed in the Prisoner's Dilemma game, the NE of a game does not need to be socially optimal. A similar argument applies to the "Modified Battle of the Sexes" game in Matrix 3.10, where (Opera, Opera) yields a higher payoff than (Football, Football) for every player, yet (Football, Football) can be sustained as one of the NEs of the game.

---

[7] In fact, when we study how to find mixed strategy NE in Chapter 5, we show that coordination and anticoordination games have one equilibrium where players randomize, implying that these games have a total of three NEs (two in pure strategies and one in mixed strategies).

		Wife	
		Football	Opera
Husband	Football	8, 8	6, 6
	Opera	4, 4	10, 10

Matrix 3.10. Modified Battle of the Sexes game.

Appendix: Equilibrium Selection

Coordination and anticoordination games give rise to two NEs, and a similar argument applies to other games where more than one NE arises. In games with more than one NE, a typical question is whether we can select one of them as being more "natural" or "reasonable" to occur according to some criteria. The literature offers several equilibrium selection criteria and we present some of the most popular below. Throughout our presentation, let $s^* = (s_i^*, s_{-i}^*)$ denote an NE strategy profile.

Pareto Dominance

Definition 3.4. **Pareto dominated NE**. An NE strategy profile s^* Pareto dominates another NE strategy profile $s' \neq s^*$ if $u_i(s^*) > u_i(s')$ for every player i.

We can easily apply Pareto dominance in the technology coordination game between firms 1 and 2 in Matrix 3.11. Every firm must simultaneously and independently choose between technology A or B, exhibiting similar incentives as in coordination games. For easier reference, the matrix underlines best response payoff for each player, helping us identify two NEs: (A, A) and (B, B). The former Pareto dominates the latter, since every player (firm) earns a higher profit at (A, A) than at (B, B), i.e., $10 > 5$.

		Firm 2	
		Tech. A	Tech. B
Firm 1	Tech. A	<u>10</u>,<u>10</u>	0, 2
	Tech. B	2, 0	<u>5</u>,<u>5</u>

Matrix 3.11. Technology coordination game.

Pareto dominance, however, does not help us rank NEs in all games. Consider the Battle of the Sexes game, for instance, where the husband prefers (F, O) to (O, F); but his wife prefers otherwise. In that context, we cannot select (F, O) over (O, F)

or vice versa. A similar argument applies in the Chicken game, where player 1 prefers (*Stay, Swerve*) to (*Swerve, Stay*) while player 2 prefers otherwise.

Risk Dominance

Pareto dominance helps us compare NEs where, for instance, both firms simultaneously change their technology decision from B to A, yielding (A, A). But what if each firm unilaterally deviates from an NE? If (B, B) is played, every firm's payoff loss when unilaterally switching to A is $5 - 0 = 5$; but if (A, A) is being played, every firm losses $10 - 2 = 8$ when unilaterally deviating to B. Informally, at (A, A) unilateral deviations are more risky than at (B, B), as we present below.

Definition 3.5. **Risk dominated NE.** An NE strategy profile $s^* = (s_1^*, s_2^*)$ risk dominates another NE strategy profile $s' = (s_1', s_2')$ if the total payoff loss of moving from s' to s^*,

$$\left[u_1(s_1^*, s_2^*) - u_1(s_1', s_2^*)\right] \times \left[u_2(s_1^*, s_2^*) - u_2(s_1^*, s_2')\right],$$

exceeds that of moving from s^* to s',

$$\left[u_1(s_1', s_2') - u_1(s_1^*, s_2')\right] \times \left[u_2(s_1', s_2') - u_2(s_1', s_2^*)\right].$$

In Matrix 3.11, we can say that (B, B) risk dominates (A, A), because the total payoff loss of moving from (B, B) to (A, A),

$$[u_1(A, A) - u_1(B, A)] \times [u_2(A, A) - u_2(A, B)]$$
$$= (10 - 2) \times (10 - 2) = 64,$$

exceeds that of moving from (A, A) to (B, B),

$$[u_1(B, B) - u_1(A, B)] \times [u_2(B, B) - u_2(B, A)]$$
$$= (5 - 0) \times (5 - 0) = 25.$$

Exercises

3.1. **More general Prisoner's Dilemma game.**[A] Consider the following Prisoner's Dilemma game where every player simultaneously chooses between confess (F) and not confess (NC). Payoffs satisfy $a_i > b_i > 0$ for every player $i = \{1, 2\}$, implying that everyone earns a higher payoff at (NC, NC) than at (C, C). In addition, $b_i > d_i$ and $c_i > a_i$.

		Player 2	
		C	NC
Player 1	C	b_1, b_2	c_1, d_2
	NC	d_1, c_2	a_1, a_2

a. Find the best responses of each player.
b. Find the NEs of this game.
c. Are your results affected if player 1 enjoys a larger payoff gain from confessing when player 2 does not confess, $c_1 - a_1 > c_2 - a_2$? Interpret.

3.2. **More general Pareto coordination game.**[A] Consider the following Pareto coordination game where every firm simultaneously chooses between technology A or B, and payoffs satisfy $a_i > b_i > 0$ for every firm $i = \{1, 2\}$, implying that both firms regard technology A as superior. However, the premium that firm i assigns to technology A, as captured by $a_i - b_i$, can be different between firms 1 and 2.

		Firm 2	
		Tech. A	Tech. B
Firm 1	Tech. A	a_1, a_2	0, 0
	Tech. B	0, 0	b_1, b_2

a. Find the best responses of every firm.
b. Find the NEs of this game.
c. Are your results affected if firm 1 assigns a larger premium to technology A than firm 2 does, $a_1 - b_1 > a_2 - b_2$? Interpret.

3.3. **More general anticoordination game.**[A] Consider the following Game of Chicken where every player drives toward each other and simultaneously chooses between swerve and stay. Payoffs satisfy $a_i > b_i > 0$ for every player $i = \{1, 2\}$, implying that drivers suffer more when they both stay (and almost die!) than when they both swerve (and are called chickens), and $a_i > d_i > c_i > b_i$, so being the only player who stays yields a payoff of d_i for that player and $-c_j$ for his rival, where $j \neq i$.

		Player 2	
		Swerve	Stay
Player 1	Swerve	$-b_1, -b_2$	$-c_1, d_2$
	Stay	$d_1, -c_2$	$-a_1, -a_2$

a. Find the best responses of each player.
b. Find the NEs of this game.

c. Are your results affected if the players' payoffs are symmetric, implying that $a_1 = a_2 = a$, $b_1 = b_2 = b$, $c_1 = c_2 = c$, and $d_1 = d_2 = d$?

d. Are your results affected if player 1's payoffs satisfy $a_1 = a_2 + \gamma$, $b_1 = b_2 + \gamma$, $c_1 = c_2 + \gamma$, and $d_1 = d_2 + \gamma$, where parameter γ represents the asymmetry between player 1's and 2's payoffs? Interpret.

3.4. **More general stag hunt game.**[A] Consider the following Stag Hunt Game where two individuals simultaneously choose between hunt a stag or a hare. Payoffs satisfy $a_i > b_i > d_i > c_i$ for every player i, so every player's ranking of payoffs is: (1) hunting a stag with the help of his opponent; (2) hunting a hare with the help of his opponent; (3) hunting a stag without help; and (4) hunting a hare without help.

		Player 2	
		Stag	Hare
Player 1	Stag	a_1, a_2	c_1, d_2
	Hare	d_1, c_2	b_1, b_2

a. Find the best responses of each player.
b. Find the NEs of this game.
c. Are your results affected if players' payoffs are symmetric, implying that $a_1 = a_2 = a$, $b_1 = b_2 = b$, $c_1 = c_2 = c$, and $d_1 = d_2 = d$?

3.5. **Strict dominance and NE.**[B] Consider the following simultaneous-move game between player 1 (in rows) and player 2 (in columns).

		Player 2		
		L	M	R
Player 1	U	4, 2	2, 0	0, 3
	C	5, 1	3, 2	8, 4
	D	5, 2	6, 2	α, β

We are informed that player 1 finds that strategy C weakly dominates U. In addition, we are told that strategy profile (D, R) is a Nash equilibrium but (D, M) is not. Using this information, answer the following questions. Before we start, let us define $u_i(s_i, s_j)$ to be the utility function of player i when player i deploys strategy $s_i \in S_i$ and his counterpart, player j, deploys strategy $s_j \in S_j$, where $i, j = \{1, 2\}$. In this context, $S_1 = \{U, C, D\}$ and $S_2 = \{L, M, R\}$.

a. Does player 2 have a strictly dominated strategy?
b. Is strategy profile (D, R) the unique Nash equilibrium of this game?
c. Does player 1 have a strictly dominant strategy?

3.6. **Rationalizable strategies and IDSDS.**[B] Consider the following two-player game, where player 1 (in rows) chooses U, M, or D, while player 2 (in columns) chooses L or R.

	Player 2	
	L	R
U	4, 2	0, 1
M	0, 2	4, 1
D	1, 0	1, 3

Player 1

a. Assuming that players are restricted to use pure strategies, show that IDSDS has no bite. In this context, identify the strategy profiles that survive IDSDS.
b. Allowing now for players to use mixed strategies, show that rationalizability has some bite. (Hint: Use the "never a best response" approach.)

3.7. **Strict dominance and Rationalizability.**[B] Consider Exercise 2.3 again (Chapter 2), where we found the strategy profiles that survive IDSDS and those surviving IDWDS. Find now which strategy profiles survive rationalizability. Compare your equilibrium results against those under IDSDS.

3.8. **IDSDS and rationalizability—Allowing for more general payoffs.**[A] Consider the following simultaneous-move game between Anthony (in rows) and Christine (in columns), where payoff x satisfies $x \geq 0$.

	Christine	
	North	South
Up	x, 7	3, 6
Center	4, 1	9, 2
Bottom	8, 5	1, 4

Anthony

a. Identify any strategies that are strictly dominated by pure strategies.
b. Suppose that $x = 6$. Is the strategy Up rationalizable? Explain why or why not.
c. Find the minimum value of x for the strategy Up to be rationalizable. How are your results affected by an increase in x?

3.9. **NE and rationalizability.**[C] In this exercise, we formally show the relationship between NE strategy profiles and rationalizability.
a. Show that if strategy profile $s^* = (s_1^*, \ldots, s_N^*)$ is a Nash equilibrium of a N-player game, it must also survive rationalizability.

b. Show that the converse of part (a) is not necessarily true. (For this part, an example suffices.)
3.10. **If a game has an NE, it must survive IDSDS.**[B] Consider the two-player game in the following matrix, where payoffs satisfy $a \neq b \neq c \neq d$. If the game has a unique psNE, show that it must be the unique strategy profile surviving IDSDS.

		Player 2	
		A	B
Player 1	A	a, a	c, c
	B	b, b	d, d

3.11. **A unique strategy profile surviving IDSDS must also be an NE.**[B] Consider that strategy profile $s^* = (s_1^*, s_2^*, \cdots, s_N^*)$ is the unique strategy profile surviving IDSDS in a game with $N \geq 2$ players. Show that s^* must also be the unique NE of the game.
3.12. **Finding NEs in four categories of simultaneous-move games.**[A] Consider the following simultaneous-move game with payoff matrix.

		Player 2	
		A	B
Player 1	A	a, a	b, c
	B	c, b	d, d

Find the players' best responses and the psNEs in the following settings. Interpret and relate your results to common games.
   a. Payoffs satisfy $a > c$ and $b > d$.
   b. Payoffs satisfy $a > c$ and $b < d$.
   c. Payoffs satisfy $a < c$ and $b < d$.
   d. Payoffs satisfy $a < c$ and $b > d$.
3.13. **Strictly Dominant equilibrium.**[C] Recall the definition of a strictly dominant equilibrium (SDE) from Sect. 2.8.
   a. *Existence*. Provide an example of a game with an SDE and a game without an SDE.
   b. *Uniqueness*. Can a game have more than one SDE?
   c. Show that if s^{SDE} exists, s^{SDE} must survive IDSDS.
   d. Show that if s^{SDE} exists, s^{SDE} must be the only strategy profile surviving IDSDS.
   e. Show that if strategy profile s survives IDSDS, it does not need to be an SDE. An example suffices.
3.14. **Strictly Dominant equilibrium and NE.**[C] Consider the discussion about strictly dominant equilibrium (SDE) in Sect. 2.8, where $s^{SDE} = (s_1^{SDE}, s_2^{SDE}, ..., s_N^{SDE})$ denotes an SDE strategy profile.

a. Show that if s^{SDE} exists, then every player has a unique best response to his rivals' strategies.
b. Show that if strategy profile s^{SDE} is a SDE, it must also be an NE.
c. Show that if strategy profile s^{SDE} is a SDE, it must also be the unique NE of the game.

3.15. **Strict Nash equilibrium.**[B] Consider the following definition: A strategy profile $s^* \equiv (s_1^*, ..., s_N^*)$ is a *strict Nash equilibrium* (SNE) if it satisfies

$$u_i(s_i^*, s_{-i}^*) > u_i(s_i, s_{-i}^*) \text{ for every player } i, \text{ and every } s_i \in S_i.$$

You probably noticed that this definition is almost identical to the definition of Nash equilibrium (NE) strategy profiles, except for using a strict, rather than weak, inequality. In this exercise we connect both solution concepts, but first examine the relationship between a strict Nash equilibrium and IDSDS.
a. Show that a game can have more than one SNE. An example suffices.
b. Provide an example of a game with more than one NE but just one SNE.
c. Provide an example with a game where SNEs are a subset of all strategy profiles surviving IDSDS.
d. Show that if strategy profile s^* is the only strategy profile surviving IDSDS, then s^* must be an SNE.
e. Show that if strategy profile s^* is an SNE, it must also be an NE.

3.16. **Strict and weak Nash equilibria.**[B] Consider the game in the following payoff matrix.

		Player 2		
		L	C	R
	U	6, 5	7, 3	8, 4
Player 1	M	4, 3	10, 6	5, 8
	D	5, 2	11, 11	4, 10

a. Find the strategy profiles that can be supported as strict Nash equilibria and weak (non-strict) Nash equilibria in this game.
b. Which strategy profiles survive IDSDS?
c. Provide an example of a game with strict and weak Nash equilibria.

3.17. **Split or steal game.**[A] Consider the following split or steal game. In a TV show, two contestants answer trivia questions, with correct answers increasing a common pot. Once the trivia part of the game is over, assume that the pot accumulates V dollars. Every player is then asked to select independently and simultaneously (from a covered button) whether he wants to split or steal the prize. As described in the following matrix, if both players split the prize, each earns $\frac{V}{2}$; if a player is the only one to steal, he keeps all the price, V; if he is the only one to split the prize, he earns zero; and if both players steal the prize, both earn zero.[8]

		Player 2	
		$Split$	$Steal$
Player 1	$Split$	$\frac{V}{2}, \frac{V}{2}$	$0, V$
	$Steal$	$V, 0$	$0, 0$

a. Does any player have strictly dominated strategies? Does any player have weakly dominated strategies? Explain.
b. Find the NEs of the game.
c. Among all NEs of the game, is some of them strict as defined in Sect. 2.8?

References

Camerer, C. F. (2003) *Behavioral Game Theory: Experiments in Strategic Interaction* (The Roundtable Series in Behavioral Economics), Princeton University Press.
Camerer, C. F., T. Ho, and J.-K. Chong (2004) "A Cognitive Hierarchy Model of Games," The Quarterly Journal of Economics, 119(3), pp. 861–898.
Crawford, V. P., M. A. Costa-Gomes, and N. Irriberri (2013) "Structural Models of Nonequilibrium Strategic Thinking: Theory, Evidence, and Applications," Journal of Economic Literature, 51(1), pp. 5–62.
Fudenberg, D., and J. Tirole (1991) *Game Theory*, The MIT Press, Cambridge, Massachussets.
Harrington, J. E. (2014) *Games, Strategies, and Decision Making*, Worth Publishers.
Nagel, R. (1995) "Unraveling in Guessing Games: An Experimental Study," American Economic Review, 85(5), pp. 1313–1326.
Nash, J. F. (1950) "Non-Cooperative Games," Annals of Mathematics, 54(2), pp. 286–295.
Pearce, D. (1984) "Rationalizable Strategic Behavior and the Problem of Perfection," Econometrica, 52, pp. 1029–1050.
Schelling, T. C. (1960) *The Strategy of Conflict* (1st ed.), Cambridge: Harvard University Press.
Stahl, D. O., and P. W. Wilson (1995) "On Players Models of Other Players: Theory and Experimental Evidence," Games and Economic Behavior, 10(1), pp. 218–254.

---

[8] For more details, you can watch a short video of the TV show here: https://youtu.be/jJzyc6VPMcc.

Nash Equilibria in Games with Continuous Action Spaces

4.1 Introduction

In previous chapters, we examined how to find best responses when players face discrete strategy spaces (e.g., left or right, or set high or low prices). While helpful to understand the notion of best response in simultaneous-move games, many economic settings allow for continuous, rather than discrete, action spaces, such as firms choosing how much output to produce, which price to set for their products, or how much to invest in advertising. A similar argument applies to other social sciences, such as the political platform that candidates choose when they run for office, or how much each donor contributes to a charity.

In this chapter, we first present the Cournot model of quantity competition, finding every firm's best response function and their equilibrium output. In Sect. 4.3, we consider that firms compete in prices, rather than quantities, presenting the Bertrand model of price competition. While this model is often verbally presented in introductory economics textbooks, here we find each firm's best response function to identify equilibrium price pairs.

Sections 4.2.3 and 4.3.2 investigate how equilibrium results in the above models are affected when firms sell differentiated goods; as opposed to the homogeneous goods that firms produce in standard Cournot and Bertrand models. Examples of homogeneous goods are certain primary commodities, such as coal and iron ore and agricultural products such as wheat or cotton, while examples of heterogeneous goods can be found in two brands of clothing or smart phones, which are usually very differentiated. In this context, product differentiation entails that, if a firm sets a slightly higher price than its rival, it does not drive all of its customers away, still retaining a share of them, and thus ameliorating price competition relative to the setting where products are homogeneous.

In the last two sections of the chapter, we discuss how to find best responses in other contexts: (i) donations to a public good, seeking to understand players' free-riding incentives; and (ii) electoral competition, predicting whether candidates running for office have incentives to converge to the same position in the political spectrum.

4.2 Quantity Competition

4.2.1 Quantity Competition with Homogeneous Goods and Two Firms

Consider again the industry analyzed in Example 3.1, where two firms simultaneously and independently choose their output levels. This setting is commonly known as the Cournot model of quantity competition, after Cournot (1838). In Example 3.1, we showed that every firm i chooses its output q_i to maximize its profits, treating its rival's output, q_j, as given, yielding best response function

$$q_i(q_j) = \begin{cases} \frac{1-c}{2} - \frac{1}{2}q_j & \text{if } q_j < 1-c \\ 0 & \text{otherwise} \end{cases}$$

which is decreasing in q_j, implying that firms regard their output as "strategic substitutes" (see Example 3.1).

We now seek to use this best response functions (one for each firm) to find the NE of the game. By definition, at the NE output pair (q_i^*, q_j^*) firms must play mutual best responses to each other's output. In other words, both firms must choose an output that is part of their best responses. Formally, this means that we have a system of two equations—one best response function for each firm, $q_i(q_j)$ and $q_j(q_i)$—and two unknowns, q_i and q_j. We only need to simultaneously solve for q_i and q_j to obtain the equilibrium output levels. We next present two approaches to solve this system of equations.

First approach: *Inserting one firm's BRF into the other.* One way to find these output levels is by inserting firm j's best response function into firm i's, as follows

$$q_i = \frac{1-c}{2} - \frac{1}{2}\overbrace{\left(\frac{1-c}{2} - \frac{1}{2}q_i\right)}^{q_j(q_i)}$$

which is then a function of q_i alone. Rearranging, we obtain

$$q_i = \frac{1-c}{4} - \frac{1}{4}q_i,$$

or $\frac{1}{4}q_i = \frac{1-c}{4}$ which, solving for q_i, yields firm i's equilibrium output

$$q_i^* = \frac{1-c}{3}.$$

4.2 Quantity Competition

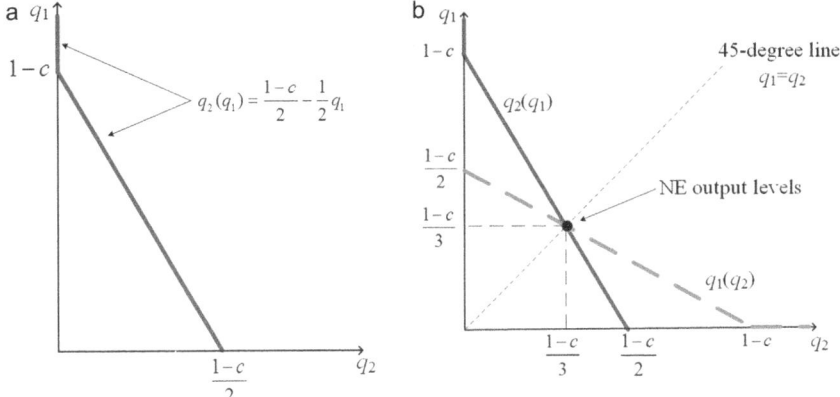

Fig. 4.1 **a** Firm 2's best response function, $q_2(q_1)$. **b** NE output in the Cournot game

Figure 4.1a depicts firm 2's best response function and Fig. 4.1b superimposes firm 1's best response function (from Fig. 3.1 in Chapter 3) and firm 2's. At their crossing point, firms are choosing output pairs that lie on both best response functions and, therefore, firms must be playing a "mutual best response," entailing that output choices must be a NE of the game.

Second approach: *Invoking symmetry.* An alternative, faster, approach is to recognize that this is a game with symmetric players (firms face the same demand and cost functions). In a symmetric equilibrium, firms must produce the same units of output, entailing $q_i^* = q_j^* = q^*$. Inserting this property into either firm's best response function, yields

$$q^* = \frac{1-c}{2} - \frac{1}{2}q^*$$

or $\frac{3}{2}q^* = \frac{1-c}{2}$. Solving for q^*, we find the same equilibrium output as with the first approach, that is,

$$q^* = \frac{1-c}{3}.$$

Generally, the second approach is useful for games with symmetric players, as it helps us identify symmetric equilibria if one exists. The first approach, while more time consuming, is however required in games with asymmetric players.

Finally, we can evaluate equilibrium profits by inserting q^* into the profit function, to obtain

$$\pi^* = (1 - q^* - q^*)q^* - cq^*$$
$$= \left(1 - \frac{1-c}{3} - \frac{1-c}{3}\right)\frac{1-c}{3} - c\frac{1-c}{3}$$

$$= \frac{(1-c)^2}{9}$$

which coincides with the square of the equilibrium output, that is, $\pi^* = (q^*)^2$.

4.2.2 Extending Quantity Competition to $N \geq 2$ Firms

The above Cournot model of quantity competition can be easily extended in many dimensions, such as considering an industry with more than two firms, allowing for firms to sell differentiated (rather than homogeneous) products, or considering different production costs, among others. (We consider some of these extensions in the end-of-chapter exercises.) In this section, we allow for more than two firms, which we denote by saying there are $N \geq 2$ firms in the industry. We show that the duopoly model with $N = 2$ firms in the previous section becomes a special case of this more general setting. To highlight the use of a similar approach to solve this game, we start again by finding every firm i's best response function.

Step 1. *Finding best response functions.* Every firm i solves

$$\max_{q_i \geq 0} (1 - q_i - Q_{-i})q_i - cq_i$$

where $Q_{-i} = \sum_{j \neq i} q_j$ denotes the aggregate output from firm i's rivals, so aggregate output in the entire industry can be written as $Q = q_i + Q_{-i}$. In the case of three firms, the aggregate output from firm 1's rivals is $Q_{-1} = q_2 + q_3$, and similarly for $Q_{-2} = q_1 + q_3$, and $Q_{-3} = q_1 + q_2$.

We now differentiate with respect to q_i, since this is the only output that firm i controls. (Technically, we say that q_i is firm i's "choice variable," and we write it below the "max" operator in firm i's maximization problem.) Differentiating firm i's profit with respect to q_i, we find

$$1 - 2q_i - Q_{-i} = c.$$

And solving for q_i, we obtain firm i's best response function

$$q_i(Q_{-i}) = \begin{cases} \frac{1-c}{2} - \frac{1}{2}Q_{-i} & \text{if } Q_{-i} < 1 - c \\ 0 & \text{otherwise.} \end{cases}$$

thus being almost identical to the expression we found in a duopoly (same vertical intercept), except for the last term, where we had $\frac{1}{2}q_j$ rather than $\frac{1}{2}Q_{-i}$. Needless to say, in a duopoly the aggregate output produced by firm i's rivals, Q_{-i}, satisfies $Q_{-i} = q_j$, because firm i has only one rival (firm j).

Step 2. *Finding NE output.* In a symmetric equilibrium, every firm produces the same output level, that is, $q_i^* = q_j^* = q^*$ for every two firms i and j. Therefore,

4.2 Quantity Competition

$Q_{-i}^* = (N-1)q^*$. Inserting this property into the above best response function, yields

$$q^* = \frac{1-c}{2} - \frac{1}{2}(N-1)q^*$$

which is only a function of q^*. Solving for q^*, we can find the equilibrium output of each firm in this industry, that is,

$$q^*(N) = \frac{1-c}{N+1}.$$

which is decreasing in the number of firms, N. In the case of a duopoly, $N = 2$, it becomes $q^*(2) = \frac{1-c}{3}$, thus coinciding with the expression we found in Sect. 4.2.1. In this context, aggregate output is $Q^* = N \times q^*(N) = \frac{N(1-c)}{N+1}$, and equilibrium price $p^* = 1 - Q^*$ is

$$p^* = 1 - \frac{N(1-c)}{N+1} = \frac{1+Nc}{N+1}$$

which is decreasing in the number of firms, N, since

$$\frac{\partial p^*}{\partial N} = -\frac{1-c}{(N+1)^2} < 0$$

given that $c < 1$ by assumption. Therefore, equilibrium profit for every firm becomes

$$\begin{aligned}\pi^* &= p^* q^* - cq^* \\ &= \frac{1+Nc}{N+1} \frac{1-c}{N+1} - c\frac{1-c}{N+1} \\ &= \frac{(1-c)^2}{(N+1)^2}\end{aligned}$$

or the square of equilibrium output, that is, $\pi^* = (q^*)^2$.

Special cases. The above results also embody other market structures as special cases:

1. When only one firm operates in the industry, $N = 1$, we obtain an equilibrium output of $q^* = \frac{1-c}{2}$, with associated price $p^* = \frac{1+c}{2}$, and equilibrium profits of $\pi^* = \frac{(1-c)^2}{4}$, thus coinciding with the standard results in a monopoly market with inverse demand curve $p(q) = 1 - q$ and constant marginal cost c.
2. Similarly, when infinitely many firms compete (i.e., $N \to +\infty$), we can take the limit of our above results, finding that the equilibrium output of every firm i approaches $q^* = 0$, and the equilibrium price converges to $p^* = c$ (marginal cost pricing), entailing that every firm makes zero economic profits in equilibrium—as in a perfectly competitive market.

4.2.3 Quantity Competition with Heterogeneous Goods

Consider again the industry of Sect. 4.2.1, where two firms compete in quantities, but assume that each firm i faces the following inverse demand function

$$p_i(q_i, q_j) = 1 - q_i - dq_j,$$

where parameter d satisfies $1 \geq d \geq 0$. The last property means that own-price effects dominate cross-price effects, or more intuitively, that firm i's price is more sensitive to changes in its own output, q_i, than to changes in its rival's output, q_j.

Step 1. *Finding best response functions.* Every firm i solves

$$\max_{q_i \geq 0} (1 - q_i - dq_j)q_i - cq_i$$

Differentiating with respect to q_i, yields

$$1 - 2q_i - dq_j = c.$$

or, after rearranging, $1 - c - dq_j = 2q_i$. Solving for q_i, we obtain firm i's best response function

$$q_i(q_j) = \begin{cases} \frac{1-c}{2} - \frac{d}{2}q_j & \text{if } q_j < \frac{1-c}{d} \\ 0 & \text{otherwise.} \end{cases}$$

Before moving to Step 2, where we combine firm i's and j's best response functions to find the NE of the game, let us build some intuition of $q_i(q_j)$ by looking at the extreme cases. When parameter d increases, eventually approaching one, $d = 1$, products become more homogeneous (i.e., firm i's price is similarly affected by a marginal increase in the sales of i and j). In this case, firm i's best response function becomes steeper; as depicted in Fig. 4.2 by the downward rotation of $q_i(q_j)$. Intuitively, this means that firm i's output choice is more sensitive to a given increase in q_j, indicating that firms' competition is more intense.

In contrast, when $d = 0$, products become more differentiated and firm i's price is, essentially, unaffected by its rival's output decision. In this case, its best response function becomes almost flat (as shown in the upward rotation in Fig. 4.2), approaching to $q_i(q_j) = \frac{1-c}{2}$, which implies that firm i's output choice does not depend on its rival's, as if firms competed in two monopoly markets.

Step 2. *Finding NE output.* In a symmetric equilibrium, every firm produces the same output level, that is, $q_i^* = q_j^* = q^*$ for every two firms i and j. Inserting this property in the above best response function, yields

$$q^* = \frac{1-c}{2} - \frac{d}{2}q^*$$

Fig. 4.2 Firm i's best response function

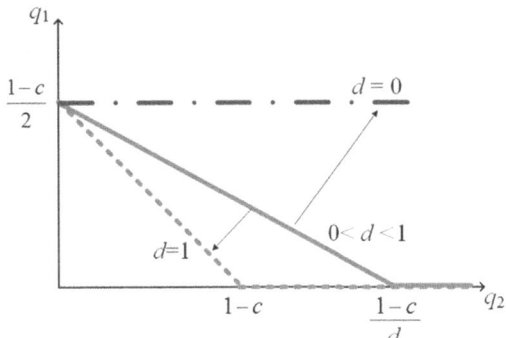

Fig. 4.3 NE output levels with heterogeneous products

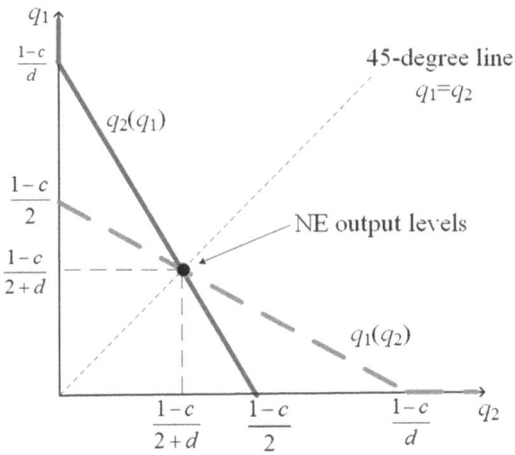

which is only a function of q^*. Solving for q^*, we find the equilibrium output of each firm in this industry, that is,

$$q^* = \frac{1-c}{2+d}.$$

Figure 4.3 illustrates both firms' best response functions, and their crossing point, which identifies the NE of this Cournot game of quantity competition with heterogeneous products.

As expected, this equilibrium output is decreasing in parameter d. This result indicates that, as products become more homogeneous (higher d), individual output decreases. In the extreme case where products are undifferentiated, $d \to 1$, this equilibrium output becomes $q^* = \frac{1-c}{3}$, as in a standard Cournot model where firms sell homogenous goods. In contrast, in the case where products are extremely differentiated, $d \to 0$, the equilibrium output is $q^* = \frac{1-c}{2}$, as in a monopoly, since the two goods are unrelated (each firm serving different markets).

In this setting, equilibrium price, $1 - q_i - dq_j$, is

$$p^* = 1 - \underbrace{\frac{1-c}{2+d}}_{q_i^*} - d\underbrace{\frac{1-c}{2+d}}_{q_j^*}$$
$$= \frac{1 + c(1+d)}{2+d}$$

which is decreasing in the degree of product homogeneity (parameter d), since

$$\frac{\partial p^*}{\partial d} = -\frac{(1-c)}{(2+d)^2} < 0$$

given that $0 < c < 1$ by assumption. Finally, equilibrium profits for every firm i become

$$\pi^* = p^*q^* - cq^*$$
$$= \frac{1+c(1+d)}{2+d} \frac{1-c}{2+d} - c\frac{1-c}{2+d}$$
$$= \frac{(1-c)^2}{(2+d)^2}$$

which are decreasing as products become more homogeneous (higher d). Equilibrium profits can, more compactly, be represented as $\pi^* = (q^*)^2$.

4.3 Price Competition

4.3.1 Price Competition with Homogeneous Goods

Consider now two firms competing in prices and selling a homogeneous good. This form of competition is often known as "à la Bertrand," after Bertrand (1883). Since the product that firms sell is undifferentiated, if firm i sets the lowest price, $p_i < p_j$, it attracts all customers while its rival makes no sales, $q_j = 0$. If both firms set the same price, $p_i = p_j$, they evenly share the market demand. We can, therefore, summarize the demand of firm i as follows

$$q_i(p_i, p_j) = \begin{cases} 1 - p_i & \text{if } p_i < p_j \\ \frac{1-p_i}{2} & \text{if } p_i = p_j \\ 0 & \text{if } p_i > p_j \end{cases}$$

For simplicity, we assume that both firms face the same marginal production cost c, where $1 > c \geq 0$. As in other games where players have a continuous action space (a price $p_i \geq 0$ in this context), we first identify the best response function of each firm, and then find the price pair where both best response functions intersect, obtaining a NE of the Bertrand game.

4.3 Price Competition

Step 1. *Finding best response functions.* In this context, every firm i's best response function is depicted in Fig. 4.4, and can be formally expressed as

$$p_i(p_j) = \begin{cases} p^m & \text{if } p_j > p^m \\ p_j - \varepsilon & \text{if } c < p_j \leq p^m \\ c & \text{if } p_j \leq c \end{cases}$$

where $p^m = \frac{1+c}{2}$ denotes the monopoly price.[1] We can interpret each of the segments of this best response function as follows:

- First, when firm j sets a price above the monopoly price p^m, firm i best responds by setting a price $p_i = p^m$ since, by doing that, it captures all sales and maximizes its profits. This is illustrated on the flat horizontal segment of the best response function, at the northeast of Fig. 4.4.
- Second, when firm j sets a price between p^m and c, firm i optimally responds by slightly undercutting its rival's price. This is illustrated in the figure by the segment slightly below the 45-degree line, so that $p_i = p_j - \varepsilon$, where ε is a small number, i.e., $\varepsilon > 0$ and $\varepsilon \to 0$.
- Finally, when firm j sets a price below the marginal cost, c, firm i responds by setting a price $p_i = c$. This price guarantees no sales since $p_i = c > p_j$, but at least generates no losses. This is depicted in Fig. 4.4 by the horizontal segment in the southwest of the figure.[2]

Figure 4.4 illustrates that firms' prices are (weak) "strategic complements": as firm j increases its price by \$1, firm i responds by increasing its price by \$1 too (in the intermediate segment between c and p_m); or keeping its price unchanged (when $p_j \leq c$ or $p_j > p^m$). Generally, when best responses are strategic complements, a player increases (decreases) her strategy when her rival chooses a higher (lower) strategy or, more compactly, players' strategies move in the same direction.

Firm j's best response function is symmetric, that is,

$$p_j(p_i) = \begin{cases} p^m & \text{if } p_i > p^m \\ p_i - \varepsilon & \text{if } c < p_i \leq p^m \\ c & \text{if } p_i \leq c \end{cases}$$

as depicted in Fig. 4.5.

Step 2. *Finding NE prices.* Figure 4.6 depicts the best response function of firms i and j, so we can find the price pair where they intersect. At this price pair, both

---

[1] The monopoly price solves $\max_{p_i \geq 0} (p_i - c)(1 - p_i)$. Differentiating with respect to p_i, yields $1 - 2p_i + c = 0$ and, solving for p_i, we obtain the monopoly price $p^m = \frac{1+c}{2}$.

[2] Technically, any price $p_i \geq c$ is a best response to firm j setting a price $p_j < c$, all of them yielding zero sales for firm i. For illustration purposes, however, we focus here on a specific best response from firm i, namely, setting $p_i = c$.

Fig. 4.4 Firm i's best response function when competing in prices

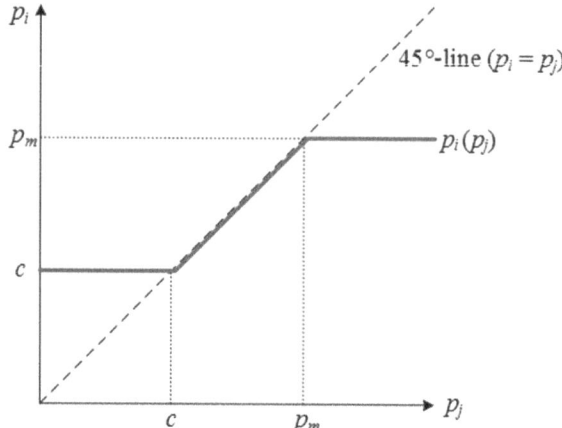

Fig. 4.5 Firm j's best response function when competing in prices

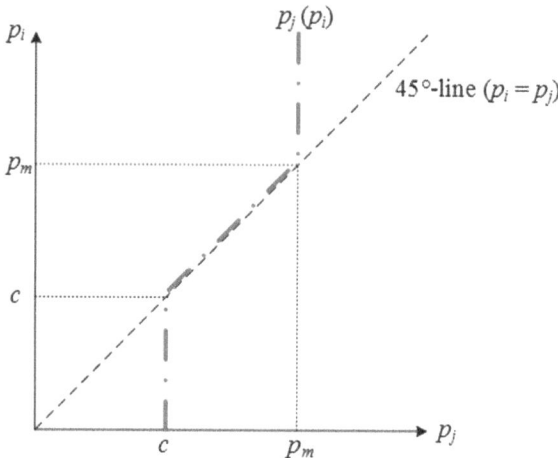

firms are choosing best responses to each other's strategies, thus becoming a NE of the Bertrand game of price competition when firms sell homogeneous products.

Therefore, the NE prices are

$$p_i^* = p_j^* = c$$

or, alternatively, the NE price pair is $\left(p_i^*, p_j^*\right) = (c, c)$, which entails zero profits for both firms, $\pi_i^* = \pi_j^* = 0$. In other words, firms make no economic profits when selling a homogeneous product. This is an extremely competitive result: with only two firms, we found that they behave as firms operating in a perfectly competitive market. This finding remains unaffected when we allow for more than two firms, but, as we explore in the next section, it is ameliorated when firms sell heterogeneous products.

Fig. 4.6 Price equilibrium

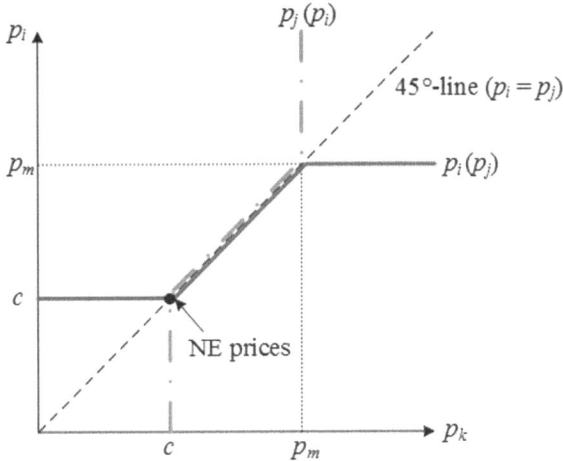

4.3.2 Price Competition with Heterogeneous Goods

Assume now that firms compete in prices but sell a horizontally differentiated product.[3] This can be the case of clothing, certain electronics, food, or music. The demand function of every firm i is

$$q_i(p_i, p_j) = 1 - p_i + \gamma p_j$$

where $j \neq i$, and parameter γ satisfies $\gamma \in (0, 2)$. When $\gamma \to 0$, firm j's price does not affect firm i's sales; when $\gamma = 1$, firm j's price affects firm i's sales at a similar rate as changes in its own price; and when $\gamma > 1$, an increase in p_j increases firm i's sales more than a similar decrease in its own price. As in the previous section, we assume that both firms face the same marginal cost c, where $1 > c > 0$.

To solve this game, we follow a similar approach as in its homogeneous-good version, starting with finding best response functions and then identifying the price pair where both functions cross each other.

Step 1. *Finding best response functions.* Every firm i chooses its price p_i to solve the following profit maximization problem,

$$\max_{p_i > 0} \pi_i(p_i) = (p_i - c)(1 - p_i + \gamma p_j)$$

---

[3] Recall that vertical product differentiation refers to goods that, if their prices coincide, all consumers would purchase one good over another (all consumers regard one good to be of superior quality). In horizontal differentiation, however, some consumers may purchase each of the goods even if they are both equally priced, thus allowing for consumers to exhibit different preferences for each good. We focus on this type of product differentiation in this section.

Differentiating with respect to p_i, we obtain

$$1 - 2p_i + c + \gamma p_j = 0$$

Solving for p_i, we find firm i's best response function:

$$p_i(p_j) = \frac{1+c}{2} + \frac{\gamma}{2} p_j$$

which originates at $\frac{1+c}{2}$ and increases in p_j at a rate of $\frac{\gamma}{2}$, as depicted in Fig. 4.7. Firm i, then, regards in this context its rival's price, p_j, as a "strategic complement": as firm j increases its price by \$1, firm i responds by increasing its own price, p_i, by only \$$\frac{\gamma}{2}$, which is less than \$1 since $\gamma < 2$ by assumption. In other words, firm i responds by increasing its price, but less than proportionally, stealing some of j's customers and still maximizing its profit.

It is straightforward to notice that, as firm i's sales become less affected by firm j's price (lower γ), the vertical intercept is unaffected, but the slope decreases, making the best response function flatter, thus softening price competition.

Step 2. *Finding NE prices.* In a symmetric equilibrium, both firms set the same price, $p = p_1 = p_2$. Inserting this property in the above best response function, we obtain

$$p = \frac{1+c}{2} + \frac{\gamma}{2} p.$$

Solving for p, we find the equilibrium price for every firm i, as follows:

$$p^* = \frac{1+c}{2-\gamma}$$

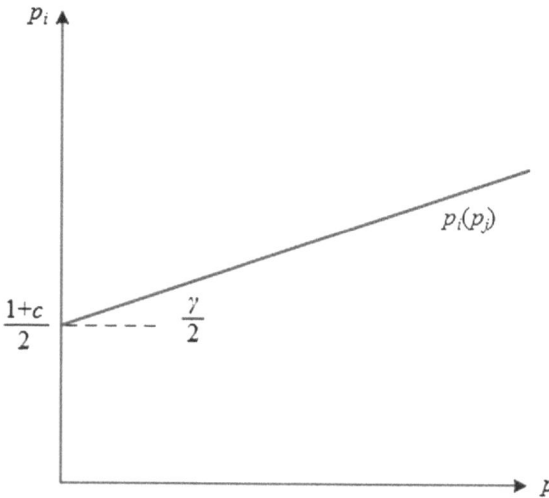

Fig. 4.7 Firm i's best response function, $p_i(p_j)$, with heterogeneous goods

4.3 Price Competition

Fig. 4.8 Equilibrium price pair with heterogeneous goods

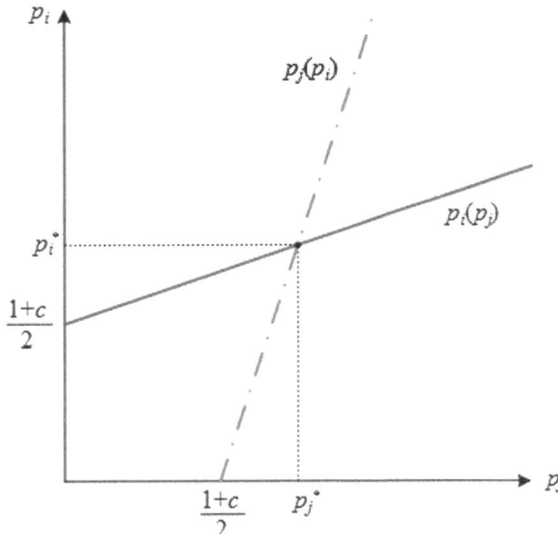

which is positive since $\gamma < 2$; as depicted in Fig. 4.8. The equilibrium price, in addition, satisfies $p^* > c$, implying that differentiated products help firms make a positive margin; as opposed to our results under homogeneous products where, in equilibrium, $p^* = c$. Furthermore, price p^* increases in γ since $\frac{\partial p^*}{\partial \gamma} = \frac{1+c}{(2-\gamma)^2} > 0$. Intuitively, as firm i's sales are more positively affected by its rival's price, p_j, firm i can charge a higher price for its product.[4]

Figure 4.8 depicts this equilibrium price pair.

Substituting the equilibrium price p^* into the demand function, we can find equilibrium output

$$q^* = 1 - p_1^* + \gamma p_2^*$$
$$= 1 - (1-\gamma)\frac{1+c}{2-\gamma}$$
$$= \frac{1 - (1-\gamma)c}{2-\gamma}$$

We can now insert equilibrium output q^* into the profit function, to find equilibrium profit, as follows:

$$\pi^* = \left(p^* - c^*\right)q^*$$

---

[4] Graphically, an increase in γ does not change the intercepts of the best response functions in Fig. 4.8, but it produces a clockwise rotation in $p_j(p_i)$ and a counterclockwise rotation of $p_i(p_j)$. The new crossing point between $p_i(p_j)$ and $p_j(p_i)$ happens to the northeast of the crossing point in Fig. 4.8, indicating that both firms set higher prices.

$$= \left(\frac{1+c}{2-\gamma} - c\right)\frac{1-(1-\gamma)c}{2-\gamma}$$

$$= \left(\frac{1-(1-\gamma)c}{2-\gamma}\right)^2$$

which, as expected, exceeds the zero profit in the Bertrand model of price competition with homogeneous products. For compactness, note that we can express equilibrium profits as $\pi^* = (q^*)^2$.

4.4 Public Good Game

Consider now a public project, such as a public library or a highway, where every individual i simultaneously and independently chooses her contribution x_i. For simplicity, consider only two players (1 and 2), both benefiting from the aggregate contributions to the public good, $X = x_1 + x_2$. This property of public goods is often known as "non-rivalry," namely, that my enjoyment of the public good does not reduce your enjoyment of this good (e.g., we can both visit the Seaquest State Park, in Washington State, without affecting each other's utility). Every player i's utility function is

$$u_i(x_i, x_j) = (w_i - x_i)\sqrt{mX}$$

where $w_i > 0$ is player i's wealth, and $m \geq 0$ represents the return from aggregate contributions, X. This information is common knowledge.

Step 1. *Finding best response functions.* Every player i chooses her donation x_i to solve the following utility maximization problem,

$$\max_{x_i > 0} u_i(x_i, x_j) = (w_i - x_i)\sqrt{m(x_i + x_j)}$$

where $X = x_i + x_j$, so the objective function is explicitly a function of x_i. Differentiating with respect to x_i, we obtain

$$-\sqrt{m(x_i + x_j)} + \frac{m(w_i - x_i)}{2\sqrt{m(x_i + x_j)}} = 0$$

which, after rearranging, simplifies to

$$\frac{m(w_i - 2x_j - 3x_i)}{2\sqrt{m(x_i + x_j)}} = 0$$

which holds if $3x_i = w_i - 2x_j$. Solving for x_i, we find player i's best response function:

$$x_i(x_j) = \begin{cases} \frac{w_i}{3} - \frac{2}{3}x_j & \text{if } x_j < \frac{w_i}{2}, \text{ and} \\ 0 & \text{otherwise,} \end{cases}$$

4.4 Public Good Game

Fig. 4.9 Player i's best response function, $x_i(x_j)$

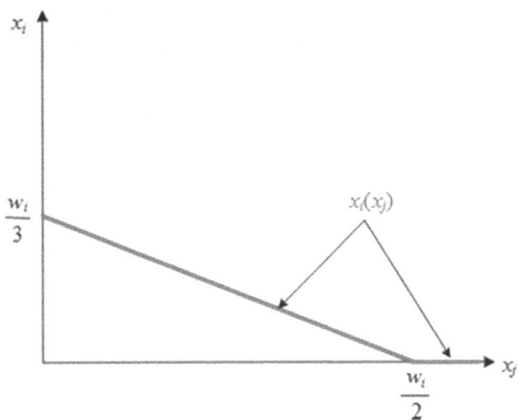

which originates at $\frac{w_i}{3}$ and decreases in player j's contributions, x_j, at a rate of $\frac{2}{3}$, as depicted in Fig. 4.9. Therefore, players regard donations to the public good as strategic substitutes.

In addition, as player i's wealth increases (higher w_i), the best response function shifts upward in a parallel fashion. Intuitively, this indicates that player i makes a larger contribution as she becomes richer, for a given donation of her rival.

Step 2. *Finding NE contributions.* We now use the above best response function to find the NE contribution of each player. For completeness, we first consider the case in which players have the same wealth, $w_i = w_j = w$, so players' best response functions become symmetric, and then analyze a setting with asymmetric wealth levels.

Symmetric wealth. In a symmetric equilibrium, each player makes the same contribution, $x_i = x_j = x$. Inserting this property in the above best response function, yields

$$x = \frac{w}{3} - \frac{2}{3}x.$$

or $3x = w - 2x$. Solving for x, we find the equilibrium contribution of every player i:

$$x^* = \frac{w}{5}$$

which means that every player donates a fifth of her wealth to the public good, leaving four-fifths of her wealth for private consumption. As expected, this equilibrium contribution is increasing in the player's wealth, w.

Asymmetric wealth. We now consider how our equilibrium results are affected when every player holds different wealths, $w_i \neq w_j$. In this setting, we cannot invoke symmetry (as we did when players' wealth levels coincide). Instead, we insert one best response function into the other, as follows

$$x_i = \frac{w_i}{3} - \frac{2}{3}\underbrace{\left(\frac{w_j}{3} - \frac{2}{3}x_i\right)}_{x_j}.$$

which simplifies to $3w_i = 2w_j + 5x_i$. Solving for x_i, yields individual i's equilibrium contribution,

$$x_i^* = \frac{3w_i - 2w_j}{5}$$

which is positive if $3w_i > 2w_j$ or, after rearranging, $w_i > \frac{2}{3}w_j$.

Therefore, individual j's equilibrium contribution is $x_j^* = \frac{3w_j - 2w_i}{5}$ which is positive if $3w_j > 2w_i$ or, after rearranging, $w_i < \frac{3}{2}w_j$. Figure 4.10 depicts cutoffs $\frac{2}{3}w_j$ and $\frac{3}{2}w_j$, both as a function of w_j (on the horizontal axis), which helps us identify that three cases emerge:

1. When player i's wealth is relatively low, $w_i < \frac{2}{3}w_j$, player j is the only one making positive contributions to the public good.
2. When player i's wealth is intermediate, $\frac{2}{3}w_j \leq w_i < \frac{3}{2}w_j$, both players submit a positive contribution. This setting embodies, as a special case, that of wealth symmetry, $w_i = w_j$, as depicted in the 45^0-line of Fig. 4.10.
3. When player i's wealth is relatively high, $w_i > \frac{3}{2}w_j$, she is the only one making positive contributions.[5]

Therefore, every player i's equilibrium contribution increases in her own wealth, w_i, but decreases in her rival's, w_j. This occurs because, as one's rival becomes richer, she makes a larger contribution, which all players benefit from (as we deal with a public good), ultimately inducing player i to reduce her own contribution (i.e., "slack").

In this setting, aggregate contributions to the public good, in equilibrium, are

$$\begin{aligned}X^* &= x_i^* + x_j^* \\ &= \frac{3w_i - 2w_j}{5} + \frac{3w_j - 2w_i}{5} \\ &= \frac{w_i + w_j}{5}\end{aligned}$$

---

[5] In this case, condition $w_i > \frac{2}{3}w_j$ holds too since cuotff $\frac{3}{2}w_j$ lies above $\frac{2}{3}w_j$, entailing that player i makes a positive contribution.

4.4 Public Good Game

Fig. 4.10 Contribution profiles in equilibrium

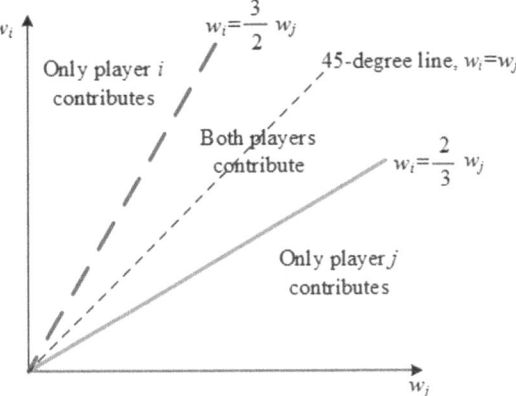

indicating that a fifth of aggregate wealth, $w_i + w_j$, is contributed to the public good. For instance, if wealth levels are $w_i = \$15$ and $w_j = \$12$, player i's equilibrium contribution is $x_i^* = \frac{3 \times 15 - 2 \times 12}{5} = \$4.2$, but only $x_j^* = \frac{3 \times 12 - 2 \times 15}{5} = \$1.2$ for player j, entailing an aggregate contribution of $\$5.4$.

4.4.1 Inefficient Equilibrium

We can show that the above equilibrium donations to the public good are Pareto inefficient, as there is another strategy profile (i.e., another pair of contributions from players i and j) that yields a higher utility level than the utility players earn in equilibrium.

Assuming that wealth levels are symmetric, $w_i = w_j = w$, for simplicity, if players choose the donation pair (x_i, x_j) that maximizes their joint utility, they would solve

$$\max_{x_i, x_j \geq 0} (w - x_i)\sqrt{mX} + (w - x_j)\sqrt{mX}$$

where X denotes aggregate donations. The above maximization problem simplifies to

$$\max_{X \geq 0} (W - X)\sqrt{mX}$$

where $W = w_i + w_j$ represents aggregate wealth, and we changed the choice variables from (x_i, x_j) to aggregate donations, X. Differentiating with respect to X, yields

$$-\sqrt{mX} + \frac{m(W - X)}{2\sqrt{mX}} = 0$$

which simplifies to $\frac{m(W-3X)}{2\sqrt{mX}} = 0$. Solving for X, we obtain the aggregate socially optimal donation

$$X^{SO} = \frac{W}{3}$$

which entails that player i's socially optimal donation is

$$x_i^{SO} = \frac{X^{SO}}{2} = \frac{W}{6}.$$

so every individual contributes a sixth of aggregate wealth to the public good. Since $W = 2w$ in this case, we can rewrite individual contributions as

$$x_i^{SO} = \frac{2w}{6} = \frac{w}{3},$$

so every individual donates a third of his wealth to the public good.

Comparing socially optimal and equilibrium results, we find that individual donations are "socially insufficient" since

$$x_i^{SO} = \frac{w}{3} > \frac{w}{5} = x_i^*$$

and, similarly, aggregate donations satisfy $X^{SO} = \frac{W}{3} > \frac{W}{5} = X^*$.

4.5   Electoral Competition

Politicians running for office compete in several dimensions, such as their stand in certain policies (education, social programs, or national defense), who endorses them, and even in their looks. For simplicity, we model their competition along one dimension. Consider two politicians running for president who must simultaneously and independently choose their position in a policy, $x_i \in [0, 1]$. Intuitively, x_i is the proportion of the budget allocated to the policy, with $x_i = 0$ meaning none and $x_i = 1$ indicating the entire budget.

We make two simplifying assumptions about the candidates: first, candidates only care about winning the election, not exhibiting political preferences for this policy; and, second, they commit to implement their announcement if they win. Therefore, voters can anticipate that if they vote for a candidate who promised a position x_i on this policy during the campaign, the candidate (now president) will deliver on her promise.[6] We also make two assumptions on voters who exhibit an

---

[6] Relaxing this assumption can significantly complicate the model and its analysis, which is beyond the scope of this book. For articles relaxing the first assumption, see Wittman (1983) and Dunaway and Munoz-Garcia (2020); and for articles relaxing the second assumption, see Aragon et al. (2007). For an accessible introduction to this literature, see the books by Grossman and Helpman (2001) and Smith (2015).

4.5 Electoral Competition

ideal point about this policy located in the unit interval [0, 1]: first, voters' ideal policies are uniformly distributed in the unit interval; and, second, they are "non-strategic," meaning that they just vote for the candidate whose announced policy (x_i or x_j) is closest to the voter's ideal.

Before we find each candidate i's best response function, $x_i(x_j)$, let us clarify one point. If the announcements satisfy $x_i < x_j$, we can anticipate that:

- All voters to the left of x_i will vote for candidate i since position x_i is closer to their ideal.
- All voters to the right of x_j will vote for candidate j because x_j is closer to their ideal.
- The voters between x_i and x_j are splitted between candidates i and j. In particular, all those between x_i and the midpoint between x_i and x_j, $\frac{x_i+x_j}{2}$, vote for candidate i; while those from midpoint $\frac{x_i+x_j}{2}$ to x_j vote for candidate j.

When does each candidate win? The above discussion is relevant to identify the winning candidate in different settings, which ultimately helps us understand best response functions in this game. If $x_i < x_j$, then candidate i wins if she receives more votes than j, which occurs when $\frac{x_i+x_j}{2} > 1 - \frac{x_i+x_j}{2}$, which simplifies to $x_i > 1 - x_j$.[7] Otherwise, candidate i loses the election.

The opposite argument applies when the announcements satisfy $x_i > x_j$, where now all voters to the right of x_i vote for candidate i; all voters to the left of x_j vote for j, those between x_j and the midpoint $\frac{x_i+x_j}{2}$ vote for j, and voters between midpoint $\frac{x_i+x_j}{2}$ and x_i vote for i. In this context, candidate i wins the election if and only if $1 - \frac{x_i+x_j}{2} > \frac{x_i+x_j}{2}$ or, after rearranging, $x_i < 1 - x_j$.

Step 1. *Finding best response functions.* Putting together the discussion in the last two paragraphs, we can identify candidate i's best response correspondence, as follows[8]:

(a) if $x_i < x_j$, then candidate i sets x_i such that $x_i > 1 - x_j$ and wins the election, as depicted in the policy announcements in Region I of Fig. 4.11; and
(b) if $x_i > x_j$, then candidate i sets x_i such that $x_i < 1 - x_j$ and wins the election, as illustrated in Region II of the figure.

---

[7] Another, more graphical, argument says that candidate i wins when the midpoint $\frac{x_i+x_j}{2}$ lies to the right side of $\frac{1}{2}$. Because every voter to the left of midpoint $\frac{x_i+x_j}{2}$ votes for candidate i, $\frac{x_i+x_j}{2} > \frac{1}{2}$ means that this candidate has more than 50% of the votes.

[8] Recall that, intuitively, a *correspondence* maps every point in the domain to potentially more than one point in the range, while a function maps every point in the domain to a single point in the range. In this context, we talk about candidate i's best response correspondence, rather than function, because, for every policy announcement of her rival, x_j, candidate i has a continuum of policies x_i that help her win the election, being indifferent among all of them.

Fig. 4.11 Candidate i's best response correspondence

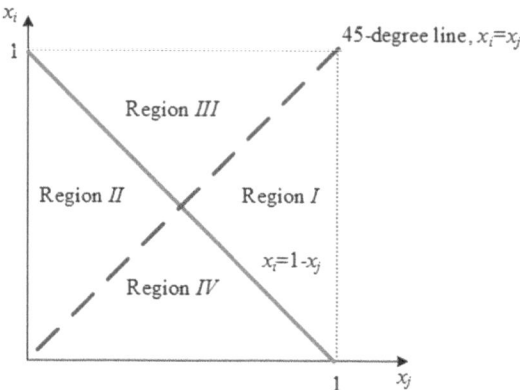

A symmetric argument applies to the best response function of candidate j:

(i) if $x_j < x_i$, then candidate j sets x_j such that $x_j > 1-x_i$ and wins the election, as depicted in Region III of Fig. 4.11; and
(ii) if $x_j > x_i$, then candidate j sets x_j such that $x_j < 1-x_i$ and wins the election, as illustrated in Region IV of the figure.

Step 2. *Finding the NE of the game.* In equilibrium, we know that both players must be choosing mutual best responses to each other's strategies. By inspection of Fig. 4.11, this only occurs at the midpoint, where

$$x_i^* = x_j^* = \frac{1}{2},$$

implying that both candidates locate at the midpoint of the political spectrum, converging in their policy announcements. Furthermore, as voters' ideal policies are uniformly distributed, $x_i^* = x_j^* = \frac{1}{2}$ coincides with the ideal policy of the median voter. This result, known as the "median voter theorem," is often used to explain why political candidates tend to converge toward similar policy proposals when competing for office. The theorem was originally described by Hotelling (1929) and formally shown by Downs (1957). For a literature review, see Congleton (2003).

4.5.1 Alternative Proof to the Electoral Competition game

An alternative approach to show the same result works by considering different policy announcements, and systematically proving that they cannot be a NE because at least one candidate has incentives to deviate. First, consider asymmetric policy announcements where each candidate promises a different policy, $x_i \neq x_j$, and, without loss of generality, assume that $x_i < x_j$. From our above discussion,

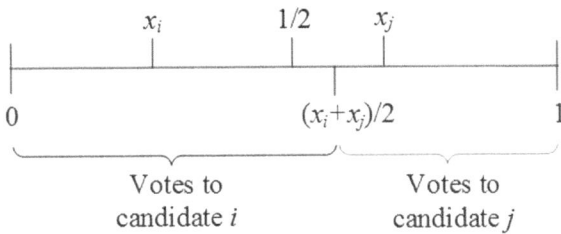

Fig. 4.12 Votes to candidate i and j when $x_i < x_j$ and $x_i > 1 - x_j$

we know that if, in addition, condition $x_i > 1 - x_j$ holds, candidate i wins the election,[9] as depicted in Fig. 4.12. In this context, candidate j would have incentives to deviate (e.g., moving x_j downward closer to x_i). If, instead, condition $x_i > 1 - x_j$ is violated, candidate i loses the election, implying that it is now this candidate who has incentives to deviate upward (moving closer to x_j). Therefore, asymmetric policy profiles cannot be sustained in equilibrium.

Second, we can consider symmetric policy announcements where both candidates announce the same policy, $x_i = x_j = x$. If $x < \frac{1}{2}$, the mass of voters to the right of x is larger than that to its right, entailing that one candidate receives more votes and wins the election, providing incentives to her rival to deviate from $x < \frac{1}{2}$. A similar argument applies, of course, if $x > \frac{1}{2}$.

However, at $x = \frac{1}{2}$, both candidates receive the same number of votes, and the winner is randomly selected. In this context, the loser of the random assignment does not have incentives to unilaterally deviate, as this deviation (above or below x) would decrease her votes below 50%, losing the election with certainty. Therefore, $x_i^* = x_j^* = \frac{1}{2}$ is a NE of the electoral competition game, because no candidate has incentives to deviate from her policy announcement. When one allows for three or more candidates running for office, however, no NE exists, as shown in Cox (1987) and Chisik and Lemke (2006).

Exercises

4.1. **Cournot duopoly with asymmetric marginal costs.**[A] Consider the Cournot duopoly in Sect. 4.2.1, but assume that firm 1 faces marginal cost c_1 while firm 2's is c_2, where $c_1 < c_2 < 1$ (so firm 1 enjoys a cost advantage relative to firm 2).
   (a) Find the best response function of firm 1 and of firm 2. Compare them.

---

[9] Recall that candidate i receives more votes than j does, thus winning the election, when $\frac{x_i + x_j}{2} > 1 - \frac{x_i + x_j}{2}$, which simplifies to condition $x_i > 1 - x_j$ or, alternatively, to $\frac{x_i + x_j}{2} > \frac{1}{2}$. In the context of Fig. 4.12, the last inequality indicates that the midpoint between x_i and x_j, $\frac{x_i + x_j}{2}$, lies to the right of the center of the line, 1/2. If, in contrast, $x_i < 1 - x_j$, the midpoint would lie to the left of 1/2, and candidate j would win the election.

(b) Insert firm 2's best response function into firm 1's to find the output that each firm produces in the Nash equilibrium of the Cournot game of quantity competition. Which firm produces a larger output?

(c) *Symmetric firms.* Assume that both firms now become cost symmetric, so that, $c_1 = c_2 = c$. Evaluate your results from part (b) at $c_1 = c_2 = c$, showing that you obtain the same results as in Sect. 4.2.1.

4.2. **Coordinated production decisions with differentiated products.**[B] Consider a duopoly where every firm i faces inverse demand function

$$p(q_i, q_j) = 1 - q_i - \gamma q_j$$

where $\gamma \in [0, 1]$ denotes how sensitive firm i's price is to its rival's sales. When $\gamma = 0$, both firms are unaffected by each other's sales, as if they were two separate monopolists; whereas when $\gamma = 1$, firms sell homogeneous goods. For simplicity, assume that both firms face the same marginal production cost, c, where $1 > c \geq 0$.

(a) *Maximizing individual profits.* Find firm i's best response function, $q_i(q_j)$.

(b) How is the best response function affected by an increase in parameter γ? Interpret.

(c) Find the equilibrium output q_i^* and profits, π_i^*.

(d) How are equilibrium output and profits affected by an increase in parameter γ? Evaluate your equilibrium output and profits in the extreme cases where $\gamma = 0$ and $\gamma = 1$. Interpret.

(e) *Maximizing joint profits.* Assume that both firms could coordinate their output decision to maximize their profits. Find their equilibrium output and profits in this context.

(f) How are output and profits that you found in part (e) affected by an increase in parameter γ? Evaluate your equilibrium output and profits in the extreme cases where $\gamma = 0$ and $\gamma = 1$. Interpret.

4.3. **Cournot competition with cost externalities.**[B] Consider a Cournot duopoly game where every firm i's inverse demand is $p(Q) = 1 - Q$, and Q denotes aggregate output. Every firm's marginal cost is, instead of c, given by $c - \alpha q_j$, where $\alpha \in [-1, 1]$ measures the degree of cost externalities that firm j's output imposes in firm i. This setting allows for cost externalities to be positive, $\alpha > 0$, if firm j's production decreases its rival's marginal cost; and negative, $\alpha < 0$, if q_j increases firm i's marginal cost.

(a) Find every firm's best response function. How is it affected by parameter α? Interpret.

(b) Find every firm's equilibrium output, aggregate output, price and profit. How are they affected by parameter α? Interpret.

4.4. **Mergers with cost externalities.**[B] Consider the setting in Exercise 4.3 about firms competing in quantities and facing cost externalities. Assume that firms merge into a single entity that maximizes joint profits.

(a) Find individual output and aggregate output in equilibrium.

(b) Does the merger reduce aggregate output under all values of parameter α? If not, when does it increase aggregate output? Interpret in terms of merged firms internalizing their cost externalities.
(c) Find equilibrium price and profits.
(d) Does the merger increase price and profits?
(e) For which values of α does the merger increase both profits and aggregate output? For which values it only increases one of them? For which values it decreases both?

4.5. **Cournot competition with only one firm benefiting from a cost advantage.**[B] Consider an industry like that in Sect. 4.2.2, namely, $N \geq 2$ firms selling a homogeneous good, competing á la Cournot, with inverse demand function $p(Q) = 1 - Q$, where Q denotes aggregate output. One of the firms (firm 1) faces marginal cost $c \in (0, 1)$, while its $N - 1$ rivals face a higher marginal cost c', where $1 > c' \geq c$. This cost advantage may be explained because firm 1 had more years of experience in the industry or operates in similar markets.
(a) Find firm 1's best response function and the best response function of each of its rivals.
(b) Find equilibrium output levels for each firm.
(c) Evaluate the equilibrium output for firm 1 and its $N - 1$ rivals when all firms are symmetric, that is, $c = c'$.
(d) Under which conditions on firm 1's cost, c, will all firms produce a positive output in equilibrium? Under which conditions will only firm 1 be active? Interpret.
(e) *Symmetric firms.* Evaluate your results in part (d) in the case that all firms are symmetric, that is, $c = c'$.

4.6. **Bertrand competition with asymmetric costs.**[A] Consider the Bertrand model with two firms competing in prices, see Sect. 4.3.1. Assume now that firm 1's marginal cost is c_1, while that of firm 2 is c_2, where $c_1 < c_2 < 1$.
(a) Find a pure-strategy NE in this game.
(b) Evaluate equilibrium profits in your results from part (a).
(c) Consider now that the industry has $N \geq 2$ firms, where firm 1's marginal cost, c_1, is the lowest, that is, $c_1 < c_2 \leq c_3 \leq \ldots \leq c_N$. Find a pure-strategy NE in this setting.

4.7. **Three firms competing in prices and selling differentiated products.**[C] Consider three firms competing à la Bertrand selling heterogeneous goods with an inverse demand function

$$p_i(q_i, q_j, q_k) = 1 - q_i - \beta(q_j + q_k),$$

where parameter β denotes the degree of product differentiation. When $\beta = 0$ products are completely differentiated, but when $\beta = 1$ products are homogeneous. For simplicity, assume that marginal costs are normalized to zero.
(a) Find every firm i's demand function. [*Hint*: Solve for q_i.]

(b) Find every firm i's best response function.
(c) Find the equilibrium price of every firm i. How does it change with β? Interpret.
(d) Find the equilibrium output of every firm i. How does it change with β? Interpret.

4.8. **Public good game with $N \geq 2$ players.**[B] Extend the public good game in Sect. 4.4 to $N \geq 2$ players. For simplicity, assume that all players have the same wealth, w.
  (a) Find player i's best response function $x_i(X_{-i})$, where $X_{-i} = \sum_{j \neq i} x_j$ denotes the aggregate donations by all player i's opponents.
  (b) Find equilibrium donation x_i^*.
  (c) How are equilibrium donations affected by a marginal increase in N? Interpret.
  (d) Evaluate your equilibrium results at $N = 2$. Show that your equilibrium results coincide with those in Sect. 4.4.
  (e) How are aggregate donations, X^*, affected by a marginal increase in N? Interpret.

4.9. **Socially optimal contributions with $N \geq 2$ players.**[B] Consider the setting in Exercise 4.8.
  (a) If players could coordinate their donations to maximize their joint utility, what would their donations be?
  (b) Compare your socially optimal results with those in equilibrium.

4.10. **Public good game with quasilinear utility.**[A] Consider the public good game in Sect. 4.4, but assuming that every player i's utility function is quasilinear, as follows

$$u_i(x_i, x_j) = (w_i - x_i) + \sqrt{mX}.$$

  (a) Find player i's best response function $x_i(x_j)$. How does it depend on player i's wealth, w_i?
  (b) Find the equilibrium donation x_i^*.
  (c) How are equilibrium donations affected by a marginal increase in w_i or w_j? Are they affected by parameter m? Interpret.
  (d) Find aggregate donations, X^*. Are they affected by a marginal increase in w_i or w_j? Interpret.
  (e) *Socially optimal outcome.* If players could coordinate their donations to maximize their joint utility, what would their donations be?

4.11. **Testing for inefficiency in public good games with asymmetric wealth.**[B] Consider our discussion in Sect. 4.4 about the equilibrium contribution in public good games being inefficient. For simplicity, that section assumed that players' wealth was symmetric. In this exercise, we relax this assumption, allowing for asymmetric wealth, $w_i \neq w_j$.

(a) Set up the joint utility maximization problem to find the socially optimal aggregate and individual contributions to the public good.
(b) Under which conditions on wealth levels w_i and w_j every player submits a strictly positive contribution? Under which conditions only one player submits a positive contribution?
(c) Compare the socially optimal contributions that you found in part (a) against the equilibrium contributions in Sect. 4.4. Which contribution is larger? Interpret.

4.12. **Impure public good.**[A] Consider the public good game in Sect. 4.4, but assume now that player i's return from aggregate donations is m_i while player j's is m_j, where $m_i, m_j \geq 0$ For simplicity, we consider symmetric wealth levels, so player i's utility function is

$$u_i(x_i, x_j) = (w - x_i)\sqrt{m_i X}.$$

The asymmetry in the return to the public good entails that its rivalry is impure because one player can benefit less from the same aggregate donations, X, than his rival does. In extreme settings, one player benefits from aggregate donations, if $m_i > 0$, while his rival does not benefit at all, if $m_j = 0$.

(a) Find player i's best response function, $x_i(x_j)$. How is it affected by a marginal increase in m_i?
(b) Find the NE contribution pair, (x_i^*, x_j^*).
(c) How is the equilibrium contribution, x_i^*, affected by a marginal increase in m_i? And by a marginal increase in m_j?

4.13. **Exploiting a common pool resource.**[A] Consider a common pool resource such as a fishing ground, a forest, or an aquifer, exploited by $N = 2$ firms. In addition, consider that every firm takes prices as given (i.e., firms sell their appropriation in an international, perfectly competitive, market), where we can normalize this price to $p = \$1$, and every firm i's cost function is

$$C_i(q_i, q_j) = \frac{q_i(q_i + q_j)}{S},$$

where $q_i \geq 0$ denotes firm i's appropriation, $q_j \geq 0$ represents its rival's appropriation, and S is the initial stock of the resource.

(a) Find every firm i's marginal cost function. Interpret.
(c) Find every firm i's best response function, $q_i(q_j)$.
(c) How is $q_i(q_j)$ affected by an increase in S? Interpret.
(d) Find the equilibrium appropriation q_i^*.
(e) Find the appropriation levels that firms choose if they could coordinate their appropriation decisions to maximize their joint profits.

4.14. **A common pool resource with more general cost externalities.**[B] Consider the setting in Exercise 4.13, but assume now that every firm i's cost function is

$$C_i(q_i, q_j) = \frac{q_i(q_i + \theta q_j)}{S},$$

where $\theta \geq 0$ denotes the severity of the cost externality. When $\theta = 0$, firm i's costs are unaffected by its rival's appropriation q_j; but when $\theta > 0$, firm i's costs are affected by q_j. Therefore, the setting in Exercise 4.12 can be interpreted as a special case of this (more general) model, where $\theta = 1$.
(a) Find every firm i's best response function, $q_i(q_j)$.
(b) How is $q_i(q_j)$ affected by parameter θ? Interpret.
(c) Find firms' appropriation in the NE of this game, q_i^*.
(d) How is q_i^* affected by parameter θ? Interpret.

4.15. **A common pool resource with asymmetric firms.**[B] Consider the setting in Exercise 4.13 again, but assume that firm i faces cost functions $C_i(q_1, q_2) = \frac{\alpha_i q_i (q_i + q_j)}{S}$, where $\alpha_1 = 1$ and $\alpha_2 \in [0, 1]$. When $\alpha_2 = 1$, both firms are cost symmetric, as in Exercise 4.13; but otherwise firm 2 benefits from a cost advantage.
(a) Find each firm's best response function.
(b) Compare firms 1 and 2's best response functions.
(c) Find the NE appropriation pair (q_1^*, q_2^*).
(d) Which firm appropriates more units in equilibrium?
(a) How is equilibrium appropriation affected by parameter α? Interpret.

References

Aragonès, E., A. Postlewaite, and T. Palfrey (2007) "Political Reputations and Campaign Promises," Journal of the European Economic Association, 5(4), pp. 846–884.

Bertrand, J. (1883) "Théorie Mathématique de la Richesse Sociale," Journal des Savants, pp. 499–508.

Chisik, R. A. and R. J. Lemke (2006) "When Winning Is the Only Thing: Pure Strategy Nash Equilibria in a Three-Candidate Spatial Voting Model," Social Choice and Welfare, 26, pp. 209–215.

Congleton, R. (2003) "The Median Voter Model." in: *The Encyclopedia of Public Choice*, eds. C. K. Rowley and F. Schneider, Kluwer Academic Press.

Cournot, A. A. (1838) *Cournot, Recherches sur les Principes Mathematiques de la Theorie des Richesses*, Hachette, Paris. Translated to English as: Researches into the Mathematical Principles of the Theory of Wealth (1927), by N. T. Bacon, translator, Macmillan.

Cox, G. W. (1987) "Electoral Equilibrium Under Alternative Voting Institutions," American Journal of Political Science, 31, pp. 82–108.

Downs, A. (1957) "An Economic Theory of Political Action in a Democracy," Journal of Political Economy, 65(2), pp. 135–150.

Dunaway, E. and F. Munoz-Garcia (2020) "Campaign Limits and Policy Convergence with Asymmetric Agents," "Public Choice, 184, pp. 429–461.

Grossman, G. M. and E. Helpman (2001) *Special Interest Politics*, The MIT Press, Cambridge, Massachussets.

References

Hotelling, H. (1929) "Stability in Competition," The Economic Journal, 39, pp. 41–57.

Smith, S.B. (2015) *Chance, Strategy, and Choice. An Introduction to the Mathematics of Games and Elections*, Cambridge University Press, Cambridge.

Wittman, D. (1983) "Candidate Motivation: A Synthesis of Alternative Theories," "The American Political Science Review, 77(1), 142–157.

Mixed Strategy Nash Equilibrium

5.1 Introduction

In previous chapters, we considered games that had at least one NE, such as the Prisoner's Dilemma, the Battle of the Sexes, and the Chicken games. But, do all games have at least one NE? If we restrict players to choose a specific strategy with certainty, some games may not have an NE. This occurs when players do not want to be *predictable*—playing in the same way every time they face the game—such as in board games and professional sports (aiming left or right in a penalty kick in soccer or serving front or back in tennis). For instance, if you are a soccer player in a penalty kick, you do not want to be known as the player who always aims right, as the goalie would anticipate your moves and, hence, you would never score! This setting also applies to tax underreporting and, more generally, to conducting illegal activities (e.g., drug trafficking). Like the kicker in penalty kicks, drug dealers do not want to use the same channels and steps, as otherwise they could be easily caught by the police.

To illustrate the above class of games, Matrix 5.1a represents the Penalty Kicks in Soccer game. Two teams reach the FIFA World Cup Final (Argentina and France's men's teams in the 2023 final) or the UEFA Champions League (with FC Barcelona and Juventus in 2015). Their scores are tied and everything boils down to one more penalty kick. If the kicker scores, her team wins. If the kicker does not, the goalie's team wins. Specifically, if the kicker aims left (right) and the goalie dives to the left (right, respectively), the kicker does not score, and both players' payoffs are zero.[1] However, when the kicker aims left (right) and

---

[1] For simplicity, we say that the goalie dives to the left, meaning the same left as the kicker (rather than to the goalie's left). A similar argument applies when the goalie dives to the right. Otherwise, it would be a bit harder to keep track of the fact that the goalie's left corresponds to the kicker's right, and vice versa.

the goalie dives in the opposite direction (right and left, respectively), the kicker scores, yielding a negative payoff of -10 to the goalie and a positive payoff of 16 for the kicker.[2]

		Kicker	
		Aim Left	Aim Right
Goalie	Dive Left	0, 0	−10, 16
	Dive Right	−10, 16	0, 0

Matrix 5.1a. Anticoordination game.

To confirm that this game has no NE where players choose a specific strategy without randomizing, let us find the best responses for the goalie. When the kicker aims left (in the left column), the goalie's best response is $BR_G(L) = L$ because her payoffs satisfy $0 > -10$. In contrast, when the kicker aims right (in the right column), the goalie's best response is $BR_G(R) = R$ because $0 > -10$. Intuitively, the goalie tries to move in the same direction as the kicker, so she prevents the latter from scoring. The kicker's best responses are not symmetric, but instead completely opposed. When the goalie dives left (at the top row), the kicker's best response is $BR_K(L) = R$ because her payoffs satisfy $16 > 0$, and when the goalie dives right (at the bottom row), the kicker's best response is $BR_K(R) = L$ because $16 > 0$. Intuitively, the kicker seeks to aim to the opposite location of the goalie to score a goal. Underlining best response payoffs, we obtain Matrix 5.1b.

		Kicker	
		Aim Left	Aim Right
Goalie	Dive Left	<u>0</u>, 0	−10, <u>16</u>
	Dive Right	−10, <u>16</u>	<u>0</u>, 0

Matrix 5.1b. Anticoordination game–underlining best response payoffs.

In summary, there is no strategy profile (no cell in the matrix) where the payoffs for all players are underlined, indicating that there is no mutual best response. As a consequence, there is no NE when we restrict players to use a specific strategy (either left or right) with 100% probability. If, instead, we allow players to

---

[2] Similar payoffs would still produce our desired result of no NE when players are restricted to using a specific strategy with 100% probability. You can make small changes on the payoffs, and then find the best responses of each player again to confirm this point. More generally, you can replace the payoff of 10 with a generic payoff $a > 0$ and -16 with a generic payoff $-b$, allowing for $b > a$ or $b < a$.

randomize (e.g., playing left with some probability, such as 1/3, and right with the remaining probability of 2/3) we can find an NE of this game. Because players in this scenario mix, this type of NE is known as a "mixed strategy NE" (msNE). In contrast, strategy profiles where players use a specific strategy with 100% probability are referred to as a "pure-strategy NE" (psNE).

Generally, the penalty kicks game illustrates contexts where players do not want to be predictable in their decisions (e.g., always diving left or always aiming right), and applies to other sports competitions, like serving in tennis. These incentives also extend nonetheless to law, like when a police department chooses where to locate most of its police patrols and, simultaneously, a criminal organization decides in which streets to run its business. As illustrated in Matrix 5.2, the police seeks to choose the same action as the criminal, while the latter seeks to miscoordinate by selecting the opposite location as the police patrol to avoid being caught.[3] A similar argument applies to firm monitoring, such as a polluting firm choosing how many emissions to abate and an environmental protection agency deciding the frequency of its inspections.

		Criminal	
		Street A	Street B
Police	Street A	$\underline{10}, 0$	$-1, \underline{6}$
	Street B	$0, \underline{8}$	$\underline{7}, -1$

Matrix 5.2. Police and criminal game.

In the following section, we formally define mixed strategy and msNE. The rest of the chapter applies the definition to different settings.

5.2 Mixed Strategy

As an introductory example, consider an individual i with a binary strategy set $S_i = \{H, L\}$ representing, for instance, a firm choosing between high and low prices. We can then define a player i's mixed strategy (or randomization) as a probability distribution over her pure strategies (H and L), as follows $\sigma_i = \{\sigma_i(H), \sigma_i(L)\}$, where

$$\sigma_i(H) = p \text{ and } \sigma_i(L) = 1 - p,$$

---

[3] It is straightforward to show that the police's best responses are $BR_P(A) = A$ and $BR_P(B) = B$, thus choosing the same strategy as the criminal; while those of the criminal are $BR_C(A) = B$ and $BR_C(B) = A$, selecting the opposite strategies as the police. As a consequence, no NE exists in pure strategies, which you can confirm by underlining best response payoffs in Matrix 5.2, seeing that no cell has both players' payoffs underlined.

indicating the probability that player i assigns to each strategy in S_i. As usual, we require that these probabilities satisfy $\sigma_i(H), \sigma_i(L) \geq 0$ and $\sigma_i(H) + \sigma_i(L) = 1$. Intuitively, probabilities must be positive or zero (but never negative) and the probabilities of all the strategies that player i can choose from (H and L in this example) must add up to 1.

We can next apply this probability distribution to larger discrete strategy sets with two or more pure strategies.

Definition 5.1. **Mixed strategy.** Consider a discrete strategy set $S_i = \{s_1, s_2, ..., s_m\}$, where $m \geq 2$ denotes the number of pure strategies. The mixed strategy

$$\sigma_i = \{\sigma_i(s_1), \sigma_i(s_2), ..., \sigma_i(s_m)\}$$

is a probability distribution over the pure strategies in S_i, with the property that:

1. $\sigma_i(s_k) \geq 0$ for every pure strategy s_k, and
2. $\sum_{k=1}^{m} \sigma_i(s_k) = 1$.

When a mixed strategy concentrates all probability weight on a pure strategy, that is, $\sigma_i(s_j) = 1$ while $\sigma_i(s_k) = 0$ for all $k \neq j$, it is commonly called a "degenerated mixed strategy" because, graphically, it collapses to a pure strategy. However, to avoid unnecessary complications, we only use the term "mixed strategy" for probability distributions over at least two pure strategies.

The above definition can also be applied to games where players choose their strategies from a continuous strategy space, e.g., an output level so that $s_i \geq 0$. In this context, the player i's probability distribution over her pure strategies in S_i can be represented with a cumulative distribution function

$$F_i : S_i \to [0, 1]$$

mapping every strategy $s_i \in S_i$ into a cumulative probability. For instance, if s_i denotes firm i's output level, the probability that this firm produces an output level equal or lower than $\overline{s}$ is $F_i(\overline{s})$ and, because $F_i(\overline{s})$ is a probability, it must satisfy $F_i(\overline{s}) \in [0, 1]$. In addition, recall that if the cumulative distribution function $F_i(\overline{s})$ has a density function $f_i(\overline{s}) = F_i'(\overline{s})$, we can interpret this function as representing the probability that firm i produces exactly an output level $s_i = \overline{s}$.

5.3 Mixed Strategy Nash Equilibrium

Using the above definition of a mixed strategy as a probability distribution over pure strategies, we can adapt the definition of the NE solution concept from Chapter 3 to a setting where players randomize over their pure strategies. Like in Chapter 3, we first define a best response in a context where players randomize, as it will be key in our equilibrium definition.

5.4 Finding Mixed Strategy Equilibria

Definition 5.2. **Best response with mixed strategies.** Player i's mixed strategy σ_i is a best response to her opponents' mixed strategy σ_{-i}, denoted as $\sigma_i \in BR_i(\sigma_{-i})$, if and only if her expected utility from σ_i satisfies

$$EU_i(\sigma_i, \sigma_{-i}) \geq EU_i(\sigma'_i, \sigma_{-i}) \text{ for all } \sigma'_i \neq \sigma_i.$$

Mixed strategy σ_i is player i's best response to her opponents' mixed strategy σ_{-i} if no other randomization σ'_i yields a higher expected utility than σ_i does, where σ'_i allows for player i to use mixed or pure (degenerated) strategies. We use *expected utility*, of course, because player i needs to compute her expected payoff from randomizing over at least two of her pure strategies and, potentially, her rivals also randomize.

Equipped with this definition of best response in a context where players can randomize, we next define a mixed strategy Nash equilibrium.

Definition 5.3. **Mixed strategy Nash equilibrium (msNE).** A strategy profile $(\sigma_i^*, \sigma_{-i}^*)$ is a mixed strategy Nash equilibrium if and only if $\sigma_i^* \in BR_i(\sigma_{-i}^*)$ for every player i.

Therefore, when player i chooses her equilibrium strategy σ_i^* she is optimally responding to her opponents' strategies, σ_{-i}^*, implying that players are choosing mutual best responses and, thus, have no incentives to unilaterally deviate. This definition is, then, analogous to that of the pure strategy NE in Chapter 3, but using mixed strategy σ_i^* rather than pure-strategy s_i^*.

5.4 Finding Mixed Strategy Equilibria

If a player randomizes, she must be indifferent between at least two of her pure strategies. In the context of the anticoordination game presented in the Introduction, and reproduced in Matrix 5.3 for easier reference, the goalie (in rows) must be indifferent between diving left and right, or more formally, her expected utility from diving left must coincide with that of diving right, that is,

$$EU_{\text{Goalie}}(\text{Left}) = EU_{\text{Goalie}}(\text{Right}).$$

If she was not indifferent, her expected utility would satisfy either $EU_{\text{Goalie}}(\text{Left}) > EU_{\text{Goalie}}(\text{Right})$, leading her to choose Left in pure strategies, or $EU_{\text{Goalie}}(\text{Left}) < EU_{\text{Goalie}}(\text{Right})$, inducing her to select Right in pure strategies. In other words, if she mixes between diving left and right, her expected utility from each pure strategy must coincide. A similar argument applies to the kicker, who must be indifferent between aiming left and right.

			Kicker	
			Prob. q	Prob. $1-q$
			Aim Left	Aim Right
Goalie	Prob. p	Dive Left	$0,0$	$-10,16$
	Prob. $1-p$	Dive Right	$-10,16$	$0,0$

Matrix 5.3. Anticoordination game—including probabilities.

At this point of our analysis, we do not know the exact probabilities with which each player randomizes, so we denote them as p and $1-p$ for the goalie (see these probabilities in Matrix 5.3, next to their corresponding rows), and q and $1-q$ for the kicker (see these probabilities on top of their corresponding column in Matrix 5.3). Intuitively, $p=1$ means that the goalie dives left with 100% probability, whereas $p=0$ indicates the opposite (she dives right with 100% probability). In contrast, when p lies strictly between 0 and 1, $0<p<1$, the goalie randomizes her diving decision. A similar argument applies to the kicker's probability of aiming left, q.

The next tool provides a step-by-step approach to find msNEs, which applies to two-player games and, more generally, to games with more players if their payoffs are symmetric.

Tool 5.1 How to find msNEs in a two-player game

1. Focus on player 1. Find her expected utility of choosing one pure strategy. Repeat for each pure strategy.
2. Set these expected utilities equal to each other.
3. Solve for player 2's equilibrium mixed strategy, σ_2^*.
   (a) If players have two pure strategies, step 2 just entails a equality (the expected utility of one pure strategy coincides with that of the other pure strategy). Solving for σ_2 in this context yields a probability, σ_2^*, which should satisfy $0<\sigma_2^*<1$. (If σ_2^* does not satisfy this property, then player 2 is not playing mixed strategies.)
   (b) If players have three pure strategies, step 2 entails several equalities (the combination of different expected utilities equal to each other), which gives rise to a system of two equations and two unknowns. The solution to this system of equations is, nonetheless, a player 2's equilibrium mixed strategy, σ_2^*, with the caveat that σ_2^* is a vector of three components describing the probability that player 2 assigns to each pure strategy.[4]

---

[4] A similar argument applies when players have $N \geq 3$ pure strategies, where the combination of expected utilities generates a system of $N-1$ equations and $N-1$ unknowns, implying that the resulting σ_2^* is a vector of N components.

5.4 Finding Mixed Strategy Equilibria

4. Focus now on player 2. Repeat steps 1–3, to obtain player 1's equilibrium mixed strategy, σ_1^*.

Example 5.1 applies this tool in the anticoordination game of Matrix 5.3, focusing on the goalie first.

Example 5.1 Finding msNEs. According to Step 1 of Tool 5.1, we need to find the goalie's expected utility from diving left, as follows,

$$EU_{\text{Goalie}}(\text{Left}) = \underbrace{q \times 0}_{\text{kicker aims left}} + \underbrace{(1-q) \times (-10)}_{\text{kicker aims right}}$$
$$= 10q - 10.$$

This expression can be understood as follows: when the goalie dives left (at the top row of Matrix 5.3), her payoff is zero when the kicker aims left (which happens with probability q, and no goal is scored) or -10 when the kicker aims right (which occurs with probability $1-q$, and a goal is scored).

Similarly, the goalie's expected utility from diving right is

$$EU_{\text{Goalie}}(\text{Right}) = \underbrace{q \times (-10)}_{\text{kicker aims left}} + \underbrace{(1-q) \times 0}_{\text{kicker aims right}}$$
$$= -10q.$$

In this case, the goalie dives right (in the bottom row of the matrix), and she either obtains a payoff of -10, which occurs when the kicker aims left and thus scores, or a payoff of 0, which happens when the kicker aims right and does not score. Moving now to Step 2 of Tool 5.1, we set the expected utilities equal to each other, $EU_{\text{Goalie}}(\text{Left}) = EU_{\text{Goalie}}(\text{Right})$, or

$$10q - 10 = -10q.$$

We can now move to Step 3 in Tool 5.1, which prescribes that we must solve for player 2's probability (q in this example). Rearranging the above indifference condition, we obtain $20q = 10$, and solving for q yields a probability $q = 10/20 = 1/2$. Therefore, the goalie is indifferent between diving left and right when the kicker aims left 50% (because we found that $q = 1/2$).

We can now repeat steps 1–3 with the kicker. First, we find her expected utility from aiming left:

$$EU_{\text{Kicker}}(\text{Left}) = \underbrace{p \times 0}_{\text{goalie dives left}} + \underbrace{(1-p) \times 16}_{\text{goalie dives right}}$$

$$= 16 - 16p.$$

In terms of Matrix 5.3, we fixed our attention on the left column because the kicker aims left. Recall that the kicker's payoff is uncertain because she does not know if the goalie will dive left, preventing the kicker from scoring (which yields a payoff of 0 for the kicker, at the top row of the matrix), or if the goalie will dive right, which entails a score and a payoff of 16 (at the bottom row of the matrix). Similarly, the kicker's expected payoff from aiming right is

$$EU_{\text{Kicker}}(\text{Right}) = \underbrace{p \times 16}_{\text{goalie dives Left}} + \underbrace{(1-p) \times 0}_{\text{goalie dives right}}$$
$$= 16p,$$

which entails the opposite payoffs than before because the kicker scores only when the goalie dives left, which occurs with probability p (the probability with which the goalie plays the top row). Moving now to Step 2 in Tool 5.1, we consider that the kicker must be indifferent between aiming left and right, $EU_{\text{Kicker}}(\text{Left}) = EU_{\text{Kicker}}(\text{Right})$, which entails

$$16 - 16p = 16p.$$

Finally, moving to Step 3, we simplify the above indifference condition to obtain $16 = 32p$. Solving for p, yields $p = 16\backslash 32 = 1/2$. Therefore, the kicker is indifferent between aiming left and right when the goalie aims left with 50% probability (because we found that $p = 1/2$).

We can now summarize that the msNE of this game as follows:

$$\text{msNE} = \left\{ \underbrace{\left(\frac{1}{2}\text{Dive Left}, \frac{1}{2}\text{Dive Right}\right)}_{\text{Goalie}}; \underbrace{\left(\frac{1}{2}\text{Aim Left}, \frac{1}{2}\text{Aim Right}\right)}_{\text{Kicker}} \right\}$$

Remember that players do not need to randomize with the same probability. They only did it in this case because payoffs are symmetric in Matrix 5.3.□

This example illustrates Tool 5.1 in a two-player game where every player has two pure strategies (see Step 3a). In subsequent sections of this chapter we consider examples where players have three pure strategies, showing the similarities with the above example in an anticoordination game, but also highlighting its differences (namely, Step 3b).

5.4 Finding Mixed Strategy Equilibria

5.4.1 Graphical Representation of Best Responses

In this section, we describe how to graphically represent the best response of each player and its interpretation. For presentation purposes, consider the goalie and kicker in Example 5.1. We separately examine the goalie and kicker's best responses next.

Goalie. From the previous analysis, the goalie chooses to dive left if her expected utility from diving left is higher than from diving right; that is,

$$EU_{\text{Goalie}}(\text{Left}) > EU_{\text{Goalie}}(\text{Right}),$$

which can be expressed as $10 + 10q > -10q$, ultimately simplifying to $q > \frac{1}{2}$. Intuitively, when the probability that the kicker aims left (as measured by q) is sufficiently high (in this case, $q > \frac{1}{2}$), the goalie responds by diving left, so she increases her chances of blocking the ball. Mathematically, this means that, for all $q > \frac{1}{2}$, the goalie chooses to dive left (i.e., $p = 1$). In contrast, for all $q < \frac{1}{2}$, the goalie responds by diving right (i.e., $p = 0$). Figure 5.1 depicts this best response function.[5] Summarizing, this best response function can be expressed as follows

$$BR_{\text{Goalie}}(q) = \begin{cases} \text{Left} & \text{if } q > \frac{1}{2}, \\ \{\text{Left}, \text{Right}\} & \text{if } q = \frac{1}{2}, \text{ and} \\ \text{Right} & \text{if } q < \frac{1}{2} \end{cases}$$

which indicates that the goalie dives left when the kicker is likely aiming left, $q > \frac{1}{2}$; dives right when the kicker is likely aiming right, $q < \frac{1}{2}$; but randomizes between left and right, as represented by $\{\text{Left}, \text{Right}\}$ otherwise.

Kicker. A similar argument applies to the kicker. From our analysis in Example 5.1, we know that she aims left if

$$EU_{\text{Kicker}}(\text{Left}) > EU_{\text{Kicker}}(\text{Right}),$$

which entails $16 - 16p > 16p$, or, after simplifying, $p < \frac{1}{2}$. Intuitively, when the goalie is likely diving right (as captured by $p < \frac{1}{2}$), the kicker aims left, increasing her chances of scoring. Formally, we can write this result by saying that, for all $p < \frac{1}{2}$, the kicker aims left (i.e., $q = 1$); whereas for all $p > \frac{1}{2}$, the kicker

---

[5] Graphically, condition "for all $q > \frac{1}{2}$" means that we are on top half of Fig. 5.1a. For these points, the goalie's best response is $p = 1$ at the top vertical line of the graph. Similarly, condition "for all $q < \frac{1}{2}$" indicates that we look at the bottom half of the figure. For these points, the goalie's best response is $p = 0$ at the vertical line on the left of the graph (overlapping a segment of the vertical axis).

Fig. 5.1 The goalie's best responses

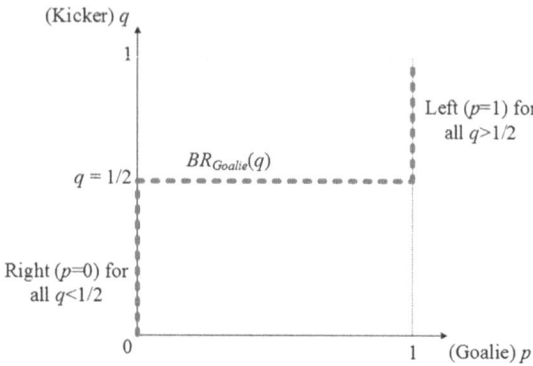

Fig. 5.2 Kicker's best responses

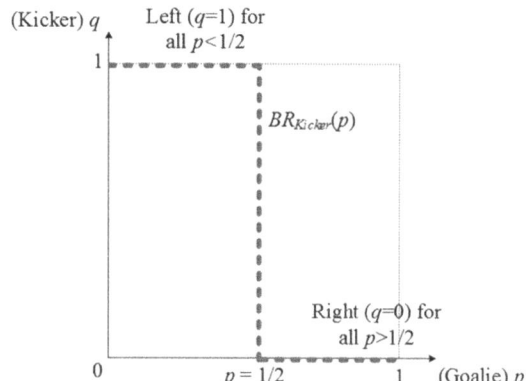

aims right (i.e., $q = 0$). Figure 5.2 illustrates the kicker's best response function.[6] Summarizing, the kicker's best response function can be expressed as follows

$$BR_{\text{Kicker}}(q) = \begin{cases} \text{Left} & \text{if } p < \frac{1}{2}, \\ \{\text{Left}, \text{Right}\} & \text{if } p = \frac{1}{2}, \text{ and} \\ \text{Right} & \text{if } p > \frac{1}{2} \end{cases}$$

which implies that the kicker aims left when the goalie is likely diving right, $p < \frac{1}{2}$; aims right when the goalie is likely diving left, $p > \frac{1}{2}$; but randomizes otherwise.

---

[6] Graphically, condition "for all $p < \frac{1}{2}$" means that we are on the left half of Fig. 5.2. For these points, the kicker's best response says that $q = 1$ at the horizontal line on the left side of the graph. Similarly, condition "for all $p > \frac{1}{2}$" indicates that we look at the right half of the figure. For these points, the kicker's best response is $q = 0$, at the horizontal line at the bottom of the graph (which overlaps part of the horizontal axis).

5.5 Some Lessons

Fig. 5.3 Both players' best responses

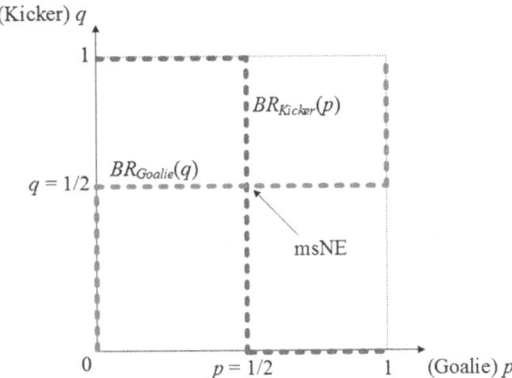

Putting best responses together. Figure 5.3 superimposes the goalie's and the kicker's best response functions, which we can do because we used the same axes in Figs. 5.1 and 5.2. The players' best responses only cross each other at one point, where $p = q = \frac{1}{2}$, as predicted in Example 5.1. Graphically, the point where both players' best responses cross each other indicates that both are using their best responses or, in other words, that the strategy profile is a *mutual* best response, as required by the definition of NE. For this example, the crossing point $p = q = \frac{1}{2}$ is the only NE of the game, an msNE, which has both players mixing. If the game had one or more psNEs, the best response functions should cross at some point in the vertices of the unit square of Fig. 5.3. Generally, if the game you analyze has more than one NE, the best responses you depict should cross at more than one point; namely, one point in the (p, q)—quadrant of Fig. 5.3 for each psNE that you find and, similarly, one for each msNE that you obtain.

5.5 Some Lessons

The above example helps us highlight some important points about msNE, which we expand on below.

- **Indifference.** If it is optimal to randomize over a collection of pure strategies, then a player receives the same expected payoff from each of those pure strategies. In other words, she must be indifferent between the pure strategies over which she randomizes. Otherwise, she would be earning a higher expected payoff from one of her pure strategies, and thus would select that pure strategy without the need to randomize.
- **When analyzing player i, ignore her probability to randomize.** This can be a typical confusion for some readers when learning about msNE for the first time. When we make player i indifferent between two (or more) of her pure strategies, we write the player i's expected utility from choosing pure strategy

s_i, $EU_i(s_i)$, as a function of *her rival's* randomization, σ_j, where $j \neq i$ (or, more generally, σ_{-i} if player i faces more than one rival); and, set $EU_i(s_i)$ equal to player i's expected utility from another pure strategy, s_i', $EU_i(s_i')$, which is also a function of her rival's randomization, σ_j. Importantly, neither $EU_i(s_i)$ or $EU_i(s_i')$ are functions of player i's own probability to randomize, σ_i. Indeed, at $EU_i(s_i)$, she chooses s_i with certainty (e.g., the goalie diving left), and at $EU_i(s_i')$ she chooses s_i' with certainty as well (e.g., dives right). Graphically, at $EU_i(s_i)$ we fix our attention to one of the rows (columns) for the row (column) player, but allow her opponent to randomize, which explains why $EU_i(s_i)$ is a function of player j's randomization, σ_j.

In contrast, when we search for randomizations that can help player i delete strictly dominated strategies (see Sect. 2.5), we compute her expected utility from two pure strategies, such as s_i and s_i', conditional on her rival choosing s_j, $EU_i(\sigma_i|s_j)$. In that setting, her expected utility was a function of her own mixed strategy, σ_i, instead of her rival's, σ_j. Graphically, this implies that we fixed our attention on a specific column (row) when analyzing the expected utility of the row (column) player's randomization.

- **Never use strictly dominated strategies.** If a pure strategy is strictly dominated (either by another pure strategy or by a mixed strategy), then such a pure strategy cannot be part of psNE or a msNE. In the context of msNEs we study in this chapter, this means that every msNE must assign a zero probability to that strictly dominated pure strategy. This comes, of course, at no surprise, because we know that a rational player would never play a strictly dominated strategy and that, by common knowledge of rationality, her rivals can anticipate she won't use this strategy.

  As an illustration, consider the Prisoner's Dilemma game, where *Not confess* is strictly dominated, and thus is not part of the psNE of the game (*Confess, Confess*). Alternatively, consider a larger payoff matrix where we can delete one or more players' strategies when applying IDSDS, but do not obtain a unique equilibrium prediction using IDSDS. In this setting, we can find the psNEs and msNE of the surviving strategy profiles (after applying IDSDS) where, importantly, the strictly dominated strategies cannot not receive any positive probability.

- **Odd number of equilibria.** In almost all finite games (games with a finite set of players and available actions), there is a finite and odd number of equilibria. As examples, consider the Prisoner's Dilemma game (with one psNE and no msNE), the Battle of the Sexes game (with two psNEs and one msNE), the Penalty Kicks in Soccer game (with one msNE and no psNEs), or the Chicken game (with two psNEs and one msNE). This is a helpful point to remember when searching for all pure and mixed NEs in a finite game because, if you find two NEs, you should keep looking for one more equilibrium (either in pure or mixed strategies).[7]

5.6 Extensions

Our above analysis helped us identify msNE in games with two players (e.g., goalie and kicker), each of them with two pure strategies (e.g., left or right). Our approach, nonetheless, extends to settings where two players randomize over more than two pure strategies and to games with more than two players. We next separately examine each of these extensions.

5.6.1 Mixed Strategy Equilibria in Games with $k \geq 3$ Pure Strategies

Consider the game in Matrix 5.4, which represents the Rock–Paper–Scissors game, which you may have played in your childhood, where R denotes Rock, P represents Paper, and S stands for Scissors. Players' payoffs are symmetric and illustrate that Rock beats Scissors, Scissors beats Paper, and Paper beats Rock. When both players display the same object (e.g., both show Paper), there is a tie, and both players earn zero.

		Player 2		
		R	P	S
	R	0, 0	−1, 1	1, −1
Player 1	P	1, −1	0, 0	−1, 1
	S	−1, 1	1, −1	0, 0

Matrix 5.4. Rock-Paper-Scissors game.

It is straightforward to show that no psNE exists in this game. Matrix 5.4a underlines best response payoffs, illustrating that no cell has both players' payoffs underlined.

---

[7] This result, due to Wilson (1971) and subsequently by Harsanyi (1973), is often known as the "Oddness Theorem," and states that almost all finite games (i.e., games that we can represent with a matrix with a finite number of rows and columns) have a finite and odd number of NEs.

Player 2

Player 1		R	P	S
	R	0, 0	−1,<u>1</u>	<u>1</u>, −1
	P	<u>1</u>, −1	0, 0	−1,<u>1</u>
	S	−1,<u>1</u>	<u>1</u>, −1	0, 0

Matrix 5.4a. Rock-Paper-Scissors game - Underlining best response payoffs.

We now examine the msNE. First, note that players' payoffs are symmetric, entailing that they randomize with the same probabilities: r, p, and $1 - r - p$, where r denotes the probability that every player i chooses Rock, p represents the probability she selects Paper, and $1 - r - p$ is the probability she plays Scissors.

Example 5.2 Finding msNE when players have three pure strategies. If player i randomizes between Rock, Paper, and Scissors, she must be indifferent between all her three pure strategies:

$$EU_i(R) = EU_i(P) = EU_i(S).$$

For presentation purposes, we separately find player 1's expected utility from each pure strategy, yielding

$$\begin{aligned} EU_1(R) &= r0 + p(-1) + (1 - r - p)1 \\ &= 1 - r - 2p. \end{aligned}$$

Graphically, when computing $EU_1(R)$, we focus our attention on the top row of Matrix 5.4a. Player 2 chooses Rock with probability r, entailing a payoff of zero for player 1; Paper with probability p, which yields a payoff of -1 for player 1; and Scissors with probability $1 - r - p$, which entails a payoff of 1 for player 1.

Player 1's expected payoff from choosing Paper (in the middle row) is

$$\begin{aligned} EU_1(P) &= r1 + p0 + (1 - r - p)(-1) \\ &= 2r + p - 1. \end{aligned}$$

Finally, player 1's expected payoff from Scissors (at the bottom row) is

$$\begin{aligned} EU_1(S) &= r(-1) + p1 + (1 - r - p)0 \\ &= p - r. \end{aligned}$$

We can now set the expected utilities from Rock, Paper, and Scissors equal to each other. First, from $EU_1(R) = EU_1(S)$, we find that

$$1 - r - 2p = p - r$$

5.6 Extensions

which rearranges to $p = \frac{1}{3}$. Second, from $EU_1(P) = EU_1(S)$, we find that

$$2r + p - 1 = p - r$$

which simplifies to $2r + r = 1$, yielding $r = \frac{1}{3}$. Therefore, the probability of playing Scissors is

$$\begin{aligned} 1 - r - p &= 1 - \frac{1}{3} - \frac{1}{3} \\ &= \frac{1}{3} \end{aligned}$$

and we can summarize the msNE of this game as that every player i randomizes according to the mixed strategy

$$(r^*, p^*, 1 - p^* - r^*) = \left(\frac{1}{3}, \frac{1}{3}, \frac{1}{3}\right)$$

which means that every player assigns the same probability weight to each of her pure strategies. □

5.6.2 Finding Mixed Strategy Equilibria in Games with $N \geq 2$ Players

We can extend the above analysis to games with $N \geq 2$ players. Consider, for instance, the so-called "snob effect" game, adapted from Watson (2013), which illustrates how individuals may choose conforming strategies when they interact in large groups but take the "risk" of adopting a non-conforming strategy in small groups. To understand these incentives, consider that every player i independently and simultaneously chooses between alternatives X and Y, where X can be interpreted as the snob option while Y is the conforming option. If she is the only player choosing X, she earns a payoff of a (she is the "cool" girl in the group) but a payoff of b otherwise (if anyone else chooses X too). When she chooses Y, she earns a payoff of c, where c satisfies $a > c > b$.

In this setting, we seek to identify a symmetric msNE where every player i chooses X with probability p. Therefore, we need to, first, find the expected utility from X and from Y; and, second, set these expected utilities equal to each other to obtain the equilibrium probability p^*.

Expected utility from X. When player i chooses X, his expected utility is

$$EU_i(X) = \underbrace{(1-p)^{N-1}}_{\text{No other player chooses } X} a + \underbrace{\left[1 - (1-p)^{N-1}\right]}_{\text{At least someone else chooses } X} b.$$

To understand the probabilities of each event, recall that the probability with which every player chooses X is p, so the probability with which she plays Y is $1 - p$, implying that all other $N - 1$ players (everyone but player i) choose alternative Y with probability

$$\underbrace{(1-p) \times (1-p) \times \cdots \times (1-p)}_{N-1 \text{ times}} = (1-p)^{N-1}.$$

If this event occurs, player i is the only one choosing X, earning a payoff of a. Otherwise, there must be a player who also chose X, which yields a lower payoff of b to player i. Importantly, this occurs with probability $1 - (1-p)^{N-1}$ or, intuitively, with the converse probability of the event yielding a payoff of a.

Expected utility from Y. When player i chooses Y, she earns a payoff of c with certainty, entailing that $EU_i(Y) = c$.

Indifference condition. If player i randomizes between X and Y, she must be indifferent, thus earning the same expected payoff from each pure strategy, $EU_i(X) = EU_i(Y)$, which means

$$(1-p)^{N-1} a + \left[1 - (1-p)^{N-1}\right] b = c.$$

Rearranging, yields

$$(1-p)^{N-1} = \frac{c-b}{a-b}.$$

After powering both sides to $\frac{1}{N-1}$, and solving for p, we obtain the equilibrium probability of choosing alternative X,

$$p^* = 1 - \left(\frac{c-b}{a-b}\right)^{\frac{1}{N-1}}.$$

For instance, when $a = 2$, $b = 0$, and $c = 1$, this probability simplifies to

$$p^* = 1 - \left(\frac{1}{2}\right)^{\frac{1}{N-1}}$$

which decreases in the number of players, N; as Fig. 5.4 depicts.[8] Intuitively, as the population grows, it is more likely that someone selects alternative X, driving each player to individually decrease her probability of choosing the snob option X; opting, instead, for the conforming option Y, with probability $1 - p^* = \left(\frac{1}{2}\right)^{\frac{1}{N-1}}$, which increases in N.

---

[8] Formally, the derivative of p^* with respect to N, yields $\frac{\partial p^*}{\partial N} = -\frac{2^{1/1-N} \ln(2)}{(N-1)^2}$, which is unambiguously negative.

Fig. 5.4 Equilibrium probability p^*

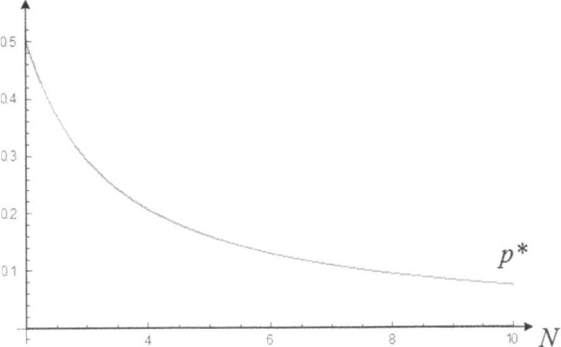

We can also examine how the above expression of p^* changes when players are more attracted to the snob option, $a - b$, which captures the payoff gain of the snob option relative to the conforming option. An increase in $a - b$ increases the probability of choosing X. In contrast, an increase in $c - b$, which measures the payoff loss of not being the only individual choosing X, decreases the probability of players choosing X.

5.7 Strictly Competitive Games

In this section, we first describe a special class of games where players' competitive pressures are particularly intense, and then find equilibrium behavior using two solutions concepts: security strategies, originally developed by von Neumann (1928), and the msNE defined in previous sections of this chapter. We finish our discussion by comparing these solution concepts and identifying in which games they yield the same equilibrium predictions.

5.7.1 Strictly Competitive Games

Definition 5.4. **Strictly competitive game.** A two-player game is strictly competitive if, for every two strategy profiles s and s', if $u_1(s) > u_1(s')$ then $u_2(s) < u_2(s')$; and if $u_1(s) = u_1(s')$ then $u_2(s) = u_2(s')$.

Intuitively, if player 1 prefers strategy profile s to s', then player 2 has the opposite preference order: preferring s' over s; and if player 1 is indifferent between s and s', player 2 must also be indifferent between these two strategy profiles. Graphically, this means that, in a game matrix, we need to check if every time that player 1 prefers one cell over another, then player 2 must prefer the opposite.

The penalty kicks game, reproduced in Matrix 5.5, provides us with an example of a strictly competitive game where we can test this definition:

1. Comparing (L, l) and (L, r), we see that the goalie prefers the former, since $0 > -10$, while the kicker prefers the latter because $0 < 16$.
2. Comparing (L, l) and (R, l), we find that the goalie prefers the former, given that $0 > -10$, while the kicker prefers the latter since $0 < 16$.
3. Comparing (L, l) and (R, r), we see that the goalie is indifferent, and so is the kicker, both players earning a payoff of zero in both strategy profiles.
4. Finally, comparing (R, l) and (L, r), we find that the goalie is indifferent between these two strategy profiles, earning -10 in both of them. A similar argument applies to the kicker, who earns a payoff of 16 in both strategy profiles.

Importantly, the above exhaustive list compares all strategy profiles in Matrix 5.5 against each other (pairwise), to confirm that the definition of strictly competitive games (i.e., opposite preferences of players 1 and 2) holds *for every* two strategy profiles, s and s', in the matrix.

		Kicker	
		Aim left (l)	Aim right (r)
Goalie	Dive Left (L)	$0, 0$	$-10, 16$
	Dive Right (R)	$-10, 16$	$0, 0$

Matrix 5.5. Anticoordination game.

Because the definition of strictly competitive games requires opposite preferences across all strategy profiles, we can show that a game is *not* strictly competitive by finding at least two strategy profiles, s and s', where players have aligned preferences, that is, both players prefer s to s' or both prefer s'. Formally, we say that a two-player game is not strictly competitive if, for at least two strategy profiles s and s', every player i's utility satisfies $u_i(s) > u_i(s')$.

Matrix 5.6 reproduces the police and criminal game from Sect. 5.1, which is not strictly competitive: comparing strategy profiles (A, A) and (B, B), along the main diagonal, we can see that the police prefers (A, A) to (B, B), since her payoff satisfies $10 > 7$; and a similar argument applies for the criminal, as her payoff satisfies $0 > -1$. Because we found that players' preferences over strategy profiles are aligned, rather than misaligned, we can already claim that the game is not strictly competitive without having to compare other pairs of strategy profiles.

		Criminal	
		Street A	Street B
Police	Street A	$10, 0$	$-1, 6$
	Street B	$0, 8$	$7, -1$

Matrix 5.6. Police and criminal game.

5.7.1.1 Constant-Sum Games

We next present two special classes of strictly competitive games where players' incentives are even more opposed.

Definition 5.5. **Constant-sum games.** A two-player game is a constant-sum game if, for every strategy profile s, players' payoffs satisfy $u_1(s) + u_2(s) = K$, where $K > 0$ is a constant.

In other words, players' payoffs add up to the same constant K (e.g., $100), in *all cells* of the matrix. If, instead, players' payoffs in at least one strategy profile add to a different number, $K' \neq K$, we can claim that the game is not a constant-sum game, although it may still be strictly competitive, such as the game in Matrix 5.7. In particular, the game is strictly competitive (check this as a practice), but is not constant-sum since players' payoffs in strategy profiles like (U, l) and (D, r) add up to $10, while those in strategy profiles (U, r) and (D, l) add up to $12.

$$
\begin{array}{c|c|c|}
 & \multicolumn{2}{c}{\text{Player 2}} \\
 & l & r \\
\hline
U & 10, 0 & 9, 3 \\
\hline
D & 9, 3 & 10, 0 \\
\hline
\end{array}
$$

Matrix 5.7. A strictly competitive game that is non constant-sum.

This class of games is, clearly, strictly competitive: for two strategy profiles s and s' where $u_1(s) > u_1(s')$, we must have that $u_2(s) < u_2(s')$ given that the sum of players' payoffs must be equal to the same constant K in both s and s'. Alternatively, we can see that condition $u_1(s) + u_2(s) = K$ can be expressed as $u_1(s) = K - u_2(s)$, implying that, if player 1's payoff increases from s' to s, player 2's payoff must decrease.

5.7.2 Zero-Sum Games

A special class of constant-sum games often mentioned in media outlets is zero-sum games, where constant K is exactly zero, as we define next.

Definition 5.6. **Zero-sum games.** A two-player game is a zero-sum game if, for every strategy profile s, players' payoffs satisfy $u_1(s) + u_2(s) = 0$.

Alternatively, condition $u_1(s) + u_2(s) = 0$ can be expressed as $u_1(s) = -u_2(s)$, which means that, every dollar that player 1 earns comes from the same dollar that player 2 losses and vice versa. These games are typically mentioned in politics or, more generally, when players seek to divide a surplus (fixed budget), where $1 going to one player means $1 less going to her opponent (e.g., one million dollars going to education means that exactly one million dollars less are allocated to other programs, such as national defense).

The Matching Pennies game that we analyzed in Chapter 3, reproduced in Matrix 5.8 for easier reference, provides an example of a zero-sum game since, indeed, condition $u_1(s) + u_2(s) = 0$ holds for every strategy profile s. Specifically, for every cell, we have that either $1 + (-1) = 0$ or $-1 + 1 = 0$. Another example is the The Rock–Paper–Scissors game in Matrix 5.4, where the payoff gain (loss) of the row player coincides with the payoff loss (gain) of the column player.

		Player 2	
		Heads	Tails
Player 1	Heads	1, −1	−1, 1
	Tails	−1, 1	1, −1

Matrix 5.8. Matching Pennies game.

5.7.3 Security Strategies

Strictly competitive games can be solved using msNE, as described in previous sections of this chapter, or using security strategies, which we define below.

Definition 5.7. **Security strategies.** In a two-player game, player i's security strategy, σ_i, solves

$$\max_{\sigma_i} \min_{\sigma_j} u_i(\sigma_i, \sigma_j).$$

To better understand the above definition, let us first consider the "worst-case scenario" from choosing strategy σ_i:

$$w_i(\sigma_i) = \min_{\sigma_j} u_i(\sigma_i, \sigma_j).$$

This expression indicates that, for strategy σ_i, player i anticipates that player j chooses her strategy σ_j to maximize her own payoff, which entails minimizing i's payoff, $u_i(\sigma_i, \sigma_j)$, because the two players are in a strictly competitive game. Player i then chooses her strategy σ_i to maximize the payoff across all worst-case scenarios. Intuitively, player i seeks to find the strategy σ_i that provides her with the "best of the worst" payoffs, as represented by the max-min problem; which also explains why security strategies are sometimes known as max-min strategies.

Tool 5.2 describes how to find security strategies and then Example 5.3 illustrates the tool.

5.7 Strictly Competitive Games

Tool 5.2 How to find security strategies in a two-player game

1. Find the expected utility of player 1's randomization, fixing player 2's strategy.
2. Repeat step 1 until you have considered all strategies of player 2, fixing one at a time.
3. *"Min" part.* Find the lower envelope of player 1's expected utility. That is, for each strategy σ_1, find the lowest expected utility that player 1 earns.
4. *"Max" part.* Find the highest expected utility of the lower envelope identified in step 3, and the corresponding strategy σ_1. This is player 1's security strategy, σ_1^{Sec}.
5. To find the security strategy for player 2, follow a similar process as in steps 1–4 above.

Example 5.3 Finding security strategies. Consider Matrix 5.9, which reproduces the strictly competitive game in Matrix 5.7.

$$\begin{array}{c} & \text{Player 2} \\ & \begin{array}{cc} l & r \end{array} \\ \text{Player 1} & \begin{array}{c} U \\ D \end{array} \begin{array}{|c|c|} \hline 10,0 & 9,3 \\ \hline 9,3 & 10,0 \\ \hline \end{array} \end{array}$$

Matrix 5.9. A strictly competitive game that is non constant-sum.

To find the security strategy for player 1, we follow the next steps:

1. We find player 1's expected utility of randomizing between U and D, with associated probabilities p and $1-p$, respectively. First, we fix player 2's strategy at column l, which yields

$$\begin{aligned} EU_1(p|l) &= p \times 10 + (1-p) \times 9 \\ &= 9 + p. \end{aligned}$$

2. We now find her expected utility of randomizing, but fixing player 2's strategy at column r, as follows

$$\begin{aligned} EU_1(p|r) &= p \times 9 + (1-p) \times 10 \\ &= 10 - p. \end{aligned}$$

3. To find the lower envelope of the previous two expected utilities, we can depict each line as a function of p, as we do in Fig. 5.5. The lower envelope is the segment $9+p$ for all $p \leq \frac{1}{2}$, but segment $10-p$ otherwise. (To see this, note that $9 + p \leq 10 - p$ simplifies to $2p \leq 1$, or $p \leq \frac{1}{2}$.)

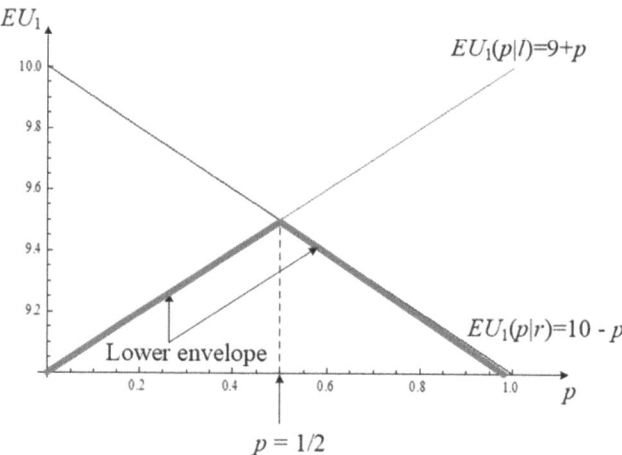

Fig. 5.5 Lower envelope and security strategies

4. Among all points in the lower envelope, player 1 enjoys the highest utility at $p = \frac{1}{2}$, which yields an expected payoff $EU_1(p|l) = 9 + \frac{1}{2} = 9.5$, as illustrated in Fig. 5.5 by the height of the crossing point between $EU_1(p|l)$ and $EU_1(p|r)$. This is player 1's security strategy, $p^{\text{Sec}} = \frac{1}{2}$.
5. Following the same steps for player 2, we find that, since payoffs are symmetric, her security strategy is $q^{\text{Sec}} = \frac{1}{2}$.

5.8 Security Strategies and NE

At this point, you may be wondering about the relationship between security strategies and msNE. In other words, if we apply security strategies to a game, does the equilibrium outcome always coincide with the msNE of the game? It does, but only for two-player strictly competitive games. Otherwise, security strategies and msNE may produce different equilibrium outcomes. Example 5.4 illustrates that both solution concepts yield the same equilibrium outcome in our ongoing strictly competitive game, and Example 5.5 shows that they yield different outcomes in a game that is not strictly competitive.

Example 5.4 Security strategies and msNE yielding the same equilibrium outcome. Consider the strictly competitive game in Matrix 5.9 again. From Example 5.3, we know that the security strategy predicts every player randomizing with probability $\frac{1}{2}$, that is, $\left(p^{\text{Sec}}, q^{\text{Sec}}\right) = \left(\frac{1}{2}, \frac{1}{2}\right)$. Let us now confirm that the msNE produces the same result. Assuming that player 1 randomizes between U and D with probabilities p and $1 - p$, respectively, and player 2 mixes between l and r with probabilities

5.8 Security Strategies and NE

q and $1 - q$, respectively, we find that player 1's expected utility from choosing U is[9]

$$EU_1(U) = q \times 10 + (1 - q) \times 9$$
$$= 9 + q.$$

Similarly, her expected utility from selecting D is

$$EU_1(D) = q \times 9 + (1 - q) \times 9$$
$$= 10 - q.$$

Therefore, player 1 randomizes between U and D when she is indifferent between these two pure strategies, $EU_1(U) = EU_1(D)$, which entails

$$9 + q = 10 - q$$

yielding $q = \frac{1}{2}$. We can follow similar steps for player 2, finding her expected utility from l and r as follows

$$EU_2(l) = p \times 0 + (1 - p) \times 3 = 3 - 3p, \text{ and}$$
$$EU_2(r) = p \times 3 + (1 - p) \times 0 = 3p.$$

Player 2 randomizes when she is indifferent between l and r, that is, $EU_2(l) = EU_2(r)$, or $3 - 3p = 3p$, which yields $p = \frac{1}{2}$. Summarizing, we can claim that the msNE of this game is $(p, q) = \left(\frac{1}{2}, \frac{1}{2}\right)$, which coincides with the security strategies we found in Example 5.3.□

Example 5.5 Security strategies and msNE yielding different equilibrium outcomes. Consider the game in Matrix 5.10. This game is not strictly competitive since we can find two strategy profiles where players' interests are aligned; both players prefer, for instance, (U, l) to (U, r).[10] Since the game is not strictly competitive, we can expect that security strategies may produce a different equilibrium prediction than msNE.

---

[9] This expected utility resembles that in Example 5.3, but describes a different randomization. In the current setting, player 1 chooses one of her pure strategies, U, fixing our attention in the top row, but player 2 randomizes between l and r, yielding player 1 with a payoff of 10 with probability q and a payoff of 9 with probability $1 - q$. In contrast, in Example 5.3, we took player 2's pure strategy as given (for instance, when computing $EU_1(p|l)$ we fixed our attention in the left column) and player 1 randomized between U and D with probability p and $1 - p$, respectively.

[10] In particular, player 1's payoffs satisfy $3 > -1$, and player 2's satisfy $5 > 1$. Other examples include both players preferring outcome (D, l) to (D, r), or both preferring (D, l) to (U, r).

	Player 2	
	l	r
U	3, 5	−1, 1
D	2, 6	1, 2

Player 1 (rows), Player 2 (columns)

Matrix 5.10. A game that is not strictly competitive.

Security strategies. When player 2 chooses l, player 1's expected payoff from randomizing between U and D with probabilities p and $1 - p$, respectively, is

$$EU_1(p|l) = p \times 3 + (1 - p) \times 2$$
$$= 2 + p.$$

And when player 2 chooses r, player 1's expected utility is

$$EU_1(p|r) = p \times (-1) + (1 - p) \times 1$$
$$= 1 - 2p.$$

Figure 5.6 plots $EU_1(p|l) = 2 + p$ and $EU_1(p|r) = 1 - 2p$, both as a function of p, showing that $EU_1(p|l)$ lies above $EU_1(p|r)$ for all $p \in [0, 1]$. This means that the lower envelope coincides with $EU_1(p|r) = 1 - 2p$ for all values of p. The highest point of her lower envelope occurs in this example at $p = 0$, which implies that player 1 assigns no probability weight to U or, alternatively, that she plays D in pure strategies, meaning that D is player 1's security strategy.[11]

Following a similar approach with player 2's expected payoff, we find that

$$EU_2(q|U) = q5 + (1 - q)1 = 1 + 4q, \text{ and}$$
$$EU_2(q|D) = q6 + (1 - q)2 = 2 + 4q.$$

Since $EU_2(q|D) > EU_2(q|U)$ for all values of q, we can claim that U is the lower envelope. We can, then, notice that the highest point of $2 + 4q$ occurs at $q = 1$, reaching a height of 6, meaning that player 2 puts full probability weight on l, which becomes her security strategy. In summary, the security strategy profile in this game is (D, l). We next show that this equilibrium prediction does not coincide with the msNE of this game.

msNE. To find the msNE of the game in Matrix 5.10, we could start finding player 1's expected utility from U and D, and then setting them equal to each other. However, we can facilitate our analysis by noticing that strategy l strictly dominates r since it yields a strictly higher payoff than r regardless of the row that

---

[11] Alternatively, the worst payoff that player 1 can earn from choosing D is 1, which is above his payoff from U, −1.

5.9 Correlated Equilibrium

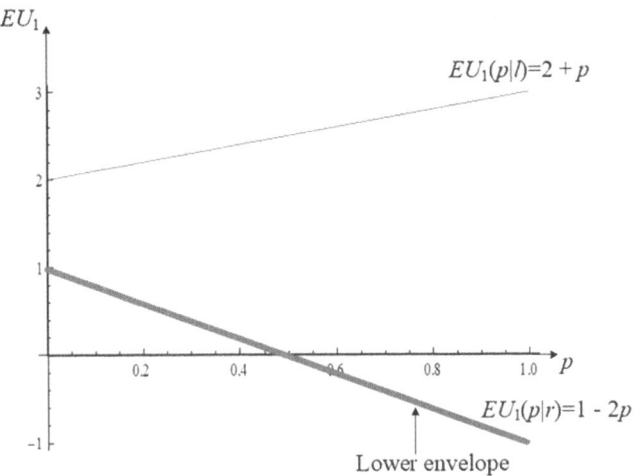

Fig. 5.6 Lower envelope and security strategies—Corner solution

player 1 chooses ($5 > 1$ and $6 > 2$). We know players put no probability weight in strictly dominated strategies, so we can delete column r from the matrix, obtaining Matrix 5.11.

$$
\begin{array}{c}
\textit{Player 2} \\
l \\
\textit{Player 1} \quad
\begin{array}{c|c}
U & 3,5 \\ \hline
D & 2,6
\end{array}
\end{array}
$$

Matrix 5.11. A game that is not strictly competitive - After deleting column r.

Turning now to player 1, we do not need to consider her randomization since, at this point, she has a clear best response to l, $BR_1(l) = U$ since $3 > 2$. Therefore, we can say that the game in Matrix 5.10 only has a psNE, (U, l), and no msNEs. As expected, this equilibrium outcome does not coincide with the security strategy profile, (D, l), found above. $\square$

5.9 Correlated Equilibrium

Consider the modified version of the Chicken game in Matrix 5.12, where both players simultaneously and independently choose whether to swerve or stay. (The game is strategically equivalent to that in Sect. 3.6.3, but increasing all payoffs to avoid negative values.) Underlining best response payoffs, one can easily show that the game has two psNEs, (*Swerve, Stay*) and (*Stay, Swerve*), with each player choosing a different strategy than her rival. The game also has a msNE, where player 1 chooses *Swerve* with probability $p = 2/3$, and the same probability applies to player 2, $q = 2/3$, since payoffs are symmetric.[12] That is, $\sigma^* = (\sigma_1^*, \sigma_2^*)$, where $\sigma_i^* = \left(\frac{2}{3} Swerve, \frac{1}{3} Stay\right)$ for every player i.

Player 2

		Swerve	Stay
Player 1	Swerve	6, 6	2, 7
	Stay	7, 2	0, 0

Matrix 5.12. Modified Chicken game.

In this msNE, player 1's expected utility in equilibrium is

$$EU_1(\sigma^*) = \underbrace{\frac{2}{3}\left(\frac{2}{3}6 + \frac{1}{3}2\right)}_{\text{Player 1 chooses } Swerve} + \underbrace{\frac{1}{3}\left(\frac{2}{3}7 + \frac{1}{3}0\right)}_{\text{Player 1 chooses } Stay} = \frac{28}{9} + \frac{14}{9} = \frac{14}{3} \simeq 4.67.$$

and, by symmetry, the same expected payoff accrues to player 2.

A natural question at this point is whether players can reach a higher expected payoff if, instead, they rely on a probability distribution, such as a coin toss, that each player privately observes before playing the game, and that informs the player about which action to choose. Intuitively, the probability distribution can be interpreted as a external "recommender" who, first, draws one strategy profile, $s = (s_i, s_{-i})$, such as one cell in Matrix 5.12.[13] Second, the recommender makes a recommendation s_i to player i, without informing her of the recommendation s_{-i} that her rivals receive. The next solution concept, due to Aumann (1974), formalizes this two-step approach, requiring that every player has no incentives to deviate from the recommendation, s_i, that she receives.

Definition 5.8. **Correlated equilibrium.** A probability distribution over strategy profiles is a correlated equilibrium if every player i follows her recommendation, s_i.

---

[12] For instance, when player 1 chooses *Swerve*, on the top row, her expected utility is $EU_1(\text{Swerve}) = q \times 6 + (1 - q) \times 2$ while her expected utility when choosing *Stay*, on the bottom row, is $EU_1(\text{Stay}) = q \times 7 + (1 - q) \times 0$. Setting them equal to each other, yields $q = 2/3$. A similar argument applies to player 2's expected payoffs, which entails $p = 2/3$.
[13] This step allows for the strategy profile to involve pure or mixed strategies but, for simplicity, most examples consider pure-strategy profiles.

5.9 Correlated Equilibrium

Intuitively, a probability distribution over strategy profiles (e.g., over cells in the matrix) is stable in the sense that every player i has no incentives to unilaterally deviate from the recommendation, s_i, that she received from the external recommender.

5.9.1 Public or Private Recommendations?

While recommendations can be privately observed by each player, they can also be public. For simplicity, we present this case first in Example 5.6. Example 5.7, then, considers the case of privately observed recommendations.

Example 5.6 Correlated equilibrium with public signals. Consider the game in Matrix 5.12 and assume that players observe a public signal that assigns probability α to one of the psNEs in this game, (*Stay*, *Swerve*), and probability $1 - \alpha$ to the other psNE, (*Swerve*, *Swerve*), as illustrated in Matrix 5.13. A public signal could be a traffic light, or any other stochastic mechanism that players agree on before starting the game, that yields the probability distribution in Matrix 5.13.

		Player 2	
		Swerve	Stay
Player 1	Swerve	0	$1 - \alpha$
	Stay	α	0

Matrix 5.13. Correlated equilibrium with public signals - Probability of each strategy profile.

In this context, player 1 does not have incentives to deviate. Indeed, upon observing (*Stay*, *Swerve*), her payoff from following the recommendation is 7, and that of unilaterally deviating to *Swerve* decreases to 6. Similarly, upon observing (*Swerve*, *Stay*), her payoff is 2, but decreases to 0 if she unilaterally deviates to *Stay* (a car crash occurs!). By symmetry, the same argument applies to player 2. As a consequence, a continuum of correlated equilibria can be sustained, where players alternate between the two psNEs of the game with probabilities α and $1 - \alpha$, respectively.□

As described above, the probability distribution of recommendations is common knowledge among players, but each recommendation s_i can be privately observed by each player. The next example considers this setting, illustrating that, upon privately observing her recommendation, every player i infers the recommendation that her rivals observed (or updates the recommendation they observed given the information that player *i* now has). In this context, for a probability distribution to be a correlated equilibrium, every player *i* must find that following her recommendation yields a higher expected utility than unilaterally deviating.

Example 5.7 Correlated equilibrium with private signals. Consider the game in Matrix 5.12 again, where the recommendations assign the same probability weight to (*Swerve, Swerve*), (*Swerve, Stay*), and (*Stay, Swerve*), as summarized in Matrix 5.14.

		Player 2	
		Swerve	Stay
Player 1	Swerve	1/3	1/3
	Stay	1/3	0

Matrix 5.14. Correlated equilibrium with private signals - Probability of each strategy profile.

Intuitively, if player 1 receives the recommendation of *Stay*, on the bottom row, she knows that the strategy profile recommended by the signal is (*Stay, Swerve*). However, if she receives the recommendation of *Swerve*, on the top row, she knows that player 2 may also receive the same recommendation, *Swerve*, with probability $\frac{\frac{1}{3}}{\frac{1}{3}+\frac{1}{3}} = \frac{1}{2}$, or the recommendation of *Stay* with the same probability, $1/2$.[14] A similar interpretation applies to the recommendations that player 2 receives.

We can now show that player 1 does not have incentives to deviate from this recommendation profile. If she receives the recommendation of *Stay*, her payoff is 7 at (*Stay, Swerve*), which she cannot improve by unilaterally deviating to *Swerve* (earning only 6). If, instead, she receives the recommendation of *Swerve*, her expected payoff is $\frac{1}{2}6 + \frac{1}{2}2 = 4$, which she cannot improve by unilaterally deviating to *Stay*, as that only yields $\frac{1}{2}7 + \frac{1}{2}0 = 3.5$. Since payoffs are symmetric, a similar argument applies to player 2, making the above recommendation profile stable. We, then, say that the recommendation profile can be sustained as a correlated equilibrium, with an expected payoff

$$\frac{1}{3}6 + \frac{1}{3}2 + \frac{1}{3}7 = 5$$

which exceeds that in the msNE of the game, 4.67. □

---

[14] Note that this is a direct application of Bayes' rule, which we discuss in more detail in subsequent chapters. Intuitively, player 1 knows that (*Swerve, Swerve*) and (*Swerve, Stay*) are each recommended with probability 1/3. Therefore, observing that her recommendation is *Swerve* means that her rival must be receiving a recommendation of *Swerve* or *Stay*, both of them being equally likely.

As a practice, you can check that a probability distribution where all four strategy profiles are equally likely cannot be supported as a correlated equilibrium. This highlights a usual and reasonable trick when finding correlated equilibria: do not assign probability weight to outcomes where both players' payoffs are the lowest, such as (*Stay*, *Stay*) in the Chicken game.

Existence. Finally, note that every psNE, $s^* = (s_i^*, s_{-i}^*)$, can be defined as a (trivial) correlated equilibrium where the probability distribution recommends player i to choose s_i^* with probability one. That is, every player is recommended to play as she would under the psNE. A similar argument applies to msNE $\sigma^* = (\sigma_i^*, \sigma_{-i}^*)$, where the probability distribution recommends player i to randomize according to the same mixed strategy she uses in the msNE of the game, σ_i^*. In other words, if σ^* is an NE, it must also be a correlated equilibrium, but the converse is not necessarily true.

$$\sigma^* \text{ is a NE} \implies \sigma^* \text{ is a correlated equilibrium}$$
$$\not\Leftarrow$$

Therefore, since an NE exists (either in pure or mixed strategies), a correlated equilibrium must also exist, i.e., at least the trivial one described above where every player is recommended to play according to her NE strategy.

5.10 Equilibrium Refinements in Strategic-Form Games (Technical)

Mixed strategies can help us discard NEs which seem fragile to small strategic mistakes, as if a player's hands could "tremble" when choosing her strategy. We next present two refinement criteria based on these strategic mistakes, or "trembles," and how they restrict the set of NEs in matrix games.

5.10.1 Trembling-Hand Perfect Equilibrium

Consider the game in Matrix 5.15, which underlines each player's best response payoffs, identifying two psNEs: (U, l) and (D, r). The second one, however, seems to be fragile to small changes in players' decisions. For instance, if player 1 deviates from D by putting a small probability weight on U, however small, player 2's best response would change to l. (A similar argument applies if player 2 deviates from r, assigning a small probability weight on l, which would change player 1's best response from D to U.)

	Player 2	
	l	r

Player 1

	l	r
U	**1,1**	0,0
D	0,0	0,0

Matrix 5.15. A game with two psNEs, but only (U, l) is THPE.

You may have noticed that the issue with (D, r) is that every player uses a weakly dominated strategy. We now seek to rule out psNEs such as (D, r), considering only those that are robust to trembles, as we define next.

Definition. **Totally mixed strategy.** Player i's mixed strategy, σ_i, is "totally mixed" if it assigns a strictly positive probability weight on every pure strategy, that is, $\sigma_i(s_i) > 0$ for all s_i.

A totally mixed strategy, therefore, allows for the trembles that we informally discussed above, where player 1, despite planning to choose U in Matrix 5.15, ends up choosing D with a positive probability, e.g., $\sigma_i(U) = 0.99$ and $\sigma_i(D) = 0.01$. We can now use this definition to present an equilibrium refinement of NE, based on Selten (1975), which selects those NEs (in pure or mixed strategies) that are robust to trembles (mistakes).

Definition 5.9. **Trembling-Hand perfect equilibrium.** A mixed strategy profile $\sigma = (\sigma_i, \sigma_{-i})$ is a Trembling-Hand Perfect Equilibrium (THPE) if there exists a sequence of totally mixed strategies for each player i, $\{\sigma_i^k\}_{k=1}^{\infty}$, that converges to σ_i, and for which $\sigma_i \in BR_i(\sigma_{-i}^k)$ for every k.

Hence, in a THPE, we can find at least one sequence of strategy trembles (totally mixed strategies) that converges to σ_i, and where σ_i is player i's best response to her opponents' strategies at all points of the sequence (for all k). While the first requirement (convergence) is easier to satisfy, the second (σ_i being player i's best response for all k) is more difficult to hold, as Example 5.9 illustrates below.

The above definition has some immediate properties, such as: (1) every THPE must be an NE; (2) every strategic-form game with finite strategies for each player has a THPE; and (3) every THPE assigns zero probability weight on weakly dominated strategies. (Exercise 5.17 asks you to prove these three properties.) Intuitively, points (1) and (2) show that THPEs are a subset of all NEs in a strategic-form game, as they were in Matrix 5.15, or, alternatively:

$$\sigma \text{ is a THPE} \implies \sigma \text{ is a NE}.$$
$$\nLeftarrow$$

Point (3) says that weakly dominated strategies, such as D for player 1 in Matrix 5.15 (and, similarly, r for player 2), cannot be sustained as a THPE.

Example 5.9 Trembling-hand perfect equilibrium. Consider Matrix 5.15 and one of the NEs, (U, l). In this setting, we study the following sequence of totally mixed strategies

$$\sigma_i^k = \left(1 - \frac{\varepsilon_k}{2}, \frac{\varepsilon_k}{2}\right) \text{ for every player } i, \text{ where } \varepsilon_k = \frac{1}{2^k}.$$

As an example, note that when $k = 1$, $\varepsilon_1 = \frac{1}{2}$, and σ_i^k becomes $\sigma_i^1 = \left(\frac{3}{4}, \frac{1}{4}\right)$, indicating that every player i makes mistakes with 1/4 probability. When $k = 2$, however, $\varepsilon_2 = \frac{1}{4}$, and σ_i^k now becomes $\sigma_i^2 = \left(\frac{7}{8}, \frac{1}{8}\right)$, representing that mistakes are less likely (they only occur with 1/8 probability). A similar argument applies when $k = 5$, where $\varepsilon_2 = \frac{1}{32}$, and σ_i^k now becomes $\sigma_i^2 = \left(\frac{63}{64}, \frac{1}{64}\right)$, meaning that mistakes happen with a probability lower than 2%. Generally, as k increases, mistakes become less likely, and the above totally mixed strategy converges to the psNE (U, l); as depicted in Fig. 5.7. Specifically, the limit of σ_i^k is $\lim_{k \to +\infty} \sigma_i^k = (1, 0)$ since $\lim_{k \to +\infty} \varepsilon_k = \lim_{k \to +\infty} \frac{1}{2^k} = 0$, which implies that player 1 (2) chooses U (l, respectively) in pure strategies, as in strategy profile (U, l).

Therefore, an NE (U, l) can be supported as a THPE because: (i) the totally mixed strategy σ_1^k (σ_2^k) converges to U (l) and (ii) U (l) is the best response of player 1 (2) to her rival's totally mixed strategy, σ_2^k (σ_1^k, respectively) for all k. Indeed, as discussed above, when $k = 1$, σ_2^k becomes $\sigma_2^1 = \left(\frac{3}{4}, \frac{1}{4}\right)$, where U is player 1's best response. Similarly, when $k = 2$, σ_2^k becomes $\sigma_2^2 = \left(\frac{7}{8}, \frac{1}{8}\right)$, where U is still player 1's best response. The same argument applies to higher values of k, and to player 2's best response to σ_1^k being l for every k.

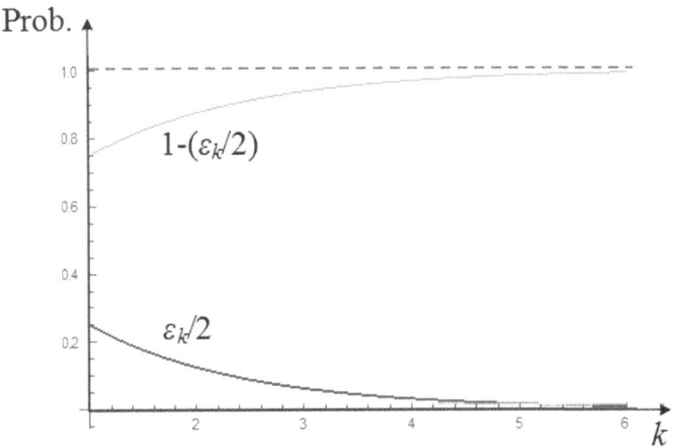

Fig. 5.7 Totally mixed strategy converging to $\sigma_i = (1, 0)$

In contrast, the other NE in Matrix 5.15, (D, r), cannot be sustained as a THPE. Although we can find sequences of totally mixed strategies that converge to (D, r) (first requirement for an NE to be a THPE), choosing D (r) is *not* player 1's (2's) best response to his rival's totally mixed strategy, σ_2^k (σ_1^k), for every k. To see this point, consider the totally mixed strategy

$$\sigma_i^k = \left(\frac{\varepsilon_k}{2}, 1 - \frac{\varepsilon_k}{2}\right) \text{ for every player } i, \text{ where } \varepsilon_k = \frac{1}{2^k}$$

which, intuitively, assigns the opposite probability weights than the totally mixed strategy we used for (U, l) in our above discussion. This totally mixed strategy converges to the psNE (D, r) since $\lim_{k \to +\infty} \sigma_i^k = (0, 1)$ given that $\lim_{k \to +\infty} \varepsilon_k = 0$, yielding strategy profile (D, r). However, U instead of D is player 1's best response to σ_2^k for every k. To understand this point, consider $k = 1$, where σ_2^k becomes $\sigma_2^1 = \left(\frac{1}{4}, \frac{3}{4}\right)$, and U is player 1's best response. Likewise, when $k = 2$, σ_2^k becomes $\sigma_2^2 = \left(\frac{1}{8}, \frac{7}{8}\right)$, where U is still player 1's best response; and similarly for higher values of k. The same argument applies to player 2's best response to player 1's totally mixed strategy, σ_1^k, which is l (not r) for every k. We can extend a similar argument to every totally mixed strategy that satisfies $\lim_{k \to +\infty} \sigma_i^k = (0, 1)$, i.e., converging to (D, r), since D (r) is *not* player 1's (2's) best response to her rival's totally mixed strategy, σ_2^k (σ_1^k, respectively), for every k.□

5.10.2 Proper Equilibrium

While THPE helps us rule out NEs that are not robust to strategic mistakes, this refinement criterion is often criticized as allowing for any trembles (totally mixed strategies), not imposing any restriction on the probability weight that strategies with high or low payoffs should receive. Myerson (1978) suggested, however, that a rational player, while playing a totally mixed strategy, should put a higher (lower) probability weight on strategies yielding high (low) payoffs. Alternatively, players are less likely to make costly trembles (mistakes) than less costly ones, as we define below.

Definition 5.10. ε**-proper equilibrium.** For any $\varepsilon > 0$, a totally mixed strategy profile $\sigma = (\sigma_i, \sigma_{-i})$ is an "ε-proper equilibrium" if, for every player i, and for every two pure strategies s_i and $s_i' \neq s_i$ such that

$$u_i(s_i, \sigma_{-i}) > u_i(s_i', \sigma_{-i}),$$

and the probabilities of playing s_i and s_i', $\sigma_i(s_i)$ and $\sigma_i(s_i')$, satisfy $\varepsilon \times \sigma_i(s_i) \geq \sigma_i(s_i')$.

Intuitively, if player i's expected payoff from choosing s_i is higher than that from s_i', then the probability of playing s_i must be at least ε times higher than the probability of playing s_i'.

5.10 Equilibrium Refinements in Strategic-Form Games (Technical)

Example 5.10 ε-**proper equilibrium.** Consider the game in Matrix 5.15 and the following mixed strategy

$$\sigma_i = \left(1 - \frac{\varepsilon}{a}, \frac{\varepsilon}{a}\right) \text{ for every player } i, \text{ where } a \geq 2 \text{ and } 0 < \varepsilon < 1.$$

This mixed strategy is an ε-proper equilibrium because: (1) it is a totally mixed strategy, assigning a positive probability weight to all players' strategies and (2) for pure strategies U and D, their expected utilities satisfy[15]

$$u_1(U, \sigma_2) = \underbrace{1\left(1 - \frac{\varepsilon}{a}\right)}_{\text{Player 2 chooses } l} + \underbrace{0\frac{\varepsilon}{a}}_{\text{Player 2 chooses } r} = 1 - \frac{\varepsilon}{a} > 0 = u_1(D, \sigma_2),$$

and the probabilities of playing U and D are

$$\varepsilon \times \sigma_1(U) = \varepsilon\left(1 - \frac{\varepsilon}{a}\right) = \frac{\varepsilon(a - \varepsilon)}{a} \text{ and}$$

$$\sigma_1(D) = \frac{\varepsilon}{a}$$

which satisfy

$$\varepsilon \times \sigma_1(U) = \frac{\varepsilon(a - \varepsilon)}{a} \geq \frac{\varepsilon}{a} = \sigma_1(D)$$

since, after rearranging, this inequality simplifies to $a \geq \varepsilon$, which holds given that $a \geq 2$ and $0 < \varepsilon < 1$ by assumption. (Since the game is symmetric, a similar argument applies to player 2's utility from choosing l and r, and its associated probabilities.)□

We can now use this notion of ε-proper equilibrium to define a proper equilibrium.

Definition 5.11. Proper equilibrium. A mixed strategy profile $\sigma = (\sigma_i, \sigma_{-i})$ is a proper equilibrium if there exists: (1) a sequence $\{\sigma_i^k\}_{k=1}^{\infty}$ that converges to σ_i for every player i; (2) a sequence $\{\varepsilon^k\}_{k=1}^{\infty}$, where $\varepsilon^k > 0$ for all k, that converges to zero; and (3) $\{\sigma_i^k\}_{k=1}^{\infty}$ is an ε_k-proper equilibrium for every k.

As you may suspect, proper equilibria are also THPE, but the converse is not necessarily true, that is,

$$\not\Leftarrow \sigma \text{ is a Proper Equilibrium} \quad \Longrightarrow \quad \sigma \text{ is a THPE}.$$

---

[15] Conditional on player 1 choosing U, at the top row of Matrix 5.14, if player 2 chooses l, player 1's payoff is 1, and if player 2 chooses r, player 1's payoff is zero. Conditional on player 1 choosing D, at the bottom row of the matrix, her payoff is zero regardless of whether player 2 chooses l or r.

Intuitively, if σ is a proper equilibrium, it must be robust to a sequence of decreasing trembles where costly mistakes are less likely to occur; while σ being THPE only requires that it is robust to any sequence of decreasing trembles. In the context of Example 5.10, NE strategy profile (U, l), which is THPE, may or may not be proper. (Example 5.11 examines whether or not (U, l) is a proper equilibrium.) However, NE strategy profile (D, r), which cannot be sustained as THPE, cannot be proper either.

Example 5.11 Proper equilibrium. Consider Matrix 5.15 again. The sequence of totally mixed strategies from Example 5.9,

$$\sigma_i^k = \left(1 - \frac{\varepsilon_k}{2}, \frac{\varepsilon_k}{2}\right) \text{ for every player } i, \text{ where } \varepsilon_k = \frac{1}{2^k},$$

is a proper equilibrium since it satisfies the three requirements in the above definition: (1) σ_i^k converges to (U, l); (2) ε^k converges to zero; and (3) σ_i^k is an ε_k-proper equilibrium for every k, as shown in Example 5.10.

Appendix—NE Existence Theorem (Technical)

This appendix provides a short proof of the existence of an NE, either in pure or mixed strategies, using Kakutani's fixed-point theorem. Therefore, we start by introducing the concept of fixed point and two standard fixed-point theorems in the literature.

Fixed-Point Theorems, an Introduction

Consider a function $f : X \to X$, mapping elements from X into X, where $X \subset \mathbb{R}^N$. We, then, say that a point $x \in X$ is a "fixed point" if $x \in f(x)$. For instance, if $X \subset \mathbb{R}$, we can define the distance function $g(x) = f(x) - x$, which graphically measures the distance from $f(x)$ to the 45-degree line, as illustrated in Fig. 5.8a and b. At points such as x' where $g(x') > 0$, we have that $f(x') > x'$, meaning that $f(x')$ lies above the 45-degree line. In contrast, at points such as $x'' > x'$ where $g(x'') < 0$, we have that $f(x'') < x''$, entailing that $f(x'')$ lies below the 45-degree line.

Since $x'' > x'$, if distance function $g(x)$ is continuous, we can invoke the intermediate value theorem to say that there must be an intermediate value, $\hat{x}$, between x' and x'' (or more than one) where $g(\hat{x}) = 0$, implying that $g(\hat{x}) = f(\hat{x}) - \hat{x} = 0$, which implies that $f(\hat{x}) = \hat{x}$, as required for a fixed point to exist.[16] (Note that

---

[16] As a reference, recall the intermediate value theorem: let f be a continuous function defined in the interval $[a, b]$, and let c be a number such that $f(a) < c < f(b)$. Then, there exists some

Appendix—NE Existence Theorem (Technical)

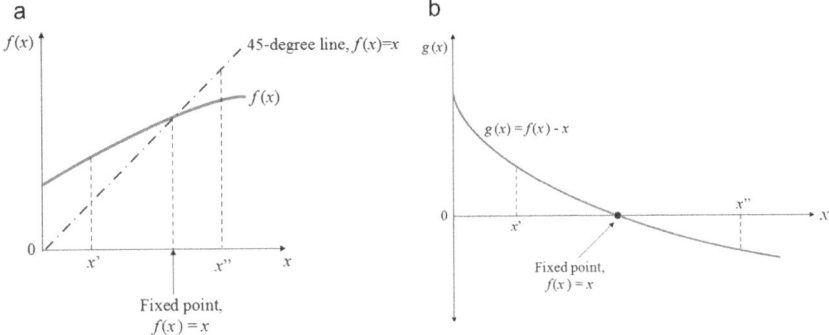

Fig. 5.8 (a) Function f(x) and the 45-degree line (b)

if $f(x)$ was not continuous, then $g(x)$ would not be continuous either, allowing for $g(x') > 0$ and $g(x'') < 0$ to occur, yet we could not guarantee the existence of an intermediate point between x' and x'' for which $g(x) = 0$.) The Brouwer's fixed-point theorem formalizes this result.

Definition 5.12. **Brouwer's fixed-point theorem.** If $f : X \to X$ is a continuous function, where $X \subset \mathbb{R}^N$, then it has at least one fixed point, that is, a point $x \in X$ where $f(x) = x$.

While Brouwer's fixed-point theorem is useful when dealing with best response functions, it does not apply to best response correspondences, where player i is, for instance, indifferent between two or more of her pure strategies when her opponent chooses strategy s_j. The following theorem generalizes Brouwer's fixed-point theorem to correspondences. For a more detailed presentation on fixed-point theorems, see Border (1985).

Definition 5.13. **Kakutani's fixed-point theorem.** A correspondence $F : X \to X$, where $X \subset \mathbb{R}^N$, has a fixed point $x \in X$ where $F(x) = x$, if these conditions hold:

1. X is a compact, convex, and non-empty set.
2. $F(x)$ is non-empty.
3. $F(x)$ is convex.
4. $F(x)$ has a closed graph.

Nash's (1950) existence theorem is a direct application of Kakutani's fixed point theorem, as the next section describes.

$x \in [a, b]$ such that $f(x) = s$. As a corollary of this theorem, if $f(a) > 0$ and $f(b) < 0$, there must exist some $x \in [a, b]$ such that $f(x) = 0$ (that is, there must be at least one root of function f).

Nash Existence Theorem

First, define player i's pure-strategy set, S_i, to be finite, i.e., a discrete list of pure strategies, and denote a mixed strategy for this player as σ_i, where $\sigma_i \in \sum_i$, meaning that player i chooses her randomization among all possible mixed strategies available to her. Therefore, the Cartesian product

$$\sum_i \times \sum_{-i} = \sum$$

denotes the set of all possible mixed strategy profiles in the game, so that every strategy profile $\sigma = (\sigma_i, \sigma_{-i})$ satisfies $\sigma \in \sum$. Second, let us define player i's best response correspondence to her rivals choosing σ_{-i} as $\sigma_i \in BR_i(\sigma_{-i})$.

We now define the joint best response correspondence $BR(\sigma)$, as the product of $BR_i(\sigma_{-i})$ and $BR_{-i}(\sigma_i)$, that is,

$$BR(\sigma) \equiv BR_i(\sigma_{-i}) \times BR_{-i}(\sigma_i).$$

Importantly, if BR has a fixed point, then, an NE exists. Therefore, we next check if BR satisfies the four conditions on Kakutani's fixed-point theorem, as that would guarantee the existence of an NE. Before doing that, we identify X in Kakutani's fixed-point theorem with the set of all possible mixed strategy profiles, $\sum$; and correspondence F with BR.

1. $\sum$ *is a non-empty, compact, and convex set.*
   (a) The set $\sum$ is non-empty as long as players have some strategies, so we can identify pure or mixed strategy profiles.
   (b) Recall that if a set is closed and bounded it is compact. The set of all possible mixed strategy profiles is closed and bounded, thus satisfying compactness.
   (c) Convexity is satisfied since, for any two strategy profiles, σ and σ', their linear combination, $\lambda\sigma + (1-\lambda)\sigma'$ where $\lambda \in [0, 1]$, is also a mixed strategy profile, thus being part of $\sum$.
2. $BR(\sigma)$ *is non-empty.* Since every player i's payoff, $u_i(\sigma_i, \sigma_{-i})$, is linear in both σ_i and σ_{-i}, she must find a maximum (a best response to her rivals choosing σ_{-i}) among her available strategies, $\sum_i$, which we know it is a compact set from point 1b. Because $\sigma_i \in BR_i(\sigma_{-i})$ and $\sigma_{-i} \in BR_i(\sigma_i)$ are non-empty (a best response exists), then their product, $BR(\sigma)$, must also be non-empty.
3. $BR(\sigma)$ *is convex.* To prove this point, consider two strategies for player i, σ_i and σ'_i, that are best responses to her rivals choosing σ_{-i}, that is, $\sigma_i, \sigma'_i \in BR_i(\sigma_{-i})$. Because both σ_i and σ'_i are best responses, they must both yield the same expected payoff; otherwise, one of them cannot be a best response. Therefore, a linear combination of σ_i and σ'_i, $\lambda\sigma_i + (1-\lambda)\sigma'_i$ where $\lambda \in [0, 1]$, must yield the same expected payoff as σ_i and σ'_i, thus being a best response as well, that is, $\lambda\sigma_i + (1-\lambda)\sigma'_i \in BR_i(\sigma_{-i})$.

Alternatively, this result can be proven by contradiction. Consider strategies σ_i and σ'_i, both being best responses to σ_{-i}, that is, $\sigma_i, \sigma'_i \in BR_i(\sigma_{-i})$; but suppose that strategy $\sigma''_i = \lambda \sigma_i + (1 - \lambda)\sigma'_i$ is not a best response. In particular, assume that there is a value of $\lambda \in [0, 1]$ for which strategy σ''_i yields a strictly higher payoff than σ_i and σ'_i. Then, σ''_i must be a best response to σ_{-i}, i.e., $\sigma''_i \in BR_i(\sigma_{-i})$. This result, however, contradicts our premise that strategies σ_i and σ'_i are best responses to σ_{-i}. Therefore, if strategies σ_i and σ'_i are best responses to σ_{-i}, their linear combination $\sigma''_i = \lambda \sigma_i + (1 - \lambda)\sigma'_i$ is a best response to σ_{-i} as well.

4. $BR(\sigma)$ *has a closed graph.* This property states that the set $\{(\sigma_i, \sigma_{-i}) | \sigma_i \in BR_i(\sigma_{-i})\}$ is "closed," meaning that every player i's best response correspondence has no discontinuities. The best responses depicted in this chapter, for instance, showed no discontinuities. Because every player i's payoff, $u_i(\sigma_i, \sigma_{-i})$, is continuous and compact, the set $\{(\sigma_i, \sigma_{-i}) | \sigma_i \in BR_i(\sigma_{-i})\}$ is closed.

The previous result guarantees that an NE exists when players face finite strategy spaces (i.e., a list of pure strategies). What if players choose from a continuous strategy space, as when firms set their prices or output levels? Glicksberg (1952) extended the above result to settings with continuous strategy spaces, where $S_i \subset \mathbb{R}^N$, showing that, if every player i's strategy space, S_i, is compact and her utility function, $u_i(\cdot)$, is continuous, an NE exists, in pure or mixed strategies. (For a generalization of this result to non-continuous utility functions, see Dasgupta and Maskin (1986).)

Exercises

5.1 **General Battle of the Sexes game.**[A] Consider the following Battle of the Sexes game where every player simultaneously chooses between going to the football game (F) or the opera (O). Payoffs satisfy $a_i > b_i > 0$ for every player $i = \{H, W\}$, implying that the husband (Wife) prefers that both players are together at the football game (opera house, respectively), and attending any event is preferred to being alone anywhere. However, the premium that player i assigns to his or her most preferred event, as captured by $a_i - b_i$, can be different between husband and wife.

		Wife	
		F	O
Husband	F	a_H, b_W	0, 0
	O	0, 0	b_H, a_W

(a) Find the msNE of this game.

(b) How are players' randomization affected by an increase in payoffs a_H, a_W, b_H, and b_W? Interpret.

(c) How are your results affected if the husband earns a larger payoff from the football game than his wife earns from the opera, $a_H > a_W$? How are your findings affected otherwise?

5.2 **General Pareto coordination game.**[A] Consider the following Pareto coordination game where every firm simultaneously choosing between technology A or B, and payoffs satisfy $a_i > b_i > 0$ for every firm $i = \{1, 2\}$, implying that both firms regard technology A as superior. However, the premium that firm i assigns to technology A, as captured by $a_i - b_i$, can be different between firms 1 and 2.

		Firm 2	
		Tech. A	Tech. B
Firm 1	Tech. A	a_1, a_2	0, 0
	Tech. B	0, 0	b_1, b_2

(a) Find the msNE of this game.

(b) How are players' randomization affected by an increase in payoffs a_1, a_2, b_1, and b_2? Interpret.

(c) How are your results affected if firm 1 earns a larger payoff from technology A than firm 2 does, $a_1 > a_2$? How are your findings affected otherwise?

5.3 **General anticoordination game.**[A] Consider the following Game of Chicken where every player drives toward each other and simultaneously chooses between swerve and stay. Payoffs satisfy $a_i > b_i > 0$ for every player $i = \{1, 2\}$, implying that drivers suffer more when they both stay (and almost die!) than when they both swerve (and are called chickens), and $a_i > d_i > c_i > b_i$, so being the only player who stays yields a payoff of d_i for that player and $-c_j$ for his rival, where $j \neq i$.

		Player 2	
		Swerve	Stay
Player 1	Swerve	$-b_1, -b_2$	$-c_1, d_2$
	Stay	$d_1, -c_2$	$-a_1, -a_2$

(a) Find the msNE of this game.

(b) How are players' randomization affected by an increase in payoffs a_1 through d_1? How are they affected by a marginal increase in a_2 through d_2? Interpret.

5.4 **Asymmetric Matching Pennies game.**[A] Consider the following Matching Pennies game where every player simultaneously chooses Heads (H) or Tails (T) and payoff a satisfies $a > 0$.

	Player 2	
Player 1	H	T
H	$a, 0$	$0, 1$
T	$0, 1$	$1, 0$

(a) Find the msNE of this game.
(b) How are the equilibrium probabilities of each player affected by an increase in payoff a? Interpret your results.

5.5 **Anticoordination in the Stop–Go game.**[A] Consider the following anticoordination game where two drivers meet at a road intersection. Each driver i chooses to stop (waiting for the other driver to continue) or to go (not waiting for the other driver). Parameter c denotes the cost of stopping while the other driver goes, where $1 > c > 0$. If both drivers stop, they both earn 1; if they both go, they crash their cars, earning zero; and if driver i stops while his opponent goes, driver i earns $1 - c$ while his opponent earns 3.

	Driver 2	
Driver 1	Stop	Go
Stop	$1, 1$	$1 - c, 3$
Go	$3, 1 - c$	$0, 0$

(a) Find the best responses of each driver.
(b) Find the psNEs of this game.
(c) Find the msNE of this game.

5.6 **Anticoordination in the job application game.**[A] Consider two college students applying for a job at either of the two companies (A or B). Firm B offers a more attractive salary than firm A, $w_B > w_A$, but both students working for firm A would be better-off than only one of them working for firm B, that is, $2w_A > w_B$. Every student simultaneously and independently applies for a job to one, and only one, company. If only one student applies to a firm, he gets the job. If both apply to the same firm, each of them gets the job with probability of 1/2.

(a) Depict the game's payoff matrix.
(b) Find the best responses of each player.
(c) Find the psNEs of this game.
(d) Find the msNE of this game.

(e) How are your equilibrium results in the msNE affected by a marginal increase in w_A? And by a marginal increase in w_B? Interpret.

5.7 **Finding msNE in the salesman's game.**[B] Consider a salesman who, after returning from a business trip, can be honest about his expenses (H) or lie about them (L) claiming more expenses than he actually had. His boss can either not check the salesman (NC) or check whether his claimed expenses are legit (C). If the salesman is honest, he is reimbursed for his expenses during the trip, entailing a payoff of zero for him, and a payoff of zero for his boss when he does not check on the salesman but $-c$ otherwise (we can interpret c as the cost that the boss incurs when checking). If the salesman lies, he earns $a > 0$ when his boss does not check on him, but when he does, the salesman's payoff is $-b$ from his boss detects his cheating (which occurs with probability p) and a when his boss does not detect his cheating (which happens with probability $1-p$). If the boss checks when the salesman lies, the boss' payoff is $-c + [p0 + (1-p)(-\beta)] = -c - (1-p)\beta$, which embodies his (certain) cost from checking and his expected cost of not detecting the lie, which occurs with probability $1-p$.
(a) Depict the game's payoff matrix.
(b) Find the best responses of each player.
(c) Find the psNEs of this game.
(d) Find the msNE of this game.

5.8 **Quantal Response Equilibrium—An introduction.**[C] Consider again the asymmetric Matching Pennies game. Luce (1959) suggests that every player $i = \{1, 2\}$ chooses Heads based on its "intensity," defined as follows

$$\sigma_{iH} = \frac{u_{iH}}{u_{iH} + u_{iT}}$$

where u_{iH} denotes player i's utility from choosing H, while u_{iT} represents his utility from selecting T. A similar intensity applies to his probability of playing Tails, $\sigma_{iT} = \frac{u_{iT}}{u_{iH} + u_{iT}}$, which satisfies $\sigma_{iT} = 1 - \sigma_{iH}$.
(a) Show that, in the asymmetric matching pennies game, utilities u_{iH} and u_{iT} are themselves functions of probabilities σ_{iH} and σ_{iT}.
(b) Find the equilibrium probabilities σ_{iH}^{QRE} and σ_{iT}^{QRE}, where the superscript QRE denotes "quantal response equilibrium."
(c) How are the equilibrium probabilities of each player affected by an increase in payoff a? Interpret your results.
(d) Compare σ_{iH}^{QRE} and σ_{iT}^{QRE} against the mixed strategy equilibrium, σ_{iH}^* and σ_{iT}^*, found in the previous exercise.
(e) Consider now that the intensity function is

$$\sigma_{iH} = \frac{(u_{iH})^\lambda}{(u_{iH})^\lambda + (u_{iT})^\lambda}$$

where $\lambda \geq 0$ is known as the responsiveness or precision parameter. Find the equilibrium probabilities in this context. Then, evaluate your results at $\lambda = 0$, showing that strategy H is unresponsive to expected payoffs, implying that players mix between H and T with equal probability. Finally, evaluate your results at $\lambda = 4$, and interpret your findings.

5.9 **Stag Hunt game.**[B] Consider the following Stag Hunt Game where two individuals simultaneously choose between hunt a stag or a hare. Payoffs satisfy $a_i > b_i > d_i > c_i$ for every player i, so every player's ranking of payoffs is: (1) hunting a stag with the help of his opponent; (2) hunting a hare with the help of his opponent; (3) hunting a stag without help; and (4) hunting a hare without help.

		Player 2	
		Stag	Hare
Player 1	Stag	a_1, a_2	c_1, d_2
	Hare	d_1, c_2	b_1, b_2

(a) Find the msNE of this game.
(b) How are players' randomization affected by an increase in payoffs a_1 through d_1? Interpret.

5.10 **Finding msNEs in four categories of simultaneous-move games.**[B] Consider the following simultaneous-move game with payoff matrix

		Player 2	
		A	B
Player 1	A	a, a	b, c
	B	c, b	d, d

Find the msNE in the following settings. Interpret and relate your results to common games.
(a) Payoffs satisfy $a > c$ and $b > d$.
(b) Payoffs satisfy $a > c$ and $b < d$.
(c) Payoffs satisfy $a < c$ and $b < d$.
(d) Payoffs satisfy $a < c$ and $b > d$.
(e) In the setting of part (d), discuss how your results are affected if the difference $b - d$ increases. What if the difference $c - a$ increases? Interpret.

5.11 **Mixed strategies do not assign probability to dominated pure strategies.**[B] Consider the following Prisoner's Dilemma game between players 1 and 2.

		Player 2	
		Confess	Not confess
Player 1	Confess	−4, −4	0, −8
	Not confess	−8, 0	−2, −2

Assume that player i randomizes with a mixed strategy σ_i that puts positive probability on Not Confess, despite being strictly dominated.

(a) Find the probability of player j, q, that makes player i indifferent between Confess and Not Confess. Show that probability q would satisfy $q \notin (0, 1)$. Interpret.

(b) Let us now take a more general approach. Consider a two-player game (not necessarily the Prisoner's Dilemma game) where player i randomizes with a mixed strategy σ_i that puts positive probability on a strictly dominated strategy. Show that σ_i is dominated by another mixed strategy σ_i' that puts zero probability weight on a strictly dominated strategy. For generally, let player i have L pure strategies $S_i = \{s_{i1}, ..., s_{iL}\}$ and let s_{ik} be a pure strategy which is strictly dominated by $s_{ik'}$, that is, $u_i(s_{ik'}, s_{-i}) > u_i(s_{ik}, s_{-i})$ for every s_{-i}.

5.12 **Pure and mixed NEs with weakly dominant strategies.**[A] Consider the payoff matrix in Exercise 2.4 again.
(a) Find all pure-strategy Nash equilibrium of this game.
(b) Can you identify a mixed strategy Nash equilibrium? Explain.
(c) Characterize and plot the best response functions you found in part (c). Interpret.

5.13 **Finding msNEs in a general game.**[B] Consider the following two-player game, based on Fudenberg and Tirole (1991, page 480). Answer the following questions allowing for psNEs and msNEs.

		Player 2	
		L	R
Player 1	U	1, 1	a, a
	D	0, 0	b, b

(a) If $0 < a < 1$ and $b \leq 0$, show that the game has one psNE, and find the condition for msNE.
(b) If $a = 0$ and $b < 0$, show that the game has only one NE.
(c) If $a = 0$ and $b > 0$, show that the game has three NEs.

5.14 **Small perturbations and NE.**[B] Consider the following two-player game.

Exercises

		Player 2	
		L	R
Player 1	U	1,1	1,1
	D	1,1	0,0

(a) Find the set of psNE and msNE.
(b) If both players' payoffs in (U, L), at the top left cell of the matrix, increase from 1 to $1 + \varepsilon$, where $\varepsilon > 0$, how are your results in part (a) affected?
(c) How does your answer to part (b) change if, instead, $\varepsilon < 0$?

5.15 **IDSDS, NE, and security strategies.**[B] Consider the following simultaneous-move game where player 1, in rows, chooses A, B, or C; while player 2, in columns, chooses between a and b.

		Player 2	
		a	b
	A	3, −1	2, 1
Player 1	B	−1, 7	1, 3
	C	4, −3	−2, 9

(a) Does player 1 or 2 have a strictly dominated strategy? What about weakly dominated strategies?
(b) Find the strategy profile/s surviving IDSDS. Find the strategy profile/s surviving IDWDS.
(c) Find player 1's best responses to each of player 2's strategies. Then, find player 2's best responses to each of player 1's strategies.
(d) Find the psNEs of this game.
(e) Find the security strategy profile (maxmin) of this game.
(f) Compare your equilibrium results according to IDSDS and IDWDS (from part b), psNE (from part d), and security strategies (from part e). Interpret your results.

5.16 **Trembling-hand perfect equilibrium—Examples.**[B] Consider the following payoff matrix.

		Player 2		
		a	b	c
	A	14, 14	4, 4	2, 2
Player 1	B	4, 4	4, 4	2, 4
	C	2, 2	4, 2	2, 2

(a) Find all pure-strategy NEs of this game.
(b) Show that only one of the NEs found in part (a) can be supported as a THPE.

5.17 **Trembling-hand perfect equilibrium-Proofs.**[C] Show that:
(a) Every THPE must be an NE.
(b) Every strategic-form game with finite strategies for each player has a THPE.
(c) Every THPE assigns zero probability weight on weakly dominated strategies.

5.18 **Proper equilibrium—Example.**[B] Consider the Hawk–Dove game in the matrix below (an example of an anticoordination game), where player 1 chooses between Hawk and Dove in rows (H or D), and player 2 chooses between these two pure strategies in columns (h and d).

		Player 2	
		h	d
Player 1	H	6, 6	0, 12
	D	12, 0	0, 0

(a) Find the pure-strategy NEs in this game.
(b) Check with of the NEs found in part (a) are proper.

5.19 **Finding msNE and security strategies with players have three available actions.**[B] Consider the simultaneous-move game between players 1 and 2 in the following payoff matrix.

		Player 2		
		L	M	R
Player 1	U	4, 0	6, 6	0, 4
	D	2, 6	1, 0	6, 2

(a) Find the NEs involving pure strategies.
(b) Find the msNE of this game.
(c) Find the security strategies of this game.

References

Aumann, R. (1974) "Subjectivity and Correlation in Randomized Strategies," Journal of Mathematical Economics, 1(1), pp. 67–96.

Border, K. C. (1985) *Fixed Point Theorems with Applications to Economics and Game Theory*, Cambridge University Press, Cambridge.

Dasgupta, P., and E. Maskin (1986) "The Existence of Equilibrium in Discontinuous Economic Games, II: Applications," Review of Economic Studies, 46, pp. 27–41.

References

Fudenberg, D., and J. Tirole (1991) *Game Theory*, The MIT Press, Cambridge, Massachussets.

Glicksberg, I. L. (1952) "A Further Generalization of the Kakutani Fixed Point Theorem with Application to Nash Equilibrium Points," Proceedings of the National Academy of Sciences, 38, pp. 170–174.

Harsanyi, J.C. (1973) "Oddness of the Number of Equilibrium Points: A New Proof," International Journal of Game Theory, 2, pp. 235–250.

Luce, R. D. (1959) *Individual Choice Behavior*, John Wiley, New York.

Myerson, R. (1978) "Refinements of the Nash Equilibrium Concept," International Journal of Game Theory, 7, pp. 73–80.

Selten, R. (1975) "Reexamination of the Perfectness Concept for Equilibrium Points in Extensive Games," International Journal of Game Theory, 4, pp. 301–324.

Von Neumann, J. (1928) "Zur Theorie der Gesellschaftsspiele," Mathematische Annalen, 100, pp. 295–320.

Watson, J. (2013) *Strategy: An Introduction to Game Theory*, Worth Publishers. Third edition.

Wilson, R. (1971) "Computing Equilibria in N-person Games," SIAM Journal of Applied Mathematics, 21, pp. 80–87.

Subgame Perfect Equilibrium

6.1 Introduction

In this chapter, we switch our attention to sequential-move games. While we can still use the NE solution concept to predict equilibrium behavior in these games, we show that this solution concept would yield several equilibrium outcomes, thus not being very precise. More importantly, several NEs in this type of game are based on beliefs that cannot be credible in a dynamic setting.

Therefore, we need a more appropriate solution concept for sequential-move games than NE. This solution concept must consider that every player maximizes her payoff given the stage of the game when she is called to move and given the information she observes at that point—which we denote as players being "sequentially rational." If every player is sequentially rational, we can be rest assured that, when called to move at a node (or information set), she chooses the action that maximizes her payoff, yielding a Subgame Perfect Equilibrium (SPE) of the game.

We find the SPE in different game trees with discrete strategy spaces (e.g., high or low prices). We study settings where every player observes the actions of her rivals in previous stages and, then, we examine contexts where at least one player does not observe the choices of her rival. Finally, we find SPE in sequential-move games where players face continuous strategy spaces, such as the Stackelberg game of sequential output competition or the sequential-move version of the public good game discussed in Chapter 4.

Before we start our journey into sequential-move games, we must present some properties that game trees need to satisfy, which we informally denote as "tree rules."

6.2 Tree rules

To facilitate our analysis of equilibrium behavior in sequential-move games, and following Watson (2013), we consider the following properties in all game trees.

1. *Every node is the successor of the initial node.* Figure 6.1 illustrates this property: the game tree in the left panel satisfies it since there is only one initial node, while the game tree in the right panel violates it. If a modeler considers two different players acting simultaneously at the beginning of the game, then the game tree should have a single initial node where, after player 1 selects B or C, player 2 chooses between c or d without observing player 1's choice (as they are simultaneous).

   If, instead, the modeler considers the same player in two initial nodes (e.g., player 1), there must be a mistake in the game tree (right panel of Fig. 6.1), because player 1 cannot simultaneously choose between A and B (on the top of the figure) and between C and D (on the bottom), as if she had multiple personalities! In this case, one of player 1's decisions should happen at a subsequent stage.

2. *Every node has only one immediate predecessor, and the initial node has no predecessors.* The left panel in Fig. 6.1 satisfies this property, since no nodes happen before the initial node, whereas Fig. 6.2 depicts a game tree that violates it. If we allowed for a node to have more than one predecessors, we could run into misunderstandings, where we do not know if, for instance, player 4 is

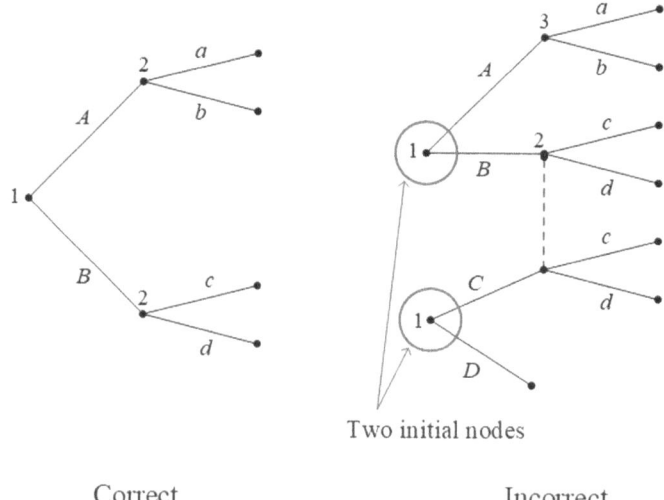

Fig. 6.1 Games with one or two initial nodes

6.2 Tree rules

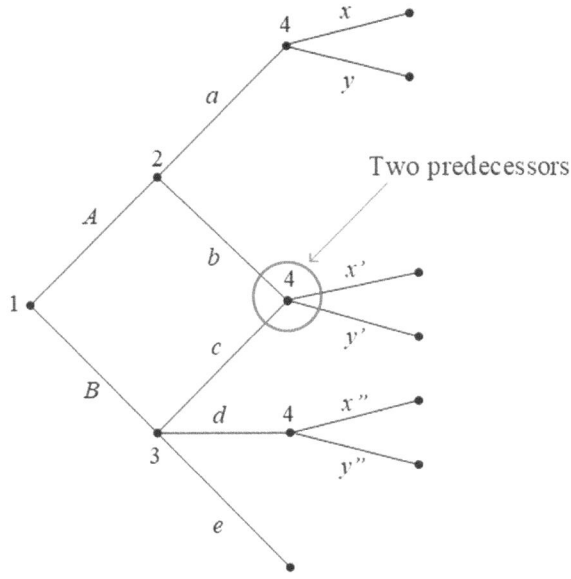

Fig. 6.2 Game with two predecessors

called to move after player 2 chose *b* (at the top of the figure) or after player 3 chose *c* (at the bottom of the figure).

3. *Multiple branches extending from the same node have different action labels.* The left panel in Fig. 6.3 satisfies this property while the right panel violates it because label *A* is on two branches for player 1. If the modeler seeks to represent that player 1 chooses *A* in both the top and middle branches that stem from the initial node, then these two branches should be collapsed into a single branch. If, instead, the modeler tries to represent that player 1 has two different actions in each of these branches, then they should be labeled differently to avoid misunderstandings.

4. *Each information set contains nodes for only one player.* The left panel in Fig. 6.4 satisfies this property but the right panel violates it. Intuitively, if player 2 is called to move at the top node of the game tree in the right panel, she can infer that player 1 selected *A*, leaving her no uncertainty about player 1's choices in the previous stage. A similar argument applies to player 3: if she is called to move at the bottom node, she knows that player 1 selected *B*. Therefore, the information set connecting the top and bottom nodes is incorrect because players 2 and 3, if called to move, know which strategy player 1 chose in the previous stage.

5. *Same number of branches and labels.* All nodes in a given information set have the same number of branches stemming from them (i.e., the same number of immediate successors). In addition, these branches must coincide across all nodes connected by the same information set. Figure 6.5a depicts a game tree that violates the first requirement because only two branches stem from the top node while three branches stem from the bottom node. If the modeler seeks to

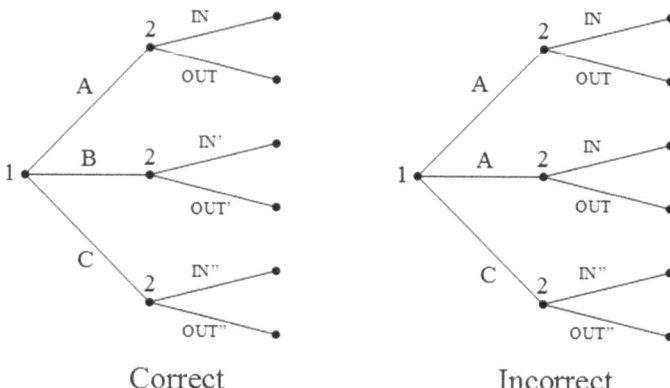

Fig. 6.3 Games with same/different action labels

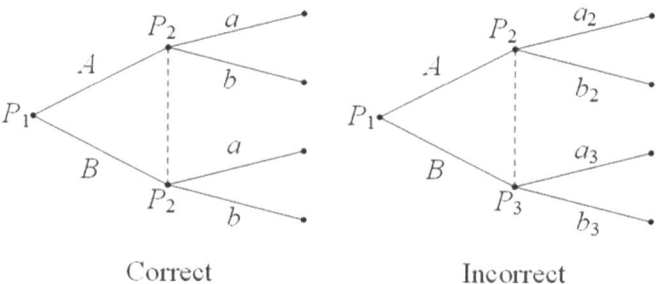

Fig. 6.4 Information sets with one or more players

represent a game where player 2 has only two available strategies after player 1 chooses A but three available strategies when player 1 selects B, then we do not need to draw an information set connecting the top and bottom nodes. Indeed, player 2 would be able to infer whether player 1 chose Invest or Not invest by just observing the number of available strategies: player 1 must have chosen Invest if player 2 has two available strategies, but she must have selected Not invest if player 2 has three available strategies.

Figure 6.5b illustrates a game tree which violates the second requirement: every node connected by the same information set has the same number of branches (i.e., two branches stem from each node), but their action labels do not coincide. By a similar argument as in Fig. 6.5a, the information set would be unnecessary in this setting, as player 2 would be able to infer that player 1 must have chosen Invest if she must select between A and B, but must have chosen Not invest if she must select between A and C.

Fig. 6.5 a. Different number of immediate successors. **b.** Different immediate successors

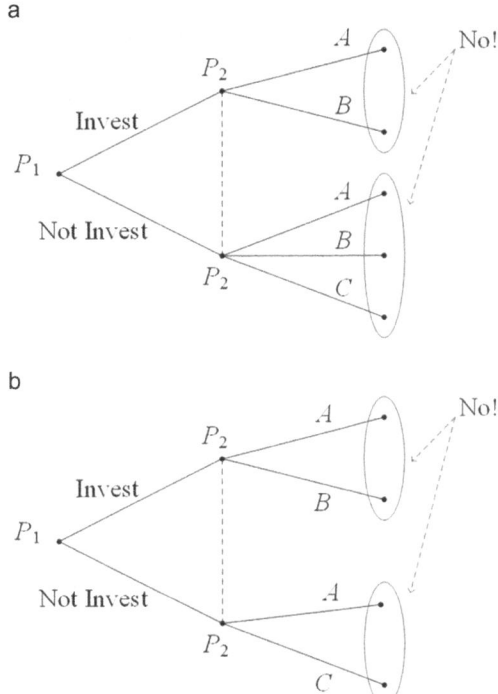

6.2.1 Actions vs. Strategies

From our above discussion, you may have noticed we distinguish between a player's *action* at any node (or information set) of the game tree where she is called to move and a player's *strategy*, which describes the specific list of actions that she chooses along the game tree. Graphically, an action is just a specific branch that player i chooses when it is her turn to move, whereas a strategy is the list of all the branches that she would choose along the game tree, both in nodes that are reached in equilibrium and those that are not reached. In other words, a pure strategy identifies a complete contingent plan, as we formally define next.

An action for player i is often denoted as a_i, or $s_i(h_i)$ to emphasize that it is one action among those available in information set h_i, $A_i(h_i)$. (Recall that information set h_i can be a singleton, thus including a single node, if player i perfectly observes previous players' actions, but $A_i(h_i)$ must include at least two available actions to choose from.) In addition, let H_i be the set of all information sets in a game tree, so a specific information set h_i satisfies $h_i \in H_i$.

Definition 6.1. **Pure strategy.** In a sequential-move game, a pure strategy for player i is a mapping

$$s_i : H_i \to A_i$$

that assigns an action $s_i(h_i) \in A_i(h_i)$ at information set $h_i \in H_i$, where $A_i(h_i)$ denotes the set of available actions at information set h_i.

Importantly, the above definition applies to *every* information set, $h_i \in H_i$, describing how player i behaves once she reaches information set h_i. In contrast, an action describes how player i behaves in a specific information set h_i. In Fig. 6.3, for instance, *In* denotes player 2's action at the top node of the game tree, but (In, Out', In'') represents her strategy (which indicates she responds with *In* after player 1 chooses A, with Out' after player 1 selects B, and with In'' after player 1 chooses C).

Our above discussion ignored the possibility that player i randomizes her actions, or her strategies, but we return to this possibility in the appendix at the end of the chapter.

6.3 Why Don't We Just Find the Nash Equilibrium of the Game Tree?

Consider the entry game in Fig. 6.6, where a firm (potential entrant) chooses whether to enter into an industry where an incumbent firm operates as a monopolist or stay out of the market. If the potential entrant stays out (at the bottom of the figure), the incumbent remains a monopolist, earning $10 while the potential entrant earns zero. However, if the entrant joins the industry, the incumbent observes this decision and responds either by accommodating the entry (e.g., setting moderate prices) which yields a payoff of $4 for each firm, or fighting it (e.g., starting a price war against the entrant) leading to a payoff of −$2 for each firm. (All payoffs are in millions of dollars.)

To find the NEs in this game tree, we first represent the game in its matrix form (see Matrix 6.1). The potential entrant has only two available strategies, *In*

Fig. 6.6 The entry game

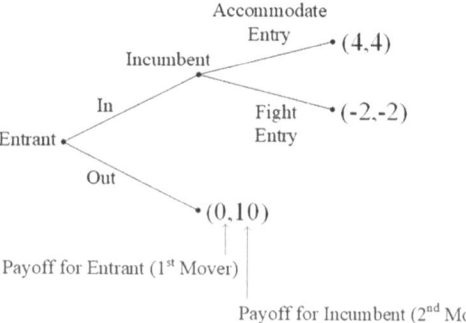

6.3 Why Don't We Just Find the Nash Equilibrium of the Game Tree?

and *Out*, and thus the matrix has two columns. Similarly, the incumbent has two strategies at its disposal: *Accommodate* or *Fight Entry*.

		Potential entrant	
		In	Out
Incumbent	Accommodate	4, 4	10, 0
	Fight entry	−2, −2	10, 0

Matrix 6.1. The Entry game in matrix form.

Matrix 6.1a underlines best response payoffs, as we did in Chapter 4, to find the NEs of the game. Specifically, for the incumbent, we find that its best response to *In* (in the left column of Matrix 6.1a) is $BR_{inc}(In) = Acc$ because $4 > -2$, while its best response to *Out* is $BR_{inc}(Out) = \{Acc, Fight\}$ because both yield a profit of 10.

		Potential entrant	
		In	Out
Incumbent	Accommodate	<u>4</u>,<u>4</u>	<u>10</u>, 0
	Fight entry	−2, −2	<u>10</u>,<u>0</u>

Matrix 6.1a. Finding NEs in the Entry game.

For the potential entrant (the column player), we find that its best responses are $BR_{ent}(Acc) = In$, because $4 > 0$, and $BR_{ent}(Fight) = Out$, because $0 > -2$. Intuitively, if the entrant believes that the incumbent will choose to accommodate after its entry, then it should enter, obtaining a profit of 4, rather than staying out, with a payoff of 0. In contrast, if the entrant believes that the incumbent will respond by fighting (starting a price war), it should remain outside the industry. In summary, we found two NEs in this game

$$NE = \{(Acc, In), (Fight, Out)\}.$$

Therefore, in the first NE, entry occurs and the incumbent responds by accommodating, whereas in the second NE, the entrant does not enter because it believes that the incumbent will start a price war. But, is this belief credible? Of course not! Once entry has already occurred, the incumbent takes entry as given and responds by accommodating entry, as that strategy yields a larger payoff ($4) than starting a price war (−$2). Therefore, we can say that the entrant's belief that a price war will ensue upon entry is not sequentially rational, because the incumbent would never have incentives to start a price war after entry has occurred. Alternatively, if the incumbent promises to start a price war upon entry, the entrant can anticipate that this is just an empty threat or, informally, that the incumbent is bluffing.

In contrast, the first NE, where the entrant joins the industry and the incumbent responds by accommodating is sequentially rational: upon entry, we know that the incumbent's best response is to accommodate. Anticipating this accommodation, the potential entrant compares its payoff from staying out (zero) and from joining the industry and being accommodated ($4), choosing to join the market.

6.4 Subgames

At this point, you may wonder:

> Ok, I see that in the Entry game we find NEs that are sequentially irrational, but how can we find only those equilibria that are sequentially rational?

We can guarantee sequential rationality by using a new solution concept, Subgame Perfect Equilibrium (SPE). Before presenting this solution concept, however, we must define subsets of the entire game tree, called "subgames," where we will require players to behave optimally.

Definition 6.2. **Subgame.** A subgame is a tree structure defined by a node and all its successors.

This definition means that if nodes a and b are connected with an information set (so player i does not know whether she is at node a or b), both nodes must be part of the same subgame. Graphically, if we draw a rectangle (or a circle) around a part of the game tree to identify one of its subgames, the circle *cannot break any players' information set*. Figure 6.7a depicts the Entry game, identifying only two subgames: (1) that initiated when the incumbent is called on to move; and (2) the game as a whole.

A similar argument applies to the game in Fig. 6.7b, where we can identify 4 subgames: (i) one initiated after player 2 is called to move after player 1 chooses *Up*; (ii) another when player 2 is called to move after player 1 chooses *Down*; (iii) another after player 2 responds to *Down* with *C*, where player 3 is called to move; and (iv) the game as a whole.

6.4.1 What If the Game Tree Has Information Sets?

The game trees in Fig. 6.7a and b have no information sets, so how does our search for subgames change if the game tree has at least one information set? Figure 6.8 depicts such a game tree, where the smallest subgame must include player 2's information set (otherwise we would be breaking it!). The presence of information sets, therefore, reduces the number of subgames we can identify in a game tree.

6.4 Subgames

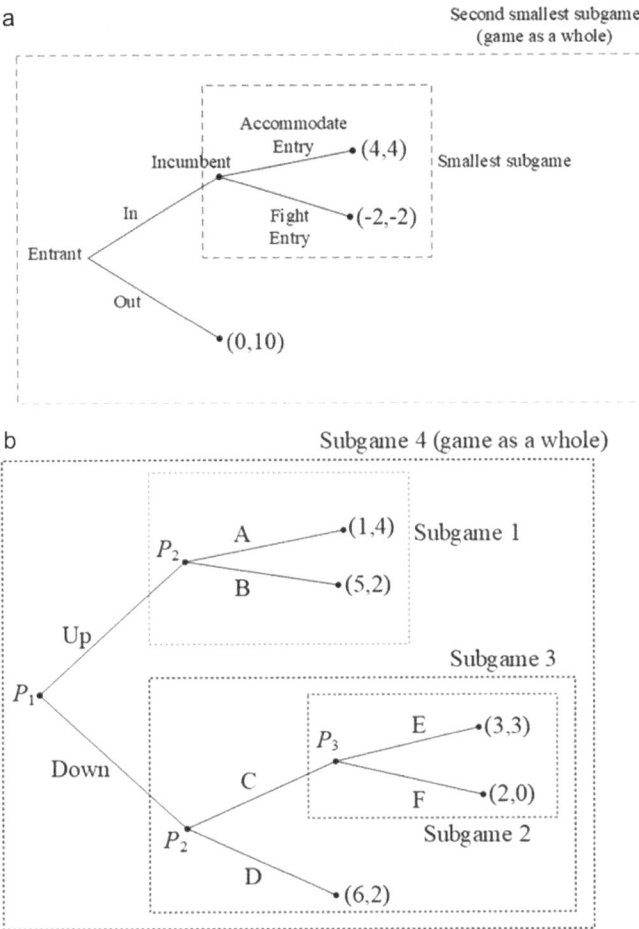

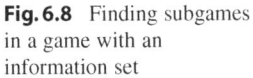

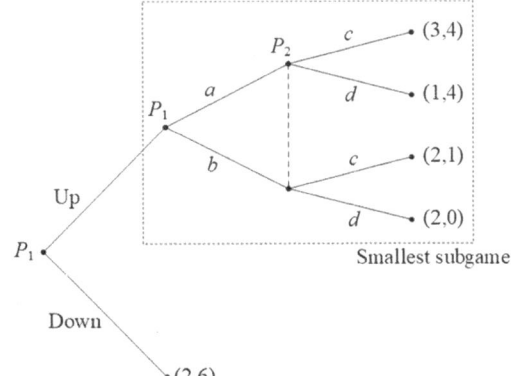

Fig. 6.7 a. Subgames in the entry game. b. Four subgames in a game tree

Fig. 6.8 Finding subgames in a game with an information set

Intuitively, a subgame captures the information that the player called to move at that point of the game tree observes. In the Entry game, the incumbent observes whether the potential entrant joined the industry or not. However, in the game tree of Fig. 6.8, player 2 does not observe whether player 1 selected a or b. This entails that player 2, when choosing whether to respond with c or d, operates as if player 1 was selecting a or b at the same time, being equivalent to a setting where players *interact in a simultaneous-move game*.

This is an important point when solving games like that in Fig. 6.8, since we can represent the subgame where players 1 and 2 interact as a simultaneous-move game (a 2×2 matrix), and find the NE in that matrix by underlining best response payoffs. We can, then, move to the previous stage of the game and find which action the player called to move in that stage chooses, anticipating her equilibrium payoff in the subsequent subgames.

6.5 Subgame Perfect Equilibrium

We are now ready to use the definition of subgame, as a part of the game tree, to characterize a new solution concept in sequential-move games.

Definition 6.3. **Subgame perfect equilibrium (SPE).** A strategy profile $\left(s_i^*, s_{-i}^*\right)$ is a Subgame Perfect Equilibrium if it specifies an NE in each subgame.

To find SPEs in a sequential-move game, we apply the notion of "backward induction," described in Tool 6.1.

Tool 6.1. Applying backward induction

1. Go to the farthest right side of the game tree (where the game ends), and focus on the last mover.
2. Find the strategy that yields the highest payoff for the last mover.
3. Shade the branch that yields the highest payoff for the last mover.
4. Go to the next-to-last mover and, following the response of the last mover that you found in step 3, find the strategy that maximizes her payoff.
5. Shade the branch that yields the highest payoff for the next-to-last mover.
6. Repeat steps 4–5 for the player acting before the previous-to-the-last mover, and then for each player acting before her, until you reach the first mover at the root of the game.

Intuitively, we start from the terminal nodes at the end of the game tree, finding optimal actions for the player called to move in the last subgame of the game tree. Then, we move to the second-to-last mover who, anticipating equilibrium behavior in the last subgames, chooses her optimal actions. We can repeat this process by moving one step closer to the initial node, successively finding equilibrium

6.5 Subgame Perfect Equilibrium

behavior in subgames as each player can anticipate how players behave in the subgames that unfold in subsequent stages.

The equilibrium that results from applying backward induction is also referred to as "rollback equilibrium" because the backward induction procedure looks like rolling back the game tree from its branches on the right side to the game's root on the left side. Backward induction then helps us find the equilibrium strategy of each player when she is called on to move at any point along the tree (the NE in each subgame), yielding the SPE of the game.

In the next subsections, we apply backward induction to solve different types of game trees, first, in relatively straightforward games where no player faces an information set (i.e., every player observes the actions that other players chose in previous stages) and, second, in more involved game trees where at least one player faces an information set.

6.5.1 Finding SPEs in Games Without Information Sets

Example 6.2 applies backward induction to the Entry game.

Example 6.2. Applying backward induction in the Entry game Consider the Entry game again. As discussed in the previous section, there are only two subgames: (1) that initiated after the potential entrant joins the industry; and (2) the game as a whole.

Applying backward induction, we first focus on the last mover, the incumbent. Comparing its payoff from accommodating entry, 4, and starting a price war, -2, we find that its best response to entry is to accommodate. We shade the branch corresponding to *Accommodate* in Fig. 6.9 to keep in mind the optimal response of the incumbent in this subgame.

We now move to the first mover, the entrant, who anticipates that, if it enters, the incumbent will respond by accommodating. Graphically, this means that the entrant expects that, upon entry, the game will proceed through the shaded branch of accommodation in Fig. 6.9, ultimately yielding a payoff of 4 from entering. If,

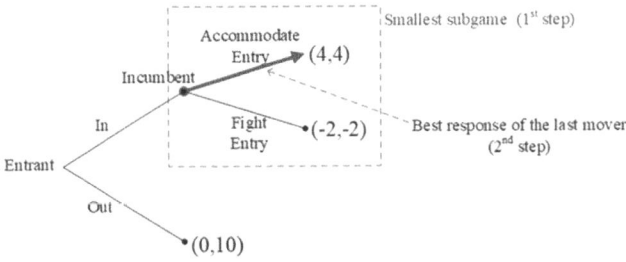

Fig. 6.9 Applying backward induction in the entry game—last mover

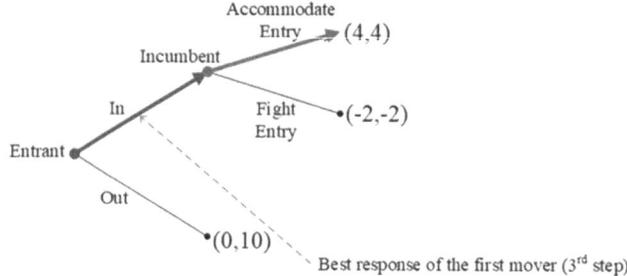

Fig. 6.10 Applying backward induction in the entry game—first mover

instead, the entrant stays out, its payoff is only 0. As a consequence, the optimal strategy for the entrant is to enter. We then conclude that the SPE after applying backward induction is

$$\text{SPE} = \{(In, \ Acc)\},$$

which indicates that the first mover (entrant) chooses to enter, and the second mover (incumbent) responds by accommodating, entailing equilibrium payoffs of (4, 4).

Figure 6.10 shades the branches that players choose in equilibrium, known as the "equilibrium path," starting at the initial node where the entrant chooses *In*, and proceeding to the second stage where the incumbent responds with *Acc*. The equilibrium path of play is a visual tool to understand how players behave in equilibrium, from the initial node to one of the terminal nodes in the tree, but does not coincide with the SPE in more involved games, as we show in Example 6.3 below. □

Recall that $(In, \ Acc)$ was also one of the NEs in the Entry game, as shown in example 6.1. Then examples 6.1 and 6.2 illustrate that every SPE must be an NE, since (In, Acc) is both an SPE and one of the NEs in the Entry game, but the converse is not necessarily true. More formally, for a strategy profile s^*,

$$s^* \text{ is a SPE} \implies s^* \text{ is a NE}$$
$$\Longleftarrow\!\!\!\!/$$

Alternatively, the set of strategy profiles that can be supported as SPEs of a game is a subset of those strategy profiles that can be sustained as NE, as depicted in Fig. 6.11.

Example 6.3. Applying backward induction in the Modified Entry game Consider, for extra practice, the modified version of the Entry game depicted in Fig. 6.12. The top part of the game tree coincides with that in the original Entry game. However, if the entrant chooses to stay out of the industry, the incumbent can

6.5 Subgame Perfect Equilibrium

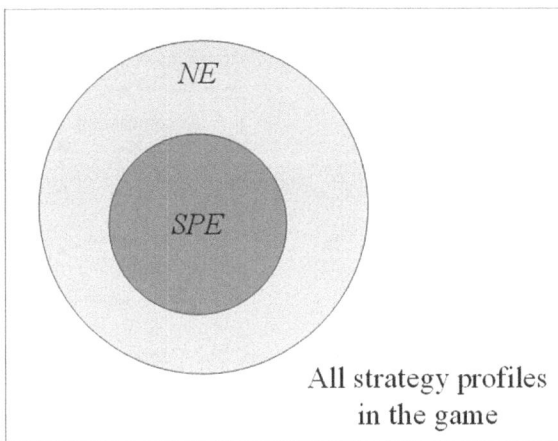

Fig. 6.11 SPEs as a subset of NEs

now respond by investing in a new technology or not, with associated payoffs (0, 12) and (0, 10), respectively.

Figure 6.13 shows that there are three subgames: (1) one initiated after the entrant joins the industry; (2) another initiated after the entrant remains out; and (3) the game as a whole.

Solving subgame 1, we find that the incumbent's best response to entry is still to accommodate, as in the original Entry game since the incumbent's payoffs in this subgame have not changed. Solving subgame 2, we obtain that the incumbent's best response upon no entry is to invest given that $12 > 10$. Anticipating *Acc* upon entry and *Invest* upon no entry, the entrant can, in the initial node, compare its payoff from entry, 4, and from no entry 0, thus choosing to enter. The SPE is, therefore,

$$SPE = \{(In, Acc/Invest)\}$$

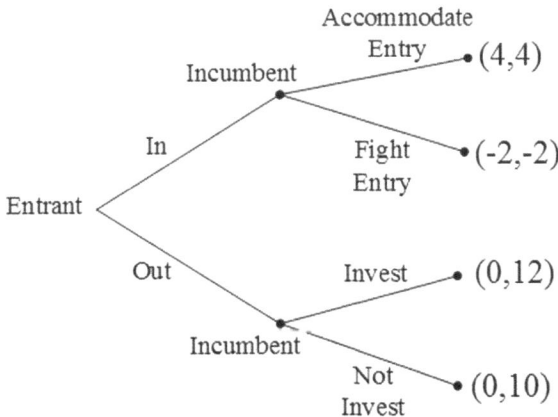

Fig. 6.12 Modified entry game

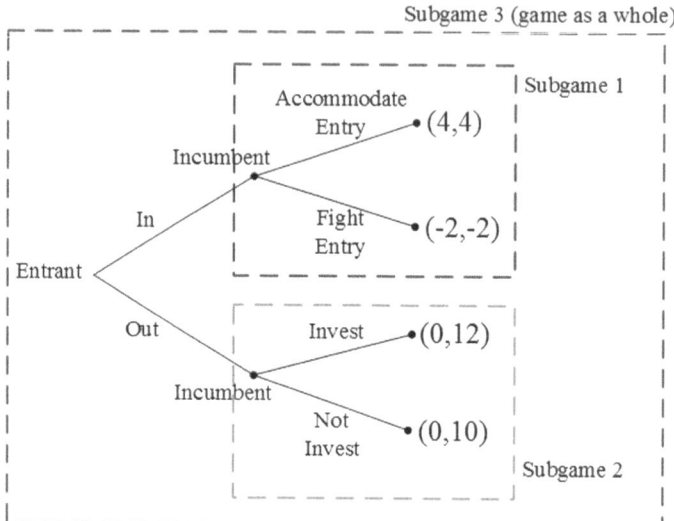

Fig. 6.13 Modified entry game—subgames

which denotes that the entrant chooses *In* and the incumbent responds with *Acc* after *In*, but with *Invest* after *Out*. □

Equilibrium path vs. SPE. Figure 6.14 shades the branches that players choose in equilibrium. One can see that the equilibrium path starts at the initial node, with the entrant joining the industry, and follows with the incumbent accommodating entry in the second stage. The equilibrium path, then, coincides with that in the original Entry game. The SPE, however, is more intricate because it specifies the incumbent's equilibrium behavior both in the node that it reaches in equilibrium (when the entrant joins the market) and in the node that the incumbent does not even reach in equilibrium (when the entrant stays out).

This is an important detail to remember when describing the SPE of a sequential-move game since it must specify equilibrium behavior for every player *at every node* where she is called to move, even in nodes that may not be reached in equilibrium.

6.5.2 Finding SPEs in Game Trees with Information Sets

Consider the game tree in Fig. 6.15, where firm 1 acts as the first mover, choosing either *Up* or *Down*. If firm 1 selects *Down*, the game is over, with firm 1 obtaining a payoff of 2, while firm 2 earns a payoff of 6. However, if firm 1 chooses *Up*, this firm gets to play again, choosing between *A* and *B*. Firm 2 is then asked to respond,

6.5 Subgame Perfect Equilibrium

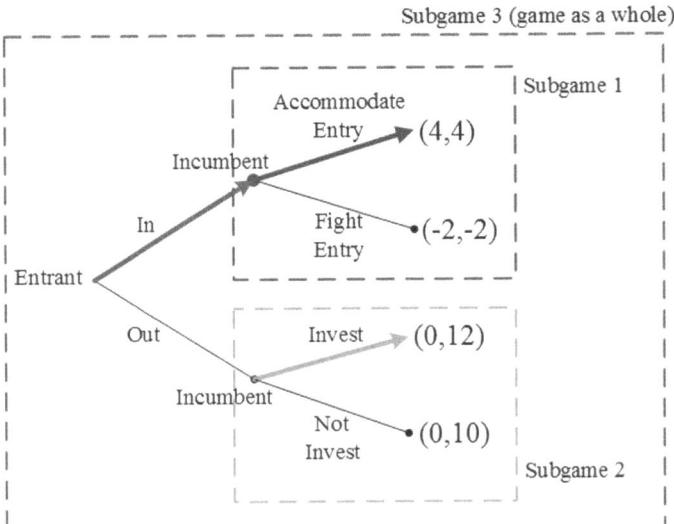

Fig. 6.14 Equilibrium path vs. SPE

but without seeing whether firm 1 chose A or B. Firm 2's uncertainty is graphically represented by the dotted line connecting the end of the branches that it doesn't distinguish, A and B. This dotted line is formally known as an "information set" of firm 2, because this firm doesn't know which of these two actions was chosen by firm 1.[1]

Before applying backward induction to this game, a usual trick is to find all the subgames (i.e., circling the portions of the tree that do not break any information set). Starting from the last mover (firm 2), the smallest subgame that we can circle is one initiated after firm 1 chooses Up, which is labeled "Subgame 1" in Fig. 6.16a. If we now move to the lower part of the game tree, note that we cannot circle any part of the tree without breaking firm 2's information set. Circles that break firm 2's information set are included in Fig. 6.16b as a reference. As a result, the only subgame that we can identify in this tree, besides subgame 1, is the game as a whole. You may be wondering: "These are nicely drawn rectangles, but why should we care about the subgames in a game tree?" The answer is simple: we can next apply backward induction by just focusing on the two subgames we found.

---

[1] Firm 2 has the same available strategies when firm 1 chooses A and when it chooses B, i.e., firm 2 must select either X or Y in both cases. If, instead, firm 2 had to choose between X and Y when firm 1 chooses A, but between a different pair of actions, X' and Y', when firm 1 chooses B, firm 2 would be able to infer which action firm 1 selected by just observing its own available actions.

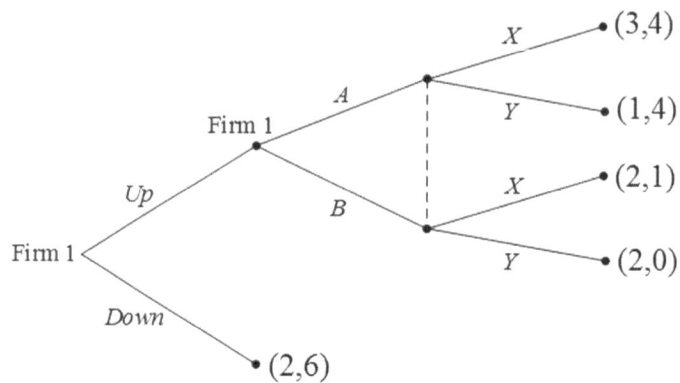

Fig. 6.15 A more involved game tree

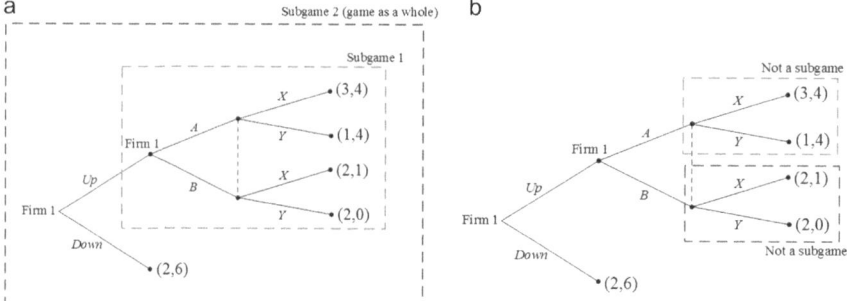

Fig. 6.16 a. Finding subgames; b. Not subgames

Subgame 1 Let us start by analyzing subgame 1. In this subgame, firm 2 does not observe which action firm 1 chose (either A or B).[2] Therefore, subgame 1 can be represented using Matrix 6.2, with firm 1 in rows and 2 in columns.

	Firm 2	
Firm 1	X	Y
A	3, 4	1, 4
B	2, 1	2, 0

Matrix 6.2. Representing subgame 1 in matrix form.

---

[2] Even if firm 2 acts a few hours (or days) after firm 1 chooses between A and B, firm 2 cannot condition its response (i.e., whether to respond with X or Y) on the specific action selected by firm 1. In this sense, firm 2 acts as if selecting its action at the same time as firm 1 chose its own, making the analysis of subgame 1 analogous to a simultaneous-move game.

6.5 Subgame Perfect Equilibrium

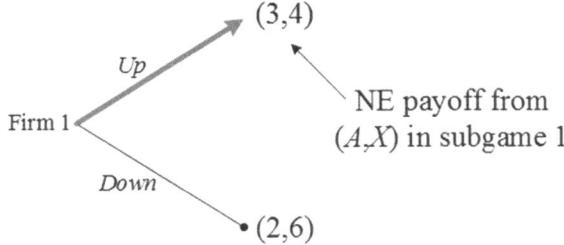

Fig. 6.17 The reduced game tree from Fig. 6.16a

We can now find the NE of subgame 1 by underlining best response payoffs, as discussed in Tool 3.1 and Sect. 3.3 in Chapter 3. Matrix 6.3 reproduces Matrix 6.2, but underlining best response payoffs.[3] As discussed in Chapter 3, the cell in which both firms' payoffs are underlined constitutes the NE of subgame 1, (A, X), with corresponding payoffs $(3, 4)$.

$$\begin{array}{c|c|c} & \multicolumn{2}{c}{Firm\ 2} \\ & X & Y \\ \hline A & \underline{3},\underline{4} & 1,\underline{4} \\ \hline B & 2,\underline{1} & \underline{2},0 \\ \end{array}$$

Firm 1

Matrix 6.3. Finding the NE of subgame 1.

The game as a whole We can now study the game as a whole. Firm 1 must choose between *Up* and *Down*, anticipating that if it chooses *Up*, subgame 1 will start. From our previous analysis, firm 1 can anticipate equilibrium behavior in subsequent stages of the game; that is, the NE of subgame 1 is (A, X), with payoffs $(3, 4)$. Firm 1 can then simplify its decision problem to the tree depicted in Fig. 6.17, where we insert the equilibrium payoffs from subgame 1, $(3, 4)$, if firm 1 were to select *Up*. Therefore, firm 1 only needs to conduct the following payoff comparison: if it chooses *Down*, the game is over and its payoff is 2, whereas if it chooses *Up*, subgame 1 is initiated, anticipating a payoff of 3. Because $3 > 2$, firm 1 prefers to choose *Up* rather than *Down*, as illustrated by the thick arrow on the branch corresponding to *Up*.

---

[3] This is a good moment to practice best responses. Recall that the underlined payoffs in Matrix 6.3 illustrate that firm 1's best response to firm 2 selecting X (in the left column) is A because $3 > 2$, and to firm 2 choosing Y (in the right column) is B, given that $2 > 1$. Similarly, firm 2's best response to firm 1 selecting A (in the top row) is $\{X, Y\}$ because firm 2 receives a payoff of 4 in both X and Y, and its best response to firm 1 choosing B (in the bottom row) is X because $1 > 0$.

Summarizing, after applying backward induction, the SPE of this game is $(Up, (A, X))$, which yields an equilibrium payoff of 3 for firm 1 and 4 for firm 2. □

6.6 Evaluating SPE as a Solution Concept

In previous chapters we evaluated solution concepts, such as IDSDS or NE, according to a list of desirable properties. We now follow a similar approach with SPE:

1. **Existence? Yes.** When we apply backward induction in any game tree, we find that at least one equilibrium exists; which is often known as "Zermelo's theorem," after Zermelo's (1913) article on chess, and later on extended by Kuhn (1953). Intuitively, this means that every player selects one strategy at every node where she is called to move. If a player is indifferent between two available actions a and b, for instance, then her best response is to choose either (or randomize between them), but every player selects one strategy at every node, yielding an SPE strategy profile.
2. **Uniqueness? No.** While the examples in this chapter display game trees that produce a unique SPE, we cannot guarantee that applying backward induction will always induce a unique SPE in *every* game tree, as required by this criterion. For instance, if in the Entry game of Example 6.2 the incumbent's payoff from both accommodating and fighting entry was 4, then this firm would be indifferent between *Acc* and *Fight*, leading to two SPEs:

$$\text{SPE} = \{(In, Acc), (In, Fight)\}.$$

   As a remark, in this case the strategy profiles that are SPEs would exactly coincide with those that are NEs; as opposed to other examples where the former is a strict subset of the latter.
   This example, however, suggests an interesting point: if players are not indifferent about their actions at any of the nodes (or information sets) where they are called to move, equilibrium behavior in each subgame must specify a unique strategy for every player in that subgame, ultimately implying that the SPE must be unique. Most games in economics and social sciences often share this feature—players are not indifferent about which action to choose at the point they are called to move—identifying a unique SPE in a large class of games.
3. **Robust to small payoff perturbations? Yes.** If we change the payoff of one of the players by a small amount (e.g., 0.001, but generally, for any ε that approaches zero), backward induction still provides us with the same equilibrium outcome, implying that SPE is robust to small payoff perturbations. This is due to the fact that, if a strategy s_i yields a strictly higher payoff than another strategy s_i', it must still yield a strictly higher payoff than s_i' after we apply a small payoff perturbation (remember that ε is infinitely small), meaning that

player i still chooses s_i in the subgame where this action was optimal. As a consequence, the SPE is unaffected.

4. **Socially optimal? No.** The application of backward induction does not necessarily produce a socially optimal outcome. This goes in line with our evaluation of IDSDS and NE, which did not always yield socially optimal outcomes.

6.7 Applications

The above examples assumed that players had a discrete action space. However, we can apply backward induction in games where players face a continuous action space, like an output decision, $q \geq 0$, or a donation to a charity, $x \geq 0$. As in previous examples, we still solve these sequential-move games by using backward induction: starting from the smallest subgames closer to the terminal nodes (i.e., in the last stages of the game), finding equilibrium behavior in those subgames; and then moving up to previous stages, identifying equilibrium choices in those stages too. Unlike previous examples, we now need to write down the maximization problem of every player at the point she is called to move.

6.7.1 Stackelberg Game of Sequential Quantity Competition

Consider the Stackelberg game of sequential quantity competition. Two firms produce a homogeneous good facing an inverse linear demand function $p(Q) = 1-Q$, where $Q = q_1 + q_2$ denotes aggregate output. For simplicity, we assume that all firms have constant marginal cost of production c, where $1 > c \geq 0$. Firms interact in a sequential-move game:

1. In the first stage, firm 1 (the industry leader) chooses its output q_1.
2. In the second stage, firm 2 (the industry follower) observes the leader's output q_1 and responds with its own output level q_2.

Solving the game by backward induction, we start solving the second stage.

Second stage, follower In this stage, the follower observes a given output level from the leader, q_1, and responds by choosing the output level q_2 that maximizes its profits, as follows

$$\max_{q_2 > 0} \pi_2 = (1 - q_1 - q_2)q_2 - cq_2$$

In other words, this problem says that the follower treats the leader's output level, q_1, as a parameter (because the follower cannot alter it). Differentiating with respect to q_2, we obtain

$$1 - 2q_2 - q_1 - c = 0.$$

Rearranging, we find $2q_2 = 1 - c - q_1$ and, solving for q_2, yields

$$q_2 = \frac{1-c}{2} - \frac{1}{2}q_1,$$

which is positive for all $\frac{1-c}{2} - \frac{1}{2}q_1 \leq 0$ or, after solving for q_1, we obtain $q_1 \leq 1-c$. Therefore, firm 2's best response function is

$$q_2(q_1) = \begin{cases} \frac{1-c}{2} - \frac{1}{2}q_1 & \text{if } q_1 \leq 1-c \\ 0 & \text{otherwise.} \end{cases}$$

Intuitively, the follower produces an output of $q_2 = \frac{1-c}{2}$ when the leader is inactive ($q_1 = 0$), but decreases its output in q_1 at a rate of $1/2$. When the leader's output is sufficiently large, $q_1 > 1-c$, the follower responds staying inactive, $q_2 = 0$. (This best response function coincides with that under the Cournot model of simultaneous quantity competition in Chapter 4.)

First stage, leader The leader anticipates the follower's best response function, $q_2(q_1)$, in the next stage of the game, so its profit-maximization problem in the first stage is

$$\max_{q_1 \geq 0} \pi_1 = [1 - q_1 - q_2(q_1)]q_1 - cq_1$$

which is evaluated at $q_2(q_1)$. Inserting firm 2's best response function, we obtain

$$\max_{q_1 \geq 0} \pi_1 = \left[1 - q_1 - \overbrace{\left(\frac{1-c}{2} - \frac{1}{2}q_1\right)}^{q_2(q_1)} \right] q_1 - cq_1$$

$$= \left[\frac{1+c}{2} - \frac{1}{2}q_1 \right] q_1 - cq_1$$

which is only a function of the leader's choice variable (its output level q_1). Differentiating the leader's profit with respect to q_1, we find

$$\frac{1+c}{2} - q_1 - c = 0.$$

Rearranging and solving for q_1, we obtain the leader's equilibrium output

$$q_1^{Seq} = \frac{1-c}{2}.$$

6.7 Applications

Therefore, we can report the SPE in this Stackelberg game of quantity competition as follows

$$\text{SPE} = \left(q_1^{Seq}, q_2(q_1)\right) = \left(\frac{1-c}{2}, \frac{1-c}{2} - \frac{1}{2}q_1\right)$$

where firm 2 (the follower) responds with its best response function, $q_2(q_1)$, for any output that the leader chooses (both its equilibrium output, $q_1 = q_1^{Seq}$, and any off-the-equilibrium output, $q_1 \neq q_1^{Seq}$). In equilibrium, we can claim that the follower observes the leader's output, q_1^{Seq}, and inserts it into its best response function, to obtain the follower's equilibrium output

$$q_2^{Seq} = \frac{1-c}{2} - \frac{1}{2}\overbrace{\frac{1-c}{2}}^{q_1^{Seq}}$$
$$= \frac{1-c}{4}.$$

Importantly, we *do not say* that the SPE of this Stackelberg game with two symmetric firms is $\left(q_1^{Seq}, q_2^{Seq}\right) = \left(\frac{1-c}{2}, \frac{1-c}{4}\right)$. This output vector only describes firms' output along the equilibrium path. In contrast, the SPE must specify each firm's output decision, both in- and off-the-equilibrium path.

Stackelberg vs. Cournot Recall that the equilibrium output in the Cournot model of simultaneous quantity competition was $q_i^{Sim} = \frac{1-c}{3}$ for every firm i (see Sect. 4.2.1), where we assumed the same indirect demand function $p(Q) = 1 - Q$ and marginal cost c for both firms. Therefore, the leader produces more units than when firms choose their output simultaneously, $\frac{1-c}{2} > \frac{1-c}{3}$, while the follower produces fewer units, $\frac{1-c}{4} < \frac{1-c}{3}$. Intuitively, the leader exercises its "first-mover advantage" by increasing its output, relative to that under Cournot competition, anticipating that the follower will respond to this increase in q_1 by decreasing its own output, q_2, given that its best response function $q_2(q_1)$ is negatively sloped.

Figure 6.18 depicts the equilibrium output level in the Cournot and Stackelberg games, showing that the output pair moves from the 45^0-line (where both firms produce the same output) to above this line, which indicates that firm 1 produces more units than firm 2. Graphically, firm 1 anticipates firm 2's best response function, $q_2(q_1)$, and chooses the point along this line that yields the highest profit.

We can confirm that, as expected, the leader's profit when firms compete sequentially,

$$\pi_1^{Seq} = \left[1 - q_1^{Seq} - q_2^{Seq}\right]q_1^{Seq} - cq_1^{Seq}$$
$$= \left[1 - \frac{1-c}{2} - \frac{1-c}{4}\right]\frac{1-c}{2} - c\frac{1-c}{2}$$

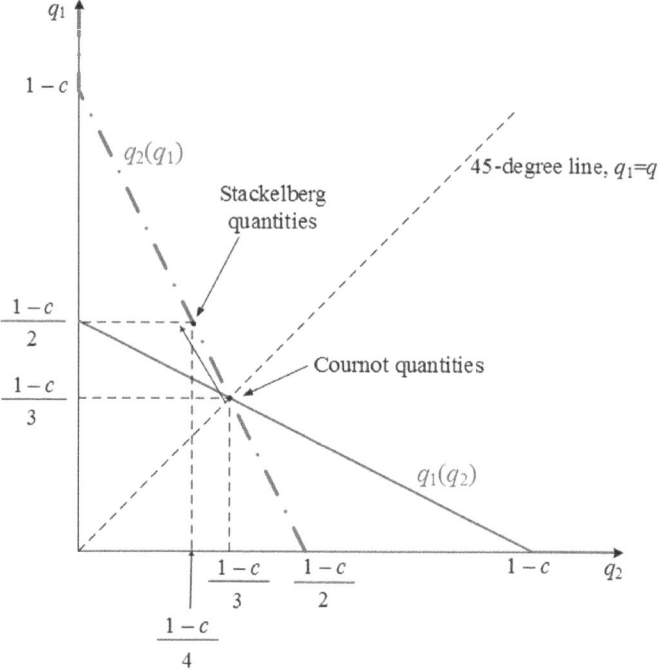

Fig. 6.18 Equilibrium quantities in Cournot vs. Stackelberg games

$$= \frac{(1-c)^2}{8},$$

is larger than its profit when they compete simultaneously, $\pi_1^{Sim} = \frac{(1-c)^2}{9}$. In contrast, the follower's profits when firms compete sequentially,

$$\pi_2^{Seq} = \left[1 - q_1^{Seq} - q_2^{Seq}\right] q_2^{Seq} - c q_2^{Seq}$$
$$= \left[1 - \frac{1-c}{2} - \frac{1-c}{4}\right] \frac{1-c}{4} - c \frac{1-c}{4}$$
$$= \frac{(1-c)^2}{16},$$

is lower than its profits under simultaneous quantity competition, $\pi_2^{Sim} = \frac{(1-c)^2}{9}$.

6.7.2 Sequential Public Good Game

Let us now return to the public good game we examined in Sect. 4.4 but, rather than assuming that players simultaneously choose their contributions to the public

6.7 Applications

good, we now consider that player 1 makes her contribution x_1 in the first period and, after observing this contribution, player 2 responds with her contribution x_2 in the second period. Because this is a sequential-move game of complete information, we solve it by backward induction, focusing on player 2 first, and then moving on to player 1.

Player 2, follower Player 2 takes player 1's contribution, x_1, as given, and chooses her contribution x_2 to solve the following utility maximization problem.

$$\max_{x_2 > 0} u_2(x_1, x_2) = (w_2 - x_2)\sqrt{m(x_1 + x_2)}.$$

Differentiating with respect to x_2, we obtain

$$-\sqrt{m(x_1 + x_2)} + \frac{m(w_2 - x_2)}{2\sqrt{m(x_1 + x_2)}} = 0$$

which, after rearranging, simplifies to

$$\frac{m(w_2 - 2x_1 - 3x_2)}{2\sqrt{m(x_1 + x_2)}} = 0$$

which holds when $3x_2 = w_2 - 2x_1$. Solving for x_2, we find that $x_2 = \frac{w_2}{3} - \frac{2}{3}x_1$, which is positive when $\frac{w_2}{3} > \frac{2}{3}x_1$, or $x_1 < \frac{w_2}{2}$. Therefore, player 2's best response function is

$$x_2(x_1) = \begin{cases} \frac{w_2}{3} - \frac{2}{3}x_1 & \text{if } x_1 < \frac{w_2}{2}, \text{ and} \\ 0 & \text{otherwise.} \end{cases}$$

This best response function coincides with that found in Sect. 4.4, originating at $x_2 = \frac{w_2}{3}$ when player 1 does not donate to the public good, but decreasing at a rate of 2/3 for every dollar that player 1 contributes. (For more details about this game, see Sect. 4.4.)

Player 1, leader In the first period, player 1 anticipates player 2's best response function in the subsequent period, $x_2(x_1)$, and inserts it into her utility maximization problem, as follows

$$\max_{x_1 \geq 0} u_1(x_1, x_2) = (w_1 - x_1)\sqrt{m[x_1 + x_2(x_1)]}$$

$$= (w_1 - x_1)\sqrt{m\left[x_1 + \underbrace{\left(\frac{w_2}{3} - \frac{2}{3}x_1\right)}_{x_2(x_1)}\right]}.$$

which simplifies to

$$\max_{x_1 \geq 0} (w_1 - x_1)\sqrt{m\frac{w_2 + x_1}{3}}$$

thus being a function of her own contribution, x_1, alone. Differentiating with respect to x_1, we obtain

$$\frac{m(w_1 - x_1)}{2\sqrt{3}\sqrt{m(w_2 + x_1)}} - \frac{\sqrt{m(w_2 + x_1)}}{\sqrt{3}} = 0$$

which simplifies to

$$\frac{m(w_1 - 2w_2 - 3x_1)}{2\sqrt{3}\sqrt{m(w_2 + x_1)}} = 0$$

and, solving for x_1, we find player 1's equilibrium contribution in this sequential-move version of the public good game

$$x_1^{Seq} = \frac{w_1 - 2w_2}{3}.$$

Therefore, the SPE of the game is

$$\text{SPNE} = \left(x_1^{Seq}, x_2(x_1)\right) = \left(\frac{w_1 - 2w_2}{3}, \frac{w_2}{3} - \frac{2}{3}x_1\right).$$

As a consequence, players contributions evaluated in equilibrium are $x_1^{Seq} = \frac{w_1 - 2w_2}{3}$ for player 1 and $x_2^* = x_2(x_1^{Seq})$ for player 2, that is,

$$x_2(x_1^{Seq}) = \frac{w_2}{3} - \frac{2}{3}\left(\frac{w_1 - 2w_2}{3}\right)$$

$$= \frac{7w_2 - 2w_1}{9}.$$

Simultaneous vs. Sequential donations When players simultaneously submit their donations, we found that their equilibrium contributions are $x_i^{Sim} = \frac{3w_i - 2w_j}{5}$ for every player i (see Sect. 4.4). Comparing our results, we obtain that the leader contributes less to the charity than when players submit donations simultaneously,

$$x_1^{Seq} = \frac{w_1 - 2w_2}{3} < \frac{3w_1 - 2w_2}{5} = x_1^{Sim},$$

because this inequality simplifies to $-w_2 < w_1$, which always hold given that wealth levels are positive by assumption. The follower, however, contributes more in the sequential- than simultaneous-move game since the following inequality

$$x_2^{Seq} = \frac{7w_2 - 2w_1}{9} > \frac{3w_2 - 2w_1}{5} = x_2^{Sim}$$

simplifies to $w_2 > -w_1$, which also holds for all wealth levels. Intuitively, the leader anticipates that her decreased contributions will be responded by the follower with an increase in her own donation. In other words, the leader exploits her first-mover advantage, which in this case means decreasing her own contribution to free-ride the follower's donation.

6.7.3 Ultimatum Bargaining Game

Bargaining is prevalent in many economic situations where two or more parties negotiate how to divide a certain surplus, such as the federal budget, a salary increase, or the price of a house. These strategic settings can be described as a sequential-move game where one player is the first mover in the game, proposing a "split," or division, of the surplus among the players.

For presentation purposes, we first examine the shortest bargaining setting, the so-called "ultimatum bargaining game" depicted in Fig. 6.19a, where player 1 makes a division of the surplus to player 2, d, where $d \in [0, 1]$ can be interpreted as a share of the total surplus and, observing the offer d, player 2 only has the ability to accept it or reject it. (The "arc" at the top of the game represents the continuum of offers that player 1 can make to player 2, where $d \in [0, 1]$.)

If player 2 rejects the offer, both players earn a payoff of zero, but if she accepts the offer, she receives d while player 1 earns the remaining surplus, $1 - d$. (For instance, if player 1 offers 20% of the surplus to player 2, $d = 0.2$, which if accepted, means that player 1 keeps the remaining 80% of the surplus, $1 - d = 0.8$.) This bargaining game is, then, equivalent to a take-it-or-leave-it offer, or an ultimatum, from player 1, explaining the game's name.

The smallest subgame that we can identify in the above game tree is that initiated after the responder (player 2) observes the proposer's offer, d, as depicted in Fig. 6.19b, and the game as a whole.

Applying backward induction to solve this bargaining game, we can start with the subgame, where the responder accepts any offer d if her utility from doing so, d, exceeds that from rejecting the offer, zero. More compactly, we can say that, upon receiving an offer d, the responder accepts it if d satisfies $d \geq 0$.

Moving to the first stage, the proposer anticipates the responder's decision rule, $d \geq 0$, and makes an offer that maximizes her payoff conditional on that offer being accepted. Formally, the proposer solves

$$\max_{d \geq 0} 1 - d$$

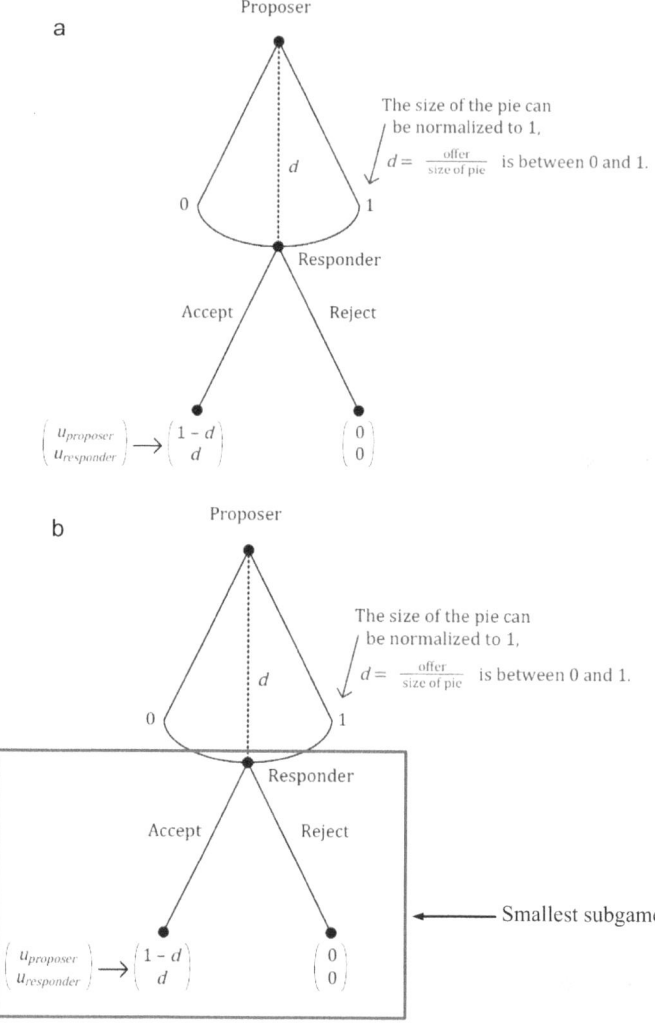

Fig. 6.19 a. Ultimatum bargaining game. b. Ultimatum bargaining game—Smallest subgame

which considers the constraint $d \geq 0$ to induce the responder to accept the offer. (Otherwise, the responder would not accept, leaving the proposer with a zero payoff.) Differentiating with respect to d, yields -1, indicating a corner solution. Intuitively, the proposer seeks to reduce d as much as possible (since a higher division d decreases the proposer's payoff, $1 - d$), while making sure that the responder accepts the offer (which happens as long as $d \geq 0$). Therefore, the proposer reduces d all the way to $d^* = 0$, which still satisfies acceptance, making the responder indifferent between accepting and rejecting the offer.[4]

Therefore, the SPE of the ultimatum bargaining game is

$$\text{SPE} = \{d^* = 0, d \geq 0\}$$

which indicates that the proposer makes an offer $d^* = 0$ and the responder accepts any offer d that satisfies $d \geq 0$.[5] The equilibrium path prescribes that the proposer offers $d^* = 0$ and the responder accepts that offer, yielding equilibrium payoff $(1-d, d) = (1, 0)$, which implies that the proposer keeps all the surplus.

6.7.4 Two-Period Alternating-Offers Bargaining Game

Figure 6.20 extends the above bargaining, allowing the responder (player 2) to make a counteroffer in the event that she rejects the division that player 1 offers, d^1. The time structure is, then, the following: player 1 makes an offer d^1 to player 2, who accepts or rejects it. If player 2 accepts d^1, the game is over, and player 2 earns d^1 while player 1 earns the remaining surplus, $1 - d^1$. If player 2 rejects, however, she has the opportunity to make an offer d^2 to player 1 (which we can interpret as a counteroffer). Observing d^2, player 1 responds by accepting or rejecting it. As in previous stages, both players earn zero if the game ends without a division being accepted but otherwise player 1 (the responder in the second stage) earns d^2, which has a discounted value of $\delta_1 d^2$ in today's terms. Discount factor $\delta_1 \in [0, 1]$ indicates the relative importance that player 1 assigns to future payoffs.[6] Player 2 earns, in this case, the remaining surplus $1 - d^2$, whose discounted value in today's terms is $\delta_2(1 - d^2)$.

In this context, we can find four subgames, initiated when: (1) player 1 receives offer d^2; (2) player 2 rejects offer d^1; (3) player 2 receives offer d^1; and (4) the game as a whole. Operating by backward induction, we start from the smallest subgame, as described in subgame (1).

Subgame (1) At that point of the game, player 1 accepts any offer d^2 from player 2 if and only if $d^2 \geq 0$.

---

[4] For simplicity, we assume that the responder accepts offers that make her indifferent. If, instead, she rejects this type of offers, the proposer could offer her an extremely small division of the surplus, $d \to 0$, still yielding similar equilibrium results as above.
[5] Note that the SPE does not just say that the responder accepts the equilibrium offer $d^* = 0$. Instead, the SPE describes the responder's decision (accept/reject) after receiving any offer d from the proposer.
[6] Recall that when $\delta_1 = 0$, player 1 assigns no value to future payoffs, which can be alternatively understood as if she only cares about today's earnings. In contrast, when $\delta_1 = 1$, player 1 assigns the same value to today's and tomorrow's payoffs.

Fig. 6.20 Two-period alternating-offers bargaining game

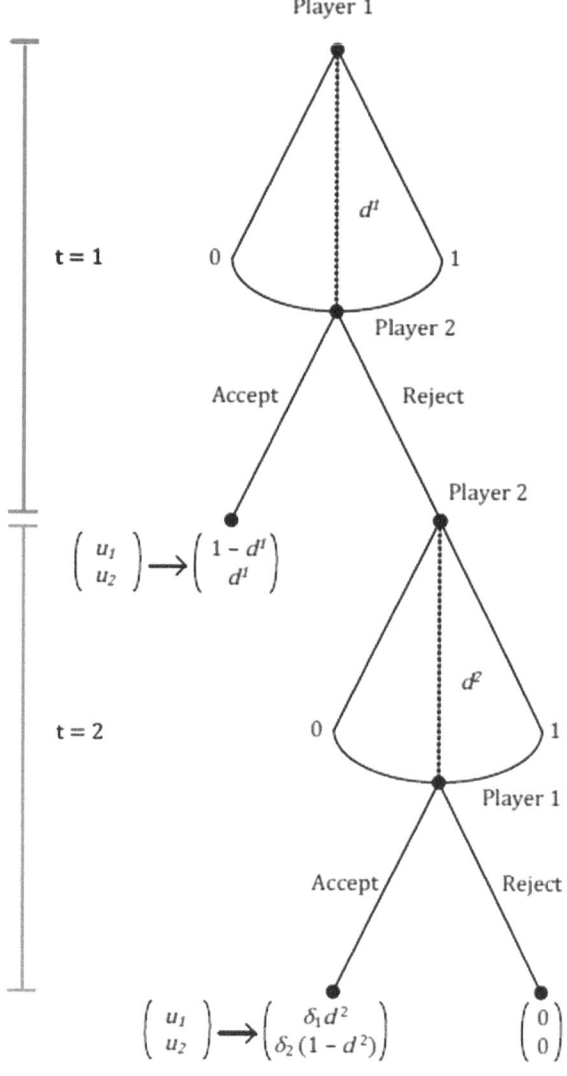

Subgame (2) In subgame (2), player 2 anticipates this best response by player 1, making an offer d^2 that solves

$$\max_{d^2 \geq 0} 1 - d^2$$

which is analogous to the proposer's problem in the ultimatum bargaining game of Sect. 6.7.3. Intuitively, when players reject offers and reach their final period of interaction, they face a strategic setting similar to the ultimatum bargaining game since not reaching an agreement at that point will lead to a zero payoff for everyone,

6.7 Applications

as in the ultimatum bargaining game. As shown in the previous section, the proposer (player 2 in the current game) offers the lowest division that guarantees acceptance from player 1, $d^2 = 0$, earning a payoff of $1 - d^2 = 1 - 0 = 1$.

Subgame (3) Moving now to subgame (3), player 2 chooses whether to accept the offer from player 1, earning d^1, or reject it and become the proposer tomorrow, where we know that her equilibrium payoff in subgame (2) is 1, with discounted value $\delta_2 \times 1 = \delta_2$ into today's terms. Therefore, player 2 accepts offer d^1 if and only if $d^1 \geq \delta_2$.

Game as a whole Finally, in the initial node, player 1 makes an offer d^1 that maximizes her payoff and guarantees player 2's acceptance, $d^1 \geq \delta_2$, that is,

$$\max_{d^1 \geq \delta_2} 1 - d^1$$

which yields an equilibrium offer $d^1 = \delta_2$. Therefore, the SPE of the two-period alternating-offers bargaining game is

$$\text{SPE} = \{\underbrace{(d^1 = \delta_2, d^2 \geq 0)}_{\text{Player 1}}, \underbrace{(d^2 = 0, d^1 \geq \delta_2)}_{\text{Player 2}}\}$$

meaning that player 1 offers a division $d^1 = \delta_2$ in the first period and accepts any offer $d^2 \geq 0$ in the second period; while player 2 offers a division $d^2 = 0$ in the first period and accepts any offer $d^1 \geq \delta_2$ in the second period.

This SPE implies that, while players have two periods to negotiate how to divide the surplus, they reach an agreement in the first period: player 1 offers a division $d^1 = \delta_2$, which player 2 accepts since it satisfies her decision rule $d^2 \geq 0$, and the game is over (see left side of Fig. 6.20). In equilibrium, player 1 earns a payoff $1 - d^1 = 1 - \delta_2$, while player 2 earns a payoff equal to the division she accepted from player 1, $d^1 = \delta_2$, that is, equilibrium payoffs are

$$(u_1^*, u_2^*) = (1 - \delta_2, \delta_2).$$

Comparative statics. As player 2 becomes more patient (higher δ_2), her equilibrium payoff increases while that of player 1 decreases. Indeed, when δ_2 increases, player 1 understands that player 2 does not discount future payoffs significantly, meaning that she can reject player 1's offer today, becoming the proposer tomorrow. If, instead, δ_2 approaches zero, player 1 anticipates that player 2 severely discounts future payoffs, having a stronger preference for today's payoffs. In this case, player 1 can exploit player 2's impatience by making a less generous offer.[7] Interestingly,

---

[7] In the extreme case where $\delta_2 = 0$, player 1 keeps the whole surplus making an offer $d^1 = 0$ to player 2, as if there was no second period of potential negotiations. In contrast, when $\delta_2 = 1$,

player 1's patience, as captured by her discount factor δ_1, does not affect the offer she makes to player 2 in equilibrium, $d^2 = 0$, or players' equilibrium payoffs, $(u_1^*, u_2^*) = (\delta_2, 1 - \delta_2)$; which are only affected by player 2's patience, δ_2. In games with more stages, player 1's discount factor δ_1 affects equilibrium payoffs, as we show in the next section.

6.7.5 Some Tricks About Solving Alternating-Offer Bargaining Games

You probably noticed a common theme in our above reasoning. Equilibrium payoffs in the last stage of the game are (0, 1), where the proposer in that stage makes a zero offer, which is accepted by the proposer, helping the proposer keep the whole surplus. This holds true when the game has only one period of possible negotiations (ultimatum bargaining game), two periods (as shown in this section), or more than two periods. Players then anticipate that, if the game reaches the last period, their equilibrium payoffs will be (0, 1) if player 2 is the proposer in that period or (1, 0) if player 1 is the proposer. (Without loss of generality, assume that player 2 is the proposer in the last period.) The proposer in the previous-to-last period (player 1), must offer a division d^1 that makes player 2 indifferent between accepting d^1 today or rejecting it to become the proposer tomorrow, earning a payoff of 1 with discounted value $\delta_2 \times 1 = \delta_2$ today. This means that player 1 offers exactly $d^1 = \delta_2$, keeping the remaining surplus $1 - \delta_2$ in equilibrium.

This approach can be applied in longer alternating-offer bargaining games with more than two stages, helping us find the SPE relatively fast. For instance, if players can negotiate for four periods, starting with player 1 making an offer d^1, we can operate by backward induction as follows:

1. *Fourth period.* In the last period, player 2 is the proposer, so the equilibrium payoff is (0, 1).
2. *Third period.* In the third period, player 1 makes an offer, so she must make player 2 indifferent between her offer today or keeping the whole surplus tomorrow, worth δ_2 today. Equilibrium payoffs are, then,

$$(1 - \delta_2, \delta_2).$$

3. *Second period.* In the second period, player 2 is the proposer, so she must make player 1 indifferent between her offer today and payoff $1 - \delta_2$ tomorrow (when she becomes the proposer), with discounted value $\delta_1(1 - \delta_2)$ today. Therefore,

---

player 1 must offer player 2 the whole surplus, $d^1 = 1$, because any offer lower than that is rejected by player 2, who can wait to earn the whole surplus when she becomes the proposer in subgame (3).

6.7 Applications

Table 6.1 An approach to solve alternating-offers bargaining games

Player	Period	Player 1's payoff	Player 2's payoff
Player 1	$t=1$	$1-\delta_2[1-\delta_1(1-\delta_2)]$	$\delta_2[1-\delta_1(1-\delta_2)]$
Player 2	$t=2$	$\delta_1(1-\delta_2)$	$1-\delta_1(1-\delta_2)$
Player 1	$t=3$	$1-\delta_2$	δ_2
Player 2	$t=4$	0	1

equilibrium payoffs are

$$(\delta_1(1-\delta_2), 1-\delta_1(1-\delta_2)),$$

where $1 - \delta_1(1 - \delta_2)$ represents the remaining surplus that player 2 earns, after offering $\delta_1(1 - \delta_2)$ to player 1.

4. *First period.* In the first period of interaction, player 1 is the proposer, and she must make player 2 indifferent between her current offer and payoff $1 - \delta_1(1 - \delta_2)$ tomorrow, with discounted value $\delta_2[1 - \delta_1(1 - \delta_2)]$ today. As a consequence, equilibrium payoffs are

$$(1 - \delta_2[1 - \delta_1(1 - \delta_2)], \delta_2[1 - \delta_1(1 - \delta_2)])$$

where player 1's equilibrium payoff denotes the remaining surplus after she offers $\delta_2[1 - \delta_1(1 - \delta_2)]$ to player 2.

Table 6.1 summarizes our approach to solve this alternating-offer bargaining game, starting in the last period. Vertical arrows indicate that a player must be indifferent between her payoff in the period when she is a proposer and her payoff in the previous period when she is the responder. Horizontal arrows, however, denote that the proposing player's equilibrium payoff is the remaining surplus, i.e., the share not offered to her rival. Graphically, we move up and sideways, and then up and sideways again until reaching the first period. Arrows resemble a ladder, explaining why students often refer to this approach as the "ladder."

6.7.6 Alternating-Offer Bargaining Game with Infinite Periods

In previous sections, players bargain over a finite number of periods (e.g., one, two, or four). How would our results change if, instead, players can bargain for as many periods as they need? This, of course, does not mean that they will negotiate for infinite periods, since they discount future payoffs and thus prefer to reach an agreement as soon as possible. In this infinite-period bargaining game, player 1 makes offers to player 2, d^1, in period 1, 3, 5,... (every odd-numbered periods) while player 2 makes offers to player 1, d^2, in period 2, 4, 6,... (every even-numbered periods).

Interestingly, players' behavior is *stationary* in equilibrium, meaning that every player's offer coincides in every period when she makes proposals (e.g., player 1's offer in period 1 and 3 coincides). A similar argument applies to her decision rule in every period she acts as the responder (e.g., player 1 accepts offers according to the same rule in periods 2 and 4). Then, we can separately analyze equilibrium behavior in odd- and even-numbered periods, as follows.

Odd-numbered periods At every odd period, player 2 compares the payoff she gets from accepting the offer from player 1, d^1, against the payoff that she can earn tomorrow when she becomes the proposer. In particular, player 2 offers d^2 tomorrow, keeping the remaining surplus $1 - d^2$, with discounted value $\delta_2(1 - d^2)$ in today's terms. Therefore, player 2 accepts player 1's offer today, d^1, if and only if $d^1 \geq \delta_2(1 - d^2)$. Player 1 anticipates this decision rule and minimizes her offer, d^1, making player 2 indifferent,

$$d^1 = \delta_2(1 - d^2).$$

Even-numbered periods Similarly, at every even period, player 1 compares her payoff from accepting player 2's offer, d^2, and the payoff she can earn tomorrow when she becomes the proposer. Specifically, player 1 offers d^1 to player 2 tomorrow, keeping the remaining $1 - d^1$ of the surplus, with discounted value $\delta_1(1 - d^1)$. Therefore, player 1 accepts d^2 today if and only if $d^2 \geq \delta_1(1 - d^1)$, where player 2 reduces d^2, ultimately making player 1 indifferent between accepting and rejecting, that is,

$$d^2 = \delta_1(1 - d^1).$$

Solving the game We have then found one equation from every odd period and one equation from every even period, and we have two unknowns, offers d^1 and d^2. To solve for these two offers in equilibrium, we can insert one indifferent condition into the other, yielding

$$d^1 = \delta_2(1 - \underbrace{\delta_1(1 - d^1)}_{d^2})$$

which rearranges to

$$d^1 = \delta_2 - \delta_1\delta_2 + \delta_1\delta_2 d^1$$

6.7 Applications

and solving for d^1, yields

$$d^1 = \frac{\delta_2(1-\delta_1)}{1-\delta_1\delta_2}.$$

Inserting this result into indifference condition $d^2 = \delta_1(1-d^1)$, we obtain

$$\begin{aligned}
d^2 &= \delta_1\left(1 - \frac{\delta_2(1-\delta_1)}{1-\delta_1\delta_2}\right) \\
&= \delta_1 \frac{(1-\delta_1\delta_2 - \delta_2 + \delta_1\delta_2)}{1-\delta_1\delta_2} \\
&= \frac{\delta_1(1-\delta_2)}{1-\delta_1\delta_2}.
\end{aligned}$$

We can then conclude that, in the first period, player 1 makes an offer d^1 to player 2, who immediately accepts it, and the game is over, yielding equilibrium payoffs

$$\left(1-d^1, d^1\right) = \left(\frac{1-\delta_2}{1-\delta_1\delta_2}, \frac{\delta_2(1-\delta_1)}{1-\delta_1\delta_2}\right)$$

where

$$\begin{aligned}
1 - d^1 &= 1 - \frac{\delta_2(1-\delta_1)}{1-\delta_1\delta_2} \\
&= \frac{1-\delta_1\delta_2 - \delta_2 + \delta_1\delta_2}{1-\delta_1\delta_2} \\
&= \frac{1-\delta_2}{1-\delta_1\delta_2}.
\end{aligned}$$

Therefore, player 2's equilibrium payoff, d^1, increases in her own discount factor, δ_2, since

$$\frac{\partial d^1}{\partial \delta_2} = \frac{1-\delta_1}{(1-\delta_1\delta_2)^2} \geq 0$$

implying that, as player 2 assigns more weight to her future payoff (i.e., she becomes more patient), she can reject player 1's offer and wait to become the proposer, which forces player 1 to make a more generous offer today. In contrast, player 2's equilibrium payoff decreases in her rival's discount factor, δ_1, because

$$\frac{\partial d^1}{\partial \delta_1} = \frac{\delta_2(\delta_2 - 1)}{(1-\delta_1\delta_2)^2} < 0$$

which intuitively means that, as player 1 assigns more weight to her future payoffs, player 2 must offer her a larger surplus share in future periods (when player 2 becomes the proposer). This, in turn, reduces player 2's future payoffs, making her less attracted to reject the offer from player 1 today.

Our above results are common in bargaining games, both when players interact for finite and infinite periods. First, if you are more patient (higher discount factor), you will not accept low offers from your opponent today, since you can wait until the next period to become the proposer and your future payoffs are not heavily discounted. Second, a more patient opponent is "more difficult to please" with low offers since she can just wait until the next period, when she becomes the proposer, forcing you to make more generous offers today to induce acceptance. Overall, the more patient you are (higher δ_i) and the less patient your rival is (lower δ_j), the larger the surplus share that you can keep, and the lower the share that she keeps, in the SPE of the game.

Appendix—Mixed and Behavioral Strategies

Recall that, in a sequential-move game, a *pure strategy* is a complete plan of action specifying what player i does at every node (or, more generally, at every information set) where she is called to move, both those that are reached in equilibrium and those that are not. We next extend this definition to allow for players to randomize over her pure strategies.

Definition 6.4. **Mixed strategy.** In a sequential-move game, a mixed strategy is a probability distribution over all player i's pure strategies $s_i \in S_i$.

In Fig. 6.21, the set of pure strategies is

$$S_1 = \{Ac, Ad, Bc, Bd\}$$

and a mixed strategy is a randomization over all these pure strategies. Intuitively, player i rolls a dice, before the game starts, and the outcome of this roll determines the path of play that she will follow (i.e., the pure strategy that she chooses). For instance, a mixed strategy for player 1 could be $\left(0Ac, 0Ad, \frac{1}{2}Bc, \frac{1}{2}Bd\right)$, which puts no probability weight on pure strategies Ac and Ad, but assigns 50% probability on Bc and Bd.

We next define a behavioral strategy where, essentially, player i randomizes over actions available at each information set, rather than over the whole complete contingent plan that a pure strategy specifies.

Definition 6.5. **Behavioral strategy.** A behavioral strategy is a mapping $b_i : H_i \rightarrow \Delta A_i(h_i)$, where $b_i(s_i(h_i))$ is the probability that player i selects action $s_i(h_i)$ at information set h_i.

In the context of Fig. 6.21, an example of a behavioral strategy for player 1 specifies that, at the first node where she is called to move, she randomizes between the two available actions in this node, A and B, such as $pA + (1 - p)B$ where $p \in [0, 1]$. Similarly, in the second node where she is called to move, she randomizes between actions c and d, e.g., $qc + (1 - q)d$, where $q \in [0, 1]$. In that context, an example of a behavioral strategy for player 1 is $(p, q) = \left(\frac{1}{3}, \frac{1}{4}\right)$, so she assigns

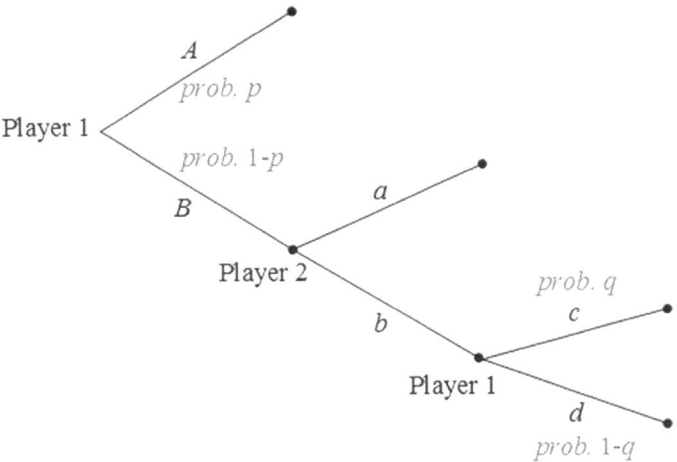

Fig. 6.21 Mixed and behavioral strategies in a game tree

a probability of $\frac{1}{3}$ to action A in the first node where she is called to move, and $\frac{1}{4}$ to action c in the second node where she plays.

Equivalence Are behavioral strategies equivalent to a corresponding mixed strategy? Yes, we can construct a one-to-one correspondence between behavioral and mixed strategies. In our above example, for instance, if probabilities p and q in the behavioral strategy satisfy $p = 0$ and $q = \frac{1}{2}$, then the behavioral strategy becomes

$$\left(B, \frac{1}{2}c + \frac{1}{2}d\right)$$

which is equivalent to the mixed strategy

$$\left(0Ac, 0Ad, \frac{1}{2}Bc, \frac{1}{2}Bd\right)$$

since, in one of them, player 1 starts choosing B in pure strategies, and then randomizes between c and d with equal probability. Otherwise, the behavioral and mixed strategy do not produce the same outcome.

"Manuals and libraries." Following Luce and Raiffa's (1957) analogy, a pure strategy $s_i \in S_i$ can be understood as an "instruction manual" in which each page tells player i which action to choose when she is called to move at information set h_i.[8]

---

[8] Such an instruction manual has as many pages as the number of information sets that player i has on the game tree.

In this analogy, strategy space S_i is a "library" with all possible instruction manuals. A mixed strategy, then, randomly chooses a specific manual s_i from the library S_i (i.e., a specific complete contingent plan), which the player considers for the rest of the game. In contrast, a behavioral strategy chooses pages from different manuals (actions in each information set) with positive probability.

Exercises

6.1 **Sequential Prisoner's Dilemma game.**[A] Consider the following Prisoner's Dilemma game where every player simultaneously chooses between confess (F) and not confess (NC). Payoffs satisfy $c > a > b > d$.

		Player 2	
		C	NC
Player 1	C	b, b	c, d
	NC	d, c	a, a

   (a) *Simultaneous-move version.* If players interact in a simultaneous-move game, what are their best responses? What is the NE of the game?
   (b) *Sequential-move version.* If player 1 acts first and, observing his action, player 2 responds, what is the SPE of the game?
   (c) *Comparison.* Compare the NE of the simultaneous version of the game against the SPE of the sequential version of the game. Interpret.

6.2 **Sequential Pareto coordination game.**[A] Consider the following Pareto coordination game where every firm simultaneously choosing between technology A or B, and payoffs satisfy $a > b > 0$, implying that both firms regard technology A as superior.

		Firm 2	
		Tech. A	Tech. B
Firm 1	Tech. A	a, a	0, 0
	Tech. B	0, 0	b, b

   (a) *Simultaneous-move version.* If firms interact in a simultaneous-move game, what are their best responses? What is the NE of the game?
   (b) *Sequential-move version.* If firm 1 acts first and, observing his action, firm 2 responds, what is the SPE of the game?
   (c) *Comparison.* Compare the NE of the simultaneous version of the game against the SPE of the sequential version of the game. Interpret.

Exercises

6.3 **Sequential anticoordination game.**[A] Consider the following Game of Chicken where every player drives toward each other and simultaneously chooses between swerve and stay. Payoffs satisfy $a > b > 0$, implying that drivers suffer more when they both stay (and almost die!) than when they both swerve (and are called chickens), and $a > d > c > b$, so being the only player who stays yields a payoff of d for that player and $-c$ for his rival.

		Player 2	
		Swerve	Stay
Player 1	Swerve	$-b, -b$	$-c, d$
	Stay	$d, -c$	$-a, -a$

(a) *Simultaneous-move version.* If players interact in a simultaneous-move game, what are their best responses? What is the NE of the game?
(b) *Sequential-move version.* If player 1 acts first and, observing his action, player 2 responds, what is the SPE of the game?
(c) *Comparison.* Compare the NE of the simultaneous version of the game against the SPE of the sequential version of the game. Interpret.

6.4 **More general version of the entry game.**[B] Consider the entry game in Fig. 6.22. The entrant can choose to stay out or enter into the market, and the incumbent can decide whether to accommodate or fight entry. The game is identical to that in Fig. 6.6, but allows for more general payoffs when the entrant stays out, or when entry is accommodated.
(a) What will the incumbent choose in the second stage of the game?
(b) What will the entrant do in the first stage of the game?
(c) Characterize the SPEs of this game. Interpret.

6.5 **NE vs. SPE in a natural resource game.**[A] Consider a situation where two players are vying for control of a natural resource with value 5. Each player has the opportunity to invest in this property, increasing its value by 2 for an investment of 1. If neither player invests, player 1 receives the natural

Fig. 6.22 More general version of the entry game

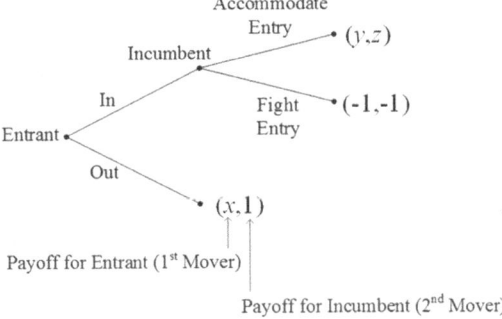

resource (player 1 receives a payoff of 5, and player 2 receives a payoff of zero). If one player invests, player 2 receives the natural resource (For example, if only player 1 invests, player 1 receives a payoff of -1, and player 2 receives a payoff of 7). If both players invest, player 1 receives the natural resource.

(a) *Simultaneous-move game.* Suppose players make their decisions simultaneously. Find all NEs in both pure and mixed strategies.

(b) *Sequential-move game.* Suppose instead that players interact in the sequential version of this game. In particular, player 1 chooses whether to Invest or Not Invest. Observing player 1's decision, player 2 responds by choosing to Invest or Not Invest. Depict the extensive form of this game and find all SPEs.

(c) Is there first or second-mover's advantage in this game? If so, identify which exists and why.

(d) Show that the SPE that you found in part (b) is also an NE of the sequential version of the game.

6.6 **Sequential product development.**[B] Consider the sequential-move game where, in the first stage, firm 0 chooses whether to launch a new technology or not (such as a new computer operating system). If it does not launch the technology, the game is over, and all players earn a zero payoff. If it launches the technology, then two other companies (1 and 2) choose whether to develop compatible products for the new technology. In this case, payoffs resemble a Pareto coordination game: (1) if they both develop the product, they both earn a payoff of a while firm 0 earns a payoff of $4 - F$, where F denotes the cost of launching the new technology; (2) if only one firm develops the product, both firms earn zero and firm 0 earns $2 - F$; and (3) if no firm develop the product, they both earn a payoff of b, where $b < a$, and firm 0 earns $1 - F$.

(a) Operating by backward induction, find the NEs in the subgame that ensues after firm 0 launches the new technology. (Consider both psNEs and msNE.) Find players' payoffs in each NE.

(b) Anticipating the NE of the subgame in part (a), find the SPEs of the game.

(c) *Numerical example.* If $a = 2$, $b = 4$, and $F = 3$, find the SPE of the game.

6.7 **Prisoner's Dilemma game with a preplay offer.**[B] Consider the following matrix, where every player simultaneously and independently chooses between high and low price.

		Firm 2	
		h	l
Firm 1	H	3,3	0,5
	L	5,0	2,2

In the first stage, firm 1 commits to pay $x to firm 2 if it chooses high price (h, on the left column). If instead firm 2 chooses low price (l, on the right column) firm 1 does not pay firm 2. We next solve the game by backward induction, starting from the second stage.

(a) Represent the above sequential-move game with a game tree
(b) Solve the game by backward induction to find the SPNEs. How are your results affected by payment x?
(c) Find each player's strategy space, S_1 and S_2, construct the simultaneous version of the game, and find the NEs.
(d) Compare the SPNEs, found in part (b), and the NEs, found in part (c). Which NEs are sequentially rational? Interpret.
(e) *Allowing for commitment.* Assume now that firm 1 can credibly commit to the payment before the game starts (e.g., signing a contract that forces firm 1 to pay x to firm 2 when firm 2 chooses a high price). Find the SPE in this context.
(f) *Comparison.* Comparing the SPNEs when firm 1 can commit, found in part (e), and when it cannot, found in part (a), how much is firm 1 willing to pay firm 2, as captured by x?

6.8 **Backward induction with three moves.**[A] Consider the sequential-move game in Fig. 6.23 where, first, player 1 chooses between L and R; observing his action, player 2 responds with A or B after L or with C, D, or E after R. Finally, player 1 gets to choose again in the third period. As in similar games, player 1's payoff is the top number at the terminal nodes and player 2's payoff is the bottom number.

(a) Operating by backward induction, find player 1's responses in the last stage.
(b) Find player 2's best responses in the second stage.
(c) Find player 1's optimal strategy in the first stage and describe the SPE of the game.

6.9 **Backward induction with information sets.**[A] Consider the sequential-move game in Fig. 6.24. As in similar games, the top number in the terminal nodes' payoffs denote player 1's payoff, whereas the bottom number represents player 2's payoff.

(a) Operating by backward induction, find player 1's responses in the last stage.
(b) Find player 2's best responses in the second stage.
(c) Find player 1's optimal strategy in the first stage and describe the SPE of the game.

6.10 **Battle of market shares.**[A] Consider the following "battle of market shares" game. In the first stage, the entrant (firm 1) can choose to stay out (S) or enter (E) into the market. If the entrant decides not to enter, then the incumbent (firm 2) and the entrant earns a payoff of 5 and 2, respectively. However, if the entrant decides to enter, then they can be tough (T) or feeble (F) in

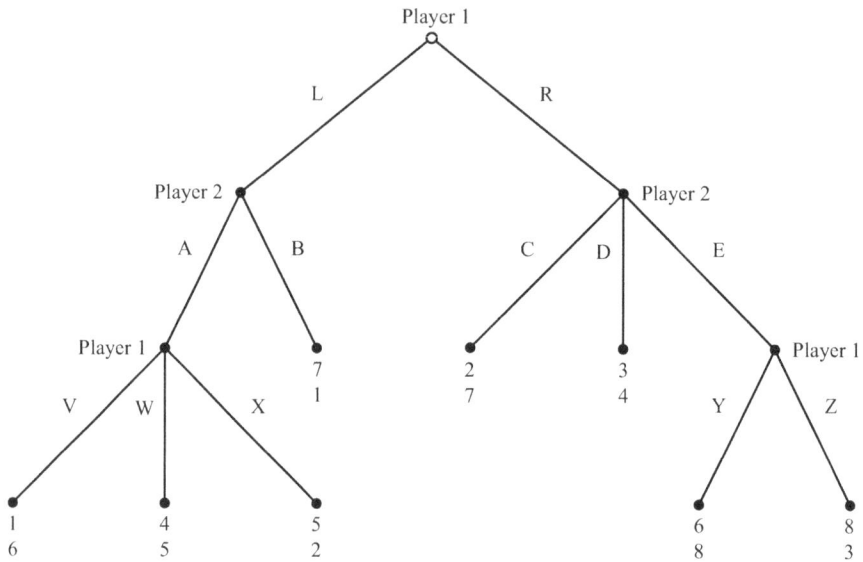

Fig. 6.23 Backward induction with three moves

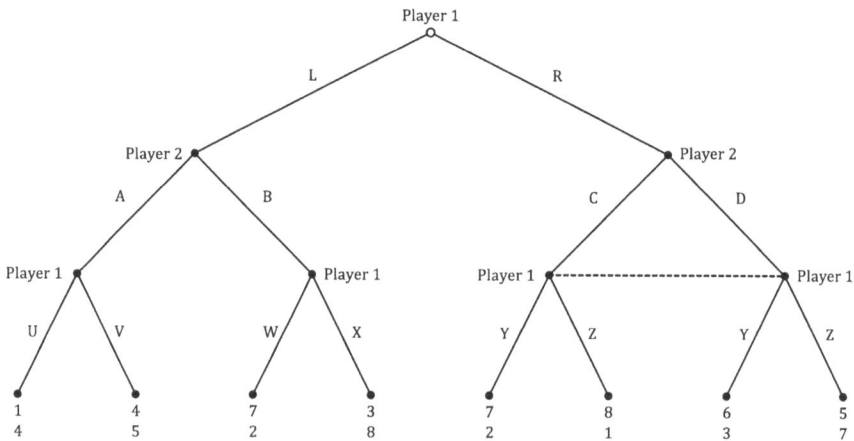

Fig. 6.24 Backward Induction with information sets

competition with one another. If both firms are tough (feeble), then every firm earns a payoff of 0. However, if one firm is tough and the other firm is feeble, the tough (feeble) firm earns a payoff of 3 (1, respectively).
(a) Depict this game in extensive form and identify all subgames.
(b) Solve for all SPE of this game with associated payoffs.

Exercises 191

6.11 **Battle of market shares, allowing for mistakes.**^{B} Consider the "battle of market shares" game again, but now assume that when the entrant decides to stay out, it can err on entering into the market with probability ε, where $0 < \varepsilon < 1$.
   (a) Identify the strategy sets of the incumbent and the entrant.
   (b) Represent this game in normal form.
   (c) Find all pure-strategy Nash equilibrium (psNE) of this game. Interpret.
   (d) How is equilibrium behavior affected by the entrant's probability of making mistakes (higher ε)? Compare your results to a setting with no mistakes, $\varepsilon = 0$.

6.12 **Burning money game.**^{B} Consider the "battle of market shares" game again, but now allow for the entrant to burn money x, where $0 < x < 2$ can be deemed as promotion and advertising expenses, if this firm decides to enter into the market in the first stage of the game. The incumbent perfectly observes how much money was burnt. In the second stage, firms simultaneously choose whether to be tough or feeble.
   (a) Write down the strategy sets of the incumbent and the entrant.
   (b) Represent this game in normal form.
   (c) Find all pure-strategy Nash equilibrium of this game. Interpret your results.

6.13 **Ultimatum bargaining game with altruistic players.**^{B} Consider the ultimatum bargaining game in Sect. 6.7.3. If an agreement is reached, the responder earns the division that the proposer offered, d, while the proposer earns the remaining surplus, $1 - d$. Assume, however, that the proposer's utility from this payoff distribution is

$$(1 - d) + \alpha_P d,$$

where $\alpha_P \in [0, 1]$ represents his degree of altruism. Similarly, the responder's utility from this payoff distribution is

$$d + \alpha_R (1 - d),$$

where $\alpha_R \in [0, 1]$ denotes his degree of altruism. Intuitively, every player cares about his own payoff (first component of every utility function) but also enjoys seeing the other player receive a higher payoff (second component of the utility function). When $\alpha_i = 0$, player $i = \{P, R\}$ does not exhibit altruistic preferences; when $\alpha_i \in (0, 1)$ he cares more about his own payoff than his rival's; and when $\alpha_i = 1$, he assigns the same weight to his payoff and those of his rival. For generality, we allow for altruistic preferences to satisfy $\alpha_P > \alpha_R$, $\alpha_P < \alpha_R$, or $\alpha_P = \alpha_R$.
   (a) Find the responder's best response. How is it affected by his altruism parameter, α_R?
   (b) Find the proposer's offer in equilibrium. How is it affected by his altruism parameter, α_P? How is it affected by the responder's altruism parameter, α_R? Interpret.

(c) How are your results affected if the proposer exhibits stronger altruistic preferences than the responder, $\alpha_P > \alpha_R$? What if, instead, the responder has stronger altruistic preferences?

(d) Evaluate your equilibrium results at $\alpha_P = \alpha_R = 0$. Interpret.

(e) Repeat parts (a)–(c) assuming that altruism parameters satisfy $\alpha_i \geq 0$ for every player i, thus not being restricted to $\alpha_i \in [0, 1]$.

6.14 **Ultimatum bargaining game with envious players.**[B] Consider the ultimatum bargaining game in the previous exercise. Assume, instead, that the proposer's utility function is

$$u_P = \begin{cases} (1-d) - \beta_P[d - (1-d)] & \text{if } 1-d < d, \text{ and} \\ 1-d & \text{otherwise.} \end{cases}$$

where $\beta_P \geq 0$ represents his disutility from disadvantageous payoff distributions, where he earns less than the responder ($1 - d < d$), thus capturing his disutility from envy. For simplicity, we assume that, if he earns more than the responder, he does not suffer from guilt, just considering $1 - d$ as his utility. Similarly, the responder's utility function is

$$u_R = \begin{cases} d - \beta_R[(1-d) - d] & \text{if } 1-d > d, \text{ and} \\ d & \text{otherwise.} \end{cases}$$

where $\beta_R \geq 0$ represents his disutility from envy.

(a) Find the responder's best response. How is it affected by his envy parameter, β_R?

(b) Find the proposer's offer in equilibrium. How is it affected by his envy parameter, β_P? How is it affected by the responder's envy parameter, β_R? Interpret.

(c) How is the equilibrium offer found in part (b) affected by the responder's envy parameter, β_R? How is it affected by the proposer's envy parameter, β_P? Interpret.

(d) Evaluate your equilibrium results at $\beta_R = \beta_R = 0$. Interpret.

6.15 **Gift exchange game.**[B] Consider the following sequential-move game. In the first stage, player 1 chooses the division of surplus, d_1, where $d_1 \in [0, 1]$. that he would like to deposit in a bank account, keeping the rest of the surplus for private consumption. In the second stage, d_1 is augmented to $m \times d_1$, where $m > 1$ indicates the return of the bank account. In the second period, player 2 is the only individual who can access the bank account, and makes an offer to player 1, d_2, where $d_2 \in [0, m \times d_1]$, keeping the remaining balance in the account. In the third stage, player 1 chooses whether to reject d_2, earning only $1 - d_1$, or accept d_2, earning $1 - d_1 + d_2$. For simplicity, assume no payoff discounting.

(a) Starting with the third stage, find player 1's acceptance rule, for a given division from player 2, d_2.

(b) Continuing with the second stage, find the offer that player 2 makes to player 1 as a function of player 1's investment, $d_2(d_1)$.

(c) Moving to the first stage, find player 1's equilibrium investment d_1^*.
(d) Is the above SPE efficient? To answer the question, think about the investment, d_1, and the division, d_2, that would maximize players' joint payoffs.

6.16 **Centipede game.**^{C} Consider a game where players 1 and 2 share a bank account, with a balance of B. In the first period, player 1 chooses whether to keep the money in the account or withdraw it. If he keeps the money, the balance grows to $B + r$, where r denotes the return on the account and satisfies $r > 0$. If, instead, he withdraws the money from the account, the game is over, and he earns $\frac{B}{2}$ while player 2 earns $\frac{B}{2} - f$, where parameter f denotes the fee that player 2 needs to pay to receive his share of the balance (e.g., amount $\frac{B}{2}$ needs to be wired to player 2 while player 1 receives it in cash without any fees) and satisfies $\frac{B}{2} > f > \frac{r}{2}$.

In the second period, if player 1 kept the money in the account, it is player 2's turn to choose whether to continue with the account or withdraw the money, earning $\frac{B+r}{2}$ while player 1 earns $\frac{B+r}{2} - f$. If he keeps the money in the account, the balance grows to $B+2r$ in the third period, and it is player 1's turn again to choose between continuing the account or withdrawing the money, with payoffs $\frac{B+2r}{2}$ for player 1 and $\frac{B+2r}{2} - f$ for player 2. The game continues for 100 periods, with alternating turns between players 1 and 2. (Player 1 plays in odd-numbered periods and player 2 plays in even-numbered periods.) If the bank account continues during all periods, every player earns $\frac{B+100r}{2} - f$.

Graphically, this game tree resembles a centipede, where every branch representing withdrawing in one of the periods looks like one of the centipede legs. For simplicity, assume no discounting of future payoffs.
(a) Starting in the last period, find player 2's response.
(b) Moving to the previous to last periods, find player 1's response.
(c) For any period t, where $1 \leq t \leq 98$, find player i's response in that period.
(d) Characterize the SPE of the game. Is it efficient?
(e) How are your results in part (d) affected if the game lasts for $T \geq 2$ periods.

6.17 **Bargaining with infinite periods and $N \geq 2$ players.**^{C} Consider the alternating-offer bargaining game with infinite periods from Sect. 6.7.6, but allowing for $N \geq 2$ players, all exhibiting the same discount factor $\delta \in (0, 1)$. Player 1 is the proposer in period 1, period $N+1$, period $2N+1$, and so on. Similarly, player 2 is the proposer in period 2, period $N+2$, period $2N+2$, and so on. A similar argument applies to any player who becomes the proposer for the first time in period k, becoming again the proposer in period $N+k$, period $2N+k$, etc.

A division of the surplus at period t is a vector describing the share that each player receives $(d_t^1, d_t^2, ..., d_t^N)$ satisfying $d_t^i \in [0, 1]$ for every player i and $\sum_{i=1}^{N} d_t^i = 1$. Assume that a division must be approved by all other $N-1$ players for it to be accepted (i.e., it requires unanimity). Focus on stationary equilibrium offers, implying that the equilibrium payoff that every player earns must only depend on who is the player making offers at that period (himself, the player making offers in the next period, the player making offers in two periods from now, etc.)

(a) Find the SPE of this game.

(b) *Two players.* Evaluate your results at $N = 2$ and show that they coincide with those in Sect. 6.7.6.

(c) *Three players.* Evaluate your results at $N = 3$. Compare them with those in part (b).

6.18 **Stackelberg competition with asymmetric firms.**[B] Consider the Stackelberg game of quantity competition in Sect. 6.7.1, but assume that the leader benefits from a cost advantage relative to the follower. In particular, the leader's marginal cost is c_1 while the follower's is c_2, where $1 > c_1 \geq c_2 > 0$.

(a) Find the follower's best response function, $q_2(q_1)$.

(b) Find the leader's equilibrium output, q_1^*, and report the SPE of the game.

(c) Find firm 2's output level in equilibrium.

(d) Under which conditions of marginal costs c_1 and c_2 only one firm produces a positive output level? Interpret.

6.19 **Stackelberg game when firms compete in prices-Homogeneous goods.**[A] Consider the Stackelberg game of quantity competition in Sect. 6.7.1, but assume that firms compete in prices rather than quantities. The inverse demand function is still $p_i(q_i, q_j) = 1 - q_i - q_j$.

(a) Find the follower's best response function, $p_2(p_1)$.

(b) Find the leader's equilibrium price, p_1^*, and report the SPE of the game.

(c) Does the leader enjoy a first-mover advantage? Compare your results with those of the Stackelberg game where firms compete in quantities. Interpret.

6.20 **Stackelberg game when firms compete in prices-Heterogeneous goods.**[B] Consider the Stackelberg game of quantity competition in Sect. 6.7.1, but assume that firms compete in prices rather than quantities. The inverse demand function is

$$p_i(q_i, q_j) = 1 - q_i - dq_j,$$

where parameter d satisfies $1 > d > 0$. A similar inverse demand function applies to firm j, $p_j(q_i, q_j) = 1 - q_j - dq_i$. Simultaneously solving for q_i and q_j in these inverse demand functions, we find the direct demand function

$$q_i(p_i, p_j) = \frac{1 - d - p_i + dp_j}{1 - d^2}$$

for every firm i.

(a) Using $q_i(p_i, p_j)$, find the follower's best response function, $p_2(p_1)$.

(b) Find the leader's equilibrium price, p_1^*, and report the SPE of the game.

(c) How is the SPE affected by a marginal change in parameter d? Interpret.

(d) Does the leader enjoy a first-mover advantage? Compare your results with those of the Stackelberg game where firms compete in quantities. Interpret.

6.21 **Stackelberg competition with heterogeneous goods.**[B] Consider the Stackelberg game of quantity competition in Sect. 6.7.1, but assume that the inverse demand function is

$$p_i(q_i, q_j) = 1 - q_i - d q_j,$$

where parameter d satisfies $1 > d > 0$.
(a) Find the follower's best response function, $q_2(q_1)$.
(b) Find the leader's equilibrium output, q_1^*, and report the SPE of the game.
(c) How is the SPE affected by a marginal change in parameter d? Interpret.
(d) Evaluate equilibrium output for the leader and follower at $d = 0$. Then evaluate these output levels at $d = 1$. Compare your results.

6.22 **Stackelberg with one leader and two followers.**[C] Consider an industry with three firms, facing inverse demand function $p(Q) = 1 - Q$, and constant marginal cost c, where $c \in [0, 1]$. Firm 1 is the industry leader, choosing its output q_1 in the first stage. In the second stage, firms 2 and 3 observe q_1 and simultaneously and independently respond by choosing their output levels q_2 and q_3.
(a) Find firms 2 and 3's best response functions. [*Hint*: They should be a function of q_1.]
(b) Find the leader's equilibrium output, q_1.
(c) Describe the SPE of the game and the output triplet (q_1, q_2, q_3) that arises in equilibrium.
(d) Compare your results in part (c) against those of a more standard Stackelberg game where each firm acts in a different stage (firm 1 chooses q_1, followed by firm 2, and then followed by firm 3). Interpret your findings.
(e) Consider now that firms 2 and 3 can coordinate their output levels q_2 and q_3 in order to maximize their joint profits in the second stage of the game. Find their output decisions (still as a function of q_1), and the leader's output in equilibrium. How are your results in part (c) affected?

6.23 **Stackelberg with two leaders and one follower.**[C] Consider an industry with three firms, facing inverse demand function $p(Q) = 1 - Q$, and constant marginal cost c, where $c \in [0, 1]$. Firms 1 and 2 are the industry leaders, independently and simultaneously choosing their output (q_1 and q_2) in the first stage. In the second stage, firm 3 observes the output pair q_1 and q_2, and responds by choosing its output level q_3.
(a) Find firm 3's best response function.
(b) Find every leader i's equilibrium output, q_i, where $i = \{1, 2\}$.
(c) Describe the SPE of the game and the output triplet (q_1, q_2, q_3) that arises in equilibrium.
(d) Compare your results in part (c) against those in a Stackelberg game with three firms, each of them choosing its output in a different period. Interpret your findings.

6.24 **Sequential-move version of public good game-I.**[B] Consider the simultaneous-move public good game in Sect. 4.4, where every player i's utility is given by

$$u_i(x_i, x_j) = (w_i - x_i)\sqrt{m(x_i + x_j)}$$

where $w_i > 0$ denotes player i's wealth, x_i represents his contribution to the public good, x_j is player j's contribution, where $j \neq i$, and m indicates the return from the aggregate donations to the public good, $X = x_i + x_j$. Consider now that player 1 is the leader, submitting his contribution to the public good, x_1, in the first period. In the second period, player 2 observes x_1 and responds with his contribution, x_2.

(a) Operating by backward induction, find player 2's best response function, $x_2(x_1)$.
(b) Moving to the first period, find player 1's equilibrium contribution, x_1^*.
(c) Under what conditions of wealth levels w_1 and w_2 will both players submit a positive contribution? Under what conditions will only one of them submit a positive contribution?
(d) Compare equilibrium contributions in the simultaneous-move and sequential-move games. Is there a first-mover advantage? Interpret.

6.25 **Sequential-move version of public good game-II.**[B] Consider a public good game where every player i has the following quasilinear utility function

$$u_i(x_i, x_j) = (w_i - x_i) + \sqrt{m(x_i + x_j)}$$

where $w_i > 0$ denotes player i's wealth, x_i represents his contribution to the public good, x_j is player j's contribution, where $j \neq i$, and m indicates the return from the aggregate donations to the public good, $X = x_i + x_j$. Consider that player 1 is the leader, submitting his contribution to the public good, x_1, in the first period. In the second period, player 2 observes x_1 and responds with his contribution, x_2.

(a) Operating by backward induction, find player 2's best response function, $x_2(x_1)$.
(b) Moving to the first period, find player 1's equilibrium contribution, x_1^*.
(c) Under what conditions of wealth levels w_1 and w_2 will both players submit a positive contribution? Under what conditions will only one of them submit a positive contribution?
(d) Compare equilibrium contributions in the simultaneous-move and sequential-move games. Is there a first-mover advantage? Interpret.

6.26 **Exploiting a common pool resource over two periods.**[B] Consider a common pool resource with an initial stock $S > 0$, exploited by two firms over two consecutive periods. For simplicity, prices are normalized to $p = \$1$, and assume a common discount factor δ, where $\delta \in (0, 1)$. In the first period, every firm i independently and simultaneously chooses its appropriation level, x_i, facing cost function

$$C_i^{1st}(x_i, x_j) = \frac{x_i(x_i + x_j)}{S},$$

which is increasing and convex in firm i's appropriation, x_i, increasing in its rival's appropriation, x_j, but decreasing in the abundance of the stock, S. In the second period, this cost function becomes

$$C_i^{2nd}(q_i, q_j) = \frac{q_i(q_i + q_j)}{S(1 + g) - X},$$

where $g \geq 0$ denotes the growth rate of the initial stock, S, and $X \equiv x_i + x_j$ represents aggregate first-period appropriation. Intuitively, when $g = 0$, the stock does not regenerate and the net stock available at the beginning of the second period is $S - x$; as captured by the denominator of the cost function. In contrast, when $g = \frac{X}{S}$, the stock is fully recovered, so the initial stock S is available again at the beginning of the second period. In this case, the second-period cost function is symmetric to that in the first period, simplifying to $\frac{q_i(q_i+q_j)}{S}$.

(a) *Second period.* Find every firm i's best response function, $q_i(q_j)$. How is $q_i(q_j)$ affected by an increase in the growth rate g? Interpret.
(b) Find every firm's second-period equilibrium appropriation $q_i^*(x)$, as a function of first-period appropriation X. How is $q_i^*(x)$ affected by an increase in the growth rate g? Interpret.
(c) *First period.* Find every firm i's first-period equilibrium appropriation, x_i^*.
(d) *Comparative statics.* Evaluate the first- and second-period appropriation in equilibrium. How are these results affected by an increase in the growth rate g? And by an increase in the discount factor δ? Interpret.

6.27 **Hotelling game with symmetric firms.**[A] Consider two firms, 1 and 2, selling the same product, and located at the endpoints of a unit line, with firm 1 located at point 0 and firm 2 located at 1 (the opposite end of the line). Consumers seek to buy only one unit, are uniformly distributed in the unit line, enjoy a utility r from product, and must pay a transportation cost $t \times d^2$, where $d \in [0, 1]$ denotes the distance that the consumer must travel when purchasing from one of the two firms, and t represents the per unit transportation cost. For simplicity, assume that r is high enough to guarantee all consumers in the unit line purchase the product.

Firms play a sequential-move game where, in the first stage, every firm i simultaneously and independently chooses its price p_i; and, in the second stage, observing the vector of prices (p_i, p_j), every consumer at location $x \in [0, 1]$ responds by buying one unit from firm 1 or 2. Consider that both firms face the same marginal cost per unit of output, $c > 0$

(a) *Second stage - Finding demand.* Given prices from the first stage, find the demand that each firm has in the second stage.

(b) *First stage - Prices.* Anticipating second stage demands, find the price that each firm sets in the first stage.

6.28 **Hotelling game with asymmetric firms.**[B] Consider the Hotelling game in the previous exercise, but assume that firm 1 faces a different marginal cost than firm 2, $c_1 \neq c_2$.

(a) *Second stage—Finding demand.* Given prices from the first stage, find the demand that each firm has in the second stage.

(b) *First stage—Prices.* Anticipating second stage demands, find the price that each firm sets in the first stage.

6.29 **Hotelling game with non-uniformly distributed consumers, based on Azar (2015).**[B] Consider the Hotelling game in Exercise 6.25, but assume that consumers are distributed in the unit line according to a cumulative distribution function $F(x)$, where $F(0) = 0$ and $F(1) = 1$, with density function $f(x) = F'(x)$ in all locations $x \in [0, 1]$. If $F(x) = x$, consumers are uniformly distributed, but in this exercise we extend our analysis to allow for other, non-uniform distributions as well.

(a) *Second stage—Finding demand.* Given prices from the first stage, find the demand that each firm has in the second stage.

(b) *First stage—Prices.* Anticipating second stage demands, find the price that each firm sets in the first stage.

(c) Assume that $F(x) = x^{1/2}$, which indicates that there is a larger mass of consumers near firm 1 than near firm 2. Use your results from part (b) to find equilibrium prices in this context. For simplicity, you can assume that $c = t = 1$.

(d) Assume that $F(x) = x^2$, which indicates that there is a larger mass of consumers near firm 2 than near firm 1. Use your results from part (b) to find equilibrium prices in this context. For simplicity, you can assume that $c = t = 1$.

(e) Compare your results in parts (c) and (d). Interpret.

6.30 **Endogenous location game.**[C] Consider the game of horizontally differentiated products with two firms. In the first stage, every firm i chooses its location, l_i, in the unit line. In the second stage, every firm, observing the location pair (l_1, l_2) that firms chose in the first stage, responds by setting its price p_i. In the third stage, given firms' location and prices, consumers buy one unit of the good from either firm 1 or 2, enjoying a utility $r > 0$ from the good. For simplicity, r is sufficiently high to induce consumers in all positions of the unit line to buy one unit of the good. Consumers are uniformly distributed in the unit line, face quadratic transportation costs, and both firms' marginal production cost is $c > 0$.

(a) *Third stage—Finding demand.* Given locations from the first stage, and prices from the second stage, find the demand that each firm has in the third stage.

(b) *Second stage—Prices.* Given locations from the first stage, find the price that each firm sets in the second stage.

(c) *First stage—Equilibrium location.* Anticipating equilibrium behavior in the second and third stages, find the equilibrium location choice of each firm in the first stage of the game.

6.31 **Cost-reducing investment followed by Cournot competition-I.**[B] Consider a duopoly market with two firms selling a homogeneous product, facing inverse demand curve $p(Q) = 1 - Q$, where $Q = q_1 + q_2$ denotes aggregate output, and facing marginal cost c, where $1 > c \geq 0$. In the first stage of the game, every firm i chooses its investment in cost-reducing technology, z_i; and, in the second stage, observing the profile of investment levels (z_i, z_j), firms compete à la Cournot.

Investment z_i reduces firm i's marginal cost, from c to $c - \frac{1}{4}z_i$; and the cost of investing z_i in the first stage is $\frac{1}{2}z_i^2$. For simplicity, assume no discounting of future payoffs.

(a) *Second stage.* Operating by backward induction, find firm i's best response function $q_i(q_j)$ in the second stage. How is it affected by a marginal increase in z_i? And by a marginal increase in z_j?

(b) Find equilibrium output and profits in the second stage as a function of investment levels (z_i, z_j). Are they increasing in z_i? Are they increasing in z_j? Interpret.

(c) *First stage.* Find the equilibrium investment levels that firms choose in the first stage, z_i^* and z_j^*.

(d) *Joint venture.* If, in the first stage, firms could coordinate their investment levels (z_i, z_j) to maximize their joint profits, what would their investment levels be? This investment decision resembles a "joint venture," where firms coordinate their R&D activities, or any other decision, and then compete in a subsequent stage (in this case, à la Cournot). Compare your results with those in part (c).

6.32 **Cost-reducing investment followed by Cournot competition-II.**[C] Consider the setting in the above exercise, but assume that firms sell a heterogeneous product, so every firm i faces inverse demand curve $p_i(q_i, q_j) = 1 - q_i - \frac{1}{2}q_j$. Firms still face the same marginal cost and time structure of the game.

(a) *Second stage.* Operating by backward induction, find firm i's best response function $q_i(q_j)$ in the second stage. How is it affected by a marginal increase in z_i? And by a marginal increase in z_j?

(b) Find equilibrium output and profits in the second stage as a function of investment levels (z_i, z_j). Are they increasing in z_i? Are they increasing in z_j? Interpret.

(c) *First stage.* Find the equilibrium investment levels that firms choose in the first stage, z_i^* and z_j^*.

(d) Compare your results with those in the previous exercise. Do firms invest more in cost-reducing technologies when they sell homogeneous or heterogeneous products?

(e) *Joint venture.* If, in the first stage, firms could coordinate their investment levels (z_i, z_j) to maximize their joint profits, what would their investment levels be? This investment decision resembles a "joint venture," where firms coordinate their R &D activities, or any other decision, and then compete in a subsequent stage (in this case, à la Cournot). Compare your results with those in part (c).

6.33 **Cost-reducing investment followed by Bertrand competition.**[B] Consider the setting in the previous exercise, but assume that firms compete, instead, in prices in the second stage (à la Bertrand).
  (a) *Second stage.* Operating by backward induction, find firm i's best response function $p_i(p_j)$ in the second stage. How is it affected by a marginal increase in z_i? And by a marginal increase in z_j?
  (b) Find equilibrium price and profits in the second stage as a function of investment levels (z_i, z_j). Are they increasing in z_i? Are they increasing in z_j? Interpret.
  (c) *First stage.* Find the equilibrium investment levels that firms choose in the first stage, z_i^* and z_j^*.
  (d) Compare your results with those in the previous exercise. Do firms invest more in cost-reducing technologies when they compete in quantities or in prices?
  (e) *Joint venture.* If, in the first stage, firms could coordinate their investment levels (z_i, z_j) to maximize their joint profits, what would their investment levels be? This investment decision resembles a "joint venture," where firms coordinate their R &D activities, or any other decision, and then compete in a subsequent stage (in this case, à la Bertrand). Compare your results with those in part (c).

6.34 **Contracts when effort is observable.**[A] Let us analyze the contract between a firm and a worker, where the firm perfectly observes the worker's effort. This is, of course, a simplifying assumption that we relax in future chapters, where we consider that the firm cannot observe the worker's effort on the job and must design the contract accordingly.

Assume that the worker's utility function is $u(w) = \sqrt{w}$, where $w \geq 0$ denotes her salary. The worker experiences disutility from exerting effort, e, measured by the linear function $g(e) = e$, and that there are only two effort levels the worker can exert, $e_H = 5$ and $e_L = 0$. His reservation utility is, for simplicity, $\bar{u} = 0$, which denotes his utility from rejecting this firm's contract and, perhaps, working for another company.

When the worker exerts a high effort, $e_H = 5$, the firm's sales are $0 with probability 0.1, $100 with probability 0.3, and $400 with probability 0.6. However, when the worker exerts low effort, $e_L = 0$, the firm's sales are $0 with probability 0.6, $100 with probability 0.3, and $400 with probability 0.1. Intuitively, low effort makes it more likely that $0 sales occur, while high effort increases the probability of $400 in sales.

Exercises

In the first stage, the firm offers a contract (w_H, w_L), which specifies a salary if a high or low effort is observed, respectively. In the second stage, the worker observes the contract (w_H, w_L), and responds by accepting or rejecting the contract. If he rejects the contract, he earns his reservation utility $\bar{u} = 0$, and if he accepts the contract he chooses his effort level, e_H or e_L.

(a) *Second stage.* Starting from the second stage, find the salary that induces the worker to accept a high-effort contract.

(b) Still in the second stage, find the salary that induces the worker to accept a low-effort contract.

(c) *First stage.* Find whether the firm chooses a high- or low-effort contract.

6.35 **Entry deterring investment.**[B] Consider an incumbent (firm 1) who can choose to invest or not invest in a technology. This investment is observable to the entrant (firm 2) who decides whether to enter into the industry or not. If firm 2 does not enter, the game ends and firm 1 enjoys monopoly profits. If firm 2 enters, firms engage in simultaneous competition, setting high (H), medium (M), or low (L) prices, with associated payoffs illustrated in Fig. 6.25.

(a) Find the SPE of this game.

(b) Suppose firm 2 does not observe firm 1's investment decision. Characterize the SPE and compare your results to part (a).

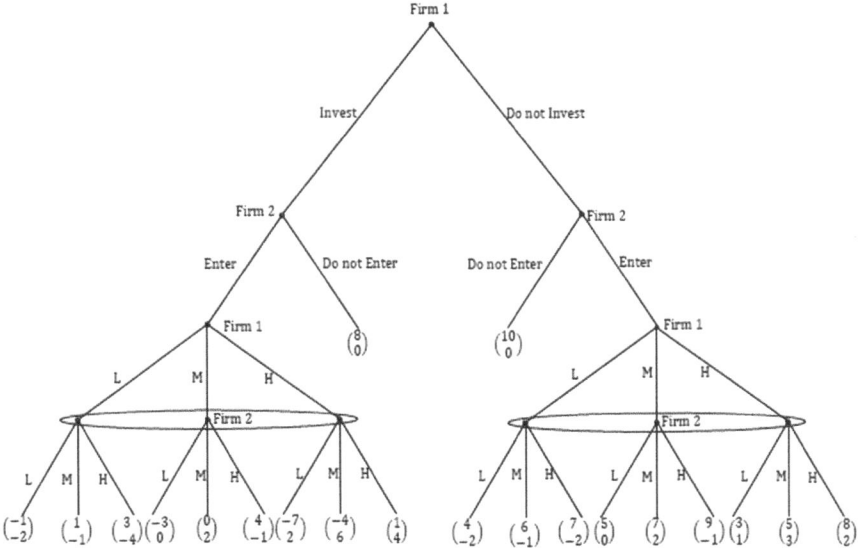

Fig. 6.25 Entry deterring investment

References

Azar, O. (2015) "A Linear City Model with Asymmetric Consumer Distribution," PLoS ONE, 10(6), e0129068.

Kuhn, H. W. (1953) ""Extensive games and the problem of information"," in: *Contri-butions to the Theory of Games II*, eds. H. W. Kuhn and A. W. Tucker, Annals of Mathematics Studies, 28, Princeton University Press, pp. 193—216.

Luce, R. D., and H. Raiffa (1957) *Games and Decisions: Introduction and Critical Survey*, John Wiley & Sons, Inc., New York.

Watson, J. (2013) *Strategy: An Introduction to Game Theory*, Worth Publishers. Third edition.

Zermelo, E. (1913) ""On an Application of Set Theory to the Theory of the Game of Chess," in U. Schwalbe and P.Walker ""Zermelo and the Early History of Game Theory," Games and Economic Behavior, 34, pp. 123–137.

Repeated Games   7

7.1   Introduction

In this chapter, we consider situations in which players interact in one of the strategic settings previously analyzed and they play the game for several rounds, which are known as "repeated games." In real life, agents often interact for long periods of time, facing the same institutional or industry characteristics in each period (i.e., the same "rules of the game"). In this respect, repeated games can help us describe many observed settings and predict strategic behavior in real-life scenarios.

We are particularly interested in identifying if the game's repetition provides players with more incentives to cooperate, relative to the unrepeated version of the game. To this end, we consider settings where cooperation cannot be sustained in the unrepeated version of the game (also known as the "stage game"), such as the Prisoner's Dilemma game. We first study its finitely repeated version, where players face the same game several times but know that their interaction will end after a fixed number of repetitions. In this context, we show that cooperation cannot be supported in equilibrium, and that players behave as in the unrepeated version of the Prisoner's Dilemma game, namely, confessing every time they interact.

We then extend players' interaction to an infinitely repeated game, which can be intuitively understood as players facing the same game and knowing that the game will continue for one more round with a positive probability. While an infinite interaction is technically possible in this context, the probability of players interacting during hundreds of rounds is very low. In this setting, we demonstrate that cooperation can emerge in equilibrium if players assign a sufficiently high value to future payoffs, while non-cooperative outcomes can also be supported regardless of players' intertemporal preferences. For instance, cooperation can be sustained when players use the so-called grim-trigger strategy, where they start cooperating in the first period of interaction, and continue to do so unless one of the players deviates (i.e., is uncooperative), which triggers a punishment phase in which every player behaves according to the NE of the stage game. In this type of strategy, every player, then, compares the instantaneous payoff gain from cheating today—deviating from cooperation while all other players cooperate—against the future loss from triggering the punishment phase.

We then apply the above tools to different settings, such as collusion in oligopoly models, where every firm chooses its output level in each period of interaction, and to stage games with more than one NE. Finally, we study modified versions of the grim-trigger strategy, such as allowing for permanent or temporal punishment phases, imperfect monitoring, or short but severe punishment; examining in each case whether cooperation is facilitated or hindered.

7.2 Repeating the Game Twice

Consider the Prisoner's Dilemma game we discussed in previous chapters, reproduced in Matrix 7.1 for easier reference.[1]

		Player 2	
		Confess	Not confess
Player 1	Confess	2, 2	8, 0
	Not confess	0, 8	4, 4

Matrix 7.1. The Prisoner's Dilemma game.

Let us repeat the game twice and find its SPE since players now interact in a sequential-move game. In the first stage, every player i simultaneously and independently chooses whether to Confess or Not confess. In the second stage, observing the outcome of the first stage (i.e., which cell was played), every player i selects again simultaneously and independently whether to Confess or Not Confess. This sequential-move game is depicted in Fig. 7.1.

The game tree includes information sets in the first stage to illustrate that player 2 does not observe which action player 1 chose (as players act simultaneously), and a similar argument applies to each of the information sets in the second stage. At the beginning of the second stage, however, players perfectly observe the strategy profile played in the first stage, indicated by the nodes at the end of the first stage not being connected by information sets. Payoff pairs in the terminal nodes are just a sum of the payoffs in the first and second stage, assuming no discounting. For instance, if (C,C) is played in the first stage but (NC,C) occurs in the second stage,

---

[1] We modified some of the payoffs, to avoid negative numbers, but Confess is still a strictly dominant strategy for every player since $2 > 0$ when her opponent confesses and $8 > 4$ when she does not.

7.2 Repeating the Game Twice

player 1 earns $2 + 0 = 2$, while player 2 earns $2 + 8 = 10$. A similar argument applies to other payoffs and we invite you to calculate a few of them as a practice.

When solving for the SPE of this game tree, we first identify the proper subgames, as we did in Fig. 7.1. There are clearly five subgames, one initiated after players choose (C,C) in the first stage, another initiated after (C,NC), after (NC,C), and after (NC,NC), and the game as a whole. Operating by backward induction, we can then solve each of these subgames, starting with subgames 1-4. Since these subgames are simultaneous-move games, we can solve each of them by transforming them to its matrix form.

Second stage. Matrices 7.2a–d present these four subgames and underline players' best response payoffs to identify the NE of the subgame, one at a time.

Subgame 1. Matrix 7.2a represents subgame 1, which starts after (C,C) is played in the first stage, showing that (C,C) is the NE in the second stage too.

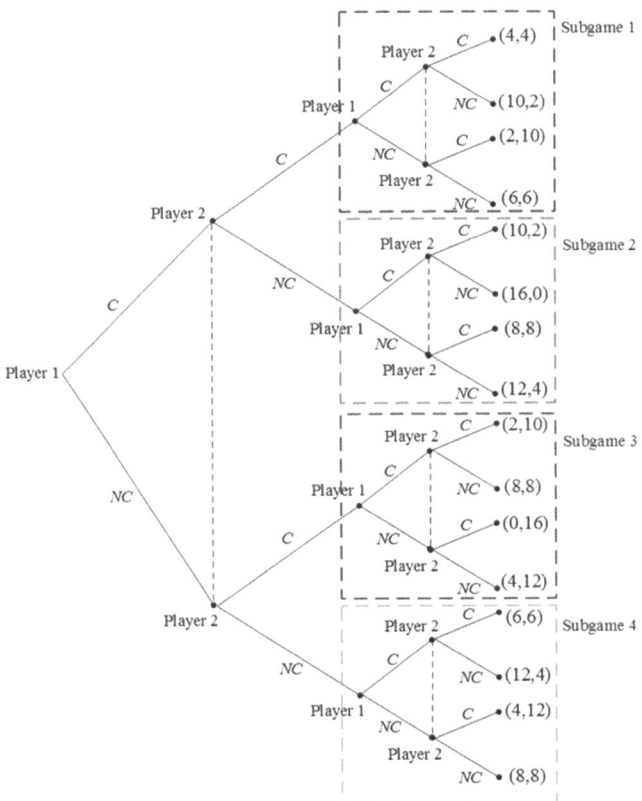

Fig. 7.1 Twice-repeated prisoner's dilemma game

		Player 2	
		Confess	Not confess
Player 1	Confess	<u>4</u>,<u>4</u>	<u>10</u>, 2
	Not confess	2,<u>10</u>	6, 6

Matrix 7.2a. Twice-repeated Prisoner's Dilemma game - Subgame 1.

Subgame 2. Matrix 7.2b analyzes subgame 2, which is initiated after players choose (C,NC) in the first stage, demonstrating that (C,C) is the NE in this subgame too.

		Player 2	
		Confess	Not confess
Player 1	Confess	<u>10</u>,2	<u>16</u>, 0
	Not confess	8, <u>8</u>	12, 4

Matrix 7.2b. Twice-repeated Prisoner's Dilemma game - Subgame 2.

Subgame 3. Matrix 7.2c analyzes subgame 3, which starts after players choose (NC,C) in the first stage. Strategy profile (C,C) is then the NE of subgame 3 as well.

		Player 2	
		Confess	Not confess
Player 1	Confess	2,<u>10</u>	<u>8</u>, <u>8</u>
	Not confess	0,<u>16</u>	4, 12

Matrix 7.2c. Twice-repeated Prisoner's Dilemma game - Subgame 3.

Subgame 4. Matrix 7.2d analyzes subgame 4, which is initiated after players choose (NC,NC) in the first stage, implying that (C,C) is the NE of this subgame too. In summary, strategy profile (C,C) is the NE of subgames 1–4 or, to put it differently, in the second stage players behave as in the unrepeated version of the Prisoner's Dilemma game regardless of how they behaved in the previous stage.

		Player 2	
		Confess	Not confess
Player 1	Confess	<u>6</u>,<u>6</u>	<u>12</u>, 4
	Not confess	4,<u>12</u>	8, 8

Matrix 7.2d. Twice-repeated Prisoner's Dilemma game - Subgame 4.

7.2 Repeating the Game Twice

Fig. 7.2 Twice-repeated prisoner's dilemma game—First stage

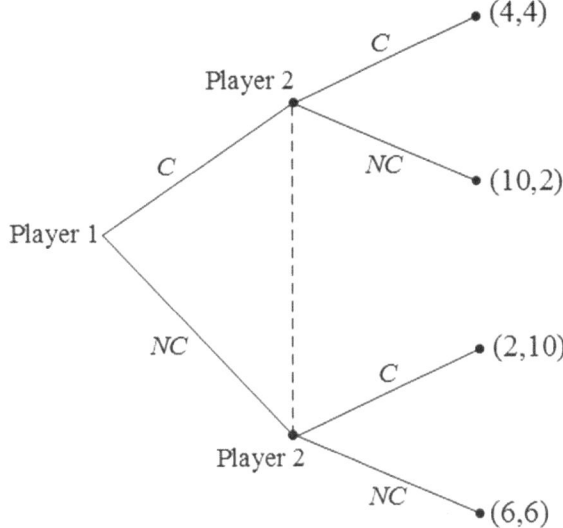

First stage. Operating by backward induction, we move now to the first stage, where every player can anticipate equilibrium behavior in subgames 1–4 and her equilibrium payoff in each subgame. Figure 7.2 insert our equilibrium payoffs from every subgame 1–4 into the nodes at the end of the first stage.

This matrix represents, of course, a simultaneous-move game that players play in the first stage, which can be represented with Matrix 7.2e. Underlining best response payoffs, we find that the NE of the first-period game is (C,C) too.

		Player 2	
		Confess	Not confess
Player 1	Confess	<u>4</u>,<u>4</u>	<u>10</u>, 2
	Not confess	2, <u>10</u>	6, 6

Matrix 7.2e. Twice-repeated Prisoner's Dilemma game - Game as a whole.

Summary. We are now ready to report the SPE of the twice-repeated Prisoner's Dilemma game:

- Every player $i = \{1, 2\}$ chooses to play Confess in the first stage, and
- Players Confess in the second stage regardless of the strategy profile played in the first stage.

Intuitively, players anticipate that they will both play Confess in the last stage independently of how cooperative or uncooperative they were in the previous stage. Informally, every player could think:

> It does not matter what I do today, Tomorrow we will all confess. I can then treat tomorrow's game as independent from today's play, as if I played two separate Prisoner's Dilemma games.

That reasoning would be correct, leading every player to choose Confess in every period. In other words, players cannot cooperate today (choosing Not confess) because there is no future punishment from not being cooperative today. Indeed, we know that, regardless of how we play today, we will all be choosing Confess tomorrow. Importantly, this result holds independently on how players discount their future payoffs (for all discount rates, δ).

In summary, repeating the game twice did not help us sustain cooperation in the SPE. Can cooperation emerge if we repeat the game for more periods, generally, for $T \geq 2$ periods? Spoiler alert: the answer is no, but let us formally show why that happens.

7.3 Repeating the Game $T \geq 2$ Times

When the game is repeated for $T \geq 2$ periods, Matrix 7.3 represents the game that players face when interacting in the last round (period T). All payoffs coincide with those in the unrepeated Prisoner's Dilemma game, but we added P_1^{T-1} to player 1's payoff (first component of every cell) to denote the discounted sum of all payoffs that this player earned in the previous $T - 1$ periods. Similarly, P_2^{T-1} denotes the discounted sum of player 2's payoffs until period $T - 1$.

		Player 2	
		Confess	Not confess
Player 1	Confess	$2 + P_1^{T-1}, 2 + P_2^{T-1}$	$8 + P_1^{T-1}, 0 + P_2^{T-1}$
	Not confess	$0 + P_1^{T-1}, 8 + P_2^{T-1}$	$4 + P_1^{T-1}, 4 + P_2^{T-1}$

Matrix 7.3. Finitely-repeated Prisoner's Dilemma game - Subgames at period T.

This presentation helps us describe the subgame that players face at period T for any previous history of play, that is, for any stream of strategy profiles occurring in previous periods (e.g., some rounds when all or some players choose Confess, followed by rounds when they choose Not Confess, or any other combinations) since the payoffs from these previous periods are included in term P_1^{T-1} for player 1

7.3 Repeating the Game $T \geq 2$ Times

and in term P_2^{T-1} for player 2. To illustrate the generality of this matrix representation, evaluate terms P_1^{T-1} and P_2^{T-1} in the special case where the game is only repeated $T = 2$ times. In this context, P_1^{T-1} just captures the payoff that player 1 earned in the first period of interaction, and similarly with P_2^{T-1} for player 2. For instance, if (C,C) emerged in the first stage of the game, $P_1^{T-1} = P_2^{T-1} = 2$; while if (C,NC) occurred, then $P_1^{T-1} = 8$ and $P_2^{T-1} = 0$.

If the Prisoner's Dilemma can produce four different strategy profiles in each period, there are 4^{T-1} nodes at the beginning of period T, when players face Matrix 7.3. In the context of the twice-repeated Prisoner's Dilemma game, for instance, there are $4^{2-1} = 4$ nodes at the beginning of the second period, as shown in the previous section (see Fig. 7.1). More generally, if the unrepeated version of the game has $k > 2$ different strategy profiles and the game is repeated T times, there are k^{T-1} nodes at the beginning of period T, only k^{T-2} at the beginning of period $T-1$, and similarly in previous periods.

Period T. Underlining best response payoffs, we can see that (C,C) is an NE in the T-period subgame in Matrix 7.3. To see this, let us find player 1's best responses. (Player 2's best responses are analogous since players' payoffs are symmetric.) When player 2 chooses Confess, in the left column, player 1's best response is to Confess because $2 + P_1^{T-1} > 0 + P_1^{T-1}$ simplifies to $2 > 0$. Similarly, when player 2 selects Not confess, in the right column, player 1's best response is to Confess since $8 + P_1^{T-1} > 4 + P_1^{T-1}$ simplifies to $8 > 4$.[2]

You may have noticed an important feature of the above result: it did not depend on players' payoffs in previous periods, as captured by P_1^{T-1} and P_2^{T-1}, which implies that every player finds Confess to be a best response to her rival's strategy *regardless* of the previous history of play (e.g., both when they cooperated in all previous rounds, when they did not in all previous periods, and any combination of the two).

Period $T-1$. Operating by backward induction, thus moving now to period $T-1$, the above results emerge again. Indeed, players face subgames that can be represented with Matrix 7.4, which coincides with Matrix 7.3 except for the terms measuring the discounted sum of payoffs in the previous $T-2$ periods, which are now P_1^{T-2} and P_2^{T-2}.

		Player 2	
		Confess	Not confess
Player 1	Confess	$2 + P_1^{T-2}, 2 + P_2^{T-2}$	$8 + P_1^{T-2}, 0 + P_2^{T-2}$
	Not confess	$0 + P_1^{T-2}, 8 + P_2^{T-2}$	$4 + P_1^{T-2}, 4 + P_2^{T-2}$

Matrix 7.4. Finitely-repeated Prisoner's Dilemma game - Subgames at period $T-1$.

---

[2] Our results then show that Confess strictly dominates Not confess for both players 1 and 2.

Underlining best response payoffs, we can show that (C,C) is again the unique NE in the subgame of period $T - 1$ represented in Matrix 7.4. The intuition is identical to that in the twice-repeated game: players anticipate that everyone will play Confess in the last round of interaction (period T), thus having no incentives to cooperate today (in period $T - 1$) as misbehavior will not be disciplined in the following period.

Period $T - 2$. A similar argument applies if, operating by backward induction, we consider period $T - 2$, where players face a matrix exactly like 7.4 but with terms P_1^{T-3} and P_2^{T-3} to represent the discounted sum of their payoffs in the previous $T - 3$ periods. We do not include such a matrix, but you can confirm, as a practice, that underlining best response payoffs still yields that (C,C) is the unique NE in this subgame.

Extending that reasoning to all previous periods, we find that (C,C) is the unique NE in *every subgame* and in *every period*. This means that the SPE is as uncooperative as one can imagine: every player i chooses Confess in every period, regardless of the previous history of play. And, once again, we encountered the motives behind this uncooperative behavior in the twice-repeated game, namely, players anticipate that none of them will cooperate in the last round of play (period T); which means that uncooperative conduct in period $T - 1$ will not be punished in period T, thus providing incentives to Confess in period $T - 1$. Moving to period $T - 2$, again, players anticipate that they will both play Confess in all subsequent periods $T - 1$ and T, regardless of how they behave at $T - 2$, hindering any cooperative behavior at period $T - 2$. Extending this argument to all previous periods, we reach the first period, where every player anticipates (C,C) to be played in all subsequent periods, which leads her to choose Confess today.

7.4 Repeating the Game Infinitely Many Times

We could not support cooperation by repeating the game twice or $T \geq 2$ times, but our above results can point us to why uncooperative outcomes emerge: the presence of a period in which players know that the game will end. At that stage, both players behave as in the unrepeated version of the game, regardless of how the game was played in previous rounds, and using backward induction they extend this behavior to all previous interactions. Therefore, if we seek to sustain cooperation, players cannot know, with certainty, when the game will end. They interact today and will keep playing with each other tomorrow with a probability p, where $p \in [0, 1]$. For instance, if $p = 0.7$, the chances that players interact for more than 20 rounds is less than 1 percent, but it is still positive for all rounds. In other words, players could be playing the game for infinitely many periods, but this probability is close to zero.

7.4.1 Uncooperative Outcome

We first show that the uncooperative outcome, (C,C) in every period, is still one of the SPEs in the infinitely repeated game. To show this, note that if a player chooses C at every period t, anticipating that her rival will choose C as well, she obtains a sequence of 2's, with discounted value

$$2 + 2\delta + 2\delta^2 + ... = 2\left[1 + \delta + \delta^2 + ...\right]$$
$$= 2\frac{1}{1-\delta}$$

because

$$1 + \delta + \delta^2 + ... = \delta^0 + \delta^1 + \delta^2 + ...$$
$$= \sum_{t=0}^{\infty} \delta^t$$
$$= \frac{1}{1-\delta}$$

is the sum of an infinite geometric series and $\delta \in (0, 1)$ denotes the discount factor. Recall that, intuitively, this discount factor represents how much players care about future payoffs: when $\delta \to 1$, players assign a similar value to current and future payoffs (they are patient), while when $\delta \to 0$, players assign almost no value to future payoffs, only caring about today (they are impatient). In addition, we assume that δ satisfies $\delta < 1$ because, if we allowed for $\delta = 1$, the sum of an infinite sequence of payoffs would be unbounded (from above).

If, instead, she unilaterally deviates to Not confess for one period, she earns[3]

$$\underbrace{0}_{\text{Not Confess}} + \underbrace{2\delta + 2\delta^2 + ...}_{\text{Confess}}$$

which simplifies to

$$0 + 2(\delta + \delta^2 + ...) = 0 + 2\delta(1 + \delta + \delta^2 + ...)$$
$$= 0 + 2\frac{\delta}{1-\delta}$$

given that $1 + \delta + \delta^2 + ... = \frac{1}{1-\delta}$. Hence, this player earns a higher stream of payoffs choosing Confess than Not confess because

$$2\frac{1}{1-\delta} \geq 2\frac{\delta}{1-\delta}$$

---

[3] We assume that player i deviates to Not confess for only one period, yielding a lower stream of payoffs than Confess. You can check that longer deviations to Not confess (for two or more periods) would be even more unprofitable.

simplifies to $1 \geq \delta$, which holds given that $\delta \in (0, 1)$ by assumption. (Note that we multiplied both sides of the above inequality by $1 - \delta$, a common trick we use to rearrange discounted payoff streams in infinitely repeated games.)

7.4.2 Cooperative Outcome

We now turn to the main question in this chapter: can we sustain cooperation as a SPE of the infinitely repeated game? For that, we need players to cooperate and keep cooperating as long as every player cooperated in previous periods, but otherwise threat to move the game to a "grim" outcome where players earn lower payoffs than by cooperating. The following Grim-Trigger Strategy (GTS) helps us achieve exactly that (we consider other forms of GTS in subsequent sections):

1. In period $t = 1$, choose Not confess.
2. In period $t \geq 2$,
   (a) keep choosing Not confess if every player chose Not confess in all previous periods, or
   (b) choose Confess thereafter for any other history of play (i.e., if either player chose Confess in any previous period).

The name of the GTS strategy should be clear at this point: any deviation from cooperation triggers a grim punishment thereafter, where every player reverts to the NE of the unrepeated version of the game (often known as "Nash reversion").

To show that the GTS is a SPE of the infinitely repeated game, we need to demonstrate that it is an optimal strategy for each player at every subgame where she is called to move. This means that players find the GTS to be optimal at *any* period t, and after *any* previous history of play (e.g., after some rounds of cooperation and some rounds of non-cooperative behavior by either or both players). A formidable task? Not really, because the GTS restricts our attention to only two possible histories of play: one in which all players cooperated in all previous periods (as in point 2a) and all other histories meaning that at least one player did not cooperate, choosing Confess, in some previous period (as in point 2b). We next analyze each of these cases.

Case (1). No cheating history. At any period t, if no previous history of cheating occurs, the GTS prescribes that every player keeps cooperating (choosing Not confess), which yields a payoff of 4 for every player i, entailing a payoff stream of

$$4 + 4\delta + 4\delta^2 + \ldots = 4(1 + \delta + \delta^2 + \ldots)$$
$$= 4 \frac{1}{1 - \delta}.$$

If, instead, player i chooses Confess today (which we interpret as her deviating from the above GTS, or more informally as "cheating"), her current payoff

7.4 Repeating the Game Infinitely Many Times

increases from 4 to 8. Importantly, this is a *unilateral deviation* as only player i chooses Confess while her opponent still plays according to the GTS. Essentially, to prove that the GTS can be sustained in equilibrium, we need to confirm that no player has incentives to deviate from the GTS if her rival behaves as prescribed by the GTS.

In the following period, however, her deviation is detected and triggers and infinite punishment to (C,C) thereafter, with a payoff of only 2 to every player. Therefore player i's discounted payoff stream from deviating from the GTS is

$$\underbrace{8}_{\text{Deviation to Confess}} + \underbrace{2\delta + 2\delta^2 + \ldots}_{\text{Punishment thereafter}}$$

which simplifies to

$$8 + 2(\delta + \delta^2 + \ldots) = 8 + 2\delta(1 + \delta + \delta^2 + \ldots)$$
$$= 8 + 2\frac{\delta}{1-\delta}.$$

We can then say that, after a history with no previous cheating episodes, every player chooses to cooperate (playing Not confess) if

$$4\frac{1}{1-\delta} \geq 8 + 2\frac{\delta}{1-\delta}.$$

Multiplying both sides by $(1-\delta)$, yields $4 \geq 8(1-\delta) + 2\delta$. Rearranging, we find $4 \geq 8 - 6\delta$ and, after solving for δ, we obtain

$$\delta \geq \frac{2}{3}.$$

Case (2). Some cheating history. At period t, if one or both players cheated in previous periods, then the GTS dictates that every player responds with Confess thereafter, earning a discounted stream of payoffs

$$2 + 2\delta + 2\delta^2 + \ldots = 2(1 + \delta + \delta^2 + \ldots)$$
$$= 2\frac{1}{1-\delta}.$$

Player i could, instead, unilaterally deviate to Not confess, while her rival plays Confess as part of the punishment in the GTS. You may suspect that such a deviation is not profitable, as player i will be the only one choosing Not confess, decreasing her payoff in that period, rather than increasing it. Your intuition is correct, but we can confirm it by writing player i's discounted stream of payoffs from this deviation

$$\underbrace{0}_{\text{Deviation to Not confess}} + \underbrace{2\delta + 2\delta^2 + \ldots}_{\text{Punishment thereafter}}$$

In period t, players observe (NC,C) or (C,NC) being played, implying that a deviation from the fully cooperative outcome (NC,NC) occurred, triggering an infinite punishment of (C,C) thereafter, with a payoff of 2 to every player. This payoff stream simplifies to

$$2(\delta + \delta^2 + ...) = 2\delta(1 + \delta + \delta^2 + ...)$$
$$= 2\frac{\delta}{1-\delta}.$$

Comparing our payoff streams from sticking with the GTS and unilaterally deviating, we find that, after a history of cheating, every player i prefers to implement the punishment prescribed by the GTS if

$$2\frac{1}{1-\delta} \geq 2\frac{\delta}{1-\delta}$$

which, after multiplying both sides by $(1-\delta)$, simplifies to $2 \geq 2\delta$, which holds because $\delta \in (0, 1)$ by definition. As a result, every player i has no incentives to unilaterally cooperate, playing Not confess, when she anticipates that her rival will be choosing Confess thereafter.

Summary. Combining our results from Cases (1) and (2), we found only one condition restricting the value of δ, $\delta \geq \frac{2}{3}$, for us to sustain the GTS as a SPE of the infinitely repeated Prisoner's Dilemma game. Intuitively, when players assign a sufficiently high weight to future payoffs, they start cooperating in the first period and keep cooperating in all subsequent periods, yielding outcome (C,C) in every round of interaction.

Figure 7.3 illustrates the trade-off that player i experiences when deciding whether to cooperate, playing Not confess; or cheat, playing Confess, after a history of cooperation (see Case 1). If she sticks to the GTS, she earns a payoff of 4 thereafter, as depicted in the flat dashed line in the middle of the figure. If, instead, she cheats, her current payoff increases from 4 to 8 for one period, but her rival detects her cheating, triggering a punishment to play (C,C) thereafter, with associated payoff 2. Relative to what she earns by cooperating (4), cheating then provides her with an instantaneous gain (as depicted in the left square) and a future payoff loss due to the punishment (represented by the right rectangle).

7.4.3 Cooperative Outcome—Extensions

Figure 7.3 helps us understand how the condition on the minimal discount factor supporting cooperation (e.g., $\delta \geq \frac{2}{3}$) is affected when we change the GTS in different ways.

1. **Temporary punishments**. The GTS in the above example assumes an infinite reversion to the NE of the stage game (unrepeated version of the game).

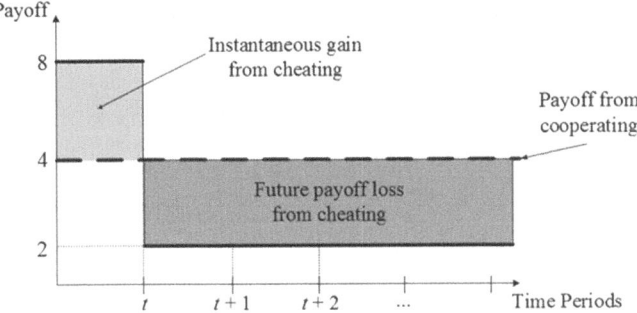

Fig. 7.3 Incentives from cheating in the infinitely repeated prisoner's dilemma game

But what if players only revert to this NE during a finite number of periods, N, moving back to cooperation once every player observes that both players implemented the punishment during N periods? Figure 7.4 depicts this setting, where the right rectangle would be narrower than that in Fig. 7.3. Intuitively, every player has stronger incentives to cheat because the punishment is less threatening (temporary rather than permanent), thus requiring a more demanding condition on δ for the GTS to be sustained as a SPE of the infinitely repeated game. In summary, shortening the punishment, while keeping its severity, shrinks the parameter values where we can sustain cooperation. Exercise 7.5 provides a formal proof of this result.

2. **More attractive cheating.** If every player i earns 10 rather than 8 when she chooses Confess while her opponent plays Not confess, the height of the instantaneous gain from cheating increases, as depicted in Fig. 7.5. This makes cheating more attractive. Formally, this increases the minimal discount factor sustaining cooperation, $\underline{\delta}$, where the GTS can be sustained as a SPE if $\delta \geq \underline{\delta}$. Because the range of δ's sustaining cooperation satisfies $\delta \in [\underline{\delta}, 1)$, an

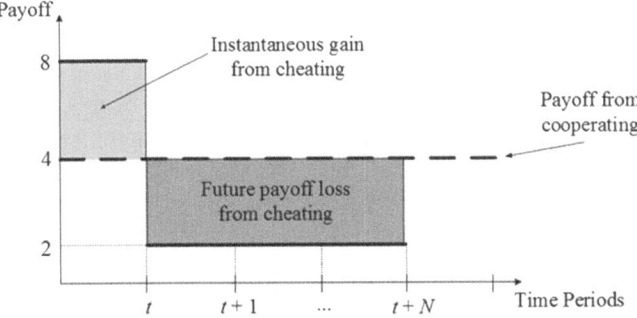

Fig. 7.4 Temporary punishments

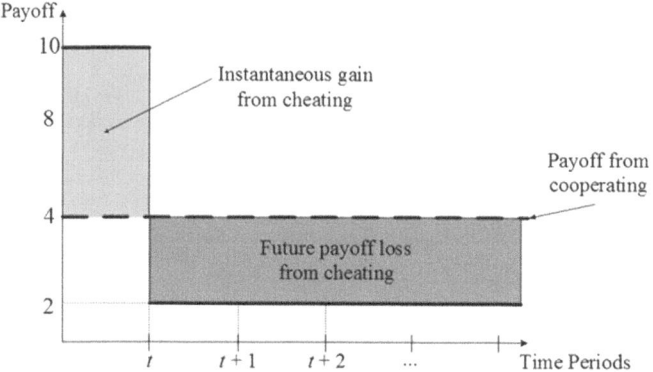

Fig. 7.5 More attractive cheating

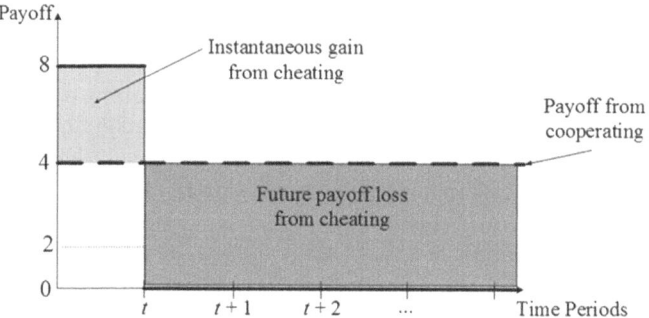

Fig. 7.6 More severe punishment

increase in cutoff $\underline{\delta}$ entails that cooperation emerges in a more restricted range of δ's. For compactness, the literature says that an increase in cutoff $\underline{\delta}$ "hinders cooperation in equilibrium," while a decrease in $\underline{\delta}$ facilitates such cooperation. Exercise 7.6 proves this result in detail.

3. **More severe punishments**. If every player earns 0 at the NE of the stage game, rather than 2, the right rectangle in Fig. 7.3 becomes deeper, as depicted in Fig. 7.6, indicating a larger future payoff loss from cheating today. Intuitively, cheating becomes less attractive, which decreases the minimal discount factor sustaining cooperation, $\underline{\delta}$. The GTS can, then, be sustained for a larger range of δ's, i.e., $\delta \in [\underline{\delta}, 1)$, implying that cooperation is facilitated. Exercise 7.7 provides a formal proof of this result.

4. **Lag in detecting cheating**. Our above example assumed, for simplicity, that every player detects cheating (any player choosing Confess) immediately when this cheating occurs. However, in some real-life examples, players may detect cheating $k \geq 0$ periods after it happened. If $k = 0$, we would still have

7.5 Folk Theorem

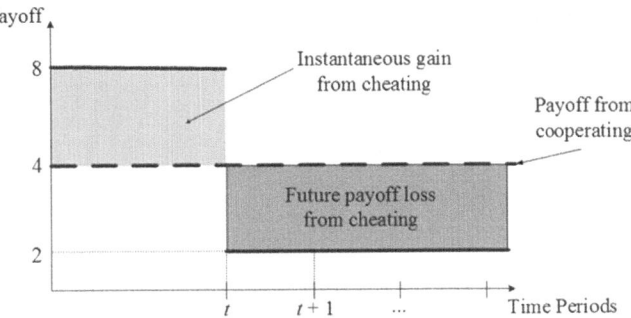

Fig. 7.7 Lag at detecting cheating

immediate detection, while $k > 0$ entails a lag in detecting cheating episodes. Graphically, this lag widens the left square representing the gain from cheating in Fig. 7.7, which is not instantaneous in this setting, as player i enjoys her cheating payoff during $t = 1 + k$ periods. Of course, this makes cheating more attractive, increasing the minimal discount factor sustaining cooperation, $\underline{\delta}$, which hinders cooperation in equilibrium. Exercise 7.8 provides a more detailed analysis of this finding.

5. **Lag in starting punishments**. A similar argument as in point (4) applies if players, despite detecting cheating immediately after it occurs, need k periods to revert to the NE of the stage game. This happens, for instance, when "cooperation" means producing few units (as when two firms collude, agreeing to reduce their output, driving prices up) while "cheating" indicates more intense competition, producing a significantly larger number of units. If the cheated firm needs several periods to expand its production process, the beginning of the punishment phase is, essentially, delayed. Graphically, this lag expands the payoff gain from cheating in the left square of Fig. 7.7 to $1 + k$ periods. A lag in starting punishments, therefore, has similar effects as the lag in detecting cheating (point 4), hindering cooperation in equilibrium.

7.5 Folk Theorem

In previous sections, we showed that we can sustain both the completely uncooperative outcome, where (C,C) occurs in every period, and the fully cooperative outcome, where (NC,NC) arises in every period of the game. Can we also support other, partially cooperative outcomes, where, for instance, players choose Confess in only some periods? More generally, which other outcomes can be sustained as SPEs of the infinitely repeated game? In particular, we seek to identify the *per-period payoffs* that players earn at different SPEs of this game. We define per-period payoffs below.

Definition 7.1. **Per-period payoff**. If player i's present value from an infinite payoff stream is defined as $PV_i = \sum_{t=0}^{\infty} \delta^t v_i^t$, her per-period payoff $\overline{v}_i$ is the constant payoff that solves

$$PV_i = \frac{\overline{v}_i}{1-\delta}$$

or, after solving for $\overline{v}_i$, $\overline{v}_i = (1-\delta)PV_i$.

Intuitively, when player i receives the constant sum $\overline{v}_i$ in every period, she is indifferent between that (flat) payoff stream and her (potentially fluctuating) stream of payoffs, as they both yield the same present value, PV_i.

7.5.1 Feasible and Individually Rational Payoffs

We now seek to find all SPEs that can be sustained in the infinitely repeated version of the game and, as a consequence, predict which per-period payoffs players earn in each SPE. To answer this question, we first need to define the set of payoffs in the unrepeated version of the game (feasible payoffs). Afterwards, we restrict this set of payoffs to those where every player earns a higher payoff than at the NE of the stage game (individually rational payoffs). As in previous sections of this chapter, we focus on simultaneous-move games of complete information, repeating it infinitely many times.

Definition 7.2. **Feasible payoffs (FP)**. A feasible payoff vector $v = (v_1, v_2, ..., v_N)$ can be achieved by convex combinations of any two payoff vectors v' and v''.

Therefore, in payoff vector v player i earns v_i, where v_i is found by a convex combination $v_i = \alpha v_i' + (1-\alpha)v_i''$ and $\alpha \in [0,1]$ represents the weight on v_i'. Intuitively, the FP set captures all possible payoff vectors (e.g., payoff pairs in a two-player game) that players can earn if they play the game in different ways. The next example illustrates the construction of the FP set in the Prisoner's Dilemma game.

7.5 Folk Theorem

Example 7.1. Finding FP in the Prisoner's Dilemma game. Consider the Prisoner's Dilemma of Matrix 7.5.

		Player 2	
		Confess	Not confess
Player 1	Confess	2, 2	8, 0
	Not confess	0, 8	4, 4

Matrix 7.5. The Prisoner's Dilemma game.

Figure 7.8 depicts the FP set. First, each vertex depicts one of the payoff pairs that players earn by playing one of the pure-strategy profiles (one of the four cells in the matrix) every period. For instance, if they play (C,C), players earn (2,2) in every period, yielding a per-period payoff of 2 for each player. A similar argument applies to (C,NC), yielding a per-period payoff pair of (8,0); to (NC,C), which yields (0,8); and to (NC,NC), which yields per-period payoffs of (4,4).

But as the figure suggests, FP includes more than just the four vertices. Convex combinations of these vertices can yield other, still feasible, per-period payoffs. For instance, if players alternate between (C,C) and (NC,NC), the per-period payoff is $\frac{1}{2}2 + \frac{1}{2}4 = 3$ to each player, graphically positioned halfway in the diagonal connecting points (2,2) and (4,4). A similar argument applies if, instead, players alternate between (C,C) and (NC,C), the per-period payoff for player 1 is 1 while that of player 2 is 5. Operating similarly, one can produce any pair inside the FP set. □

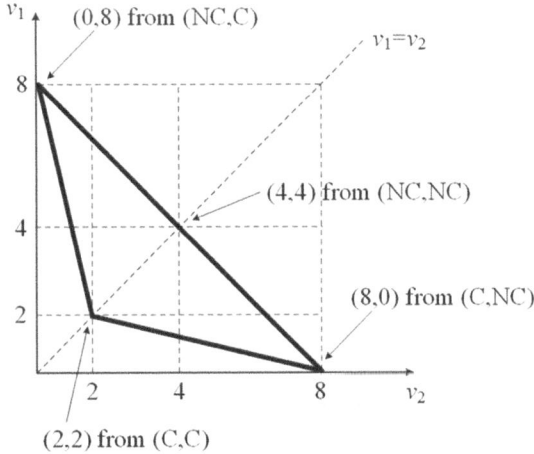

Fig. 7.8 FP set in the Prisoner's Dilemma game

Recall that the FP set does not mean that players earn any payoff pair in the FP every period. Instead, it means that players can play the game in such a way that, even if their payoffs vary across time, their per-period payoff (the constant amount they would need to be indifferent) would lie inside the FP.

Among all payoffs in the FP set, however, you may suspect that players may not want to play according to a GTS that induces per-period payoffs below those they would earn by playing the NE of the stage game. Why bother playing a GTS if they are worse off than in the NE of the stage game, right? We next restrict the FP set to those payoffs that make both players weakly better-off than in the NE of the stage game.

Definition 7.3. **Individually rational payoffs (IR)**. An individually rational (IR) payoff vector $v = (v_1, v_2, ..., v_N)$ satisfies $v_i \geq v_i^{NE}$ for every player i, where v_i^{NE} denotes her NE payoff in the unrepeated version of the game.

Combining the set of FP and IR payoffs, we are interested in payoff vectors that are both feasible and individually rational, which we denote as FIR. The next example restricts the FP set in Example 7.1, to identify the set of FIR payoffs, making both players better-off than at the NE where (C,C) holds.

Example 7.2. Finding FIR payoffs. In the Prisoner's Dilemma game of Example 7.1, a per-period payoff for player i is individually rational if $v_i \geq 2$, since every player earns 2 in the NE of the stage game. Figure 7.9 depicts a vertical line representing $v_2 \geq 2$ for player, which holds for all payoffs to the right-hand side of 2; and a horizontal line capturing $v_1 \geq 2$ for player, which occurs for all payoffs above 2.

The FP diamond is, then, restricted to a portion in the northeast of the figure (see shaded area), indicating that both players earn a higher per-period payoff than at the NE of the stage game.□

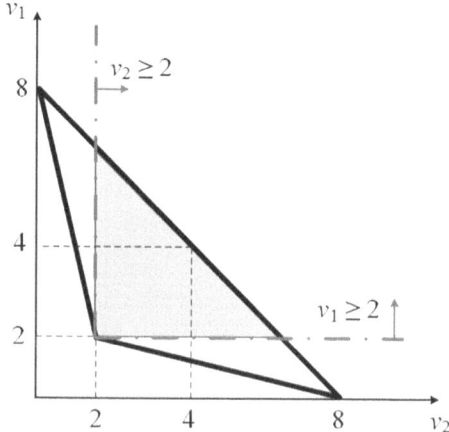

Fig. 7.9 FIR set in the prisoner's dilemma game

7.5 Folk Theorem

Table 7.1 Modified GTS inducing partial cooperation

	Outcome	Payoffs
Period 1	NC, C	0, 8
Period 2	C, NC	8, 0
Period 3	NC, C	0, 8
Period 4	C, NC	8, 0
...		

7.5.2 Folk Theorem and Cooperation

We are now ready to present the Folk theorem. This theorem, originally due to Friedman (1951), identifies equilibrium behavior in infinitely repeated games. As a curiosity, it was called the Folk theorem because it was widely known among game theorists before its publication, being part of the "folk," even though no one had formalized it.

Definition 7.4. **Folk theorem.** Every per-period payoff vector $v = (v_1, v_2, ..., v_N)$ in the FIR set can be sustained as the SPE of the infinitely repeated game for a sufficiently high discount factor, δ, where $\delta \geq \underline{\delta}$.

Graphically, the Folk theorem says that any point in the FIR diamond (on the edges or strictly inside) can be supported as a SPE of the infinitely repeated game as long as players care enough about their future payoffs (high δ).

We know that the uncooperative outcome, where (C,C) emerges in every period, can be sustained for all values of δ, yielding per-period payoffs (2,2), as depicted on the southwest corner of the FIR diamond. Similarly, the fully cooperative outcome, where (NC,NC) arises in every period, can be supported if $\delta \geq \frac{2}{3}$, as shown in Sect. 7.4.2, and illustrated on the northeast corner of the FIR diamond. But what about partially cooperative outcomes? Example 7.3 examines this possibility with our ongoing Prisoner's Dilemma game.

Example 7.3. Supporting partial cooperation. Consider the following modified GTS where players alternate between (NC,C) and (C,NC) over time, starting with (NC,C) in the first period. If either or both players deviates, both players revert to the NE of the stage game, (C,C), forever. This modified GTS means that player 1 earns a payoff of 0 in the first period, 8 in the second, 0 in the third period, and so on. Player 2 earns the opposite stream of payoffs: 8 in the first period, 0 in the second period, 8 in the third period, and so on. Table 7.1 summarizes the actions that each player takes in each period and their corresponding payoffs.[4]

---

[4] When players alternate between outcomes (NC,C) and (C,NC), they reach the midpoint of the diagonal line that connects the two vertexes of the FP diamond in Fig. 7.9, which lies inside the shaded FIR set.

To determine whether this modified GTS can be sustained as a SPE, we must show that no player benefits from unilaterally deviating (cheating). While this GTS prescribes different actions for each player, it calls player 1 (2) to choose NC (C) in every odd-numbered (even-numbered) period, yielding a payoff of 0 (8, respectively). Therefore, we only need to separately analyze the GTS in odd- and even-numbered periods for a generic player i.

Odd-numbered periods. When player 1 cooperates in this GTS, her stream of discounted payoffs starting at an odd-numbered period (e.g., period 1) is

$$0 + \delta 8 + \delta^2 0 + \delta^3 8 + ... = 0(1 + \delta^2 + ...) + 8(\delta + \delta^3 + ...)$$
$$= 8\delta(1 + \delta^2 + ...)$$
$$= \frac{8\delta}{1 - \delta^2}.$$

since $1 + \delta^2 + ...$ is the sum of an infinite geometric series, which can be rewritten as $\sum_{t=1}^{\infty} \delta^{2t} = \delta^0 + \delta^2 + \delta^4 + ... = \frac{1}{1-\delta^2}$. (A similar argument applies to player 2 in even-numbered periods.)

If, instead, player 1 deviates to C in an odd-numbered period (recall that she is supposed to choose NC in this period), her stream of discounted payoffs becomes

$$\underbrace{2}_{\text{Deviation}} + \underbrace{\delta 2 + \delta^2 2 + \delta^3 2 + ...}_{\text{Reversion to NE}} = 2(1 + \delta + \delta^2 + ...) = \frac{2}{1 - \delta}$$

since player 1's current payoff increases, from 0 under (NC,C) to 2 under (C,C). Her deviation, however, triggers an infinite reversion to the NE of the stage game, with a payoff of 2 thereafter. Comparing player 1's payoff streams, we find that she sticks to the GTS in every odd-numbered period if

$$\frac{8\delta}{1 - \delta^2} \geq \frac{2}{1 - \delta}$$

since $1 - \delta^2 = (1-\delta)(1+\delta)$, we can rearrange this inequality to obtain $8\delta \geq 2(1+\delta)$. Simplifying this expression, and solving for δ, yields $\delta \geq \frac{1}{3}$. Intuitively, player 1 must put a sufficiently important weight on future payoffs to suffer a payoff of 0 in every odd-numbered period, where (NC,C) occurs, instead of deviating toward C, which would increase her current payoff to 2.

Even-numbered periods. If player 1 chooses C while her rival selects NC, as prescribed by this GTS, it yields a payoff of 8 in the current period. Therefore, player 1's payoff stream is

$$8 + \delta 0 + \delta 8 + \delta^3 0 + ... = 8(1 + \delta^2 + ...) + 0(\delta + \delta^3 + ...)$$
$$= 8(1 + \delta^2 + ...)$$

$$= \frac{8}{1-\delta^2}.$$

A similar argument applies to player 2 in odd-numbered periods.

If, instead, player 1 unilaterally deviates, from C to NC, her current payoff actually decreases, from 8 under (C,NC) to 4 under (NC,NC). In addition, her deviation triggers an infinite reversion to the NE of the stage game. Hence, player 1's payoff stream from deviating in an even-numbered period is

$$\underbrace{4}_{\text{Deviation}} + \underbrace{\delta 2 + \delta^2 2 + \delta^3 2 + \ldots}_{\text{Reversion to NE}} = 4 + 2\delta(1 + \delta + \delta^2 + \ldots) = 4 + \frac{2\delta}{1-\delta}.$$

Therefore, player 1 sticks to the GTS instead of deviating at even-numbered periods if

$$\frac{8}{1-\delta^2} \geq 4 + \frac{2\delta}{1-\delta}$$

and since $1 - \delta^2 = (1-\delta)(1+\delta)$, we can rearrange this inequality to obtain $8 \geq 4(1-\delta)(1+\delta) + 2\delta(1+\delta)$ or, alternatively, $2\delta^2 - 2\delta + 4 \geq 0$. This inequality holds for all admissible values of $\delta \in [0, 1)$, meaning that player 1 sticks to the GTS in every even-numbered period regardless of her discount factor.[5]

In summary, players cooperate in both odd- and even-numbered periods if $\delta \geq \frac{1}{3}$. The only condition on δ that we found comes from odd-numbered periods for player 1 (even-numbered periods for player 2). In these periods, player 1 (2) must cooperate, choosing NC, while her rival selects C, indicating that player 1 (2, respectively) has the strongest incentives to cheat.□

The Folk theorem provides us with a positive and a negative result. On one hand, it helps us sustain cooperation even in settings such as the Prisoner's Dilemma game, where we know that cooperation cannot be supported in the unrepeated version of the game or if we repeat it a finite number of times. In other words, players can reach Pareto-improving outcomes, relative to those in the stage game. On the other hand, however, the Folk theorem says that we can reach *any* per-period payoff in the FIR diamond, essentially meaning that there is an extremely large set of SPEs in the infinitely repeated version of the game, thus limiting our predictive power.

---

[5] To see this point, notice that inequality $2\delta^2 - 2\delta + 4 \geq 0$ can be alternatively expressed as $4 \geq 2\delta(1-\delta)$, where the right side is smaller than 2. Graphically, equation $2\delta^2 - 2\delta + 4$ originates at a height of 4 when $\delta = 0$, decreases in δ reaching its minimum at $\delta = \frac{1}{2}$, with a height of 3.5, and then increases in δ, reaching a height of 4 at $\delta = 1$.

7.6 Application to Collusion in Oligopoly

The Folk theorem is often applied to the study of collusive agreements, where firms seek to coordinate their production decisions to raise prices and profits for cartel participants. Real-life examples include the Organization of Petroleum-Exporting Countries (OPEC), vitamin B_2 producers, steel, diamonds, and lysine, as described in Levenstein and Suslow (2006) and Harrington (2006).[6] We next examine how to model collusion in an infinitely repeated game.

Consider the duopoly with two firms competing à la Cournot that we examined in Sect. 4.2.1. Firms face the same marginal cost of production c, where $1 > c \geq 0$, and inverse demand function is $p(Q) = 1 - Q$, where $Q \geq 0$ denotes aggregate output. In Sect. 4.2.1, we found that, when firms compete in quantities, every firm i produces $q_i = \frac{1-c}{3}$ units, earning $\pi_i = \frac{(1-c)^2}{9}$ in profits. This stage game, then, has a unique NE, $(q_i, q_j) = (\frac{1-c}{3}, \frac{1-c}{3})$. As in the Prisoner's Dilemma game, we study next if a GTS can be sustained as a SPE of the infinitely repeated version of this Cournot competition game.

But which output should firms produce in the "cooperative" outcome? The most cooperative outcome is that maximizing firms' joint profits, which occurs when they perfectly coordinate their production decisions (or, without euphemisms, when firms collude).

Collusion. If firms collude to maximize their joint profits, they choose their output levels, q_1 and q_2, to solve

$$\max_{q_1, q_2 \geq 0} \pi(q_1, q_2) = \underbrace{[(1 - q_1 - q_2)q_1 - cq_1]}_{\pi_1} + \underbrace{[(1 - q_1 - q_2)q_2 - cq_2]}_{\pi_2}$$

where the first term denotes firm 1's profits and the second term represents firm 2's profits. This problem can be rearranged, more compactly, as follows

$$\max_{q_1, q_2 \geq 0} (1 - q_1 - q_2)(q_1 + q_2) - c(q_1 + q_2)$$

and since $Q = q_1 + q_2$ by definition, the above problem further simplifies to

$$\max_{Q \geq 0} \pi(Q) = (1 - Q)Q - cQ.$$

Intuitively, when firms collude, they can just choose the aggregate output Q that maximizes their joint profits, which are equivalent to the profits of a monopolist. Differentiating with respect to Q, yields

$$1 - 2Q - c = 0.$$

---

[6] For a humorous account of real-life case of collusion in the lysine industry, you can watch the movie *The Informant* (2009), featuring Matt Damon in the role of Mark Whitacre, vice-president of agribusiness giant Archer Daniels Midland Company, who turned government informant in the 1990s.

7.6 Application to Collusion in Oligopoly

Solving for Q provides an aggregate output level

$$Q^C = \frac{1-c}{2}.$$

where superscript C denotes collusion. Because both firms face the same marginal costs and demand, they each produce half of this aggregate output, that is,

$$q_i^C = \frac{Q^C}{2} = \frac{1-c}{4}.$$

with collusive price

$$p^C = 1 - Q^C = 1 - \frac{1-c}{2} = \frac{1+c}{2}.$$

Finally, we insert our equilibrium price and aggregate output into the profit function to find firm i's equilibrium profits under collusion:

$$\pi_i^C = (p^C - c)q_i^C = \left(\frac{1+c}{2} - c\right)\left(\frac{1-c}{4}\right) = \frac{(1-c)^2}{8}.$$

We are now ready to specify the GTS that we seek to sustain as a SPE in this game:

1. In period $t = 1$, every firm i chooses the collusive output $q_i^C = \frac{1-c}{4}$.
2. In all periods $t \geq 2$, every firm i chooses the collusive output $q_i^C = \frac{1-c}{4}$ if both firms produced q_i^C in every previous period. Otherwise, every firm i reverts to the NE of the stage game, choosing $q_i = \frac{1-c}{3}$, thereafter.

To show that this GTS can be sustained in equilibrium, we need to show that firms have no incentives to deviate from it, at any period t, and after any previous history of play. We separately analyze whether firms have incentives to deviate from the GTS after observing that both firms cooperated in previous periods and after observing some uncooperative episodes.

After a history of cooperation. If both firms cooperated in all previous periods, choosing q_i^C, every firm i can collude in period t, as specified by the GTS, or deviate from this collusive output. We find its profits from each option below.

Profits from collusion. If firm i colludes, it earns $\pi_i^C = \frac{(1-c)^2}{8}$ in every period, entailing a discounted present value of

$$\pi_i^C + \delta \pi_i^C + \delta^2 \pi_i^C + \ldots$$
$$= \pi_i^C (1 + \delta + \delta^2 + \ldots)$$
$$= \pi_i^C \left(\frac{1}{1-\delta}\right)$$

$$= \frac{(1-c)^2}{8(1-\delta)}$$

Profits from deviating. If, instead, firm i deviates from the collusive output q_i^C, we must first determine firm i's most profitable unilateral deviation. In this setting, firm i solves

$$\max_{q_i \geq 0} \ (1 - q_i - q_j^C)q_i - cq_i$$

where we fixed $q_j = q_j^C$ since we are considering unilateral deviations by firm i. Differentiating with respect to q_i, and assuming an interior solution, we obtain

$$1 - 2q_i - q_j^C - c = 0$$

which, solving for q_i, yields the standard best response function, evaluated in this case at q_j^C,

$$q_i(q_j^C) = \frac{1-c}{2} - \frac{1}{2}q_j^C.$$

Since $q_j^C = \frac{1-c}{4}$, firm i's most profitable deviation when its rival colludes is

$$q_i^{Dev} = \frac{1-c}{2} - \frac{1}{2}\left(\frac{1-c}{4}\right)$$
$$= \frac{3(1-c)}{8}.$$

Therefore, the price in this setting becomes

$$p^{Dev} = 1 - q_i^{Dev} - q_j^C$$
$$= 1 - \frac{3(1-c)}{8} - \frac{1-c}{4}$$
$$= \frac{3+5c}{8}$$

and firm i's profits from deviating are

$$\pi_i^{Dev} = (p^{Dev} - c)q_i^{Dev}$$
$$= \left(\frac{3+5c}{8} - c\right)\left(\frac{3(1-c)}{8}\right)$$
$$= \frac{9(1-c)^2}{64}.$$

7.6 Application to Collusion in Oligopoly

Therefore, if firm i deviates at any period t, it earns deviation profits $\pi_i^{Dev} = \frac{9(1-c)^2}{64}$ in that period, but Cournot profits $\pi_i = \frac{(1-c)^2}{9}$ in every period afterward as both firms revert to the NE of the stage game, yielding a discounted profit stream of

$$\pi_i^{Dev} + \delta\pi_i + \delta^2\pi_i + \ldots$$
$$= \pi_i^{Dev} + \delta\pi_i(1 + \delta + \delta^2 + \ldots)$$
$$= \pi_i^{Dev} + \pi_i\left(\frac{\delta}{1-\delta}\right)$$
$$= \frac{9(1-c)^2}{64} + \frac{\delta(1-c)^2}{9(1-\delta)}.$$

Calculating the minimal discount factor, $\underline{\delta}$. From our above calculations, we can say that at any period t after a history of collusion, firm i keeps colluding (as prescribed by the GTS) if

$$\pi_i^C\left(\frac{1}{1-\delta}\right) \geq \pi_i^{Dev} + \pi_i\left(\frac{\delta}{1-\delta}\right)$$
$$\frac{(1-c)^2}{8(1-\delta)} \geq \frac{9(1-c)^2}{64} + \frac{\delta(1-c)^2}{9(1-\delta)}.$$

Rearranging, we find

$$\frac{(1-c)^2}{8(1-\delta)} \geq \frac{9(1-c)^2}{64} + \frac{\delta(1-c)^2}{9(1-\delta)}$$
$$\iff \frac{1}{8} \geq \frac{9(1-\delta)}{64} + \frac{\delta}{9}$$

Solving for δ, we obtain the minimal discount factor sustaining collusion,

$$\delta \geq \underline{\delta} \equiv \frac{9}{17}.$$

Therefore, collusion can be sustained if firms assign a sufficiently high value to future profits. Otherwise, collusion cannot be sustained.

After a history of no cooperation. If one firm deviated from collusive output $q_i^C = \frac{1-c}{4}$ in any previous period, the GTS prescribes that every firm i reverts to the NE of the stage game. To confirm that the GTS is a SPE of the infinitely repeated game, we need to show that every firm i has incentives to, essentially, implement this punishment. If, upon observing a deviation, firm i behaves as prescribed by the GTS, producing the Cournot output $q_i = \frac{1-c}{3}$, it earns profits $\pi_i = \frac{(1-c)^2}{9}$ in every period afterward, yielding a discounted profit stream

$$\pi_i + \delta\pi_i + \delta^2\pi_i + \ldots = \pi_i(1 + \delta + \delta^2 + \ldots) = \frac{\pi_i}{1-\delta}.$$

If, instead, firm i unilaterally deviates from the Cournot output, producing $q_i \neq \frac{1-c}{3}$, while firm j behaves as prescribed by the GTS, producing $q_j = \frac{1-c}{3}$ units, firm i's best response to $q_j = \frac{1-c}{3}$ is, from above calculations,

$$q_i\left(\frac{1-c}{3}\right) = \frac{1-c}{2} - \frac{1}{2}\frac{1-c}{3}$$
$$= \frac{1-c}{3}.$$

In other words, even if firm i wanted to deviate from the GTS (upon observing uncooperative behavior in some previous period), its best response would be to behave as prescribed by the GTS, producing $q_i = \frac{1-c}{3}$ in every subsequent period. As a result, every firm i has incentives to implement the punishment prescribed by the GTS.

7.6.1 Minimal Discount Factor Supporting Collusion

Our above analysis considers that firms compete à la Cournot, sell a homogeneous good, and face the same marginal cost of production, c. In a more general setting, however, we could allow for firms to compete in prices, sell differentiated products, and/or face different production costs. The above steps to identify the minimal discount factor sustaining collusion would, nonetheless, apply to these contexts. In particular, after a history of cooperation, every firm i's present value from cooperating is

$$\pi_i^C + \delta \pi_i^C + \delta^2 \pi_i^C + \ldots = \pi_i^C(1 + \delta + \delta^2 + \ldots)$$
$$= \pi_i^C\left(\frac{1}{1-\delta}\right)$$

where π_i^C denotes its collusion profits in a single period. If, instead, firm i deviates, it earns π_i^{Dev} during that period (which entails finding firm i's optimal unilateral deviation) but earns NE profit, π_i^{NE}, in all subsequent periods, for a present value of

$$\pi_i^D + \delta \pi_i^{NE} + \delta^2 \pi_i^{NE} + \ldots = \pi_i^{Dev} + \delta \pi_i^{NE}(1 + \delta + \delta^2 + \ldots)$$
$$= \pi_i^{Dev} + \pi_i^{NE}\left(\frac{\delta}{1-\delta}\right).$$

Therefore, firm i cooperates if and only if

$$\pi_i^C\left(\frac{1}{1-\delta}\right) \geq \pi_i^{Dev} + \pi_i^{NE}\left(\frac{\delta}{1-\delta}\right)$$

7.6 Application to Collusion in Oligopoly

which we can rearrange as $\pi_i^C \geq (1-\delta)\pi_i^{Dev} + \delta\pi_i^{NE}$. Solving for δ, yields

$$\delta \geq \frac{\pi_i^{Dev} - \pi_i^C}{\pi_i^{Dev} - \pi_i^{NE}} \equiv \underline{\delta}$$

Minimal discount factor $\underline{\delta}$ increases in $\pi_i^{Dev} - \pi_i^C$, intuitively indicating that, as deviations become more attractive, cooperation can only be sustained for higher discount factors (firms must care more about the future). In contrast, $\underline{\delta}$ decreases in $\pi_i^{Dev} - \pi_i^{NE}$, suggesting that, when the profit loss from reverting to the NE of the stage game becomes more severe (as captured by the difference $\pi_i^{Dev} - \pi_i^{NE}$), firms have less incentives to cheat, expanding the range of discount factors sustaining cooperation.

The above ratio is convenient, as it can be readily applied to different market structures, such as markets with two or more firms, selling homogeneous or heterogeneous products, and with symmetric or asymmetric costs. Specifically, we start finding the NE in the stage game and its associated profits, π_i^{NE} (which may differ from firm j's); the collusive output (or prices) that maximize firms' profits, yielding π_i^C to firm i; and firm i's optimal deviation, fixing firm j's collusive behavior, so firm i's profits are π_i^{Dev}. Inserting profits π_i^{NE}, π_i^C, and π_i^{Dev} in the above expression for $\underline{\delta}$, we obtain the minimal discount factor sustaining collusion in the market structure we study.

7.6.2 Other Collusive GTS

We said that the above GTS with perfect collusion, where firms maximize their joint profits, can be sustained as a SPE, rather than saying "the unique SPE" of the infinitely repeated game. From the Folk theorem, we know that other per-period payoffs can be supported in SPEs too. In fact, the above GTS can be interpreted as the most northwest payoff pair in the FIR set since firms earn the highest symmetric payoff relative to what they earn in the NE of the Cournot model (which would correspond graphically to the southwest corner of the FIR set). If we focus on symmetric payoff pairs, we can characterize the line connecting profits under the NE, $\pi_i^{NE} = \frac{(1-c)^2}{9}$, and those in the above GTS, $\pi_i^C = \frac{(1-c)^2}{8}$, with the convex combination

$$\pi_i = \alpha\pi_i^C + (1-\alpha)\pi_i^{NE},$$

where weight $\alpha \in [0,1]$ can be understood as how close firms are to the perfect collusion. In particular, firm i produces

$$q_i = \alpha q_i^C + (1-\alpha)q_i^{NE}$$
$$= \alpha\frac{1-c}{4} + (1-\alpha)\frac{1-c}{3}$$

$$= \frac{(1-c)(4-\alpha)}{12}$$

implying that this firm's output coincides with the collusive output, $q_i^C = \frac{1-c}{4}$, when $\alpha = 1$; but becomes the NE output, $q_i^{NE} = \frac{1-c}{3}$, when $\alpha = 0$. The GTS is, then, similar to the one considered in this section:

1. In period $t = 1$, every firm i chooses the output $q_i = \alpha q_i^C + (1-\alpha) q_i^{NE}$.
2. In all periods $t \geq 2$, every firm i chooses output q_i if both firms produced q_i in every previous period. Otherwise, every firm i reverts to the NE of the stage game, choosing $q_i = \frac{1-c}{3}$, thereafter.

Exercise 7.9 asks you to find the minimal discount factor in this setting, which is a function of weight α, that is, $\underline{\delta}(\alpha)$, indicating that the emergence of cooperation in equilibrium depends on how collusive the GTS is. We could, of course, consider asymmetric output pairs, where each firm produces a different output level, yielding even further SPEs in this game, but we leave them as a practice.

7.7 What if the Stage Game has More than One NE?

So far we considered games with a unique NE in its unrepeated version, as in the Prisoner's Dilemma game or the Cournot game of quantity competition. Their NE was rather uncooperative, yet we showed that, cooperation can be supported as the SPE of the infinitely repeated game. In a sense, we examined these games to highlight that, even when players face intense competitive pressures, cooperation is still possible in the infinitely repeated version of the game.

A natural question is whether cooperation can be sustained in games with more than one NE in its unrepeated version, as that in Matrix 7.6. Underlining best response payoffs, we can find two NEs: (B, B) and (C, C). However, (C, C) Pareto dominates (B, B) since both players earn a higher payoff at the former than the latter.

		Player 2		
		A	B	C
	A	8, 8	−2, 10	2, 3
Player 1	B	10, −2	2, 2	1, 1
	C	1, 1	1, 1	6, 6

Matrix 7.6. A game with two NEs.

In addition, outcome (A, A) cannot be sustained as an NE of the stage game, but it would yield higher payoffs than any of the two NEs for both players. We

7.7 What if the Stage Game has More than One NE?

then wonder if the "cooperative outcome" (A, A) can be supported as a SPE of the twice-repeated game. We know that cooperative outcomes cannot arise in the twice-repeated version of games with a unique NE. In games with more than one NE, however, cooperation can be sustained in equilibrium. To show this, consider the following GTS:

1. In the first period, every player chooses A.
2. In the second period, every player chooses C if (A, A) was played in the first period. Otherwise, every player chooses B.

Informally, this GTS is known as a "stick-and-carrot" strategy because, in the second period, it prescribes the Pareto-dominant NE, (C, C), if players were cooperative in the first period; but prescribes the Pareto-dominated NE, (B, B), if either player was uncooperative. This implies that, while a GTS can specify an outcome that is not an NE of the stage game during the first period, such as (A, A), it cannot prescribe an outcome that is not an NE in the second period. Otherwise, if the GTS specified outcome (A, A) in both periods, every player would have incentives to deviate from A in the second (last) period. In summary, when designing GTSs in finitely repeated games, we can "shoot for the stars" in all but the final period, but must settle for an NE in the last round of interaction.

But we haven't yet shown that the above GTS is a SPE of the twice-repeated game in Matrix 7.6. To demonstrate that it is a SPE, we must operate by backward induction.

Second period. Technically, there are 9 outcomes in each stage, for a total of $9^2 = 81$ terminal nodes, so you are excused if you do not plan to draw its game tree and circle the 9 different subgames that arise in the second stage. Interestingly, out of these 9 subgames, there is only one that is initiated after outcome (A, A) was played in the first period, indicating cooperation, while the remaining 8 second-period subgames emerge because one or both players did not select A in the first period. Fortunately, our analysis of these 8 subgames is, essentially, identical, saving us some time.

Non-cooperative history. Upon observing any outcome different from (A, A) in the first period, player i anticipates that player j will play B in the second period. In this setting, player i's best response is B as well (see the middle column in Matrix 7.6). In other words, player i does not have incentives to deviate from the GTS in the second period.

Cooperative history. If, instead, player i observes that outcome (A, A) was played in the first period, then she expects player j to choose C in the second period. In this context, player i's best response is C as well (see right column in Matrix 7.6). Therefore, player i does not want to deviate from the GTS in the second period.

First period. If player i chooses A, as prescribed by the GTS, her payoff stream is $8 + \delta 6$, since she anticipates that (C, C) will be played in the second period (the Pareto-dominant NE, or "carrot"). If, instead, player i unilaterally deviates, her best deviation is to B, which yields a payoff of 10 today, but that triggers outcome (B, B), the Pareto dominated NE or "stick," in the second period, for a payoff stream of $10 + \delta 2$. Therefore, player i behaves as prescribed by the GTS in the first period if

$$8 + \delta 6 \geq 10 + \delta 2$$

which simplifies to $\delta \geq \frac{1}{2}$. Players can, then, sustain cooperative behavior at least during one period if players assign a sufficiently high weight to future payoffs and the unrepeated version of the game has more than one NE, using one of them as a prize for good behavior and another one as a penalty for bad behavior. Intuitively, we could not design a similar incentive scheme in previous games, such as the Prisoner's Dilemma game, because they had a unique NE, so we could not condition last-period play on whether players were cooperative or uncooperative in previous interactions.

7.8 Modified GTSs

From our discussion of the Folk theorem, we know that different GTSs induce different per-period payoffs in the FIR diamond. This implies that we can consider GTS prescribing different cooperative behaviors (not necessarily the outcome yielding the highest per-period payoff for both players) or different punishment schemes (reverting to some outcome that yields a higher/lower payoff than the NE of the stage game). We finish this chapter describing two GTS that are quite intuitive and have been often used in real life.

7.8.1 An Eye for an Eye

Consider the old adage "an eye for an eye," or "a tooth for a tooth," which prescribes that individuals start cooperating but, if an individual is uncooperative, the victim becomes uncooperative, inflicting the same damage to the cheater, and then players return to cooperation.[7] In an infinitely repeated game context, this adage means that, if a player cheats, the cheated player gets to cheat during one period while the cheating player cooperates in that same period, and afterward players return to cooperation.

To implement this cheating sequence in a GTS, consider the game in Matrix 7.7, where strategy B is strictly dominant for every player, and thus (B, B) is the unique NE of the stage game. In this NE, every player earns a payoff of 8, but

---

[7] This old adage is a paraphrase of the Code of Hammurabi, a set of 282 laws covering trade, slavery, theft, the duties of workers, liability, and divorce. It was written around 1755–1750 BC, found in the city of Susa (modern-day Iran), and influenced subsequent legal codes.

7.8 Modified GTSs

they could both improve their payoffs if they cooperated playing (C, C), which yields (10, 10). In this context, we could use a standard GTS, where:

1. In the first period, every player i chooses C.
2. In subsequent periods, every player i chooses C if (C, C) was played in all previous periods. Otherwise, every player reverts to the NE of the stage game, (B, B).

In such a setting, it is easy to show that cooperation can be sustained if δ satisfies $\delta \geq \frac{2}{3}$. (Exercise 7.17 asks you to confirm this result.)

Player 2

		A	B	C
	A	0, 0	4, 2	5, −2
Player 1	B	2, 4	8, 8	14, 3
	C	−2, 5	3, 14	10, 10

Matrix 7.7. An eye for an eye.

But what about using a GTS that follows the "an-eye-for-an-eye" adage? As in the above GTS, every player i chooses C in the first period and continues doing so if (C, C) was played in all previous periods. Unlike in the standard GTS, the punishment phase does not entail playing the NE of the stage game thereafter, but instead says

> If player j chooses B in period $t-1$, player i selects B while player j chooses C in period t.
>
> In all subsequent periods, every player chooses C.

Intuitively, the punishment phase now prescribes that the victim gets to cheat her opponent for a period, and then they return to the cooperative outcome (C, C) thereafter. As an illustration, if player 1 cheats, deviating to B while player 2 still cooperates choosing C, then Table 7.2 summarizes the sequence of outcomes.

Table 7.2 Timing of the an-eye-for-an-eye GTS

	Period $t-1$	Period t	Period $t+1$
Outcome	B, C	C, B	C, C
Payoffs	14, 3	3, 14	10, 10
Who cheats?	Player 1	Player 2	None

7.8.1.1 After a History of Cooperation

If no player has cheated in previous periods, player i's payoff stream from cooperating is

$$10 + 10\delta + 10\delta^2 + 10\delta^3 + \ldots$$

If, instead, player 1 deviates, we first find that, conditional on player 2 cooperating (playing C), player 1's best response is to play B, earning a payoff of 14. Therefore, player 1's deviation increases her current payoff from 10 to 14, but then triggers a punishment, which reduces her payoff to 3 during one period (while she is being cheated by player 2, as described in Table 7.2). In the following period, players return to cooperation, yielding a payoff of 10 thereafter. In summary, the stream of payoffs that player 1 earns when deviating is

$$\underbrace{14}_{\text{Cheat}} + \underbrace{3\delta}_{\text{Cheated}} + \underbrace{10\delta^2 + 10\delta^3 + \ldots}_{\text{Back to cooperation}}$$

Therefore, player 1 cooperates if and only if

$$10 + 10\delta + 10\delta^2 + 10\delta^3 + \ldots \geq 14 + 3\delta + 10\delta^2 + 10\delta^3 + \ldots$$

Both sides of the inequality coincide in the third term, $10\delta^2$, and for all subsequent periods, so we can simplify this inequality to

$$10 + 10\delta \geq 14 + 3\delta$$

and, solving for δ, we find that cooperation can be sustained if $\delta \geq \frac{4}{7}$.

7.8.1.2 After a History of Cheating

When one player cheats, the GTS prescribes that the cheating player cooperates for one period, playing C, while her opponent cheats in that period, choosing B. We need to show that the cheating player also follows the GTS in this case. If player 1 cheats in period $t - 1$, the GTS prescribes that (C, B) is played in period t, and (C, C) occurs in all subsequent periods, yielding

$$\underbrace{3}_{\text{Cheated}} + \underbrace{10\delta + 10\delta^2 + 10\delta^3 + \ldots}_{\text{Back to cooperation}}$$

If player 1 does not choose C in period t, being cheated, her best deviation is to B since, conditional on player 2 choosing B in period t (see Matrix 7.7), player 1's best response is B, which yields outcome (B, B) and a payoff of 8. The instantaneous gain from deviating is, then, captured by the payoff increase from 3, in outcome (C, B), to 8, in outcome (B, B). In the next period, however, players observe that (C, B) was not played, and thus they cannot return to the cooperative outcome (C, C) yet. Instead, they still need to play (C, B) for one period, yielding a payoff of 3 to the cheating player 1, and then they return to (C, C). In summary, player 1's deviation (not being allowed to be cheated in period t) yields a payoff stream

$$\underbrace{8}_{\text{Cheat}} + \underbrace{3\delta}_{\text{Cheated}} + \underbrace{10\delta^2 + 10\delta^3 + ...}_{\text{Back to cooperation}}$$

Intuitively, if the cheating player does not follow the GTS, not allowing the victim to cheat during one period, she only postpones her return to cooperation. Therefore, player 1 allows player 2 to cheat her, after player 1 cheated in the first place, if and only if

$$3 + 10\delta + 10\delta^2 + 10\delta^3 + ... \geq 8 + 3\delta + 10\delta^2 + 10\delta^3 + ...$$

Both sides of the inequality coincide in the third term, $10\delta^2$, and for all subsequent periods. We can then rewrite this inequality as

$$3 + 10\delta \geq 8 + 3\delta$$

which, solving for δ, yields $\delta \geq \frac{5}{7}$.

Summary. Comparing this condition on δ with that found after a history of cooperation, $\delta \geq \frac{4}{7}$, we see that condition $\delta \geq \frac{5}{7}$ is more demanding than $\delta \geq \frac{4}{7}$ since $\frac{5}{7} \geq \frac{4}{7}$. In other words, if condition $\delta \geq \frac{5}{7}$ holds, condition $\delta \geq \frac{4}{7}$ must also be satisfied, so $\delta \geq \frac{5}{7}$ is a sufficient condition for cooperation.

7.8.1.3 Comparing Different GTS

We can now compare the an-eye-for-an-eye GTS against the standard GTS. On one hand, we find that the punishment phase is more equitable in the former than the latter since the cheated party is compensated in the following period, when she cheats. Under the standard GTS, however, the cheated party sees its payoff decrease during the cheating period, and then both players revert to the NE of the stage game, yielding a lower per-period payoff than the cheater's. On the other hand, the minimal discount factor supporting an-eye-for-an-eye GTS, $\delta \geq \frac{5}{7}$, is higher than that under a standard GTS, $\delta \geq \frac{2}{3}$, implying that cooperation is more difficult to arise in the former than the latter since $\frac{5}{7} \geq \frac{2}{3}$. This result was predictable because the punishment phase in this GTS requires the cheating party to be cheated during one period, which is not attractive for this player, generating strong incentives to deviate from the GTS at precisely that moment.

7.8.2 Short and Nasty Punishments

Consider now a similar GTS as above, where players start cooperating in the first period and keep doing so if everyone cooperated in the past. As in the previous

Table 7.3 Timing of the short-and-nasty-punishments GTS

	Period $t-1$	Period t	Period $t+1$
Outcome	B, C	A, A	C, C
Payoffs	14, 3	0, 0	10, 10

Table 7.4 Minimal discount factors sustaining cooperation in the game in Matrix 7.7 when using different GTS

	Minimal discount factor, $\underline{\delta}$
Standard GTS, sect. 7.4.2	$\underline{\delta} = \frac{2}{3}$
An-eye-for-an-eye GTS, sect. 7.8.1	$\underline{\delta} = \frac{5}{7}$
Short-and-nasty-punishments GTS, sect. 7.8.2	$\underline{\delta} = \frac{2}{5}$

section, we also assume now that the punishment phase is temporary, lasting only one period. The punishment, however, is now (A, A), yielding the lowest symmetric payoff in the matrix, $(0, 0)$, and players return to cooperation immediately after. Specifically, we define this GTS as follows:

1. In the first period, every player i chooses C.
2. In subsequent periods:
   1. Every player i chooses C if (C, C) was played in all previous periods.
   2. Otherwise, every player chooses A during one period. If (A, A) was played in the last period, every player selects C thereafter.

As an illustration, if player 1 cheats, deviating to B while player 2 still cooperates choosing C, Table 7.3 summarizes the sequence of outcomes if players behave according to this GTS.

Exercise 7.1 asks you to find the minimal discount factor that sustains cooperation with this GTS, which holds if $\delta \geq \frac{2}{5}$. Comparing this result with that in the an-eye-for-an-eye GTS we explored in Section 7.8.1, where cooperation can be supported if $\delta \geq \frac{5}{7}$, we see that $\frac{5}{7} > \frac{2}{5}$, implying that cooperation is easier to sustain with the threat of a one-period nasty punishment.

Table 7.4 reports the minimal discount factor $\underline{\delta}$ under different GTS in the game in Matrix 7.7. Recall that, when $\underline{\delta}$ decreases, cooperation can be sustained for a wider range of discount factors, that is, the interval $[\underline{\delta}, 1)$ becomes longer.

7.8.3 Imperfect Monitoring

Previous sections considered that players can perfectly observe deviations from cooperation, known as "perfect monitoring" of each other's previous actions. In some settings, however, players may imperfectly monitor her rival's actions,

7.8 Modified GTSs

observing which action she chose with a certain probability. To illustrate this setting, consider the Prisoner's Dilemma in Matrix 7.8, where outcome (C, C) is the NE of the stage game, but (NC, NC) would increase both players' payoffs from 2 to 4.

		Player 2	
		Confess	Not confess
Player 1	Confess	2, 2	8, 0
	Not confess	0, 8	4, 4

Matrix 7.8. The Prisoner's Dilemma game.

If players had perfect monitoring, we know from Section 7.4.2 that cooperation in outcome (NC, NC) can be supported as a SPE of the infinitely repeated game with a standard GTS if $\delta \geq \frac{2}{3}$. But what if monitoring is imperfect? Before starting to check if players have incentives to deviate from a standard GTS, we need to more precisely define what it means for players to "suffer from imperfect monitoring." In particular, if player j chooses to cooperate, NC, we assume that, for simplicity, player i cannot detect a deviation to C. (Exercise 7.10 relaxes this assumption, allowing players to incorrectly observe NC when their rivals are, actually, playing C.) If, instead, player j chooses C, the probability that player i observes a deviation from NC is $p \in [0, 1]$. Intuitively, when probability $p \to 0$, deviations are rarely observed while when $p \to 1$, deviations are almost always observed, as in a context with perfect monitoring.

We are now ready to adapt the standard GTS to this context:

1. In the first period, every player chooses NC.
2. In every subsequent period, every player chooses NC if (NC, NC) was *observed* in all previous periods. Otherwise, revert to the NE of the stage game, (C, C), thereafter.

We italicized the only difference relative to the standard GTS, namely, the cooperative outcome (NC, NC) is now observed. This suggests that outcome (NC, NC) may not have been played, because at least one player deviated, but cooperation continues if such outcome is observed. The opposite argument also applies: outcome (NC, NC) may have been played, but cooperation stops if one or both players did not observe such outcome in a previous period. We return to these two options below when we check if the GTS can be supported as a SPE.

After a history of observed cooperation. If both players observe outcome (NC, NC) in all previous periods, player i's payoff stream from cooperating is

$$4 + 4\delta + 4\delta^2 + \ldots = \frac{4}{1-\delta}.$$

If, instead, player i unilaterally deviates to C, her payoff increases to 8 in the current period. In the next period, however, player j detects her deviation with probability p, reverting to the NE of the stage game with a payoff of 2 forever; or does not detect this deviation with probability $(1-p)$, which lets player i return to cooperation (choosing NC tomorrow) as if no deviation ever happened. More compactly, player i's payoff stream from unilaterally deviating to C is

$$\underbrace{8}_{\text{Current gain from dev. to } C} + \delta \left[p \underbrace{\frac{2}{1-\delta}}_{\text{Detected}} + (1-p) \underbrace{\frac{4}{1-\delta}}_{\text{Undetected}} \right].$$

Therefore, for player i to cooperate after a history of cooperation, we need that

$$\frac{4}{1-\delta} \geq 8 + \delta \left[p \frac{2}{1-\delta} + (1-p) \frac{4}{1-\delta} \right]$$

which simplifies to to $4 \geq 8(1-\delta) + \delta(4-2p)$. Solving for δ, we obtain

$$\delta \geq \frac{2}{2+p} \equiv \underline{\delta}(p)$$

Figure 7.10 depicts the minimal discount factor $\underline{\delta}(p)$, showing that it originates at a height of $\underline{\delta}(0) = 1$, where $p = 0$, and decreases in p, ultimately reaching a height of $\underline{\delta}(1) = \frac{2}{3}$ when $p = 1$. Intuitively, as deviations are more likely detected (higher p), the reversion to the NE of the stage game becomes more likely. In turn, the *expected punishment* increases, ultimately expanding the range of δ's sustaining cooperation (shaded region in the figure). In contrast, when deviations cannot be detected ($p = 0$), the minimal discount factor becomes $\underline{\delta}(0) = 1$, implying that collusion cannot be supported in equilibrium.

After a history of observed deviation. If player i observes that (NC, NC) was not played in a previous period, this means that player j did not cooperate.[8] (Player

---

[8] Recall that, when player j chooses NC, player i perfectly observes her rival choosing NC. However, when player j chooses C, player i correctly observes C with probability p but incorrectly observes NC with probability $1-p$.

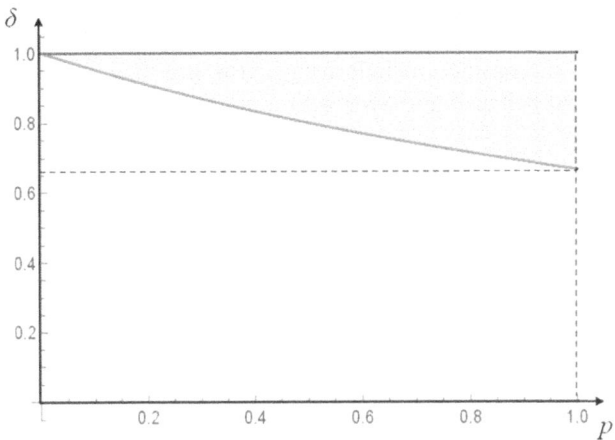

Fig. 7.10 Minimal discount factor $\underline{\delta}(p)$

i perfectly observes her own deviations from *NC*, so the probability of observing deviations only affects those originating from the other players.) If player i behaves as prescribed by the GTS, she should revert to the NE of the stage game, (C, C), thereafter, earning a payoff stream of

$$2 + 2\delta + 2\delta^2 + \ldots = \frac{2}{1-\delta}.$$

If, instead, player i deviates, choosing *NC* while player j chooses *C* (recall that this is a *unilateral* deviation from the GTS, upon observing no cooperation in a previous period), player i earns 0 in this period. Players then play (C, C) in all subsequent periods, yielding a payoff stream of[9]

$$0 + 2\delta + 2\delta^2 + \ldots = 2\delta(1 + \delta + \delta^2 + \ldots)$$
$$= \frac{2\delta}{1-\delta}.$$

Therefore, upon observing a deviation, player i prefers to behave as prescribed by the GTS than deviate from it since

$$\frac{2}{1-\delta} \geq \frac{2\delta}{1-\delta}$$

simplifies to $\delta \leq 1$, which holds by assumption since $\delta \in [0, 1)$. Overall, the only condition that we need to support this GTS as a SPE is $\delta \geq \underline{\delta}(p)$.

---

[9] We do not consider probability of detection here because this case assumes that a deviation has already been observed.

Further reading. The model on imperfect monitoring presented in this section is, needless to say, a simplification of models in the literature. For instance, Tirole (1988, Section 6.7.1) considers an industry with two firms competing à la Bertrand, but market demand collapsing to zero with a positive probability. In that setting, upon experiencing zero sales, a firm does not know whether that occurs because market demand is nil or because its rival undercuts its price, leaving the former with no sales. For a literature review, see Kandori (2002) and graduate-level students may consider the book by Mailath and Samuelson (2006).

Exercises

7.1 **Short and nasty punishments.**[A] Consider the stage game in Matrix 7.7 and the following GTS described in Sect. 7.8.2:

- In the first period, every player i chooses C.
- In subsequent periods:
  – Every player i chooses C if (C, C) was played in all previous periods.
  – Otherwise, every player chooses A during one period. If (A, A) was played in the last period, every player selects C thereafter.

(a) After a history of cooperation, under which condition for discount factor δ every player behaves as prescribed by the GTS?
(b) After observing at least one player deviating from (C, C), under which condition for discount factor δ every player behaves as prescribed by the GTS?
(c) Compare your results in parts (a) and (b), finding a sufficient condition for the GTS to be sustained as a SPE of the infinitely repeated game.

7.2 **Pareto coordination game.**[A] Consider the following Pareto coordination game where every firm simultaneously chooses between technology A or B, and payoffs satisfy $a_i > b_i > 0$ for every firm $i = \{1, 2\}$, implying that both firms regard technology A as superior. For generality, the premium that firm i assigns to technology A, as captured by $a_i - b_i$, can be different between firms 1 and 2. Firm i's discount factor is $\delta_i \in (0, 1)$.

		Firm 2	
		Tech. A	Tech. B
Firm 1	Tech. A	a_1, a_2	$0, 0$
	Tech. B	$0, 0$	b_1, b_2

(a) *Twice-repeated game.* Consider that the stage game is repeated twice. Show that the following strategy profile can be sustained as a SPE of the game: in the first period, every firm i chooses A; in the second period, if (A, A) was selected in the first period, every firm chooses A. Otherwise, every firm chooses B.

(b) Still in the twice-repeated game, can you support a strategy profile where every firm chooses A in every period of the game for any previous history of play?

(c) Still in the twice-repeated game, can you support a strategy profile where every firm chooses B in every period of the game for any previous history of play?

(d) *Infinitely repeated game.* Assume now that the game is infinitely repeated. In addition, consider a GTS where, in the first period, every firm i chooses A. In all subsequent periods, every firm chooses A if outcome (A, A) was played in all previous periods, but deviates to B otherwise. Under which conditions on δ_i can this GTS be sustained as a SPE of the infinitely repeated game?

7.3 **Anticoordination game.**[A] Consider the following Game of Chicken where every player drives toward each other and simultaneously chooses between swerve and stay. Payoffs satisfy $a_i > b_i > 0$ for every player $i = \{1, 2\}$, implying that drivers suffer more when they both stay (and almost die!) than when they both swerve (and are called chickens), and $a_i > d_i > c_i > b_i$, so being the only player who stays yields a payoff of d_i for that player and $-c_j$ for his rival, where $j \neq i$.

		Player 2	
		Swerve	Stay
Player 1	Swerve	$-b_1, -b_2$	$-c_1, d_2$
	Stay	$d_1, -c_2$	$-a_1, -a_2$

(a) *Twice-repeated game.* Consider that the stage game is repeated twice. Show that the following strategy profile can be sustained as a SPE of the game: in the first period, player 1 (2) chooses Stay (Swerve). In the second period, regardless of the outcome played in the first period, player 1 (2) chooses Swerve (Stay), switching their actions from the first period.

(b) Still in the twice-repeated game, can you sustain a SPE where every player chooses Swerve in every period?

(c) *Infinitely repeated game.* Assume that the game is infinitely repeated. Can you sustain the following strategy profile as a SPE where, player 1 (2) chooses Stay (Swerve) in every odd-numbered period and Swerve (Stay) in every even-numbered period? (Note that this strategy profile is unconditional on previous history of play, meaning that players behave as prescribed regardless of whether all players behaved according to this strategy or one/more players deviated.)

7.4 **Prisoner's Dilemma game with imperfect monitoring.**[B] Consider the following Prisoner's Dilemma game.

		Player 2	
		Confess	Not confess
Player 1	Confess	2, 2	8, 0
	Not confess	0, 8	4, 4

(a) Find the NE in the stage game.
(b) *GTS with perfect monitoring.* For the remainder of the exercise, assume that players interact in an infinitely repeated game. Find the minimal discount factor $\underline{\delta}$ sustaining the following (standard) GTS, namely, in the first period, every player chooses *NC*; in all subsequent periods, every player chooses *NC* if (*NC, NC*) was the outcome in all previous periods. Otherwise, every player reverts to the NE of the stage game, (*C, C*), thereafter. In this part of the exercise, assume that players can immediately observe deviations (perfect monitoring).
(c) *GTS with imperfect monitoring.* Assume now that players suffer from imperfect monitoring. Specifically, if player j chooses C, player i observes C with probability $p \in [0, 1]$; and if player j chooses *NC*, player i observes *NC* with probability $q \in [0, 1]$. This model embodies perfect monitoring as a special case, when $p = q = 1$; but it also includes other cases, such as every player perfectly observing if his rival chooses *NC* but imperfectly detecting deviations to C, that is, $p \in (0, 1)$ and $q = 1$; or perfectly observing if his rival chooses C but imperfectly detecting cooperation in *NC*, i.e., $p = 1$ and $q \in (0, 1)$.

In this context, the GTS described in part (b) must be rewritten as follows: In the first period, every player chooses *NC*; in all subsequent periods, every player chooses *NC* if (*NC, NC*) was the *observed* outcome in all previous periods. Otherwise, every player reverts to the NE of the stage game, (*C, C*), thereafter. Find the minimal discount factor sustaining this GTS, $\underline{\delta}(p, q)$.
(d) Evaluate $\underline{\delta}(p, q)$ at $p = q = 1$, to confirm that $\underline{\delta}(p, q) = \underline{\delta}$.
(e) Evaluate $\underline{\delta}(p, q)$ at $p \in (0, 1)$ and $q = 1$. Does $\underline{\delta}(p, 1)$ increase or decrease in p? Interpret.
(f) Evaluate $\underline{\delta}(p, q)$ at $q \in (0, 1)$ and $p = 1$. Does $\underline{\delta}(1, q)$ increase or decrease in q? Interpret.

7.5 **Prisoner's Dilemma game with temporary punishments.**[B] Consider the following Prisoner's Dilemma game.

Exercises

		Player 2	
		Confess	Not confess
Player 1	Confess	2, 2	8, 0
	Not confess	0, 8	4, 4

Players interact in an infinitely repeated game and consider the following GTS with temporary punishments: in the first period, every player chooses *NC*; in all subsequent periods, every player chooses *NC* if (*NC, NC*) was the outcome in all previous periods. Otherwise, every player reverts to the NE of the stage game, (*C, C*), during $T \geq 1$ periods, and once (*C, C*) happens for T periods, every player returns to cooperation (choosing *NC*). Assume, for simplicity, that players can immediately observe deviations (perfect monitoring).

(a) Find the minimal discount factor $\underline{\delta}$ sustaining the above GTS if the punishment lasts for only one period, $T = 1$.

(b) Find the minimal discount factor $\underline{\delta}$ sustaining the above GTS if the punishment lasts for three periods, $T = 3$.

(c) Find the minimal discount factor $\underline{\delta}$ sustaining the above GTS if the punishment lasts for T periods. Does $\underline{\delta}$ increase or decrease in T? Interpret your results.

7.6 **Prisoner's Dilemma game with more attractive cheating.**[B] Consider the following Prisoner's Dilemma game, where parameter a satisfies $a > 4$.

		Player 2	
		Confess	Not confess
Player 1	Confess	2, 2	a, 0
	Not confess	0, a	4, 4

Players interact in an infinitely repeated game and consider the following GTS with permanent punishments: in the first period, every player chooses *NC*; in all subsequent periods, every player chooses *NC* if (*NC, NC*) was the outcome in all previous periods. Otherwise, every player reverts to the NE of the stage game, (*C, C*), thereafter. Assume, for simplicity, that players can immediately observe deviations (perfect monitoring).

(a) Find the minimal discount factor sustaining the above GTS, $\underline{\delta}(a)$.

(b) Does $\underline{\delta}(a)$ increase or decrease in a? Interpret.

(c) Depict the set of feasible, individually rational payoffs (FIR diamond) in this game.

7.7 **Prisoner's Dilemma game with more severe punishments.**[A] Consider the following Prisoner's Dilemma game, where parameter b satisfies $4 > b > 0$.

	Player 2	
Player 1	Confess	Not confess
Confess	b, b	$8, 0$
Not confess	$0, 8$	$4, 4$

Players interact in an infinitely repeated game and consider the following GTS with permanent punishments: in the first period, every player chooses *NC*; in all subsequent periods, every player chooses *NC* if (*NC*, *NC*) was the outcome in all previous periods. Otherwise, every player reverts to the NE of the stage game, (*C*, *C*), thereafter. Assume, for simplicity, that players can immediately observe deviations (perfect monitoring).

(a) Find the minimal discount factor sustaining the above GTS, $\underline{\delta}(b)$.
(b) Does $\underline{\delta}(b)$ increase or decrease in b? Interpret.

7.8 **Prisoner's Dilemma game with lags at starting punishments.**[B] Consider the following Prisoner's Dilemma game.

	Player 2	
Player 1	Confess	Not confess
Confess	$2, 2$	$8, 0$
Not confess	$0, 8$	$4, 4$

Players interact in an infinitely repeated game and consider the following GTS with permanent punishments: in the first period, every player chooses *NC*; in all subsequent periods, every player chooses *NC* if (*NC*, *NC*) was the outcome in all previous periods. Otherwise, every player reverts to the NE of the stage game, (*C*, *C*), thereafter, but this reversion only starts $k \geq 1$ periods after the deviation was detected. Assume, for simplicity, that players can immediately observe deviations (perfect monitoring).

(a) Find the minimal discount factor sustaining the above GTS, $\underline{\delta}(k)$.
(b) Does $\underline{\delta}(k)$ increase or decrease in k? Interpret.
(c) Verbally discuss how your above results would coincide with those in a Prisoner's Dilemma game where players observe deviations $k \geq 1$ periods after they occurred but, once they observe the deviation, they can immediately revert to the NE of the stage game.

7.9 **Partial collusion in quantity competition.**[B] Consider an industry with two firms competing in quantities, facing inverse demand function $p(Q) = 1 - Q$, where $Q = q_1 + q_2$ denotes aggregate output; and having the same marginal cost of production, c, which satisfies $1 > c \geq 0$. In this context, consider the GTS with partial collusion, as follows:

- In period $t = 1$, every firm i chooses output $q_i = \alpha q_i^C + (1-\alpha) q_i^{NE}$, where $q_i^C = \frac{1-c}{4}$ denotes firm i's collusive output and $q_i^{NE} = \frac{1-c}{3}$ represents the NE output.
- In all subsequent periods, $t > 1$, every firm i chooses output q_i if both firms produced q_i in every previous period. Otherwise, every firm i reverts to the NE of the stage game, choosing $q_i = \frac{1-c}{3}$ thereafter.

Answer the following questions:
(a) Find the minimal discount factor sustaining the above GTS (partial collusion) as a function of parameter α, $\underline{\delta}(\alpha)$.
(b) How is $\underline{\delta}(\alpha)$ affected by an increase in α? Interpret.
(c) Evaluate $\underline{\delta}(\alpha)$ at $\alpha = 0$ and at $\alpha = 1$. Interpret.

7.10 **Collusion with imperfect monitoring (Discrete strategies).**[B] Consider the following matrix representing a simultaneous-move game between firms 1 and 2, where each firm independently chooses its output level (low, L, medium, M, or high, H).

		Firm 2		
		L	M	H
	L	20, 20	5, 13	0, 35
Firm 1	M	13, 5	15, 15	3, 23
	H	35, 0	23, 3	4, 4

(a) Find the NE in the stage game.
(b) *GTS with perfect monitoring.* For the remainder of the exercise, assume that firms interact in an infinitely repeated game. Find the minimal discount factor $\underline{\delta}$ sustaining the following (standard) GTS, namely, in the first period, every firm chooses the collusive output L; in all subsequent periods, every firm chooses L if (L, L) was the outcome in all previous periods. Otherwise, every firm reverts to the NE of the stage game, (H, H), thereafter. In this part of the exercise, assume that players can immediately observe deviations (perfect monitoring).
(c) *GTS with imperfect monitoring.* Assume now that firms suffer from imperfect monitoring.

- If firm j colludes, choosing L, we assume that, for simplicity, firm i cannot detect a deviation from the GTS, that is, if firm j colludes by producing the lowest output, firm i does not incorrectly observe its rival producing a medium or high output.
- If, instead, firm j chooses M, the probability that player i observes j deviating from L is $p \in [0, 1]$.

- Similarly, if firm j selects H, the probability that player i observes j deviating from L is $q \in [0, 1]$. In addition, we assume that $q > p$, indicating that when firm j increases its deviating output (from M to H) it is more likely that firm i detects such a deviation.

  In this context, the GTS described in part (b) must be rewritten as follows: In the first period, every firm chooses L; in all subsequent periods, every player chooses L if (L, L) was the *observed* outcome in all previous periods. Otherwise, every firm reverts to the NE of the stage game, (H, H), thereafter. Find the minimal discount factor sustaining this GTS, $\underline{\delta}(q)$, and show that it only depends on probability q.

  (d) Evaluate $\underline{\delta}(q)$ at $q = 1$ to confirm that $\underline{\delta}(1) = \underline{\delta}$.
  (e) Does cutoff $\underline{\delta}(q)$ increase or decrease in q? Interpret.

7.11 **Collusion with imperfect monitoring (Continuous strategies and firms competing à la Bertrand).**[B] Consider two firms selling a homogenous product, competing in prices (à la Bertrand), and facing a linear inverse demand function $p(Q) = 1 - Q$, where Q denotes aggregate output. For simplicity, assume that they are symmetric in their marginal production cost, c, which is normalized to $c = 0$.

  (a) *Stage game.* If the game is unrepeated, find the equilibrium price, p^C, and profit for each firm.
  (b) *GTS with perfect monitoring.* For the remainder of the exercise, assume that firms interact in an infinitely repeated game. Find the minimal discount factor sustaining cooperation if firms use a standard GTS, namely, in the first period, every firm charges the monopoly price, p^m; in all subsequent periods, every firm charges p^m if (p^m, p^m) was the outcome in all previous periods. Otherwise, every firm reverts to the NE of the stage game thereafter. In this part of the exercise, assume that firms can immediately observe deviations (perfect monitoring).
  (c) *GTS with imperfect monitoring.* Assume now that firms suffer from imperfect monitoring: if firm j deviates from the collusive price p^m, firm i observes its deviation with probability

  $$\alpha_j = \frac{p^m - p_j}{p^m},$$

  where p_j denotes firm j's price. This probability is positive when firm j deviates from the collusive price, $p_j < p^m$, and increases monotonically when firm j charges a lower price than p^m. Intuitively, firm i may not observe its rival's deviation when firm j charges barely above the collusive price p^m, but it will likely observe its deviation when firm j charges a low price. In this context, the GTS described in part (b) must be rewritten as follows: In the first period, every firm charges the monopoly price, p^m; in all subsequent periods, every firm charges p^m if (p^m, p^m) was the outcome *observed* in all previous periods. Otherwise, every firm reverts to the NE of the stage game, p^C, thereafter.

7.12 **Collusion with imperfect monitoring (Continuous strategies and firms competing à la Cournot).**[B] Consider two firms selling a homogenous product, competing (à la Cournot), and facing a linear inverse demand function $p(Q) = 1 - Q$, where Q denotes aggregate output. For simplicity, assume that they are symmetric in their marginal production cost, c, which is normalized to $c = 0$.

(a) *Stage game.* If the game is not repeated, find the equilibrium output, q^C, and profit for each firm.

(b) *GTS with perfect monitoring.* For the remainder of the exercise, assume that firms interact in an infinitely repeated game. Find the minimal discount factor sustaining cooperation if firms use a standard GTS, namely, in the first period, every firm produces half of the monopoly output, $\frac{q^m}{2}$; in all subsequent periods, every firm produces $\frac{q^m}{2}$ if $\left(\frac{q^m}{2}, \frac{q^m}{2}\right)$ was the outcome in all previous periods. Otherwise, every firm reverts to the NE of the stage game, q^C, thereafter. In this part of the exercise, assume that firms can immediately observe deviations (perfect monitoring).

(c) *GTS with imperfect monitoring.* Assume now that firms suffer from imperfect monitoring: if firm j deviates from the collusive output $\frac{q^m}{2}$, firm i observes its deviation with probability

$$p_i = \frac{q_j - \frac{q^m}{2}}{1 - \frac{q^m}{2}},$$

where q_j denotes firm j's output. This probability is positive when firm j deviates from the collusive output, $q_j > \frac{q^m}{2}$, and increases monotonically when firm j produces a larger output. Intuitively, firm i may not observe its rival's deviation when firm j produces barely above the collusive output $\frac{q^m}{2}$, but it will likely observe its deviation when firm j produces a large output. In this context, the GTS described in part (b) must be rewritten as follows: In the first period, every firm produces half of the monopoly output, $\frac{q^m}{2}$; in all subsequent periods, every firm produces $\frac{q^m}{2}$ if $\left(\frac{q^m}{2}, \frac{q^m}{2}\right)$ was the outcome *observed* in all previous periods. Otherwise, every firm reverts to the NE of the stage game, q^C, thereafter.

7.13 **Collusion with cost shocks—Public shocks.**[C] Consider two firms selling a homogenous product, competing in quantities (à la Cournot), and facing a linear inverse demand function $p(Q) = 1 - Q$, where Q denotes aggregate output. For simplicity, assume that they are symmetric in their marginal production cost, c, normalized to $c = 0$, at time $t = 1$. In subsequent periods $t \geq 2$, however, firm i's cost is $c_i^t = c + s_i^t$, where shock s_i^t can only take two values, $s_i^t = \{0, 0.02\}$. Therefore, when $s_i^t = 0$, firm i's cost is zero, entailing no cost shock; but when $s_i^t = 0.02$, its cost increases to 0.02. Assume that $s_i^t = 0$ occurs with probability p or $s_i^t = s$ with probability $1 - p$, for every firm i and for every period t. Assume that every firm can observe each

other's cost shocks, which become common knowledge in every period t, before every firm simultaneously and independently chooses its output level.
  (a) *Stage game.* If the game is unrepeated, find the equilibrium output, q^{NE}, and profit for each firm.
  (b) For the remainder of the exercise, assume that firms interact in an infinitely repeated game. Find the minimal discount factor sustaining cooperation if firms use a standard GTS as a SPE of the game, namely, in the first period, every firm produces the collusive output, $q^C(0) = \frac{Q^m(0)}{2}$, where $Q^m(0)$ denotes monopoly output when the marginal cost of both firms is 0; in all subsequent periods, every firm produces q^C if (q^C, q^C) was the outcome in all previous periods. Otherwise, every firm i reverts to the NE of the stage game, q_i^{NE}, thereafter, where output q_i^{NE} is a function of the state space (s_i^t, s_j^t).
  (c) *Special cases.* Evaluate the minimal discount factor that you found in part (a) at $p = 0$ and at $p = 1$. Interpret.

7.14 **Repeated public good game.**[C] Consider the public good game in section 4.4, where two individuals simultaneously and independently submit their contribution to the public good, x_i and x_j, with utility function

$$u_i(x_i, x_j) = (w - x_i)\sqrt{mX}$$

where m denotes the return of aggregate contributions to the public good, $X = x_i + x_j$. For simplicity, assume that both players have the same wealth, $w = 12$, that the return of the public good is $m = 2$, and they exhibit a common discount factor, $\delta \in (0, 1)$.
  (a) *Stage game.* Find the best responses and NE contributions in the stage game.
  (b) *Socially optimal outcome.* If players could coordinate their contributions to maximize their joint utility, which levels of x_i and x_j would they choose?
  (c) *Twice-repeated game.* Assume that the game is repeated twice. Analyze equilibrium contributions in the second stage, (x_i^2, x_j^2), for a given profile of contributions in the first stage, (x_i^1, x_j^1). Then, operating by backward induction, find the equilibrium contributions in the first stage. Can you sustain contributions different from those in the NE found in part (a) at every stage of the game?
  (d) *Infinitely repeated game.* If the stage game is infinitely repeated, can you sustain a GTS where, in the first period, every player chooses the cooperative (socially optimal) outcome of part (b) and, in all subsequent stages, every player keeps choosing the cooperative outcome if this outcome was played in all previous stages. Otherwise, he reverts to the NE of the stage game thereafter.

7.15 **Cooperation in a general version of the Prisoner's Dilemma game.**[B] Consider the following Prisoner's Dilemma game where payoffs satisfy $a > c$ and $b > d$, making Confess a strictly dominant strategy for every player.

Exercises

		Player 2	
		Confess	Not confess
Player 1	Confess	a, a	b, c
	Not confess	c, b	d, d

(a) Consider a standard GTS. Identify the minimal discount factor sustaining cooperation, $\underline{\delta}$.
(b) Describe how $\underline{\delta}$ is affected by a marginal increase in payoffs a through d. Interpret your results.
(c) Depict the set of feasible, individually rational payoffs (FIR diamond) in this game.

7.16 Intermittent cooperation in the Prisoner's Dilemma game.[B] Consider the following Prisoner's Dilemma game.

		Player 2	
		Confess	Not confess
Player 1	Confess	$2, 2$	$8, 0$
	Not confess	$0, 8$	$4, 4$

(a) Consider a GTS with partial cooperation, where players alternate between (C,C) and (NC,NC) over time, starting with (C,C) in the first period. If either or both players deviates, both players revert to the NE of the stage game, (C,C), forever. Find the minimum discount factor for this GTS to be sustained as a subgame perfect equilibrium of the infinitely repeated game.
(b) Allow for generic payoffs a, b, c, d (as in Exercise 6.15). Identify under which conditions for these four payoffs the GTS with partial cooperation of part (a) can be supported.

7.17 An eye-for-an-eye.[A] Consider the following payoff matrix.

		Player 2		
		A	B	C
	A	$0, 0$	$4, 2$	$5, -2$
Player 1	B	$2, 4$	$8, 8$	$14, 3$
	C	$-2, 5$	$3, 14$	$10, 10$

Show that, if players use a standard GTS, cooperation can be sustained if δ satisfies $\delta \geq \frac{2}{3}$.

7.18 **Tit-for-tat.**[A] Consider the Prisoner's Dilemma game.

		Player 2	
		Confess	Not confess
Player 1	Confess	2, 2	8, 0
	Not confess	0, 8	4, 4

(a) Find the minimal discount factor sustaining cooperation if players use a tit-for-tat strategy, where every player mimics what his rival did in the previous period. Then, a cooperative outcome in the previous period is self-perpetuating, while a non-cooperative outcome is self-perpetuating too.
(b) Find the minimal discount factor sustaining cooperation if players use a standard GTS.
(c) Compare your results in parts (a) and (b). Which strategy requires a more demanding condition on the discount factor? Interpret your results.

7.19 **War of attrition.**[B] Consider two players, 1 and 2, waiting in line. Every player $i = \{1, 2\}$ incurs a cost $c > 0$ from staying in the line for one more period. When only one player remains in the line, she receives a reward (e.g., a concert ticket) which all players value at V, where $V > c$. For simplicity, the discount factor of both players is δ.
(a) If both players remain in line until period $t-1$, and only player 1 remains in line in the next period, t, which is player 1's discounted net payoffs?
(b) Assume that players use a symmetric strategy. Show that a pure-strategy NE cannot be supported.
(c) Assume that players use a symmetric strategy. Find the mixed strategy NE where every player quits in period t with probability p.
(d) Is the probability you found in part (c) a number between 0 and 1? Is it increasing or decreasing in the reward, V?
(e) How is the probability you found in part (c) affected by c? Explain.

7.20 **Chain-store paradox.**[B] Consider the entry game in Sect. 6.3, but allowing for it to repeated $T \geq 1$ times. This can be interpreted as an incumbent firm operating as a monopolist in different geographically separated markets (as a chain store), and T different entrants sequentially deciding whether to enter each of these markets: entrant 1 chooses whether to enter market 1, entrant 2 observes 1's decision and decides whether to enter market 2, and similarly until entrant T.
(a) Show that the unique SPE prescribes every entrant i joining its market i and the incumbent responds by accommodating entry in every market.
(b) Does a larger number of markets (as captured by T) help the incumbent prevent entry relative to the setting with a single market analyzed in section 6.3?

References

Harrington, J. E. (2006) "How Do Cartels Operate?," Foundations and Trends in Microeconomics, 2(1), pp. 1–105.

Kandori, M. (2002) "Introduction to Repeated Games with Private Monitoring," Journal of Economic Theory, 102(1), pp. 1–15.

Levenstein, M. C., and V. Y. Suslow (2006) "What Determines Cartel Success?," Journal of Economic Literature, 44(1), pp. 43–95.

Mailath, G., and L. Samuelson (2006) *Repeated Games and Reputations: Long-run Relationships*, Oxford University Press, Oxford.

Tirole, J. (1988) *The Theory of Industrial Organization*, The MIT Press, Cambridge.

Bayesian Nash Equilibrium 8

8.1 Introduction

More often than not, players interact in games where at least one of them is uninformed about some relevant information, such as its rival's production costs in an oligopoly market or the available stock in a common pool resource (e.g., fishing grounds, forests, and aquifers). In other games, all players are uninformed about some piece of information, such as auctions where every bidder privately observes her valuation for the object on sale but does not observe the valuation that other bidders assign to the object.

A similar example applies to an industry where firms compete in prices, each privately observing its marginal production costs while not observing the exact cost of its rivals. In all of these cases, players may do research to find out the information they do not observe (e.g., rivals' costs) but, in most settings, this investigation only produces *estimates*, with perhaps some confidence intervals, rather than the actual value of the unobserved parameter. For instance, in the context of two firms competing in prices, each privately observing its cost, every firm knows that its rival's marginal cost is $10 with probability 2/3 or $6 otherwise. That is, while firms may not observe some parameter (e.g., its rival's cost), they know the probability distribution of this parameter. This knowledge may come, perhaps, after months of research or because the uninformed player hired a consulting company to provide more precise estimates of the parameter, each with its associated probability distribution.

We generally refer to the private information that a player observes as her "type," such as a firm observing its marginal cost being $10, or a bidder observing her valuation of a painting being $300. In contrast, every player does not observe all other players' types; otherwise, we would be in a game of complete information! In other words, for the game to qualify as an incomplete information game, there must be at least one player who does not observe the type of at least one of her rivals.

In this chapter, we essentially seek to adapt the NE solution concept, first presented in Chapter 3, to a context of incomplete information. We start by defining a player's strategy, which in this context describes the action that she chooses as a function of her type. For instance, a firm may set a price of $15 when its marginal cost is $10, but a price of $9 when its marginal cost is only $6. We then use this definition of strategy, adapted to incomplete information settings, to identify a player's best response in this context. As in Chapter 3, a best response maximizes a player's payoff against her rivals' strategies. In our current setting, however, a best response maximizes her *expected* payoff, because she cannot observe the type of at least one of her opponents, and that maximization is conditional on her type. Therefore, a player may choose a different best response against the same strategy depending on her privately observed type. (Consider, for instance, the above example about a firm setting different prices depending on its marginal costs.)

Following the same steps as in Chapter 3, we then use the definition of best response to describe an NE in the context of incomplete information, which is commonly known as a Bayesian Nash Equilibrium (BNE). As under complete information, in a BNE every player must play mutual best responses to each other's strategies, making the strategy profile stable. This stability is, in this context, understood from an ex-ante perspective, that is, given the information that every player observes when she is called to move.

We present two approaches to find BNEs: one building the Bayesian normal form of the incomplete information game, and another solving, first, for the best responses of the privately informed player/s and then moving to those of the uninformed players. We apply these approaches to different games. A typical application of BNEs, strategic bidding in auctions, is relegated to Chapter 9.

8.2 Background

8.2.1 Players' Types and Their Associated Probability

Discrete types. In a game of incomplete information, every player i observes its type, θ_i, where $\theta_i \in \Theta_i$, which may represent, for instance, a high or low production cost, $\theta_i = H$ or $\theta_i = L$, implying that the set of types in this case is $\Theta_i = \{H, L\}$. Player i, however, does not observe her rival's type: θ_j in a two-player game (players i and j) or θ_{-i} in a game with more players, where we use θ_{-i} to more compactly denote the types of player i's rivals, that is,

$$\theta_{-i} = (\theta_1, \theta_2, ..., \theta_{i-1}, \theta_{i+1}, ..., \theta_N).$$

Players, however, know the probability distribution over types, e.g., firm i knows that its rival's type is either $\theta_j = H$ with probability q, where $q \in [0, 1]$, and $\theta_j = L$ with probability $1 - q$, and this information is common knowledge. Intuitively, these probabilities can be understood as the frequency of different types of firms in an industry.

8.2 Background

More generally, in a setting where player i can have K different types, we write that her type space is

$$\Theta_i = \left\{\theta_i^1, \theta_i^2, ..., \theta_i^K\right\}$$

and the probability that her type is θ_i^1 can be expressed as $\Pr(\theta_i = \theta_i^1) = p_i^1$, the probability that her type is θ_i^2 can be written as $\Pr(\theta_i = \theta_i^2) = p_i^2$, and similarly for the probability that her type is θ_i^K, where we write that $\Pr(\theta_i = \theta_i^K) = p_i^K$. Since, in addition, probabilities of each type $\left(p_i^1, p_i^2, ..., p_i^K\right)$ satisfy $p_i^k \in [0, 1]$ for every $k = \{1, 2, ..., K\}$ and $\sum_{k=1}^{K} p_i^k = 1$, we can omit the probability of the last type, writing it, instead, as $p_i^K = 1 - \sum_{k \neq K} p_i^k$. In a context with three types, $\Theta_i = \left\{\theta_i^1, \theta_i^2, \theta_i^3\right\}$, the associated probabilities can be expressed as $\left(p_i^1, p_i^2, 1 - p_i^1 - p_i^2\right)$.

Our above discussion simplifies if players face the same set of possible types, $\Theta_i = \Theta_j = \Theta$ for every two players $i \neq j$, where we can represent the set of types as Θ, omitting the subscript. This argument extends to the probabilities of each type, which are identical across players, helping us write $\left(p^1, p^2, ..., p^K\right)$, omitting the subscripts since $p_i^k = p_j^k$ for every two players $i \neq j$. Player i's type, however, must still keep its subscript, θ_i^k, as otherwise we would not know the identity of each player (that is, we write θ_i^k rather than θ^k).

Continuous types. Our notation can be adapted to a setting where types are continuous. Following our above example of a firm privately observing its production cost, we could have that θ_i lies in an interval, such as $[0, 20]$, or more generally in any interval $[a, b]$, where $b > a$.[1]

A player i's type in this setting, θ_i, is drawn from a continuous cumulative probability distribution, that is,

$$F(x) = \Pr\{\theta_i \leq x\}.$$

Intuitively, $F(x)$ measures the probability that player i's type, θ_i, lies weakly below x. For instance, if θ_i is uniformly distributed, $\theta_i \sim U[0, 1]$, its cumulative probability distribution is $F(x) = x$. If, instead, θ_i is exponentially distributed, its cumulative distribution function is $F(x) = 1 - \exp(-\lambda x)$, where parameter $\lambda \geq 1$ represents how concave $F(x)$ is. In other words, a higher λ indicates that most of the probability weight is concentrated on low values of x.

---

[1] The length of interval $[a, b]$ is, often, normalized to one, so point $x \in [a, b]$ can alternatively be expressed as $y = \frac{x-a}{b-a}$, which implies that y lies in the unit interval, i.e., $y \in [0, 1]$. Intuitively, values of y close to 1 indicate values of the original parameter, x, close to its upper bound, while values of y close to zero entail that x is close to its lower bound.

This representation also helps us find a density function, $f(x)$, associated with the above $F(x)$, if one exists, by computing its first-order derivative because $f(x) = F'(x)$. Recall that density $f(x)$ describes the probability that player i's type, θ_i, is exactly x, that is, $f(x) = \Pr\{\theta_i = x\}$, which can be quite useful in some games. In the above example, where player i's types are uniformly distributed, $F(x) = x$, the density function is $f(x) = 1$, meaning that all types are equally likely to occur. If types are exponentially distributed, $F(x) = 1 - \exp(-\lambda x)$, its density function is $f(x) = \lambda \exp(-\lambda x)$, implying that parameter λ represents how quickly the density function decreases as we increase x, and is often known as the "decay rate." Intuitively, a higher λ means that $f(x)$ puts most probability weight on low values of x; thus following a similar interpretation as in $F(x)$.

8.2.2 Strategies Under Incomplete Information

If players operate under incomplete information, we must have that at least one player does not observe the types of at least one of her opponents. If player i observes some piece of private information (e.g., production costs), she can condition her strategy on her type, implying that her strategy in this context is a function of θ_i, which we express as $s_i(\theta_i)$. Similarly, if every player $j \neq i$ privately observes her type θ_j, we can write the strategy profile of player i's rivals as

$$(s_1(\theta_1), s_2(\theta_2), ..., s_{i-1}(\theta_{i-1}), s_{i+1}(\theta_{i+1}), ..., s_N(\theta_N)),$$

or, more compactly, as $s_{-i}(\theta_{-i})$. If we write the player i's strategy as $s_i(\theta_i, \theta_{-i})$, we mean that she can condition her choice of s_i on the observation of both her type, θ_i, and her rival's, θ_{-i}, making player i perfectly informed. In some games, we may have some perfectly informed players, who observe everyone's types because of their experience in the industry or because they get to act before everyone else; while other players privately observe their types but not their rivals.

8.2.3 Representing Asymmetric Information as Incomplete Information

Consider the game in Matrix 8.1, where player 1 observes the realization of x before playing the game, but player 2 only knows that its realization is either $x = 20$ or $x = 12$ with equal probabilities. This information is common knowledge.

		Player 2	
		L	R
Player 1	U	x, 17	5, 10
	D	10, 0	10, 17

Matrix 8.1. Simultaneous move game where player 1 privately observes x.

8.2 Background

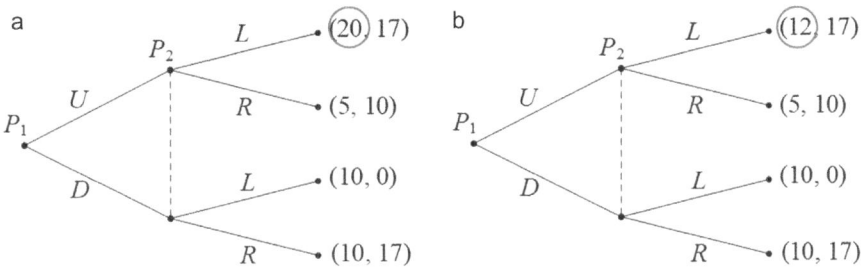

Fig. 8.1 a Simultaneous-move game where $x = 20$. b Simultaneous-move game where $x = 12$

Matrix 8.1, therefore, suggests that player 2 faces *imperfect* information because she does not observe the realization of x while player 1 does. Player 2 knows, however, that she faces either of the two games depicted in Fig. 8.1a and b, where we circled the only payoff that differs across both figures, that corresponding to (U, L).

Alternatively, we can represent the above strategic setting as a game of *incomplete* information. In this setting, player 2, instead of not observing the realization of x, does not observe the move by a fictitious player ("nature"), who determines whether $x = 20$ or $x = 12$ at the beginning of the game, as illustrated in Fig. 8.2. Graphically, this figure just connects the game trees in Figs. 8.1a and b with a move by nature, indicating that players face either the game in Fig. 8.1a or that in Fig. 8.1b, both occurring with equal probability. This useful trick—representing games of imperfect information as games of incomplete information with the help of a move by nature—is due to Harsanyi (1967) and we will often use it to facilitate our search of equilibrium behavior in games where one (or more) players privately observe some information.

8.2.4 Best Response Under Incomplete Information

We are now ready to adapt the definition of an NE to this incomplete information setting, by first defining a best response in this context.

Fig. 8.2 Combining both games of Fig. 8.1 to allow for incomplete information

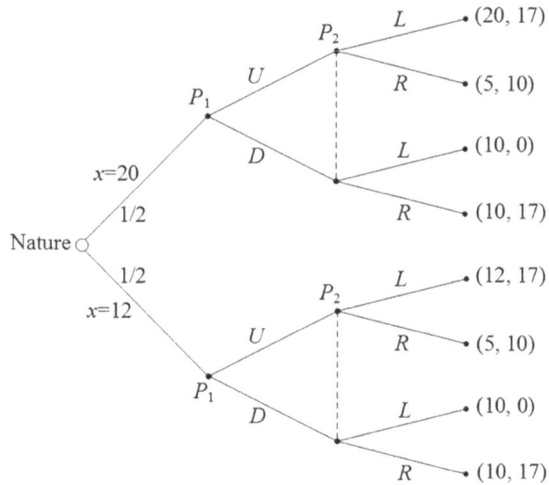

Definition 8.1. **Best response under incomplete information**. Player i regards strategy s_i as a best response to strategy profile $s_{-i}(\theta_{-i})$, if

$$EU_i(s_i(\theta_i), s_{-i}(\theta_{-i})) \geq EU_i(s'_i(\theta_i), s_{-i}(\theta_{-i}))$$

for every available strategy $s'_i(\theta_i) \in S_i$ and every type $\theta_i \in \Theta_i$.

In a two-player game, player i finds strategy $s_i(\theta_i)$ to be a best response to player j's strategy, $s_j(\theta_j)$, if

$$EU_i(s_i(\theta_i), s_j(\theta_j)) \geq EU_i(s'_i(\theta_i), s_j(\theta_j))$$

for every available strategy $s'_i(\theta_i) \in S_i$ and every type $\theta_i \in \Theta_i$. Intuitively, this means that strategy $s_i(\theta_i)$ yields a weakly higher *expected* payoff than any other strategy $s'_i(\theta_i)$ against $s_j(\theta_j)$, and this holds for all player i's types, θ_i. For instance, if $\Theta_i = \{H, L\}$, and player i's type is H, player i maximizes her expected payoff responding with $s_i(H)$ against her rival strategy; and, similarly, when i's type is L, she maximizes her expected payoff responding with $s_i(L)$.

Relative to the definition of best response in contexts of complete information (Chapter 3), this definition differs in two dimensions.

1. *Expected utility*. First, player i seeks to maximize her *expected*, instead of certain, utility level, because she can now face the uncertainty of not observing some private information (e.g., the type of some rival player). If the player is perfectly informed—and there can be some perfectly informed players in these games—she faces no uncertainty, implying that she can just consider the certain utility of each strategy. At least some player in an incomplete information game, however, will not observe everyone's types, thus having to compute the expected utility from each of her strategies.

2. *Best response as a function of types*. In addition, player i finds a strategy $s_i(\theta_i)$, as opposed to the strategy s_i in the best responses of Chapter 3. Intuitively, her choice may now depend on her privately observed type, meaning that her strategy may differ for at least two of her types, that is, $s_i(\theta_i') \neq s_i(\theta_i'')$ for at least two types θ_i' and θ_i'', where $\theta_i' \neq \theta_i''$. Player i may, instead, choose the same strategy, $s_i(\theta_i') = s_i(\theta_i'')$, for every two types $\theta_i' \neq \theta_i''$, implying that she does not condition her strategy on her private information.

Importantly, the above definition of best response concludes by saying "...and every type $\theta_i \in \Theta_i$" which means that player i must find strategy $s_i(\theta_i)$ to be optimal for each of her types. As in our above example, if her set of types is $\Theta_i = \{H, L\}$, player i's best response, is a vector of two components, $s_i(\theta_i) = \{s_i(H), s_i(L)\}$, describing the action that player i chooses when her type is H, $s_i(H)$, and when her type is L, $s_i(L)$. Similarly, in a context with K different types, player i's best response is a vector with K components, specifying the action that maximizes her expected payoff in each of her types.

8.3 Bayesian Nash Equilibrium

We next use the concept of best response to define a Bayesian Nash Equilibrium.

Definition 8.2. **Bayesian Nash Equilibrium (BNE)**. A strategy profile $\left(s_i^*(\theta_i), s_{-i}^*(\theta_{-i})\right)$ is a Bayesian Nash Equilibrium if every player chooses a best response (under incomplete information) given her rivals' strategies.

Therefore, in a two-player game, a strategy profile is a BNE if it is a *mutual* best response, thus being analogous to the definition of NE: the strategy that player i chooses is a best response to that selected by player j, and vice versa. As a result, no player has unilateral incentives to deviate. This definition assumes, of course, that the players select best responses in the sense defined in Sect. 8.2, where players seek to maximize their expected payoffs given the information they observe.

8.3.1 Ex-ante and Ex-post Stability

As under complete information, in a BNE every player must be playing mutual best responses to each other's strategies, making the strategy profile stable. This stability is, in this context, understood from an *ex-ante* perspective, that is, given the information that every player observes when she is called to move; as opposed to *ex-post*, which assumes that at the end of the game every player gets to observe her rivals' types.[2]

Informally, we can interpret that, upon playing a BNE, players could say "if I had to play the game again, given the (little) information I had, I would not have incentives to deviate. I would play in the same way I just did." Ex-post stability would, however, imply that, after playing according to a BNE, players would say "if I had to play the game again, but given the (rich) information I have now, once the game is over and I can observe everyone's types, I would still play in the same way" or, alternatively, "if I knew then what I know now, I would behave in the same way." Ex-post stability is, therefore, more demanding than ex-ante stability, as it requires that players have no incentives to deviate from their equilibrium strategies for all realizations of type profiles (i.e., regardless of their rivals' types).

8.4 Finding BNEs—First Approach: Build the Bayesian Normal Form

In this section, we present the first tool to find all pure-strategy BNEs in a game, which entails building the so-called "Bayesian normal form representation" of the game. The following tool describes the three steps we follow to build it and, then, find the BNEs. This approach is particularly useful in two-player games, where players have a discrete, and relatively small, number of available actions (such as 2–5) and few possible types (such as high, medium, and low). When players have several discrete actions to choose from, or continuous actions (such as a firm selecting its output level, $q \in \mathbb{R}_+$) and several possible types, this approach becomes rather intractable, and we will rely on a different method.

Tool 8.1. Finding BNEs using the Bayesian normal form representation (two-player game).

1. Write down each player's strategy set, S_i and S_j. Recall that privately informed players condition their strategies on their types. Similarly, players who observed other players' actions before being called to move can condition their strategy on the node or information set where they are called to move.
2. Building the Bayesian normal form representation of the game:
   a. Depict a matrix with as many rows as strategies in S_i and as many columns as strategies in S_j, leaving all cells in this matrix empty.
   b. Find the expected utility that player i earns in each cell.

---

[2] The literature often uses ex-ante, interim, and ex-post stability, meaning that players do not have incentives to deviate from their equilibrium strategies before observing their own types, after observing their own types but before observing their rivals', or after observing everyone's types, respectively. For simplicity, however, we use here "ex-ante stability" in the interim sense (players privately observe their own types) because we do not analyze the ex-ante stage, where players have not even observed their own types yet.

8.4 Finding BNEs—First Approach: Build the Bayesian Normal Form

3. Find each player's best response against her opponent's strategy and underline her best response payoffs.
4. Identify which cell/s have all players' payoffs underlined, thus being mutual best responses. This cell/s are the BNEs of the game.

The following example applies tool 8.1 to a relatively straightforward setting with two players, where one of them privately observes which game players play while her opponent does not.

Example 8.1 Finding BNEs. Consider the game in Matrix 8.2, which reproduces Matrix 8.1.

		Player 2	
		L	R
Player 1	U	$x, 17$	$5, 10$
	D	$10, 0$	$10, 17$

Matrix 8.2. Simultaneous move game where player 1 privately observes x.

First step. For simplicity, we often start with the uninformed player (player 2 in this setting), whose strategy set is $S_2 = \{L, R\}$ since she cannot condition her strategy on player 1's type. In contrast, player 1 observes her type (the realization of x) and can condition her strategy on x, that is, $s_1(x)$. Therefore, player 1's strategy set is

$$S_1 = \{U^{20}U^{12}, U^{20}D^{12}, D^{20}U^{12}, D^{20}D^{12}\}.$$

Strategy $U^{20}U^{12}$, for instance, prescribes that player 1 chooses U both after observing that $x = 20$ and that $x = 12$. In $U^{20}D^{12}$ ($D^{20}U^{12}$), however, player 1 chooses U only after observing that $x = 20$ (that $x = 12$, respectively). Finally, in $D^{20}D^{12}$, player 1 selects D regardless of her type.

Second step. From the first step, player 1 has four available strategies while player 2 only has two, yielding the Bayesian normal form representation in Matrix 8.3 with four rows and two columns. We now need to find out the expected payoff that each player earns in each strategy profile (cell of Matrix 8.3).

		Player 2	
		L	R
Player 1	$U^{20}U^{12}$		
	$U^{20}D^{12}$		
	$D^{20}U^{12}$		
	$D^{20}D^{12}$		

Matrix 8.3. Player 1 privately observes x - Bayesian normal form representation.

For instance, at strategy profile $(U^{20}U^{12}, L)$, player 1 earns

$$EU_1(U^{20}U^{12}, L) = \underbrace{\frac{1}{2}20}_{\text{if } x=20} + \underbrace{\frac{1}{2}12}_{\text{if } x=12} = 16$$

while player 2's expected utility is

$$EU_2(U^{20}U^{12}, L) = \underbrace{\frac{1}{2}17}_{\text{if } x=20} + \underbrace{\frac{1}{2}17}_{\text{if } x=12} = 17.$$

However, at strategy profile $(U^{20}D^{12}, L)$, player 1 earns

$$EU_1(U^{20}D^{12}, L) = \frac{1}{2}20 + \frac{1}{2}10 = 15$$

whereas player 2's expected utility is

$$EU_2(U^{20}D^{12}, L) = \frac{1}{2}17 + \frac{1}{2}0 = \frac{17}{2} = 8.5.$$

Operating similarly for the remaining strategy profiles, we find the expected payoffs in Matrix 8.3, obtaining Matrix 8.4.

		Player 2	
		L	R
Player 1	$U^{20}U^{12}$	16, 17	5, 10
	$U^{20}D^{12}$	15, 8.5	7.5, 13.5
	$D^{20}U^{12}$	11, 8.5	7.5, 13.5
	$D^{20}D^{12}$	10, 0	10, 17

Matrix 8.4. Player 1 privately observes x - Finding expected payoffs.

Third step. Underlining best response payoffs, we obtain Matrix 8.5. The two cells where both players' (expected) payoffs are underlined indicate the two BNEs of this game: $(U^{20}U^{12}, L)$ and $(D^{20}D^{12}, R)$.

		Player 2	
		L	R
Player 1	$U^{20}U^{12}$	<u>16, 17</u>	5, 10
	$U^{20}D^{12}$	15, 8.5	7.5, <u>13.5</u>
	$D^{20}U^{12}$	11.5	7.5, <u>13.5</u>
	$D^{20}D^{12}$	10, 0	<u>10, 17</u>

Matrix 8.5. Player 1 privately observes x - Best response payoffs.

8.5 Finding BNEs—Second Approach: Focus on the Informed Player First

From the previous section, you may anticipate that the first approach to find BNEs (building the Bayesian normal form of the game) can quickly become intractable if players have several available actions or several types. Consider, for instance, a duopoly game where firms compete in quantities, q_1 and q_2 where $q_i \geq 0$, with firm 1 observing the marginal cost of both firms (informed player), while firm 2 only observes its cost (uninformed player). For simplicity, assume that firm 1's cost is either high or low (just two types). While the setting seems stylized enough, we can immediately see that it is not easy to represent with the Bayesian normal form. Starting with step 1 in Tool 8.1, firm 2's strategy set is $S_2 = \mathbb{R}_+$ (the uninformed player, as firm 2 cannot condition its output on firm 1's type) while firm 1's is $S_1 = \mathbb{R}_+^2$ since every strategy for this firm is a pair of output levels (q_1^H, q_1^L), describing its production decision when facing high and low costs, respectively. Moving to step 2, these strategy sets would yield infinitely many rows and columns in the Bayesian normal form representation!

Is there another approach we can deploy to solve for BNEs in games that do not allow for a compact Bayesian normal form representation? Yes, focusing on the privately informed player first, we find her best response functions (one for each of her types), and then move to the uninformed player, who should have a single best response function. For simplicity, we consider two-player games where one or both players are privately informed.

Tool 8.2. Finding BNEs by focusing on the informed player first.

1. Focus on the privately informed player i. For each of her $k \geq 2$ types, find her best response function to her rival's strategies, e.g., $s_i(s_j|\theta_i)$. You must find k best response functions, one for each type, e.g., $s_i(s_j|H)$ and $s_i(s_j|L)$.
2. If all players are privately informed about their types, simultaneously solve for $s_i(\theta_i)$ in the best response function $s_i(s_j|\theta_i)$, finding k equilibrium strategies, one for each type, that is, $s_i^*(\theta_i) = \left(s_i^*(\theta_i^1), ..., s_i^*(\theta_i^k)\right)$. Therefore, $\left(s_1^*(\theta_1), s_2^*(\theta_2)\right)$ is the BNE of the game.
3. If one player is uninformed (does not privately observe some information), analyze the uninformed player j.
    a. Find which strategy gives player j the highest expected utility to her rival's strategies. This strategy is her best response function $s_j(s_i)$, which is not conditional on her rival's types.
    b. Simultaneously solve for $s_i(\theta_i)$ and s_j in best response functions $s_i(s_j|\theta_i)$, one for each of player i's types, and $s_j(s_i)$. This yields a system of $k + 1$ equations (k best response functions for the informed player plus the best response function of the uninformed player) and a BNE given by $\left(s_i^*(\theta_i), s_j^*\right)$ where the first component describes the informed player equilibrium strategy, given her type θ_i, and the last component prescribes the uninformed player's strategy, which is not conditional on player i's type, θ_i.

The next example applies this tool to a standard duopoly game.

Example 8.2 Cournot competition with asymmetrically informed firms. Consider a duopoly market where firms compete à la Cournot and face inverse demand function $p(Q) = 1 - Q$, where Q denotes aggregate output. Firms interact in an incomplete information setting where firm 2's marginal costs are either $c_H = \frac{1}{2}$ or $c_L = 0$, occurring with probability p and $1 - p$, respectively, where $p \in (0, 1)$. Firm 2 privately observes its marginal cost, but firm 1 does not. Because firm 1 has operated in the industry for a long period, all firms observe that this firm's costs are $c_H = \frac{1}{2}$.

Privately informed firm, high costs. Focusing first on the privately informed player (firm 2), we solve its profit-maximization problem when its cost is high, $c_H = \frac{1}{2}$, as follows

$$\max_{q_2^H \geq 0} \pi_2^H\left(q_2^H\right) = \left(1 - q_2^H - q_1\right)q_2^H - \frac{1}{2}q_2^H$$

Differentiating with respect to q_2^H, yields

$$1 - 2q_2^H - q_1 - \frac{1}{2} = 0$$

and solving for q_2^H, we obtain firm 2's best response function when its costs are high,

$$q_2^H(q_1) = \frac{1}{4} - \frac{1}{2}q_1$$

which originates at $\frac{1}{4}$, decreases in its rival's output at a rate of $\frac{1}{2}$, and becomes zero for all $q_1 > \frac{1}{2}$.

Privately informed firm, low costs. We keep analyzing the privately informed player (firm 2) but when its costs are low, $c_L = 0$. In this case, its profit-maximization problem is

$$\max_{q_2^L \geq 0} \pi_2^L\left(q_2^L\right) = \left(1 - q_2^L - q_1\right)q_2^L$$

Differentiating with respect to q_2^L, we find that

$$1 - 2q_2^L - q_1 = 0$$

8.5 Finding BNEs—Second Approach: Focus on the Informed Player First

and solving for q_2^L, we obtain firm 2's best response function when its costs are low,

$$q_2^L(q_1) = \frac{1}{2} - \frac{1}{2}q_1$$

which originates at $\frac{1}{2}$, decreases in its rival's output at a rate of $\frac{1}{2}$, and becomes zero for all $q_1 > 1$. Figure 8.3a compares firm 2's best response functions when its costs are high and low, where $q_2^L(q_1) > q_2^H(q_1)$ indicates that, for a given output by firm 1, firm 2 produces more units when its own costs are low than high.

Uninformed firm. The uninformed player (firm 1) does not observe firm 2's cost, which can be high or low, and thus solves the following *expected* profit-maximization problem

$$\max_{q_1 \geq 0} \pi_1(q_1) = p\underbrace{\left[\left(1 - q_2^H - q_1\right)q_1\right]}_{\text{Firm 2's costs are high}} + (1-p)\underbrace{\left[\left(1 - q_2^L - q_1\right)q_1\right]}_{\text{Firm 2's costs are low}} - \frac{1}{2}q_1$$

where the last term indicates firm 1's high costs, which are certain, and observed by all firms. Differentiating with respect to q_1, yields

$$p\left(1 - q_2^H - 2q_1\right) + (1-p)\left(1 - q_2^L - 2q_1\right) - \frac{1}{2} = 0$$

Solving for q_1, we find that firm 1's best response function is

$$q_1(q_2^H, q_2^L) = \frac{1}{4} - \frac{pq_2^H + (1-p)q_2^L}{2}$$

which originates at 1/4, and decreases at a rate of $\frac{1}{2}$ when firm 2's *expected* output increases (as captured by the term $pq_2^H + (1-p)q_2^L$). Figure 8.3b depicts firm 2's

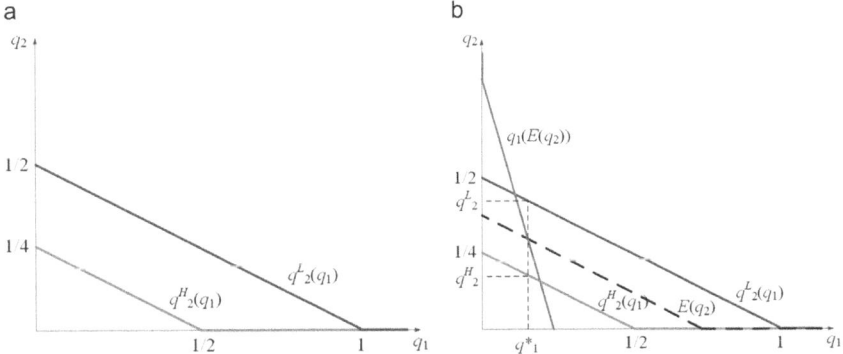

Fig. 8.3 a Firm 2—high or low costs. b Equilibrium output under incomplete information

expected output, $E(q_2) \equiv pq_2^H + (1-p)q_2^L$ in the dashed line between $q_2^H(q_1)$ and $q_2^L(q_1)$. It also plots firm 1's best response as a function of $E(q_2)$. As we next study, the point where firm 1's best response function crosses $E(q_2)$ yields firm 1's equilibrium output, q_1^*, in this incomplete information game. Afterward, inserting q_1^* into $q_2^L(q_1)$ and $q_2^H(q_1)$ will provide us with firm 2's equilibrium output for each of its types.

Finding equilibrium output in the BNE. Our above analysis produced three best response functions (one for firm 1, two for firm 2) and we have three unknowns (q_1, q_2^H, and q_2^L), so we only need to simultaneously solve for these three unknowns. A common approach is to insert best response functions $q_2^H(q_1)$ and $q_2^L(q_1)$ into $q_1(q_2^H, q_2^L)$, as follows,

$$q_1 = \frac{1}{4} - \frac{p\overbrace{\left(\frac{1}{4} - \frac{1}{2}q_1\right)}^{q_2^H(q_1)} + (1-p)\overbrace{\left(\frac{1}{2} - \frac{1}{2}q_1\right)}^{q_2^L(q_1)}}{2}$$

Rearranging, this expression simplifies to

$$q_1 = \frac{1}{4} + \frac{2q_1 + p - 2}{8}$$

which, solving for q_1, yields firm 1's equilibrium output

$$q_1^* = \frac{p}{6}.$$

Inserting q_1^* into firm 2's best response functions, we obtain

$$q_2^H(q_1^*) = \frac{1}{4} - \frac{1}{2}q_1^* = \frac{3-p}{12}, \text{ and}$$
$$q_2^L(q_1^*) = \frac{1}{2} - \frac{1}{2}q_1^* = \frac{6-p}{12}$$

where, as expected, firm 2 produces a larger output when facing low than high cost. Therefore, the BNE of this game is given by the triplet of output levels

$$\left(q_1^*, q_2^{H*}, q_2^{H*}\right) = \left(\frac{p}{6}, \frac{3-p}{12}, \frac{6-p}{12}\right).$$

Figure 8.4 depicts these three output levels, showing that firm 1's equilibrium output, q_1^*, is increasing in the probability that its rival has a high cost, p, but that firm 2's equilibrium output is decreasing in this probability, which holds both when its costs are both high and low.

As a remark, note that our results also embody those of the complete information game where firm 1 has high or low costs with certainty. When $p \to 0$ (left

Fig. 8.4 Cournot competition with asymmetrically informed firms—equilibrium output

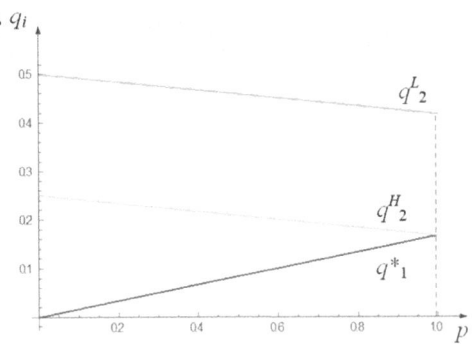

side Fig. 8.4), the privately informed firm has (almost certainly) low costs, thus being likely competitive. In this setting, the uninformed player (firm 1) remains inactive, $q_1^* \to 0$, while firm 2 produces 1/4 units when its costs are high and 1/2 when its costs are low. In contrast, when $p \to 1$ (right side of Fig. 8.4), firm 2 is, essentially, a high-cost company, inducing firm 1 to be active, producing 1/6 units. Firm 2 in this case chooses an output of 1/6 when its costs are high and 5/12 if its costs are low.

8.6 Evaluating BNE as a Solution Concept

Following our discussion in Sect. 1.6, we next evaluate BNE as a solution concept:

1. **Existence? Yes.** When we apply BNE to any game, we find that at least one equilibrium exists. Intuitively, this result is equivalent to the existence of NEs in complete information games, but extended to an incomplete information context. For this finding to hold, however, we may need to allow for mixed strategies; although, for simplicity, all applications studied in this chapter focus on pure-strategy BNEs. (Exercise 8.17, however, considers the incomplete information version of the matching pennies game, showing that a BNE in pure strategies cannot be supported, and finds a mixed strategy BNE.) Some auctions, which we examine in Chapter 3, require players to randomize their bidding decisions, giving rise to mixed strategy BNEs.
2. **Uniqueness? No.** Games of incomplete information may have more than one BNE, as illustrated in Example 8.1 where we found two BNEs. For simplicity, several strategic settings we examine in this chapter have a unique BNE (such as that in Example 8.2), giving us a unique equilibrium prediction.
3. **Robust to small payoff perturbations? Yes.** What if we change the payoff of one of the players by a small amount (i.e., for any ε that approaches zero)? As in previous solution concepts, such as IDSDS and NE, we find that BNE is also robust to small payoff perturbations. This is due to the fact that, if a strategy s_i

yields a higher expected payoff than another strategy s'_i, it must still generate a higher expected payoff than s'_i after we apply a small payoff perturbation.

4. **Socially optimal? No.** As with previous solution concepts, BNE does not necessarily yield socially optimal outcomes. Example 8.1 illustrates this property. Strategy profile $(D^{20}D^{12}, R)$ is a BNE, with equilibrium payoffs (10, 17), but it is *not* socially optimal. Indeed, we can find other strategy profiles (other cells in Matrix 8.5) where at least one player improves her payoff, such as $(U^{20}U^{12}, L)$ with payoffs (16, 17). This strategy profile is a BNE and is also socially optimal. Indeed, all other outcomes entail a payoff equal or below 16 for player 1, a payoff equal or below 17 for player 2, or both.

8.7 What If Both Players Are Privately Informed?

Previous sections considered, for simplicity, that only one player was privately informed. The tools presented in this chapter can still be applied to settings where both players are privately informed. Example 8.3 extends our analysis of Cournot competition under incomplete information, but in an industry where both firms privately observe their production costs and choose their output.

Example 8.3 Cournot competition with symmetrically uninformed firms. Consider the duopoly in Example 8.2, but assume now that every firm i observes its marginal cost, either $c_H = \frac{1}{2}$ or $c_L = 0$, occurring with probability p and $1 - p$, respectively; but does not observe its rival's cost. Using tool 8.2, we can analyze each privately informed player (firm), identify its best response function when its costs are high and low. Combining the two best responses of firm 1 and the two of firm 2, we can, then, find the equilibrium output levels in this market.

High costs. If firm i has high marginal costs, $c_H = \frac{1}{2}$, it solves the following problem:

$$\max_{q_i \geq 0} \pi_i^H(q_i) = p\underbrace{\left[\left(1 - q_j^H - q_i\right)q_i\right]}_{\text{Firm } j\text{'s costs are high}} + (1-p)\underbrace{\left[\left(1 - q_j^L - q_i\right)q_i\right]}_{\text{Firm } j\text{'s costs are low}} - \frac{1}{2}q_i$$

which is symmetric to firm 1's profit-maximization problem in Example 8.2. Differentiating with respect to q_i, and solving for q_i, we then find a similar best response function to that of firm 1 in Example 8.2, that is,

$$q_i^H(q_j^H, q_j^L) = \frac{1}{4} - \frac{pq_j^H + (1-p)q_j^L}{2}.$$

8.7 What If Both Players Are Privately Informed?

Low costs. If firm i has, instead, low marginal costs, $c_L = 0$, it solves

$$\max_{q_i \geq 0} \pi_i^L(q_i) = p\underbrace{\left[\left(1 - q_j^H - q_i\right)q_i\right]}_{\text{Firm } j\text{'s costs are high}} + (1-p)\underbrace{\left[\left(1 - q_j^L - q_i\right)q_i\right]}_{\text{Firm } j\text{'s costs are low}}$$

Differentiating with respect to q_i, yields

$$p\left(1 - q_j^H - 2q_i\right) + (1-p)\left(1 - q_j^L - 2q_i\right) = 0$$

Solving for q_i, we find that the firm i's best response function is

$$q_i^L(q_j^H, q_j^L) = \frac{1}{2} - \frac{pq_j^H + (1-p)q_j^L}{2}.$$

Relative to $q_i^H(q_j^H, q_j^L)$, it originates at a higher vertical intercept, $1/2$, but shares the same slope. Intuitively, firm i produces a larger output, for a given expected output from its rival, when its own costs are lower than high.

Finding equilibrium output in the BNE. In a symmetric BNE, firms produce the same output level when their costs coincide, that is, $q_i^H = q_j^H = q^H$ and $q_i^L = q_j^L = q^L$. (This symmetry applies because firms observe the same information, and face the same production costs: high when $q_i^H = q_j^H = q^H$ and low when $q_i^L = q_j^L = q^L$.) Inserting this property in best response functions $q_i^H(q_j^H, q_j^L)$ and $q_i^L(q_j^H, q_j^L)$, yields

$$q^H = \frac{1}{4} - \frac{pq^H + (1-p)q^L}{2} \quad \text{and} \quad q^L = \frac{1}{2} - \frac{pq^H + (1-p)q^L}{2}$$

which provides us with a system of two equations and two unknowns, q^H and q^L. Simultaneously solving for q^H and q^L, we obtain equilibrium output levels.[3]

$$q^H = \frac{1+p}{12} \quad \text{and} \quad q^L = \frac{4+p}{12}$$

which are both increasing in the probability that its rival's costs are high, p.

Comparison across information settings. A natural question at this point is how equilibrium output levels vary from those we found in Example 8.2, $(q_1^*, q_2^{H*}, q_2^{H*})$, where only one player was privately informed about its costs (firm 2) while firm 1's costs were common knowledge. Starting with firm 1's output, we

---

[3] For instance, we can solve for q^H in the first equation, to find $q^H = \frac{1-2(1-p)q^L}{2(2+p)}$. Inserting this result into the second equation yields an expression that is only a function of q^L. Solving for q^L yields the result we reported in the text.

find that it produces more units when its rival does not observe its costs (Example 8.3) than when it does (Example 8.2), because

$$q^H - q_1^* = \frac{1+p}{12} - \frac{p}{6} = \frac{1-p}{12} > 0, \text{ and}$$

$$q^L - q_1^* = \frac{4+p}{12} - \frac{p}{6} = \frac{4-p}{12} > 0$$

since $0 < p < 1$ by assumption. In contrast, firm 2's output satisfies

$$q^H - q_2^{H*} = \frac{1+p}{12} - \frac{3-p}{12} = -\frac{1-p}{6} < 0, \text{ and}$$

$$q^L - q_2^{L*} = \frac{4+p}{12} - \frac{6-p}{12} = -\frac{1-p}{6} < 0$$

indicating that firm 2 produces more output when it observes its rival's cost (Example 8.2) than when it does not (Example 8.3).

Exercises

8.1 **Bayesian normal form—An introductory example.**[A] Consider the following payoff matrix, representing an incomplete information game, where player 1 observes the realization of x before playing the game, but player 2 only knows that its realization is either $x = 20$ or $x = 3$ with equal probabilities. This information is common knowledge. We next find the set of BNEs by building the Bayesian normal form of the game.

		Player 2	
		L	R
Player 1	U	$x, 17$	$5, 10$
	D	$10, 0$	$10, 17$

  a. Find the strategy set of each player.
  b. Build the Bayesian normal form of the game with the expected utilities of each player in the appropriate cells.
  c. Underline each player's best response payoffs in the Bayesian normal form from part (b). Then, identify the BNEs in this game.

8.2 **Bayesian normal form—Allowing for more general probabilities.**[B] Consider the game in Exercise 8.1, but assume that $x = 20$ with probability p and $x = 3$ with probability $1 - p$.
  a. Find the strategy set of each player.
  b. Build the Bayesian normal form of the game with the expected utilities of each player in the appropriate cells.

c. Underline each player's best response payoffs in the Bayesian normal form from part (b). Then, identify the BNEs in this game.
d. How do your equilibrium results depend on probability p? Interpret.
e. Confirm that when the types $x = 20$ and $x = 3$ are equally likely, $p = 1/2$, you obtain the same results as in Exercise 8.1.

8.3 **Bayesian normal form—Allowing for more general type values.**[B] Consider the game in Example 8.1, but rather than assuming that random variable x takes values $x = 20$ and $x = 3$ with equal probability, let us now consider that $x = a$ and $x = b$, where $a > b > 0$, still with equal probabilities.
a. Find the strategy set of each player.
b. Build the Bayesian normal form of the game with the expected utilities of each player in the appropriate cells.
c. Underline each player's best response payoffs in the Bayesian normal form from part (b). Then, identify the BNEs in this game.
d. How do your equilibrium results depend on payoffs a and b? Interpret.
e. Confirm that when x takes value $x = 20$ and $x = 3$, you obtain the same results as in Exercise 8.1.

8.4 **Bayesian normal form—Allowing for both players to be privately informed.**[B] Consider the game in the following payoff matrix.

		Player 2	
		L	R
Player 1	U	x, y	5, 10
	D	10, 0	10, 17

where, as in Example 8.1, player 1 privately observes the realization of x, which takes values $x = 20$ and $x = 3$, with equal probabilities; but now player 2 privately observes the realization of y, which takes values $y = 15$ and $y = 5$, also with equal probabilities.
a. Find the strategy set of each player.
b. Build the Bayesian normal form of the game with the expected utilities of each player in the appropriate cells.
c. Underline each player's best response payoffs in the Bayesian normal form from part (b). Then, identify the BNEs in this game.
d. How do equilibrium results differ from those in Exercise 8.1? Interpret.

8.5 **Both tools yield the same BNEs.**[A] Consider the game in Exercise 8.1. Find the set of BNEs by using the second tool to solve for BNEs (focusing on the privately informed player first, and then moving on to the uninformed player). Confirm that you obtain the same BNEs as when using the first tool (building the Bayesian normal form of the game).

8.6 **Uninformed responder but facing two information sets.**[B] The following game tree depicts a sequential-move game where, first, nature determines whether the incumbent has a high cost (with probability $\frac{1}{3}$) or low cost (with probability $\frac{2}{3}$). The incumbent, privately observing its cost, decides to set either a high or a low price, which we denote as $\overline{p}^H$ and $\overline{p}^L$ after observing

a high cost (at the top of the game tree) and $\underline{p}^H$ and $\underline{p}^L$ after observing a low cost (at the bottom of the tree). Finally, the entrant, observing the incumbent's price (either high or low), but without observing the incumbent's type, decides whether to enter the market (E and NE after observing a low price, and E' and NE' after observing a high price).
a. Find the strategy set of each player.
b. Build the Bayesian normal form of the game with the expected utilities of each player in the appropriate cells.
c. Underline each player's best response payoffs in the Bayesian normal form from part (b). Then, identify the BNEs in this game.

8.7 **Bargaining under incomplete information.**[B] Consider a bargaining game between a seller and a buyer. The buyer values the object at v_B and the seller at v_S, where $v_S < v_B$. Intuitively, the buyer assigns a larger valuation to the object than the seller, which implies that there should be a mutually beneficial agreement between these two players. For compactness, most bargaining games normalize the buyer's valuation to $v \equiv v_B - v_S > 0$, leaving the normalized valuation of the seller to zero. In the first stage, nature determines the buyer's valuation for the object, which can be high, v^H, with probability $q \in [0, 1]$, and low, v^L, with probability $1 - q$; where $v^H > v^L > 0$. The seller does not observe the buyer's valuation, but knows that it's distributed according to the above probabilities. The seller is the first mover in this game (other than move by nature): without observing the realization of the buyer's valuation, the seller sets a price p for the good. Observing this price, but observing his valuation for the object (v^H or v^L), the buyer responds by accepting or rejecting the deal.
a. *Buyer's response.* Since the buyer is the privately informed player in this game, we start analyzing his best responses. Find his best response when his valuation is v^H and when his valuation is v^L.
b. *Seller's price.* Find the price that the seller sets in the BNE of the game.
c. How are the BNEs you identified in part (b) affected by an increase in v^H? And by an increase in v^L? Interpret.

8.8 **The "lemons" problem.**[A] Consider a used-car dealership. Let q denote the quality of a car, which is observed by the seller but not by the buyer. The buyer, however, knows that quality q is uniformly distributed in $[0, 1]$. Therefore, a buyer assigns a value q to a car whose quality is q, but the seller's value is only αq, where $\alpha \in (0, 1)$ represents the value difference between the buyer and seller. In particular, when α is close to 1, both players value the car similarly; whereas when α approaches zero, the buyer values the car more significantly than the seller does.
a. *Symmetric information.* Assume that buyer and seller can observe a car's quality, q. For which prices are both players better-off? Interpret.
b. Still in a context of symmetric information, find the subgame perfect equilibrium (SPE) of the game.
c. Are all car qualities sold?

Exercises

    d. *Asymmetric information.* Assume now that the buyer cannot observe the car's quality, q. Find the BNE of the game.

    e. *Comparison.* Compare your equilibrium results under symmetric and asymmetric information.

8.9 **Cournot competition with both firms uninformed.**[B] Consider the duopoly market in Sect. 8.5, where firms face an inverse demand function $p(Q) = 1 - Q$, and $Q \geq 0$ denotes aggregate output. Let us now assume that both firms are uninformed about each other's costs, that is, every firm i privately observes its cost, c_i, which as in Sect. 8.5 is still $c_H = \frac{1}{2}$ or $c_L = 0$, with probability p and $1 - p$, respectively. Firm i, however, does not observe its rival's cost, c_j, which is also $c_H = \frac{1}{2}$ or $c_L = 0$ that occurs with probability p and $1 - p$, respectively.

    a. Find firm i's best response function when its production cost is $c_H = \frac{1}{2}$, and denote it as $q_i^H(q_j^H, q_j^L)$. Is it increasing or decreasing in probability p? Interpret.

    b. Find firm i's best response function when its production cost is $c_L = 0$, and denote it as $q_i^L(q_j^H, q_j^L)$. Is it increasing or decreasing in probability p? Interpret.

    c. Find equilibrium output levels, q_i^H and q_i^L.

    d. How is equilibrium output q_i^H affected by a marginal increase in probability p? Interpret.

    e. How is equilibrium output q_i^L affected by a marginal increase in probability p? Interpret.

8.10 **Cournot competition under incomplete information-A twist.**[B] Consider the same industry as in Sect. 8.5, but assume that the marginal production cost of firm 1 (2) is high, $c_H = \frac{1}{2} > 0$, with probability p (q, respectively), where $p, q \in [0, 1]$. Similarly, the marginal cost of firm 1 (2) is low, $c_L = 0$, with probability $1 - p$ ($1 - q$, respectively).

    a. Find firm 1's best response function when its costs marginal costs are low, $q_1^L(q_2^H, q_2^L)$. Find firm 2's best response function when its marginal costs are low, $q_2^L(q_1^H, q_1^L)$.

    b. Find firm 1's best response function when its costs marginal costs are high, $q_1^H(q_2^H, q_2^L)$. Find firm 2's best response function when its marginal costs are high, $q_2^H(q_1^H, q_1^L)$.

    c. Use your results from parts (a) and (b) to find the BNE of the game. [*Hint*: You cannot invoke symmetry since best response functions are not symmetric in this case.]

    d. How are the equilibrium output levels $\left(q_1^H, q_1^L, q_2^H, q_2^L\right)$ affected by a marginal increase in p? And by a marginal increase in q? Interpret.

    e. *Symmetric probabilities.* Evaluate your equilibrium results in the special case where both firms' costs occur with the same probability, $p = q$. What if, in addition, these probabilities are both 1/2?

f. *Special cases.* Evaluate your equilibrium results in the special case where both firms' types are certain, as under complete information: (1) $p = q = 1$, (2) $p = 1$ and $q = 0$, (3) $p = 0$ and $q = 1$, and (4) $p = q = 0$. Interpret.

8.11 **Cournot competition with demand uncertainty.**[B] Consider a duopoly market, where every firm i's marginal production costs are c, which satisfies $\frac{1}{2} > c \geq 0$. Assume that every firm i does not observe whether the inverse demand function is high $p^H(Q) = 1 - Q$, with probability p, or low $p^L(Q) = \frac{1}{2} - Q$ with probability $1 - p$. Aggregate output is denoted as $Q \geq 0$.
   a. Find every firm i's best response function, $q_i(q_j)$.
   b. Is $q_i(q_j)$ increasing in probability p? Interpret.
   c. Find the equilibrium output levels, q_i^* and q_j^*.
   d. Is equilibrium output, q_i^*, increasing in probability p? Interpret.

8.12 **Public good game under incomplete information-I.**[B] Suppose that there are $N \geq 2$ players and that a public good is supplied if only if at least one player contributes. If the public good is supplied, every player enjoys a benefit of 1. The cost of contributing is denoted as θ_i for player i, and is independently drawn from the cumulative distribution function $F(\cdot)$ on $[\underline{\theta}, \overline{\theta}]$ where the lower and upper bounds satisfy $\underline{\theta} < 1 < \overline{\theta}$. Every player i privately observes his contribution cost, θ_i, but does not observe his rivals' costs, θ_{-i}, although he knows that they are drawing their contribution costs from $F(\cdot)$.
   a. Find the BNE of the game in this setting.
   b. *Uniformly distributed costs.* Assume that θ_i is uniformly distributed on $[\underline{\theta}, \overline{\theta}]$, where $\underline{\theta} = 0$ and $\overline{\theta} = 2$, so that $F(\theta_i) = \frac{\theta_i}{2}$, and that $N = 2$ play the game. Use your results from part (a) to find a BNE where only one player contributes to the public good.
   c. Still in the context of part (b), find the BNE when $N = 3$, $N = 4$, and $N = 10$. How are the equilibrium results affected? Interpret.

8.13 **Public good game under incomplete information-II.**[C] Consider the setting in Exercise 8.12, but assume that the public good is only provided if at least two players contribute and that $N \geq 3$.
   a. Show that there is a trivial BNE in which no player contributes.
   b. In the setting of part (a), find another BNE where every player i contributes if $\theta_i \leq \theta_i^*$.
   c. *Uniformly distributed costs.* Assume that θ_i is uniformly distributed on $[\underline{\theta}, \overline{\theta}]$, where $\underline{\theta} = 0$ and $\overline{\theta} = 2$, so that $F(\theta_i) = \frac{\theta_i}{2}$, and that $N = 4$ play the game. Use your results from part (b) to find a BNE where only one player contributes to the public good.
   d. Still in the context of part (c), find the BNE when $N = 10$. How are the equilibrium results affected? Interpret.

8.14 **Public good game under incomplete information-III.**[C] Consider the setting in Exercises 8.13 and 8.14, but assume that the public good is only supplied if at least $k \geq 2$ players contribute.
   a. Find a BNE where every player i contributes if $\theta_i \leq \theta_i^*$.

b. *Uniformly distributed costs.* Assume that θ_i is uniformly distributed on $[\underline{\theta}, \overline{\theta}]$, where $\underline{\theta} = 0$ and $\overline{\theta} = 2$, so that $F(\theta_i) = \frac{\theta_i}{2}$, that $N = 4$ players interact in this game and that $k = 2$. Use your results from part (a) to find a BNE where only one player contributes to the public good.

c. Still in the context of part (b), find the BNE when k decreases to $k = 1$. How are the equilibrium results affected? Interpret.

8.15 **Incomplete information in a simple card game.**[C] Consider the following scenario in a card game:

- First, nature selects numbers x_1 and x_2, both of them drawn from a uniform distribution, that is, $x_i \sim U[0, 1]$ for every $i = \{1, 2\}$.
- Every player i privately observes x_i, which we interpret as how good her cards are, but does not observe her rival's cards, x_j, where $j \neq i$.
- Simultaneously and independently, every player i chooses whether to fold or bid.

If both players fold, then they both lose \$1. If only one player folds, then she loses \$1, which goes to her rival. If both players bid, then player i receives \$2 if her number exceeds her rival's ($x_i > x_j$); loses \$2 if she falls below ($x_i < x_j$), and neither gains nor losses if there is a tie ($x_i = x_j$). This is a Bayesian game where every player privately observes her number, x_i, but does not observe her rival's, x_j.

a. *Complete information game.* As a benchmark, assume that every player i observes the pair (x_i, x_j). Find a pure-strategy NE in this game.

b. *Incomplete information game with binary types.* Also as a benchmark, consider a setting where x_i can only take two numbers, x_H and x_L, both equally likely. Every player i observes her number, x_i, but does not observe her rival's number, x_j. Describe every player's strategy space, build the Bayesian normal form representation of this game, and find the BNEs.

c. *Incomplete information game with uniformly distributed types.* Consider now the setting described above, where every player i draws a number $x_i \sim U[0, 1]$ for every $i = \{1, 2\}$. We seek to find a symmetric BNE where every player bids if and only if her number, x_i, is greater than a number α. Confirm that this strategy profile can be sustained as a BNE and find the equilibrium value of the number α.

8.16 **Game of chicken with uncertain payoffs, based on Tadelis (2013).**[B] Consider the game of chicken presented in Chapter 2 and reproduced in the matrix below. Two drivers simultaneously and independently choose whether to swerve and drive straight with the payoffs described in Chapter 2.

		Player 2	
		Swerve	Drive
Player 1	Swerve	$-1, -1$	$-8, 10$
	Drive	$10, -8$	$-30, -30$

We now consider an incomplete information version of this game, where, before the beginning of the game, every player privately observes the cost of repairing his car and health costs, as represented in the next matrix.

		Player 2	
		Swerve	Drive
Player 1	Swerve	$-1, -1$	$-8, 10$
	Drive	$10, -8$	$-\theta_1, -\theta_2$

where $\theta_i = \{H, L\}$, where $H > L > 10$, with probabilities 2/3 and 1/3, respectively.

a. Find the BNE (or BNEs) of this game.
b. How are your results in part (a) affected by a marginal increase in H? And by a marginal increase in L? Interpret.

8.17 **BNEs involving mixed strategies.**^{C} Consider the following Matching pennies game under incomplete information. First, nature determines whether players interact in the standard Matching pennies game (left panel, with probability 1/2) or in the same game but with "reversed" payoffs (right panel, with probability 1/2). Player 1 privately observes the move from nature (i.e., whether players interact in the left or right matrix) but player 2 does not.

		Player 2	
		h	t
Player 1	H	$1, -1$	$-1, 1$
	T	$-1, 1$	$1, -1$

		Player 2	
		h	t
Player 1	H	$-1, 1$	$1, -1$
	T	$1, -1$	$-1, 1$

a. As a benchmark, consider a complete information game where both players observe the matrix they are facing. Show that there is no NE in pure strategies.
b. Consider now the incomplete information game described above. Show that there is no BNE in pure strategies. [*Hint*: Fix player 2's strategy at h (which is type independent), find player 1's best response to h, either choosing HH, HT, TH, or TT. Then, show that player 2 has incentives to deviate. Repeat the same approach but fixing player 2's strategy at t.]
c. Find a BNE in mixed strategies. [*Hint*: For simplicity, focus on the strategy profile where player 1 chooses heads regardless of her type, HH, while player 2 mixes.]

8.18 **Hawk–Dove game with incomplete information.**^{C} Consider the following Hawk–Dove game (an anti-coordination game).

		Player 2	
		C	D
Player 1	C	$3, 3$	$2, 4$
	D	$4, 2$	$0, 0$

a. Find the NEs of this game. [*Hint*: It has two NEs in pure strategies and one in mixed strategies.]
b. Assume now that, before the game begins, every player i privately observes her cost from playing Cooperate, c_i, which is uniformly distributed in $[-C, C]$. Intuitively, parameter $C \geq 0$ represents the uncertainty around players' cost from cooperation: When $C \to 0$, there is almost no uncertainty; whereas a large C entails more uncertainty. Therefore, the payoff matrix is now changed as follows.

		Player 2	
		C	D
Player 1	C	$3-c_1, 3-c_2$	$2-c_1, 4$
	D	$4, 2-c_2$	$0, 0$

Consider a symmetric strategy profile where every player i, upon observing her cost from cooperating c_i, chooses to cooperate if and only if c_i is sufficiently low, $c_i \leq \bar{c}$, where $\bar{c}$ denotes a cutoff. Show that this strategy profile can be sustained as a BNE of this incomplete information game.

c. How is the BNE found in part (b) affected by a decrease in parameter C? Evaluate the BNE at $C \to 0$. Interpret your results.

8.19 **Market entry with incomplete information..**[B] Consider an incomplete information extension of an entry game, as depicted in the figure below. In the first stage, nature determines the incumbent's production cost, either high and low with probability of p and $1 - p$, respectively, where $p \in [0, 1]$. In the second stage, the entrant does not observe the incumbent's cost, but decides whether to enter into the industry or not. In the third stage, the incumbent, observing its own cost and whether the entrant joined the industry, responds by accommodating or fighting entry.

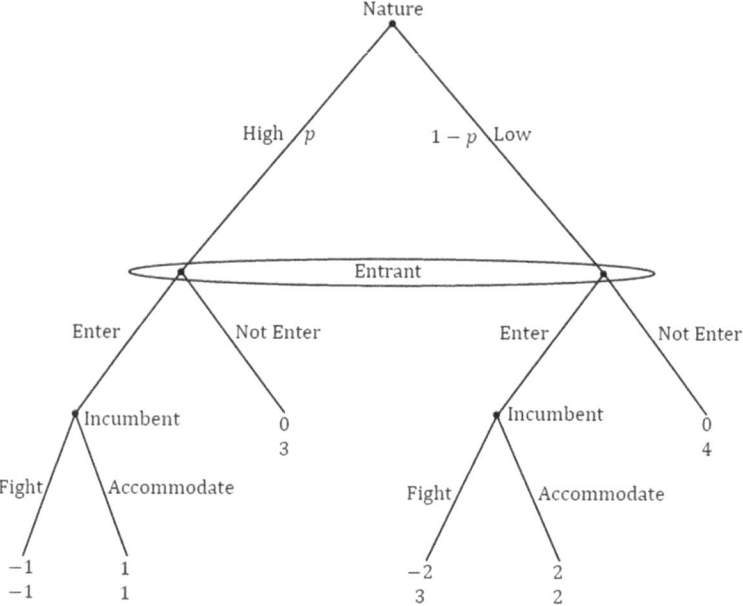

a. *Third stage.* Apply backward induction in the third stage of the game. Represent the resulting game tree.
b. *Second stage.* Find the entrant's decision.
c. Characterize the BNE of this game.

8.20 **More information may hurt.**[A] Consider a setting where player 1 has perfect information of the game but player 2 does not know whether the front or back matrix is played, as displayed below.

Front matrix, p

Player 2

		U	D
Player 1	L	2,1	3,3
	R	3,4	2,5

Back matrix, $1-p$

Player 2

		U	D
Player 1	L	3,3	1,2
	R	1,2	3,3

a. *Complete information.* As a benchmark, let us first consider a complete information setting where both players are perfectly informed about the matrix they face. Find the pure-strategy NE of each game.
b. Compare equilibrium payoffs in the NEs of part (a) and in the BNE of part (b). How does incomplete information affect player 2's payoffs?

8.21 **Jury voting game.**[B] Consider a defendant in a trial, who is guilty or innocent of a felony charge with equal probability. The judge seeks the counsel of two jurors, each of whom independently receives the signal that the defendant is guilty (innocent) when being guilty (innocent) with probability γ and innocent (guilty) when being guilty (innocent) with probability $1-\gamma$. Intuitively, parameter $\gamma \in [0, 1]$ denotes the signal precision. The defendant is convicted only when both jurors vote for conviction (i.e., unanimous vote). The payoff structure is the following: (i) if the defendant is guilty and convicted, every juror receives a payoff of 2; (ii) if the defendant is innocent but convicted, every juror receives a payoff of -3; and (iii) if the defendant is acquitted (whether being guilty or innocent) every juror receives a payoff of 0.

a. Find the joint probability of the jurors' signals.
b. Find the conditional probability of the defendant's type. That is, find the conditional probability that he is guilty given that both jurors receive G signals, and similarly for other signal pairs. Then, find the conditional probability that he is innocent given that both jurors receive G signals, and similarly for other signal pairs.
c. Characterize every juror's decision.

8.22 **Jury voting game with more general payoffs**.[B] Consider again the setting in the jury voting (Exercise 8.21). Assume now that if the defendant is innocent but convicted, every juror's payoff is not -3, but more generally $-x$, where $x > 2$. All other payoffs and probabilities are unaffected.

a. Find the BNE in this setting.
b. Evaluate your equilibrium results at $x = 3$, showing that you obtained the same results as in exercise 8.21.
c. How are your equilibrium results affected by an increase in x? Interpret.

Reference

Harsanyi, J. C. (1967) "Games with Incomplete Information Played by "Bayesian" Players, Part I–III. Part I. The Basic Model." Management Science, 14, pp. 159–182.

Auction Theory

9.1 Introduction

This chapter applies the BNE solution concept to auctions, a strategic setting where players typically face incomplete information. In particular, every player privately knows her valuation for the object being sold (e.g., a painting), but does not observe other bidders' valuations. Bids, then, are a player's strategy, and are a function of her privately observed valuation for the object.

Auctions have been historically reported as early as Babylon around 500 BC, in the Roman Empire in 193 AD, and remain popular nowadays through online auctions such as eBay, eBid, QuiBids, AuctionMaxx, DealDash, or LiveAuctioneers, among others.[1] Understanding optimal bidding behavior in different auctions, then, is not only an interesting application of BNEs but also a useful tool for online shopping.

We seek to identify equilibrium bids—the BNE of the game—in different auction formats. We start, for simplicity, with the second-price auction (SPA), where the bidder submitting the highest bid wins the auction but does not pay her bid; she pays, instead, the second-highest bid. This is an auction format for which a BNE can be identified without the use of calculus, making it is easier to describe than other formats. Afterwards, we move on to the first-price auction (FPA), where the bidder submitting the highest bid must pay the amount she submitted.

This chapter, then, evaluates the expected revenue of these two auctions formats, to identify which one generates a larger revenue for the seller. We show, however, that their expected revenues coincide. This result extends to many other auction

---

[1] More recently, purely digital non-fungible tokens (NFTs) have been gathering pace. For instance, the JPG file "Everydays, The First 5000 Days," by the artist known as Beeple, sold at Christie's for $69.3 Million in 2021, setting a new record for a digital artwork.

formats satisfying a relatively mild set of assumptions, all of them yielding the same expected revenue; a result known as the "revenue equivalence principle."

We finish the chapter considering auctions where bidders observe the value of the object, such as an oil reservoir, which is common across all bidders, but each bidder receives an estimate of this common value before submitting her bid. This auction, often referred to as "common-value auctions," are prone to the so-called "winner's curse," where the winner overpays for the object. Indeed, the winner of the auction suffers, informally, from a curse, as she receives the object but makes a loss. Intuitively, this occurs because the bidder submitting the highest bid is also receiving the highest estimate for the object's value. This overestimated value induces her to submit such a high bid that, despite helping her win the object, makes her pay a monetary amount that exceeds the object's true value. We then identify equilibrium bids that help bidders avoid the winner's curse in common-value auctions.

9.2 Auctions as Allocation Mechanisms

While we can find several auction formats, we often characterize them according to two basic criteria:

a. **Assignment rule.** Informally, the assignment (or allocation) rule in an auction just answers the question "who gets the object?" In the first-price auction, for instance, the individual submitting the highest bid receives the object for sale. A similar assignment rule applies to other auction formats, such as the second-price, third-price, kth-price, or all-pay auctions (APA), where the object still goes to the bidder submitting the highest bid. (These auctions differ, however, in the price that each bidder pays, as we describe below.) In settings where the seller offers $m \geq 2$ units of the same good (multi-unit auctions), the assignment rule in first-price auctions, for instance, determines that the first unit of the object goes to the individual submitting the highest bid, the second unit to the individual submitting the second-highest bid, and similarly for the remaining units.

In lottery auctions, however, bidder i may not receive the object even when she submits the highest bid. In particular, bidder i's probability of winning the auction is

$$\Pr(win) = \frac{b_i}{b_i + B_{-i}}$$

where $B_{-i} = \sum_{j \neq i} b_j$ denotes the aggregate bids of player i's rivals. Therefore, bidder i's probability of winning:
  i) increases in her own bid, b_i;
  ii) satisfies $\Pr(win) < 1$ if at least one of her rivals submits a positive bid, $B_{-i} > 0$; and

iii) decreases in her rivals' bids.[2]

As an example, if bidder i submits a bid of \$100 while each of her $N-1=10$ rivals submits a bid of only \$5, bidder i's probability of winning the auction is $\frac{100}{100+(10\times 5)} = \frac{2}{3}$ while that of every bidder $j \neq i$ is $\frac{5}{5+[100+(9\times 5)]} = \frac{1}{30}$.

b. **Payment rule.** This criterion answers the question "how much each bidder pays?" which allows for the winner and losers to pay a monetary amount to the seller. In the first-price auction, for instance, the winner is the only bidder paying to the seller, and she must pay the bid she submitted. In the second-price auction, the winner is again the only player paying to the seller, but in this case she pays the second-highest bid (not the bid that she submitted). A similar argument applies to kth-price auctions, where only the winning bidder pays to the seller, specifically the kth-highest bid. In the all-pay auction, however, all bidders (the winner and the $N-1$ losers) pay the bid that each of them submitted.

In all the auction formats we consider in this chapter, we assume the following setting. A seller offers an object to $N \geq 2$ bidders. Every bidder i privately observes her valuation for the object, $v_i \in [0, 1]$, but does not observe bidder j's valuation, v_j, where $j \neq i$. It is common knowledge, however, that valuations v_i and v_j are independent and identically distributed (i.i.d.),[3] and that v_j is drawn from a cumulative distribution function $F_j(v) = \Pr\{v \leq v_j\}$, with positive density in all its support, i.e., $f_j(v) = F'_j(v) > 0$ for all $v \in [0, 1]$. After observing her valuation v_i, bidder i simultaneously and independently submits her bid $b_i \geq 0$ for the object. The seller observes the profile of submitted bids,

$$b = (b_1, b_2, ..., b_N),$$

and, given the assignment and payment rules of the auction (i.e., the rules of the auction), she declares a winning bidder (e.g., the player submitting the highest

---

[2] First, note that $\frac{\partial \Pr(win)}{\partial b_i} = \frac{B_{-i}}{(b_i + B_{-i})^2}$, which is positive for all $B_{-i} > 0$ but zero otherwise. Intuitively, if at least one bidder $j \neq i$ submits a positive bid, bidder i's probability of winning the auction increases when she submits a more aggressive bid, b_i. Otherwise, if $B_{-i} = 0$, her probability of winning the auction is one, and does not increase in b_i. Secondly, her probability of winning satisfies $\frac{b_i}{b_i + B_{-i}} \leq 1$ or, after rearranging, $B_{-i} \geq 0$. Therefore, $\Pr(win) = 1$ only when no other bidder $j \neq i$ submits a positive bid, $B_{-i} = 0$; but otherwise this probability lies strictly between 0 and 1. Finally, $\frac{\partial \Pr(win)}{\partial B_{-i}} = -\frac{b_i}{(b_i + B_{-i})^2}$, which is strictly negative for all $b_i > 0$ and zero otherwise. This indicates that bidder i has a lower probability of winning the object when her rivals submit more aggressive bids (as captured by an increase in aggregate bids, B_{-i}).

[3] The "independent" part in i.i.d. means that bidders' valuations are uncorrelated. That is, after bidder i privately observes her own valuation, v_i, she cannot infer any information about her rival's valuation, v_j, where $j \neq i$. In contrast, if valuations were positively (negatively) correlated, the observation of a high value of v_i would increase (decrease) the expected value of v_j. The "identical" part in i.i.d. indicates that valuations v_i and v_j are both random variables drawn from the same probability distribution.

bid), the price that the winning bidder must pay for the object (e.g., paying the bid she submitted in the FPA, or the second-highest bid in the SPA) and potentially the prices that losing bidders must pay (e.g., as happening in APAs).

9.3 Second-price Auctions

We examine SPAs in this section. As described above, the assignment rule in SPAs prescribes that the object goes to the individual who submitted the highest bid. The payment rule, however, says that the winner must pay the second-highest bid while all other bidders pay zero. If two or more bidders submit the highest bid, then the object is randomly assigned among them with equal probabilities. (Exercise 9.2 asks you to show that equilibrium results are unaffected if we change the tie-breaking rule.)

We seek to show that submitting a bid equal to her valuation (i.e., $b_i(v_i) = v_i$) is a *weakly dominant strategy* for every player. This means that, regardless of the valuation that bidder i assigns to the object, and independently of her opponents' bids, submitting a bid equal to her valuation, $b_i(v_i) = v_i$, yields an expected profit equal to or higher than deviating to a bid lower than her valuation, $b_i(v_i) < v_i$, or higher than her valuation, $b_i(v_i) > v_i$. For presentation purposes, we separately find bidder i's payoff from each bidding strategy below.

9.3.1 Case 1: Bid Equal To Her Valuation

Let us first find bidder i's expected payoff from submitting a bid that coincides with her own valuation v_i, which we use as a benchmark for comparing the payoffs from deviations in subsequent sections. In this setting, either of the following situations can arise:

1.1 **Bidder i wins.** First, let h_i denote the highest competing bid among all bidder i's rivals, as follows

$$h_i = \max\{b_1, b_2, ..., b_{i-1}, b_{i+1}, ..., b_N\}$$

where we exclude bidder i's bid, b_i, from the list of bids.[4] If bidder i submits the highest bid, $b_i > h_i$, she wins the auction, earning a net payoff of $v_i - h_i$ because in a SPA, the winning bidder does not pay the bid she submitted, but rather the second-highest bid, h_i.

1.2 **Bidder i loses.** If, instead, the highest competing bid lies above her bid, $h_i > b_i$, then bidder i loses the auction, earning a zero payoff.

---

[4] Alternatively, we can more compactly write h_i as $h_i = \max\limits_{j \neq i}\{b_j\}$, which also finds the highest bid among all bidder i's rivals, $j \neq i$.

1.3 **Tie in bids.** If a tie occurs, where $b_i = h_i$, the object is randomly assigned, and bidder i's expected payoff becomes $\frac{1}{2}(v_i - h_i)$, where $\frac{1}{2}$ denotes the probability that bidder i receives the object. However, because $v_i = h_i$ in this case, the bidder earns a zero expected payoff.[5]

9.3.2 Case 2: Downward Deviations, Bidding Below Her Valuation

Using the above results as a reference, let us now analyze the case in which bidder i submits a bid below her valuation, shading her bid so that $b_i < v_i$ for every valuation v_i. As in the previous setting, we can identify the following three situations, 2.1–2.3, which we compare with the corresponding Cases 1.1–1.3:

2.1 If the highest competing bid h_i lies below her bid (i.e., $h_i < b_i < v_i$), bidder i still wins the auction, earning a payoff $v_i - h_i$, as when she does not shade her bid, This indicates that bidder i's payoff does not increase relative to Case 1.1, thus not having incentives to shade her bid.
2.2 If the highest competing bid h_i lies between b_i and v_i ($b_i < h_i \leq v_i$), bidder i loses, earning a payoff of zero. As a consequence, bidder i's payoff is the same (zero) when she does not shade her bid, as shown in Case 1.2, than when she does in Case 2.2.
2.3 If the highest competing bid h_i is higher than v_i (i.e., $b_i < v_i < h_i$), bidder i loses the auction, thus yielding the same outcome as when she submits a bid, $b_i = v_i$. In this case, bidder i's payoff exactly coincides when she bids according to her valuation (Case 1.3) and when she shades her bid.

Overall, when bidder i shades her bid, $b_i < v_i$ in Cases 2.1–2.2, she earns the same or lower payoff than when she submits a bid that coincides with her valuation for the object ($b_i = v_i$). In other words, she does not have strict incentives to shade her bid because her payoff would not strictly improve from doing so, regardless of the exact position of the highest competing bid, h_i.

---

[5] More generally, if $K \geq 2$ bidders are tied submitting the highest bid, the auctioneer randomly assigns the object to any of them with equal probability, $\frac{1}{K}$, implying that each of these bidders earns an expected payoff of $\frac{1}{K}(v_i - h_i)$. Since $v_i = h_i$ holds in this case, bidder i's expected payoff is still zero, regardless of the number of bidders who are tied in the highest bid, i.e., for all $K \geq 2$.

9.3.3 Case 3: Upward Deviations, Bidding Above Her Valuation

We now follow a similar approach to show that submitting a bid above her own valuation (i.e., $b_i > v_i$) does not help bidder i strictly improve her expected payoff from bidding according to her valuation (Cases 1.1–1.3). In particular, we also find three situations:

3.1 If the highest competing bid h_i lies below bidder i's valuation, v_i ($h_i < v_i < b_i$), she wins, earning a payoff of $v_i - h_i$. Her payoff in this case coincides with that when submitting a bid equal to her valuation, $b_i = v_i$, as in Case 1.1, implying that she has no strict incentives to bid above her valuation.
3.2 If the highest competing bid h_i lies between v_i and b_i ($v_i < h_i < b_i$), bidder i wins the object but earns a negative payoff because $v_i - h_i < 0$. If, instead, bidder i submits a bid equal to her valuation, $b_i = v_i$, she would lose the object, earning a zero payoff. In other words, bidder i would be better off submitting a bid $b_i = v_i$ and losing, than submitting a bid $b_i > v_i$ and winning but earning a zero or negative payoff (i.e., negative expected payoff).
3.3 If the highest competing bid h_i lies above b_i ($v_i < b_i < h_i$) bidder i loses the auction, earning a zero payoff. Therefore, bidder i has no strict incentives to deviate from bidding $b_i = v_i$.

Overall, bidder i can either earn the same payoff as in Cases 1.1–1.3, or a lower payoff, but cannot improve her payoff, thus having no strict incentives to increase her bid to $b_i > v_i$. Combining our results from Sects. 9.3.1–9.3.3, we conclude that bidder i has no incentives to shade her bid, $b_i < v_i$, or increase it to $b_i > v_i$. In short, bidding according to her valuation, $b_i = v_i$, is a weakly dominant strategy for every bidder i in the SPA.

9.3.4 Discussion

Bidding BNE vs. Dominant strategies. From our above discussion, every bidder i finds that bidding according to her valuation is not only a BNE of the SPA, it is a BNE where every player uses *weakly dominant* strategies. Technically, if bidder i finds that her bidding function $b_i^* = b_i(v_i)$ is a BNE, this means that

$$EU_i\left(b_i^*, b_{-i}^* | v_i\right) \geq EU_i\left(b_i, b_{-i}^* | v_i\right)$$

for every bid $b_i \neq b_i^*$ and every valuation v_i. Intuitively, equilibrium bid b_i^* provides bidder i with a higher expected payoff than any other bid $b_i \neq b_i^*$, conditional on her rivals selecting equilibrium bids b_{-i}^*. However, when we say that bidder i finds that her equilibrium bid b_i^* is a weakly dominant strategy, we mean that

$$EU_i\left(b_i^*, b_{-i} | v_i\right) \geq EU_i(b_i, b_{-i} | v_i)$$

for every bid $b_i \neq b_i^*$ and every valuation v_i and for every bidding profile b_{-i}. This inequality entails that bidder i's expected payoff from submitting bid b_i^* is higher than from any other bid $b_i \neq b_i^*$, *regardless* of the specific bidding profile that her rivals use, b_{-i}, that is, both when they submit equilibrium bids, $b_{-i} = b_{-i}^*$, and when they do not, $b_{-i} \neq b_{-i}^*$.

This is a strong property in the bidding strategy in SPAs, saying that bidder i, when submitting $b_i(v_i) = v_i$, can, essentially, *ignore her opponents' bids*: both when they submit equilibrium bids, $b_j(v_j) = v_j$ for all $j \neq i$, and when they do not, $b_j(v_j) \neq v_j$, bidder i cannot strictly improve her payoff by deviating from $b_i(v_i) = v_i$. In other auction formats, such as the FPA, we will find that equilibrium bid b_i^* is a BNE but it is not a weakly dominant strategy for every player i.

No bid shading. Intuitively, by shading her bid, $b_i(v_i) < v_i$, bidder i lowers the chance that she wins the auction, but does not lower the price that she pays upon winning, thus having no incentives to lower her bid. In the FPA, in contrast, bid shading gives rise to a trade-off: a lower chance of winning the auction, but paying a lower price upon winning the object. As we show in Sect. 9.4, bidders have incentives to shade their bids in FPA in most contexts.

9.4 First-Price Auctions

In the FPA, the assignment rule coincides with that in the SPA: the winner is the bidder submitting the highest bid. The payment rule in the FPA, however, differs, since the winning bidder must pay the highest bid (that is, the bid that she submitted); as opposed to the payment rule in the SPA where she pays the second-highest bid. This seemingly small difference between auction formats gives rise to bid shading in the FPA—a result that we could not sustain in equilibrium when bidders face a SPA.

For simplicity, we assume that values are independent and identically distributed (i.i.d.) according to a cumulative distribution function $F(v)$. As in Sect. 9.3, we consider that bidders are risk neutral, while Exercise 9.9 relaxes this assumption.

Step 1. *Writing bidder i's maximization problem.* Let us first write bidder i's expected utility maximization problem, as follows.

$$\max_{b_i \geq 0} \Pr(win)(v_i - b_i) - \Pr(lose)0$$

where the first term represents the net payoff that bidder i earns when winning the object, $v_i - b_i$, since she values it at v_i and pays her own bid b_i for it; while the second term just indicates that her payoff from losing the auction is zero.

At this point, we need to write the probability of winning, $\Pr(win)$, as a function of bidder i's bid, b_i. Intuitively, a more aggressive bid increases her chances of winning the object but reduces her net payoff if winning, $v_i - b_i$. To do this, note that every bidder i uses a symmetric bidding function $b_i : [0, 1] \to \mathbb{R}_+$, i.e., a function mapping her valuation $v_i \in [0, 1]$ into a positive dollar amount (her bid). If bidding functions are symmetric across players and monotonic (which we show below), bidder i wins when her bid satisfies $b_j \leq b_i$ for all $j \neq i$, indicating that her valuation satisfies $v_j \leq v_i$. This ranking between valuations v_i and v_j occurs if

$$\Pr(v_j \leq v_i) = F(v_i).$$

Therefore, when bidder i faces $N - 1$ rivals, her probability of winning the auction is the probability that her valuation exceeds that of all other $N - 1$ bidders. Since valuations are i.i.d., we can write this probability as the product

$$\Pr(v_j \leq v_i) \times \Pr(v_k \leq v_i) \times \ldots \times \Pr(v_l \leq v_i) = \underbrace{F(v_i) \times F(v_i) \times \ldots \times F(v_i)}_{N-1 \text{ times}}$$
$$= F(v_i)^{N-1}$$

where bidders $j \neq k \neq l$ represent i's rivals. As a result, we can express the above expected utility maximization problem as follows.

$$\max_{b_i \geq 0} \underbrace{F(v_i)^{N-1}}_{\Pr(win)} (v_i - b_i)$$

Using the above bidding function, we can write $b_i(v_i) = x_i$, where $x_i \geq 0$ represents bidder i's bid when her valuation is v_i. Applying the inverse $b^{-1}(\cdot)$ on both sides, yields $v_i = b_i^{-1}(x_i)$, which helps us rewrite the probability of winning, $F(v_i)^{N-1}$, as $F(b_i^{-1}(x_i))^{N-1}$. The above maximization problem, then, becomes

$$\max_{x_i \geq 0} F(b_i^{-1}(x_i))^{N-1} (v_i - x_i)$$

where we expressed the bid as x_i in the last term since $b_i(v_i) = x_i$.

Step 2. *Finding equilibrium bids.* Once the probability of winning is written as a function of bidder i's bid, x_i, we are ready to differentiate with respect to x_i to find the equilibrium bidding function in the FPA. Differentiating with respect to x_i, yields

$$-\left[F(b_i^{-1}(x_i))^{N-1}\right]$$
$$+ (N-1)F(b_i^{-1}(x_i))^{N-2} f\left(b_i^{-1}(x_i)\right) \frac{\partial b_i^{-1}(x_i)}{\partial x_i} (v_i - x_i) = 0.$$

9.4 First-Price Auctions

Since $b_i^{-1}(x_i) = v_i$, we can use the inverse function theorem[6] to obtain $\frac{\partial b_i^{-1}(x_i)}{\partial x_i} = \frac{1}{b'\left(b_i^{-1}(x_i)\right)}$. Therefore, the above expression simplifies to

$$-\left[F(v_i)^{N-1}\right] + (N-1)F(v_i)^{N-2}f(v_i)\frac{1}{b'(v_i)}(v_i - x_i) = 0.$$

Further rearranging, we obtain

$$(N-1)F(v_i)^{N-2}f(v_i)v_i - (N-1)F(v_i)^{N-2}f(v_i)x_i = F(v_i)^{N-1}b'(v_i)$$

or

$$F(v_i)^{N-1}b'(v_i) + (N-1)F(v_i)^{N-2}f(v_i)x_i = (N-1)F(v_i)^{N-2}f(v_i)v_i$$

The left-hand side is $\frac{\partial \left[F(v_i)^{N-1}b_i(v_i)\right]}{\partial v_i}$, which helps us rewrite the above expression more compactly as

$$\frac{\partial \left[F(v_i)^{N-1}b_i(v_i)\right]}{\partial v_i} = (N-1)F(v_i)^{N-2}f(v_i)v_i$$

Integrating both sides, yields

$$F(v_i)^{N-1}b_i(v_i) = \int_0^{v_i} (N-1)F(v_i)^{N-2}f(v_i)v_i dv_i. \tag{FOC}$$

Step 3. *Applying integration by parts.* We can now use integration by parts on the right-hand side of the above first-order condition. Recall that, when integrating by parts, we consider two functions $g(x)$ and $h(x)$, so that $(gh)' = g'h + gh'$. Integrating both sides, yields

$$g(x)h(x) = \int g'(x)h(x)dx + \int g(x)h'(x)dx.$$

Reordering this expression, we find

$$\int g'(x)h(x)dx = g(x)h(x) - \int g(x)h'(x)dx.$$

---

[6] Formally, this theorem says that if $f(x)$ is a continuously differentiable function with non-zero derivative at point $x = a$, $f'(a) \neq 0$, then the derivative of the inverse function at $b = f(a)$, which we denote as $(f^{-1})'(b)$, is the reciprocal of the derivative of function f at point $x = a$, that is, $(f^{-1})'(b) = \frac{1}{f'(a)} = \frac{1}{f'(f^{-1}(b))}$ since $f^{-1}(b) = a$. In our analysis of equilibrium bids in the FPA, this theorem says that $\frac{\partial b_i^{-1}(x_i)}{\partial x_i} = \frac{1}{b_i'(v_i)} = \frac{1}{b'\left(b_i^{-1}(x_i)\right)}$ since the inverse of the bidding function is bidder i's valuation, $b_i^{-1}(x_i) = v_i$.

At this point, we can apply integration by parts in our setting by defining $g'(x) \equiv (N-1)F(x)^{N-2}f(x)$ and $h(x) \equiv x$, so that $g(x) = F(x)^{N-1}$ and $h'(x) = 1$, obtaining the following result.

$$\int_0^{v_i} \underbrace{\left[(N-1)F(x)^{N-2}f(x)\right]}_{g'(x)} \underbrace{x}_{h(x)} dx = \underbrace{F(v_i)^{N-1}}_{g(x)} \underbrace{v_i}_{h(x)} - \int_0^{v_i} \underbrace{F(x)^{N-1}}_{g(x)} \underbrace{1}_{h'(x)} dx$$

Inserting this result into the right-hand side of first-order condition (denoted as FOC), yields

$$F(v_i)^{N-1} b_i(v_i) = F(v_i)^{N-1} v_i - \int_0^{v_i} F(v_i)^{N-1} dv_i.$$

We can now solve for the equilibrium bidding function $b_i(v_i)$ that we seek to find. Dividing both sides by $F(v_i)^{N-1}$, yields[7]

$$b_i(v_i) = v_i - \underbrace{\frac{\int_0^{v_i} F(v_i)^{N-1} dv_i}{F(v_i)^{N-1}}}_{\text{Bid shading}}$$

Intuitively, bidder i submits a bid equal to her valuation for the object, v_i, less an amount captured by the second term in the above expression, which is referred as her "bid shading." We can then claim that the above bidding function $b_i(v_i)$ constitutes the BNE of the first-price auction when bidders' valuations are distributed according to $F(v_i)$.

Step 4. *Checking monotonicity.* We finally check that the above bidding function $b_i(v_i)$ is monotonically increasing in bidder i's valuation, v_i; as we assumed in Step 1. A marginal increase in v_i produces the following effect in bidder i's equilibrium bidding function

$$\frac{\partial b_i(v_i)}{\partial v_i} = 1 - \frac{F(v_i)^{N-1} F(v_i)^{N-1} - (N-1)F(v_i)^{N-2} \int_0^{v_i} F(v_i)^{N-1} dv_i}{\left[F(v_i)^{N-1}\right]^2}$$

$$= \frac{(N-1)F(v_i)^{N-2} f(v_i) \int_0^{v_i} F(v_i)^{N-1} dv_i}{\left[F(v_i)^{N-1}\right]^2}$$

which is positive since $F(v_i) \in [0, 1]$, $f(v_i) > 0$ for all v_i, and $N \geq 2$ by definition. Informally, a one-dollar increase in bidder i's valuation produces a less-than-proportional increase in her bid shading (it can increase, but by less than one

---

[7] Note that, to divide both sides by $F(v_i)^{N-1}$, we need this term to be different than zero, which holds for all $v_i \neq 0$. Alternatively, the probability of winning the auction, $F(v_i)^{N-1}$, is positive for all $v_i > 0$.

dollar), ultimately driving bidder i to submit a more aggressive bid. Example 9.1 illustrates bid shading intensity and the monotonicity of the bidding function when valuations are uniformly distributed.

As a remark, note that the above bidding function is, indeed, a BNE of the FPA, because, conditional on every bidder $j \neq i$ using this bidding strategy, $b_j(v_j)$, bidder i cannot increase her expected utility by unilaterally deviating from $b_i(v_i)$. If she could, bidding function $b_i(v_i)$ would not solve her expected utility maximization problem defined in Step 1.

Example 9.1 FPA with uniformly distributed valuations. Consider, for instance, that individual valuations are uniformly distributed, $F(v_i) = v_i$. In this setting, we obtain $F(v_i)^{N-1} = v_i^{N-1}$ and $\int_0^{v_i} F(v_i)^{N-1} dv_i = \frac{1}{N} v_i^N$, producing a bidding function of

$$b_i(v_i) = v_i - \frac{\frac{1}{N} v_i^N}{v_i^{N-1}}$$

$$= v_i - \underbrace{\frac{v_i}{N}}_{\text{Bid shading}}.$$

In this context, every bidder shades her bid by $\frac{v_i}{N}$, which increases in the number of competing bidders as we explain next. In addition, the equilibrium bidding function can be rewritten as

$$b_i(v_i) = v_i \left(\frac{N-1}{N} \right),$$

which is monotonically increasing in the valuation that bidder i assigns to the object, v_i, as required.

When only two bidders compete for the object, $N = 2$, this bidding function simplifies to $b_i(v_i) = \frac{v_i}{2}$; as depicted in Fig. 9.1. When $N = 3$, equilibrium bids increase to $b_i(v_i) = \frac{2v_i}{3}$; and similarly when $N = 4$ bidders compete for the object, where $b_i(v_i) = \frac{3v_i}{4}$.[8] Intuitively, as more bidders participate in the auction, every bidder i submits more aggressive bids since she faces a higher probability that another bidder j has a higher valuation for the object and (given symmetric bidding functions) submits a higher bid than she does, leading her to lose the auction. In the extreme case where the number of bidders approaches infinity, $N \to +\infty$, the bidding function converges to $b_i(v_i) = v_i$; as illustrated in Fig. 9.1.□

---

[8] More generally, the derivative of bidding function $b_i(v_i) = v_i \left(\frac{N-1}{N} \right)$ with respect to the number of bidders, N, yields $\frac{\partial b_i(v_i)}{\partial N} = \frac{1}{N^2} v_i$, which is postive for all parameter values.

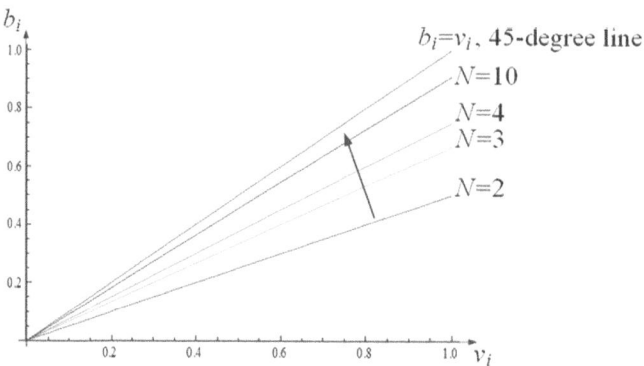

Fig. 9.1 Equilibrium bids in the first-price auction with uniformly distributed valuations

9.5 Efficiency in Auctions

An auction is deemed "efficient" when it assigns the object to the individual with the highest valuation, that is, when the assignment rule allocates the object to bidder i if and only if her valuation, v_i, satisfies $v_i > v_j$ for all $j \neq i$. Otherwise, if bidder j receives the object despite having a lower valuation than some other bidder i, these two bidders could negotiate at the end of the auction, with bidder i paying a price p that satisfies $v_i > p > v_j$, thus making both bidders better off. Indeed, bidder i only pays p, earning $v_i - p$; while bidder j sells the object for a price higher than her own valuation, v_j.

In other words, the assignment rule that allocates the object to bidder j is Pareto *inefficient*, as we can find an alternative allocation (such as bidder i receiving the object) that improves the payoff of at least one individual without making other individuals worse off. In addition, inefficient allocations do not maximize the seller's expected revenue, indicating that she could design the auction rules differently and earn a higher expected revenue.

Applying this definition of efficiency, we can easily confirm that both FPA and SPA are efficient. To see this point, first note that both auction formats assign the object to the individual submitting the highest bid and, second, the equilibrium bidding functions in both auction formats are strictly increasing in bidders' valuations. As a consequence, if bidder i wins the object in the FPA or SPA, she must have submitted the highest bid and, as a consequence, she must have the highest valuation for the object. This is an important, yet intuitive, finding:

If bidders use a *symmetric, strictly increasing, bidding function* in equilibrium, the winner of the auction must be the individual with the highest valuation, making the auction efficient.

This condition holds for most of the commonly used auction formats in real life, like the FPA, the SPA, and the APA; and even in uncommon auctions, like the third- and kth-price auctions. However, when we allow for budget constraints or risk aversion in these auction formats, auctions are not necessarily efficient.

Budget constraints. When bidder i faces a budget w_i, her bidding function is increasing in v_i for all $b_i < w_i$ (affordable bids), but becomes flat at a height of w_i for all valuations for which $b_i > w_i$ (unaffordable bids). As a consequence, budget constraints imply that bidding functions are weakly increasing in v_i, not strictly increasing, so we cannot guarantee that the auction is efficient. This occurs even if only one individual suffers from budget constraints, or if all bidders face the same budget constraint, $w_i = w$, making their bidding functions symmetric (but not strictly increasing). Exercise 9.11 examines the effect of budget constraints in the FPA.

Risk aversion. Another context where efficiency is not satisfied is that where players exhibit different degrees of risk aversion. As Exercise 9.10 shows, bidders competing in an FPA submit more aggressive bids when they become more risk averse. If both bidders are symmetric in their risk aversion, the above condition still holds (i.e., bidders use a symmetric, strictly increasing, bidding function). Therefore, the auction is still efficient even if both bidders submit more aggressive bids than when they are risk neutral. This point is illustrated in Exercise 9.9.

When bidders are asymmetric in their risk preferences, however, the FPA is not efficient. To understand this result, consider an FPA with two bidders, A and B, where bidder A exhibits more (less) risk aversion. This allows for several scenarios: bidder A can be risk averse while B is risk neutral or risk-loving, both bidders can be risk averse but A is more risk averse than B, or both are risk lovers but A is less so than B. In any of these settings, bidder A submits a more aggressive bid than B does, $b_A > b_B$, implying that A wins the auction. If bidder A values the object more than B does, $v_A > v_B$, the outcome of the auction is still efficient; but otherwise the outcome is not efficient. Generally, then, we cannot guarantee that the object goes to the individual who values the object the most, specially if bidders are relatively asymmetric in their risk aversion, entailing that the FPA is not efficient when players are risk averse.

9.6 Seller's Expected Revenue

Previous sections found equilibrium bidding functions in different auction formats. A natural question at this point is the following: anticipating these equilibrium bidding functions, but without observing players' valuations for the object, what is the seller's expected revenue? Let us first answer this question for the FPA.

9.6.1 Expected Revenue in the FPA

When finding the expected revenue in any auction format, the seller must first identify the expected payment she will receive from each winning bidder; second, take expectations over all possible valuations (because she does not observe bidders' valuations); and, third, sum across all N bidders. For presentation purposes, we separately examine each step below.

Step 1. *Finding each bidder's payment.* The seller receives a payment from bidder i if she wins the auction. In other words, bidder i's payment is

$$m(v_i) = \Pr(win) \times b_i(v_i)$$

that is, the probability of winning the auction times the bid that bidder i pays for the object upon winning. If bidder i wins (loses) the auction with certainty, her expected payment is just $b_i(v_i)$ (zero, respectively). As shown in previous sections, we know that the probability of winning is $[F(v_i)]^{N-1}$, so the expected payment can be expressed as

$$\begin{aligned} m(v_i) &= (F[v_i])^{N-1} \times b_i(v_i) \\ &= G(v_i) \times b_i(v_i) \end{aligned}$$

where, for compactness, we denote $G(v_i) = (F[v_i])^{N-1}$. At this point, recall that the equilibrium bidding function in the FPA is

$$b_i(v_i) = v_i - \frac{\int_0^{v_i} F(x)^{N-1} dx}{F(v_i)^{N-1}}$$

or, alternatively,

$$\begin{aligned} b_i(v_i) &= v_i - \frac{\int_0^{v_i} G(x) dx}{G(v_i)} \\ &= \frac{G(v_i) v_i - \int_0^{v_i} G(x) dx}{G(v_i)}. \end{aligned}$$

Applying integration by parts in the numerator, yields

$$G(v_i) v_i - \int_0^{v_i} G(x) dx = \int_0^{v_i} x g(x) dx.$$

Therefore, we can rewrite the equilibrium bidding function as follows.

$$b_i(v_i) = \frac{\int_0^{v_i} x g(x) dx}{G(v_i)}$$

Inserting this equilibrium bidding function in bidder i's expected payment, yields

9.6 Seller's Expected Revenue

$$m(v_i) = G(v_i) \times \overbrace{\left(\frac{\int_0^{v_i} xg(x)dx}{G(v_i)} \right)}^{\text{Bid, } b_i(v_i)}$$

$$= \int_0^{v_i} xg(x)dx.$$

Step 2. *Finding the expected payment.* Since the seller cannot observe bidder i's value for the object, she needs to take expectations over all possible values to find the *expected* payment from this bidder, $E[m_i(v_i)]$, as follows.

$$E[m(v_i)] = \int_0^1 m(x)f(x)dx$$

$$= \int_0^1 \underbrace{\left[\int_0^{v_i} xg(x)dx \right]}_{m(v_i)} f(x)dx$$

Third step. *Sum across all bidders.* Finally, the seller sums across all N bidders participating in the auction, which yields the expression of her expected revenue in the FPA.

$$R^{FPA} = \sum_{i=1}^N E[\pi_i(v_i)]$$

$$= N \times E[m(v_i)]$$

$$= N \int_0^1 \left[\int_0^{v_i} xg(x)dx \right] f(x)dx.$$

Example 9.2 Expected revenue in FPA with uniformly distributed valuations. When valuations are uniformly distributed, $F(v_i) = v_i$ and $f(v_i) = 1$, so that $G(v_i) = v_i^{N-1}$, which implies that $g(v_i) = G'(v_i) = (N-1)v_i^{N-2}$ and $v_i g(v_i) = (N-1)v_i^{N-1}$. Therefore, bidder i's equilibrium bid (as shown in Example 9.1) is

$$b_i(v_i) = v_i \left(\frac{N-1}{N} \right)$$

which implies that her expected payment (Step 1) is

$$m(v_i) = \int_0^{v_i} xg(x)dx$$

$$= \int_0^{v_i} \underbrace{(N-1)x^{N-1}}_{xg(x)} dx$$

$$= \frac{N-1}{N} \left[x^N \right]_0^{v_i}$$

$$= \frac{N-1}{N}\left(v_i^N - 0\right)$$
$$= \frac{N-1}{N}v_i^N.$$

Therefore, bidder i's expected payment (Step 2) is

$$E[m(v_i)] = \int_0^1 m(v_i)f(v_i)dv_i$$
$$= \int_0^1 \underbrace{\frac{N-1}{N}v_i^N}_{m(v_i)}\underbrace{1}_{f(v_i)}dv_i$$
$$= \frac{N-1}{N}\int_0^1 v_i^N dv_i$$
$$= \frac{N-1}{N}\left[\frac{v_i^{N+1}}{N+1}\right]_0^1$$
$$= \frac{N-1}{N}\left(\frac{1}{N+1} - 0\right)$$
$$= \frac{N-1}{N(N+1)}.$$

Finally, the seller sums across all N bidders to obtain the expected revenue from the FPA, R^{FPA}, as in Step 3, that is,

$$R^{FPA} = \sum_{i=1}^N E[m(v_i)]$$
$$= N \times E[m(v_i)]$$
$$= N\frac{N-1}{N(N+1)}$$
$$= \frac{N-1}{N+1}.$$

The expected revenue is increasing in the number of bidders, N, but at a decreasing rate.[9] This result goes in line with that in Example 9.1, namely, as more bidders compete in the auction, they submit more aggressive bids, i.e., $b_i(v_i)$ increases in N, increasing as a result the expected winning bid that the seller earns. For instance, when only $N = 2$ bidders participate, the seller earns a revenue of

---

[9] Indeed, $\frac{\partial R^{FPA}}{\partial N} = \frac{2}{(N+1)^2}$ is positive, but $\frac{\partial^2 R^{FPA}}{\partial N^2} = -\frac{4}{(N+1)^3}$ is negative, indicating that, graphically, the seller's expected revenue, R^{FPA}, is increasing and concave in N.

$R^{FPA} = \frac{2-1}{2+1} = \frac{1}{3}$; and increases to $R^{FPA} = \frac{3-1}{3+1} = \frac{1}{2}$ when $N = 3$ bidders compete for the object.[10]

9.6.2 Expected Revenue in the SPA

Let us now find the expected revenue in the SPA, R^{SPA}. In this auction format, the seller anticipates that every bidder i submits a bid $b_i(v_i) = v_i$, and the winning bidder pays the second-highest bid (not her own bid!). Therefore, finding the expected revenue is more direct than in the FPA, as we only need to obtain the expected second-highest bid. Since $b_i(v_i) = v_i$, the second-highest bid coincides with the second-highest valuation for the object, $v_i^{[2]}$, that is,

$$R^{SPA} = E[v_i^{[2]}]$$
$$= \int_0^1 x f^{[2]}(x) dx$$

where $f^{[2]}(x)$ is the density function of the second-highest valuation, $v_i^{[2]}$. It is straightforward to see that the only complication in the above expression is to find this density function $f^{[2]}(x)$, which requires that we first identify the cumulative distribution function $F^2(x)$. This function describes the probability that the second-highest valuation is lower than x, that is, $F^2(x) = \Pr\{v^2 \leq x\}$, which happens if either (but not both) of these two events occurs:

(1) The valuations of all N bidders are below x or, formally, $v_i \leq x$ for every bidder i. This event happens with probability

$$\Pr\{v_1 \leq x\} \times \cdots \times \Pr\{v_N \leq x\} = \underbrace{F(x) \times \cdots \times F(x)}_{N \text{ times}} = [F(x)]^N$$

(2) The valuations of $N - 1$ bidders are below x, $v_i \leq x$, but that of only one bidder j is above x, $v_j > x$. This event can occur in N different ways:
- $v_1 > x$ for bidder 1 but $v_i \leq x$ for every bidder $i \neq 1$;
- $v_2 > x$ for bidder 2 but $v_i \leq x$ for every bidder $i \neq 2$;
- ...
- $v_N > x$ for bidder N but $v_i \leq x$ for every bidder $i \neq N$.

---

[10] Furthermore, when the number of bidders grows to infinity, the expected revenue approaches 1 since $\lim_{N \to +\infty} R^{FPA} = \lim_{N \to +\infty} \left(1 - \frac{2}{N+1}\right) = 1$. In other words, the seller earns a revenue of \$1 from the bidder with the highest valuation since she wins the auction and, when N grows to infinity, bidders do not shade their bids.

Each of these N cases happens with probability

$$\underbrace{[1 - F(x)]}_{v_i > x \text{ for } i} \times \underbrace{[F(x)]^{N-1}}_{v_j \leq x \text{ for every } j \neq i},$$

where $[1 - F(x)]$ denotes the probability that $v_i > x$ for a given bidder i, while $[F(x)]^{N-1}$ represents the probability that $v_j \leq x$ for all other bidders $j \neq i$. Summing over the above N cases, we find that event (2) happens with probability

$$\sum_{i=1}^{N} [1 - F(x)][F(x)]^{N-1} = N(1 - F(x))[F(x)]^{N-1}$$

Summarizing, the cumulative distribution function of the second-highest valuation $F^{[2]}(x)$, is

$$F^{[2]}(x) = \underbrace{[F(x)]^N}_{\text{Event \#1}} + \underbrace{N(1 - F(x))[F(x)]^{N-1}}_{\text{Event \#2}}.$$

Rearranging, the above expression simplifies to

$$F^{[2]}(x) = N[F(x)]^{N-1} - (N - 1)[F(x)]^N.$$

We can now differentiate $F^{[2]}(x)$ with respect to x, to obtain the density function

$$f^{[2]}(x) = N(N-1)F(x)^{N-2} f(x) - N(N-1)F(x)^{N-1} f(x)$$
$$= N(N-1)F(x)^{N-2}[1 - F(x)]f(x).$$

Finally, inserting density function $f^{[2]}(x)$ in the seller's expected revenue defined above, yields

$$R^{SPA} = \int x f^{[2]}(x) dx$$
$$= \int x \underbrace{N(N-1)F(x)^{N-2}[1 - F(x)]f(x)}_{f^{[2]}(x)} dx.$$

Example 9.3 Expected revenue in SPA with uniformly distributed valuations. When valuations are uniformly distributed, $F(x) = x$ and $f(x) = 1$, the seller's expected revenue becomes

$$R^{SPA} = \int_0^1 x N(N-1) x^{N-2} (1 - x) dx$$
$$= (N-1) \int_0^1 N x^{N-1} dx - N(N-1) \int_0^1 x^N dx$$

$$= (N-1)x^N \Big|_0^1 - \frac{N(N-1)x^{N+1}}{N+1}\Big|_0^1$$
$$= (N-1) - \frac{N(N-1)}{N+1}$$
$$= \frac{N-1}{N+1}$$

which coincides with that in the FPA, R^{FPA}, found in Example 9.2, thus being increasing and concave in the number of bidders, N.

9.6.3 Revenue Equivalence Principle

Examples 9.2 and 9.3 illustrate that, when bidders' valuations are uniformly distributed, $R^{FPA} = R^{SPA}$, implying that the seller can expect to earn the same revenue from both auction formats. This "revenue equivalence" result extends to other auction formats, which yield the same expected revenue as the FPA and SPA, and to settings where bidders' valuations are exponentially distributed (Exercises 9.13 and 9.14 ask you to check revenue equivalence in this context). More generally, the finding extends to auctions where valuations are distributed according to a generic, but common, cumulative distribution function $F(v_i)$. While a formal analysis of this result is beyond the scope of this book, we can identify the two main requirements that two auctions must satisfy to yield the same expected revenue for the seller, namely,

1. Same allocation rule in both auction formats, e.g., the bidder submitting the highest bid receives the object, as in the FPA and SPA.
2. Same expected utility of the bidder who has the lowest valuation for the object, e.g., zero in most auction formats since this bidder loses the auction, not receiving the object.

It is straightforward to note that the comparison of FPA and SPA satisfies conditions (1) and (2), thus generating the same expected revenue; but the comparison of FPA and APA does not satisfy condition (2) because the bidder with the lowest valuation earns a zero payoff in the FPA but a negative payoff in the APA after paying her bid. Therefore, the FPA and APA do not necessarily generate the same expected revenue. Similarly, the comparison of the FPA and the lottery auction does not satisfy condition (1), since the lottery auction does not necessarily assign the object to the individual submitting the highest bid but the FPA does. As a consequence, the FPA and the lottery auction do not yield the same expected revenue for the seller.

9.7 Common-Value Auctions and the Winner's Curse

Previous sections consider auctions where bidders may enter the auction with different valuations. In some settings, however, every bidder i shares a common value for the object, v, but privately observes a noisy signal s_i about the object's value drawn from the uniform distribution $U[0, 1]$, so $F(s_i) = s_i$, where $s_i \in [0, 1]$. Based on this signal, every bidder i submits her bid, b_i. This setting is known as "common-value" auctions and are typically used to analyze the sale of oil leases, as firms exploiting the oil reservoir would earn a similar profit if they win the auction, but they have imprecise and potentially different signals of the amount of oil barrels in the reservoir. A similar argument applies to other auctions of government contracts.

Bidders participate in a first-price, sealed-bid auction, which means that if bidder i wins, her realized payoff becomes $v - b_i$, and if she loses her payoff is zero. For simplicity, we consider a setting with only two bidders and that the true value is equal to the average of bidders' signals, so that $v = \frac{s_i + s_j}{2}$.

9.7.1 Bid Shading Is a Must!

First, note that bidder i falls prey to the winner's curse if her bid exceeds the object's true value (which no bidder observes), $b_i > v$, thus earning a negative payoff, $v - b_i < 0$, from winning the auction. In particular, this occurs if

$$b_i > \frac{s_i + s_j}{2}.$$

When bidder i's bid is a function of her privately observed signal, $b_i = \alpha s_i$, where $\alpha \in [0, 1]$, this inequality becomes

$$\alpha s_i > \frac{s_i + s_j}{2}$$

which, solving for signal s_i, simplifies to

$$s_i > \frac{s_j}{2\alpha - 1}.$$

When bidder i submits a bid equal to the signal she received, $\alpha = 1$, the winner's curse occurs if $s_i > s_j$; that is, bidder i's falls prey to the winner's curse as long as her signal is larger than bidder j's. This result is quite intuitive: if every bidder submits a bid that coincides with her privately observed signal, the bidder who received the highest signal ends up submitting the highest bid and, thus, winning the auction, but suffering from the winner's curse. In other words, the fact that she won the auction means that she received an overestimated signal of the object.

9.7 Common-Value Auctions and the Winner's Curse

If, instead, bidder i submits a bid equal to 3/4 of the signal she received, $\alpha = 3/4$, the winner's curse only emerges if $s_i > \frac{s_j}{2\frac{3}{4}-1} = 2s_j$; that is, when bidder i's signal is larger than the double of bidder j's. More generally, as bidder i shades her bid more severely (decreasing α), ratio $\frac{s_j}{2\alpha-1}$ increases, and the winner's curse is less likely to occur.

9.7.2 Equilibrium Bidding in Common-Value Auctions

Our previous discussion indicates that participants in common-value auctions shade their bids to account for the possibility that the signals they receive can be higher than the true value of the object, thus avoiding the winner's curse. However, we did not explicitly find the equilibrium bid in this type of auctions. We focus on the equilibrium bid next, where we keep assuming that bidder i's bid function is $b_i = \alpha s_i$ and parameter α satisfies $\alpha \in [0, 1]$. For presentation purposes, we split our analysis in several steps.

First step. *Finding the expected utility*. Bidder i's expected payoff from participating in the auction is

$$\Pr(b_i > b_j) \times \{E[v|s_i, b_i > b_j] - b_i\}$$

where $E[v|s_i, b_i > b_j]$ is bidder i's expected valuation, conditional on her signal s_i, and on knowing that she submitted the highest bid, i.e., $b_i > b_j$.

Since bidder j uses bidding function $b_j = \alpha s_j$, this expression becomes

$$\Pr(b_i > \alpha s_j) \times \{E[v|s_i, b_i > \alpha s_j] - b_i\}$$

and solving for s_j in the probability (first term) and in the inequality inside the expectation operator (second term), we obtain that

$$\Pr\left(\frac{b_i}{\alpha} > s_j\right) \times \left\{E\left[v\middle|s_i, \frac{b_i}{\alpha} > s_j\right] - b_i\right\}$$

Next, inserting $v = \frac{s_i+s_j}{2}$ in the expectation operator, yields

$$\Pr\left(\frac{b_i}{\alpha} > s_j\right) \times \left\{E\left[\frac{s_i + s_j}{2}\middle|\frac{b_i}{\alpha} > s_j\right] - b_i\right\}$$

Recall that bidder i observes her signal, s_i, so that $E[s_i] = s_i$, but does not know her rival's, s_j, entailing that

$$\Pr\left(\frac{b_i}{\alpha} > s_j\right) \times \left\{\frac{s_i}{2} + \frac{1}{2}E\left[s_j\middle|\frac{b_i}{\alpha} > s_j\right] - b_i\right\}.$$

Using the uniform distribution on s_j, we have

$$\Pr\left(\frac{b_i}{\alpha} > s_j\right) = \frac{b_i}{\alpha}, \text{ and}$$

$$E\left[s_j \left| \frac{b_i}{\alpha} > s_j \right.\right] = \frac{b_i}{2\alpha}$$

because s_j, which is a uniformly distributed random variable, falls into the range $\left[0, \frac{b_i}{\alpha}\right]$, yielding an expected value of $\frac{b_i}{2\alpha}$. Inserting these results into bidder i's expected payoff, we obtain that

$$\frac{b_i}{\alpha}\left[\frac{s_i}{2} + \frac{1}{2}\left(\frac{b_i}{2\alpha}\right) - b_i\right]$$

Second step. *Taking first-order conditions.* Every bidder i chooses her bid b_i to maximize her expected utility, solving the following problem

$$\max_{b_i \geq 0} \frac{b_i}{\alpha}\left[\frac{s_i}{2} + \frac{1}{2}\left(\frac{b_i}{2\alpha}\right) - b_i\right]$$

Taking first-order conditions with respect to b_i, we obtain

$$\frac{1}{\alpha}\left[\frac{s_i}{2} + \frac{1}{2}\left(\frac{b_i}{2\alpha}\right) - b_i\right] + \frac{b_i}{\alpha}\left(\frac{1}{4\alpha} - 1\right) = 0$$

Simplifying, yields

$$\frac{2\alpha s_i + b - 4\alpha b_i + b_i - 4\alpha b_i}{4\alpha^2} = 0$$

which holds if $2\alpha s_i + 2b_i - 8\alpha b_i = 0$ or, after rearranging, $\alpha x_i = b_i(4\alpha - 1)$. Solving for b_i, we find that

$$b_i = \frac{\alpha}{4\alpha - 1} s_i.$$

Third step. *Finding the equilibrium bidding function.* Recall that we considered a symmetric bidding function $b_i = \alpha s_i$ for a generic value of α. The above expression, $b_i = \frac{\alpha}{4\alpha-1} s_i$, is indeed linear in signal s_i, so we can write $\alpha s_i = \frac{\alpha}{4\alpha-1} s_i$, or

$$\alpha = \frac{\alpha}{4\alpha - 1}$$

which simplifies to $4\alpha - 1 = 1$. Solving for α, we obtain

$$\alpha = \frac{1}{2}$$

In summary, a symmetric BNE has every bidder i using the bidding function

$$b_i(s_i) = \frac{1}{2} s_i$$

that is, she submits a bid equal to half of her privately observed signal s_i.[11] A natural question at this point is whether this bidding function helps bidders avoid the winner's curse. Exercise 9.16 asks you to confirm that it does.

Exercises

9.1 **How robust is the equilibrium bidding in the SPA?**[A] Consider the SPA in Sect. 9.3 where $N \geq 2$ bidders compete for an object. Every bidder i privately observes his valuation, v_i, drawn from a cumulative distribution function, $F(v_i)$, and $v_i \in [0, 1]$, which is common knowledge among the players. We now analyze how equilibrium bidding behavior in the SPA, $b_i(v_i) = v_i$, is affected by the following changes.
   a) How is equilibrium bidding in the SPA, $b_i(v_i) = v_i$, affected by an increase in the number of bidders, from N to N', where $N' > N$?
   b) How is equilibrium bidding in the SPA, $b_i(v_i) = v_i$, affected if some bidders become risk averse? What if they are risk lovers?
   c) How is equilibrium bidding in the SPA, $b_i(v_i) = v_i$, affected if valuations are uniformly distributed, $F(v_i) = v_i$? What if they are exponentially distributed, with cumulative distribution function $F(v_i) = 1 - e^{-\lambda v_i}$, where $\lambda > 0$ and $v_i \in [0, +\infty)$?

9.2 **SPA with other tiebreaking rules.**[A] In Sect. 9.3, we considered that, if any two bidders i and j submit the highest bid, the object is randomly assigned among them with equal probabilities, $p_i = p_j = 1/2$. Show that if, without loss of generality, these probabilities satisfy $p_i > p_j$, every bidder i still submits a bid equal to his valuation, $b_i = v_i$.

9.3 **SPA and the English auction.**[A] Consider an ascending-price auction (also known as the "English auction") where the seller starts announcing a zero price (or a reservation price if the auction has one) and then allows bidders to submit increasingly higher bids. The highest bid at any given moment is the "standing bid." If another bidder submits a higher bid, then that becomes the standing bid. If no competing bidder displaces the standing bid (often after a few seconds as determined by the seller), the standing bid is declared the winner of the auction, that bidder who made this bid receives the object, and pays the price where the auction was stopped.

---

[11] In a context with $N \geq 2$ bidders, one can show that this bidding function becomes $b_i(s_i) = \frac{(N+2)(N-1)}{2N^2} s_i$, where ratio $\frac{(N+2)(N-1)}{2N^2}$ is smaller than 1 since $\frac{(N+2)(N-1)}{2N^2} < 1$ simplifies to $N^2 - N + 2 > 0$, which holds for all $N \geq 2$.

As the Dutch auction will be analyzed in Chapter 2, the English auction is also an open auction since every bidder can observe other bidders' bidding behavior (by seeing whether they are raising their hand or not), as opposed to the second-price auction we considered in previous exercises of this chapter which is known as a closed (or sealed bid) auction given that bidders cannot observe other bidders' strategies until the end of the auction.

A variation of the English auction is the so-called Japanese auction, where the seller uses a price clock, starting from zero (or a reservation price) which continuously increases. Bidders do not raise their hands but, instead, decide if, given the price at any given moment, they want to leave the auction. When a single bidder remains, the clock is stopped, he is declared the winner, and pays the price where the clock was stopped.

Show that the English auction is strategically equivalent to the second-price auction, implying that every bidder submits the same equilibrium bidding function in both auction formats.

9.4 **SPA with entry fees.**[B] Consider an SPA where the seller announces an entry fee, E, that every bidder must pay in order to enter the auction room. Every bidder i's valuation is independently drawn from a uniform distribution, $F(v_i) = v_i$ where $v_i \in [0, 1]$, and there are $N \geq 2$ bidders.
   a) Find the optimal bidding function for bidder i, $b_i(v_i, E)$.
   b) How are your results in part (a) affected by the entry fee E? And how are they affected by an increase in N? Interpret.

9.5 **Equilibrium bidding in the FPA using the envelope theorem approach.**[C] Consider an FPA with $N \geq 2$ bidders, each privately drawing his valuation for the object, v_i, from a uniform distribution. In this exercise, we show that the equilibrium bidding function in Sect. 9.3 can be found using an alternative approach: instead of bidder i directly differentiating his expected utility from participating in the auction with respect to his bid b_i, we assume that he uses bidding function $b_i(z_i)$ where valuation z_i is allowed to coincide with his real valuation, $z_i = v_i$, or differ, $z_i \neq v_i$, and bidder i differentiates his expected utility with respect to valuation z_i.
   a) Find bidder i's expected utility from participating in the FPA.
   b) Differentiate the expected utility of part (a) with respect to z_i. In equilibrium, every bidder i must have no incentives to bid according a valuation different than his true valuation, implying that $z_i = v_i$. Evaluate your first-order condition at $z_i = v_i$, rearrange, and find the equilibrium bidding function $b(v_i)$. Does your result coincide with that in Sect. 9.4 (Example 9.1)?

9.6 **FPA with exponentially distributed values.**[B] Consider the equilibrium bidding function in the FPA found in Sect. 9.4. Assume that only two bidders compete for the object and their valuations are drawn from an exponential distribution $F(v_i) = 1 - \exp(-\lambda v_i)$ where $v_i \in [0, +\infty)$. Find the equilibrium bidding function $b_i(v_i)$ in this context. Evaluate your results at different values of λ, and interpret your results.

Exercises

9.7 **FPA with other distributions.**[B] Consider the equilibrium bidding function in the FPA found in Sect. 9.4, but assume that valuations are drawn from the following distribution function,

$$F(v_i) = (1+\alpha)v_i - \alpha v_i^2$$

where $v_i \in [0, 1]$, and parameter α satisfies $\alpha \in [-1, 1]$. When $\alpha = 0$, this function collapses to the uniform distribution, $F(v_i) = v_i$; when $\alpha > 0$, it becomes concave, thus putting more probability weight on low valuations; and when $\alpha < 0$, it is convex, assigning more probability weight on high valuations. For simplicity, assume that only two bidders compete for the object.
  a) Find the equilibrium bid in this setting.
  b) How does the equilibrium bidding function, $b_i(v_i)$, change in parameter α?

9.8 **Bidding one's valuation in FPAs.**[A] Consider a first-price auction (where the bidder winning the auction pays the bid he submitted). For simplicity, assume two players with valuations $\theta_1 > 0$ and $\theta_2 > 0$, both being uniformly distributed in $[0, 1]$. Without loss of generality, assume that $\theta_2 > \theta_1$. Does bidding one's valuation constitute a dominant strategy? [*Hint*: A counterexample suffices.]

9.9 **FPA with risk averse bidders.**[B] Consider a FPA where every bidder's valuation is drawn from a uniform distribution, $F(v_i) = v_i$, and $v_i \in [0, 1]$. Every bidder i is risk averse and exhibits a constant relative risk aversion (CRRA) utility function $u(x) = x^\alpha$, where x is the bidder's income, and α denotes the Arrow-Pratt coefficient of relative risk aversion, where $0 < \alpha < 1$. When $\alpha \to 1$, the utility function becomes almost linear in x, indicating that the individual's utility approaches risk neutrality; but as α decreases, the utility function becomes more concave in x, implying that the individual is more risk averse.
  a) *Two bidders.* Consider first a setting with only $N = 2$ bidders. Assuming that every bidder i uses a symmetric bidding function $b_i(v_i) = sv_i$, where $s \in (0, 1)$ denotes his bid shading, find the equilibrium bidding function in this setting.
  b) How is the equilibrium bidding function you found in part (b) affected by parameter α? Interpret.
  c) *More than two bidders.* Find the equilibrium bidding function in a setting with $N \geq 2$ bidders.
  d) How is the equilibrium bidding function affected by a marginal increase in α? And by a marginal increase in N?

9.10 **FPA with two asymmetrically risk averse bidders.**[B] Consider an FPA with two risk averse bidders, A and B. For simplicity, every bidder i's valuations are uniformly distributed, $F(v_i) = v_i$, where $v_i \in [0, 1]$. Bidder A's utility function exhibits a constant relative risk aversion (CRRA) utility function, $u_A(x) = x^\alpha$, where x is his income and $0 < \alpha \leq 1$. Bidder B's utility function is symmetric, $u_B(x) = x^\beta$, but parameter β satisfies $\beta < \alpha$.

a) *Bidding comparison.* Using the results in Exercise 7.9, compare the equilibrium bidding functions of bidders A and B, and interpret.

b) *Efficiency.* Under which combinations of v_A and v_B is the outcome of this auction inefficient? Do your results depend on parameters α and β? Explain.

c) *Symmetric players.* Show that if $\alpha = \beta$, the outcome of the auction is efficient regardless of the realizations of v_A and v_B. Interpret your results.

9.11 **FPAs with budget constrained bidders.**[B] Consider an FPA with $N \geq 2$ bidders, but assume that every bidder privately observes his valuation for the object, v_i, and his budget, w_i, both being uniformly and independently drawn from the [0, 1] interval. For simplicity, assume that if a bidder wins the auction and the winning price is above his budget, w_i, he cannot afford to pay this price, and the seller imposes a fine on the buyer, $F > 0$, for having to renege from his bid.

a) Show that bidding above his budget, $b_i > w_i$, is a strictly dominated strategy for every bidder i.

b) If bidder i's valuation, v_i, satisfies $\frac{N-1}{N} v_i \leq w_i$, show that bidding according to $b_i(v_i) = \frac{N-1}{N} v_i$ (as found in Sect. 9.4) is still a weakly dominant strategy.

c) If bidder i's valuation, v_i, satisfies $\frac{N-1}{N} v_i > w_i$, show that submitting a bid equal to his budget, $b_i = w_i$, is a weakly dominant strategy.

d) Combine your results from parts (b) and (c) to describe the equilibrium bidding function in the first-price auction with budget constraints, $b_i(v_i, w_i)$. Depict it as a function of v_i.

9.12 **Equilibrium bidding function in the APA.**[B] Consider an APA where bidders draw their valuations from a uniform distribution, $F(v_i) = v_i$ where $v_i \in [0, 1]$, and there are $N \geq 2$ bidders.

a) Assume that every bidder i uses a symmetric and strictly increasing bidding function $b(z_i)$, where $z_i \in [0, 1]$ denotes the valuation according to which he bids that can satisfy $z_i = v_i$ if he bids according to his true valuation for the object or $z_i \neq v_i$ if he does not. Write bidder i's expected utility maximization problem.

b) Instead of differentiating the expected utility expression found in part (a) with respect to bidder i's bid, differentiate it with respect to z_i. This indirect approach is often known as the "envelope theorem" approach, while directly differentiating with respect to b_i is known as the "direct approach."

c) Evaluate the first-order conditions that you found in part (b) at $z_i = v_i$ since every bidder i must have no incentives to bid according to a different valuation than v_i. Rearrange and find the equilibrium bidding function in this auction format. (For simplicity, you can assume that $b(0) = 0$, meaning that the bidder with the lowest valuation submits a bid of zero.)

d) *Comparative statics.* Evaluate the equilibrium bidding function at $N = 2$, $N = 3$, and $N = 4$. Interpret.

e) *Comparison with FPA.* Compare the equilibrium bidding function you found in part (c) against that in an FPA with uniformly distributed values. In which auction format the bidder submits more aggressive bids?

9.13 **Equilibrium bidding function in a lottery auction.**[B] Consider a lottery auction with $N \geq 2$ bidders, all of them assigning the same value to the object, v. Every bidder i's utility from submitting a bid b_i is

$$EU_i[b_i|v] = \frac{b_i}{b_i + B_{-i}} v - b_i$$

where the ratio represents bidder i's probability of winning, which compares his bid relative to the aggregate bids submitted by all players, where $B_{-i} \equiv \sum_{j \neq i} b_j$. The second term indicates that bidder i must pay his bid both when winning and losing, as in APAs. For simplicity, assume that bidders use a symmetric bidding strategy, $b(v)$.
a) Find bidder i's equilibrium bidding function.
b) *Comparative statics.* How does bidder i's equilibrium bid change with v and N?
c) *Bidding coordination.* Find equilibrium bids if bidders could coordinate their bidding decisions. Compare your results with those of part (b).

9.14 **Expected revenue in the FPA when valuations are exponentially distributed.**[C] Consider the FPA in Sect. 9.6.2, but assume $N = 2$ bidders, each with valuations drawn from an exponential distribution, $F(v_i) = 1 - \exp(-\lambda v_i)$ where $v_i \in [0, \infty)$.
a) Find the seller's expected revenue in this context. [*Hint*: You may need to apply L'Hôpital rule]
b) How does expected revenue change with the parameter λ? Interpret your results.

9.15 **Expected revenue in the SPA when valuations are exponentially distributed.**[B] Consider the SPA in Sect. 9.6.2, but assume $N = 2$ bidders with valuations drawn from an exponential distribution, $F(v_i) = 1 - \exp(-\lambda v_i)$ where $v_i \in [0, \infty)$.
a) Find the seller's expected revenue in this context. [*Hint*: You may need to apply L'Hôpital rule]
b) How does expected revenue change with the parameter λ? Interpret your results.

9.16 **Equilibrium expected utility in a common-value auction.**[A] Consider a common-value auction with two bidders.
a) Find bidder i's expected utility in equilibrium.
b) Does bidder i's fall prey to the winner's curse in equilibrium?

10 Perfect Bayesian Equilibrium

10.1 Introduction

In this chapter, we extend the incomplete information games studied in Chapter 8 to allow for sequential interaction. We also extend the sequential-move games of Chapter 6 to allow for incomplete information, that is, at least one of the players observes some information that her rivals do not observe. This class of games helps us to model several strategic settings, as in real life individuals and firms often interact sequentially and they cannot observe all information when making their decisions. Examples include an incumbent monopolist privately observing its production costs while a potential entrant only having an estimate about the incumbent's costs (e.g., $10 per unit with probability 0.6, or $5 otherwise). The incumbent chooses its price and the potential entrant, observing the price but not knowing the incumbent's cost, decides whether to enter into the industry or not. Another example would be a new car company offering a long warranty (10 years) to convince potential customers of its cars' high quality, such as when Korean automakers (Kia and Hyundai) entered the US market in the 1980s, where customers were generally uncertain about the cars' quality, and provided longer warranties than any other automakers operating in the market.

In the next section, we present the main ingredients in these games, its time structure, and how to represent them with the game trees that we already studied in Chapter 6. Section 10.3 then uses the BNEsolution concept to show that, in a sequential-move game, the set of BNEs may be too large and, more importantly, some BNEs predict players choosing actions that do not maximize their payoffs at that point of the game tree (i.e., sequentially irrational behavior).

This issue is analogous to the problem we encountered in our first approach to sequential-move games of complete information, where some NEs prescribe sequentially irrational behavior, leading us to introduce a new solution concept, Subgame Perfect Equilibrium (SPE), which guarantees that players are sequentially rational in all subgames where they are called to move. In this chapter, we follow a similar problem and solution but applied to an incomplete information context, that is, some BNEs predict sequentially irrational behavior, so we introduce a solution concept, Perfect Bayesian Equilibrium (PBE), which ensures sequentially rational behavior given players' beliefs at that point of the game tree.

An important issue in sequential-move games of incomplete information is that a player's actions can help other players acting later in the game to infer the information that she privately observed. In other words, actions can work as "signals" of a player's private information. In fact, a special class of sequential-move games of incomplete information is known as "signaling games," where the first mover is privately informed and her actions convey this information to an uninformed second mover or, instead, conceal that information from the second mover.

For presentation purposes, we first consider games of incomplete information with only one information set, where the player acting as a responder is only called to move once. Afterwards, we extend our analysis to games with two or more information sets. In each of them, we examine whether separating strategy profiles can be sustained as PBEs, where the first-mover's actions convey information about her privately observed information (her "type") to the uninformed second mover. Separating PBEs follows the old adage of "actions speak louder than words," as the first-mover's actions help the second mover infer some information. Similarly, we check if pooling PBEs can be supported where the first-mover's actions are, essentially, uninformative, concealing her privately observed information from other players. (For completeness, we also explore a blend of these two strategy profiles, the so-called semi-separating PBEs, and how to identify them in signaling games.)

The chapter ends with a list of extensions, helping the reader recognize how the previous tools and results would be affected if the first mover had more than two types, more than two available actions, or if the second mover has more than two available responses.

10.2 Sequential-Move Games of Incomplete Information—Notation

In this section, we describe the main ingredients and notation that we usually encounter in sequential-move games of incomplete information.

1. **Nature.** Nature reveals to player i a piece of private information, $\theta_i \in \Theta_i$, which we refer to as her "type." Examples include a firm's production cost, its product quality, or a worker's ability on a certain task. The realization of θ_i is privately observed by player i but not by her rivals $j \neq i$. In some games, only one player privately observes her type, θ_i, as player 1 in Fig. 10.1. In more challenging games, we can have every player i privately observing her type θ_i; or a subset of players, each privately observing her type.
2. **First mover (sender).** Player i privately observes θ_i and chooses an action $s_i(\theta_i)$. Afterwards, her opponents observe $s_i(\theta_i)$ but do not observe θ_i.

10.2 Sequential-Move Games of Incomplete Information—Notation

1. *Separating strategy.* If action $s_i(\theta_i)$ is type dependent,

$$s_i(\theta_i) \neq s_i(\theta_i') \text{ for every two types } \theta_i \neq \theta_i',$$

   we say that player i uses a "separating" strategy, as different types choose different strategies. This would be the case if, for instance, the incumbent firm uses a different price as a function of its cost.

2. *Pooling strategy.* If action $s_i(\theta_i)$ is type independent,

$$s_i(\theta_i) = s_i(\theta_i') \text{ for every two types } \theta_i \neq \theta_i',$$

   we say that player i employs a "pooling" strategy, since different types choose the same strategy (i.e., they pool into the same strategy). Following with the above example, the incumbent would set the same price regardless of its production cost.

3. *Partially separating strategy.* If action $s_i(\theta_i)$ satisfies $s_i(\theta_i) \neq s_i(\theta_i')$ for some pairs of types θ_i and θ_i', but $s_i(\theta_i'') = s_i(\theta_i''')$ for other pairs of types, we say that player i uses a "partially separating" strategy. In our ongoing example, the incumbent firm in this type of strategy would choose, for instance, price p when its production cost is high and medium, $p = s_i(H) = s_i(M)$, but a different price p' when its cost is low, $p' = s_i(L)$, where $p' \neq p$.

4. **Second mover (receiver).** Player j observes action s_i, but does not know player i's type, θ_i. We assume, however, that every player knows the prior probability distribution over types, $\mu(\theta_i)$ for every type $\theta_i \in \Theta_i$. This probability distribution is well behaved, meaning that it satisfies:

   a. $\mu(\theta_i) \in [0, 1]$, that is, the probability of player i's type being θ_i is a number between zero and one.

   b. $\sum_{\theta_i \in \Theta_i} \mu(\theta_i) = 1$ if player i's type space is discrete, e.g., $\Theta_i = \{L, H\}$; and, similarly, $\int_{\theta_i \in \Theta_i} \mu(x)dx = 1$ if her type space is continuous. In other words, when we sum (or integrate) across all possible realizations of θ_i, the probability is 1.

   For example, when player i has two types, $\Theta_i = \{L, H\}$, the prior probability can be more compactly expressed as $\mu(L) = q$ and $\mu(H) = 1 - q$, where $q \in [0, 1]$. When player i's types are uniformly distributed over $[0, 1]$, then $\Pr\{\theta \leq \theta_i\} = F(\theta_i) = \theta_i$, with corresponding density $f(\theta_i) = 1$.

3. **Second mover updates her beliefs.** Upon observing action s_i, player j updates her beliefs about player i's type being θ_i, which we write as $\mu(\theta_i|s_i)$. These beliefs are often known as "posterior beliefs" as opposed to the prior probability distribution, $\mu(\theta_i)$, which is referred to as "prior beliefs." For compactness, we will generally use "priors" and "posteriors," or "prior probability" and "beliefs." Like priors, posteriors satisfy $\mu_j(\theta_i|s_i) \in [0, 1]$ and $\sum_{\theta_i \in \Theta_i} \mu_j(\theta_i|s_i) = 1$ when types are discrete and $\int_{\theta_i \in \Theta_i} \mu_j(x|s_i)dx = 1$ when types are continuous. As we

discuss below, the PBE solution concept requires players to update their beliefs according to Bayes' rule, as follows,

$$\mu_j(\theta_i|s_i) = \frac{\mu(\theta_i)\Pr(s_i|\theta_i)}{\Pr(s_i)}$$

Intuitively, the conditional probability that, upon observing action s_i, player i's type is θ_i, as captured by $\mu_j(\theta_i|s_i)$, is equal to the (unconditional) probability that player i's type is θ_i and she chooses action s_i, $\mu(\theta_i)\Pr(s_i|\theta_i)$, given that she chose s_i, $\Pr(s_i)$. When $\mu_j(\theta_i|s_i)$ is higher (lower) than the prior probability $\mu(\theta_i)$, the observation of s_i increases (decreases) the chances that player i's type is θ_i, indicating that action s_i provided information that player j did not have in prior probability $\mu(\theta_i)$

In the extreme case that $\mu_j(\theta_i|s_i) = 1$, player j becomes "convinced" that player i's type is θ_i after observing player i choosing s_i. Similarly, if $\mu_j(\theta_i|s_i) = 0$, player j is sure that player i's type cannot be θ_i. In contrast, if $\mu_j(\theta_i|s_i) = \mu(\theta_i)$, player j does not change her beliefs upon observing s_i, meaning that s_i did not provide her with additional information. In this case, we say that the signal was uninformative.

4. **Second mover responds.** After player j updates her beliefs to $\mu_j(\theta_i|s_i)$, she responds by choosing an action s_j. Player j may have observed her type, θ_j, right before choosing her action s_j. In most settings, player j's type is publicly observable, becoming common knowledge. Alternatively, player j may privately observe her type, $\theta_j \in \Theta_j$, as player i did.[1]

After concluding the above five steps, most of the sequential-move games with incomplete information that we consider in this chapter will be over, distributing payoffs to each player. However, we can continue the game, allowing, for instance, player i to observe player j's action s_j in step 5, update her beliefs about j's type being θ_j, $\mu_i(\theta_j|s_j)$, and respond with an action a_i, where we use a_i to distinguish it from s_i. (If player j does not have privately observed types, then player i could just respond with action a_i, without having to update her beliefs about θ_j.)

For illustration purposes, we next present two examples that we will explore in this chapter.

Example 10.1. Two sequential-move games of incomplete information. Consider the game tree in Fig. 10.1, where player 1 privately observes whether a business opportunity is beneficial for player 2 (in the upper part of the game tree, which happens with probability p) or futile (bottom part of the tree, with probability $1 - p$), where $p \in [0, 1]$. In this setting, we say that player 1's types are *Beneficial* or *Futile*. If player 1 does not make an investment offer to player 2, N (moving

---

[1] For simplicity, we assume that players' types are independently distributed (no correlation), which entails that, when a player observes her type, she cannot update the prior probability of her opponent's type. In other words, every player updates her beliefs about her rival's type based on her rival's actions, but not based on the player's privately observed type.

10.2 Sequential-Move Games of Incomplete Information—Notation

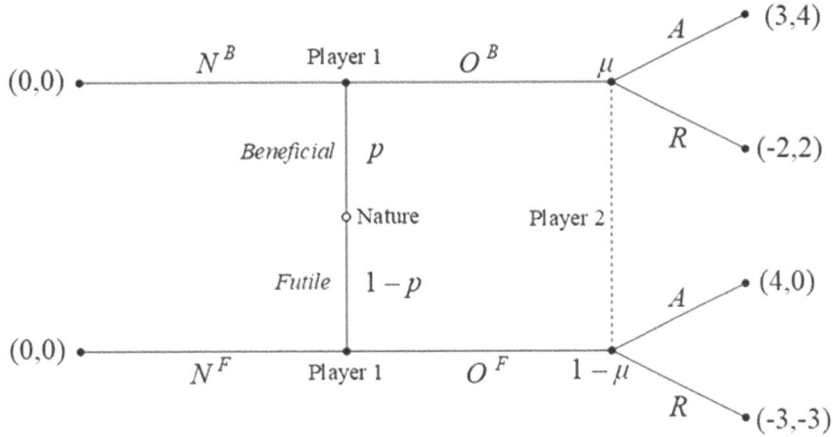

Fig. 10.1 Sequential-move game with incomplete information—One information set

leftward), the game is over and both players earn a payoff of zero. If, instead, player 1 makes an offer to player 2, O (moving rightward), player 2 receives this offer but does not observe whether it is beneficial (at the upper part of the tree) or not (at the bottom part). However, upon receiving the offer, player 2 must respond by accepting (A) or rejecting it (R).

The game in Fig. 10.1, then, has only one information set where the receiver (player 2) is called to move. Figure 10.2, however, depicts the game in which a receiver is called to respond after both messages from a sender, thus yielding two information sets. In particular, the game represents Spence's (1973) labor market signaling game, where a worker (sender) privately observes her job productivity (her type), which is either high or low, and then decides whether to pursue an advanced education degree (such as a Master's program) that she might use as a signal about her productivity level to the employer (receiver). Upon observing whether or not the applicant acquired education, but without observing her true productivity, the employer updates her beliefs about the worker's type, and responds by hiring her as a manager (M) or as a cashier (C).

Observing the worker's payoffs, one can infer that the cost of acquiring education is only \$4 for the high-productivity worker. Indeed, when she is hired as a manager, at the top of the figure, the difference in her payoff when she avoids or acquires education is $10 - 6 = 4$. (Compare M', where her payoff is 10, and M, where her payoff is only 6.) A similar argument applies when she is hired as a cashier, where the payoff difference is $4 - 0 = 4$ (in this case, compare C', where her payoff if 4, against C, where her payoff is 0).

However, the cost of acquiring education increases to \$7 for the low-productivity worker. In this case, when she is hired as a manager, her payoff difference is $10 - 3 = 7$ (comparing her payoff at M' and M). Similarly, when she

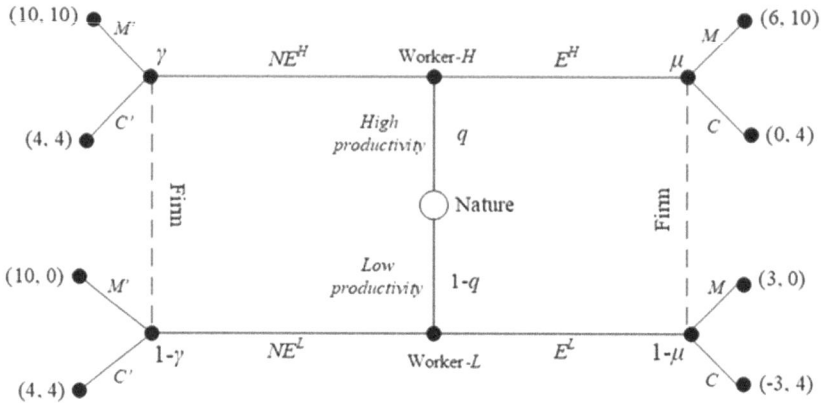

Fig. 10.2 Labor market signaling game

is hired as a cashier, the payoff difference is $4 - (-3) = 7$ (where we compare C' and C).

The firm's payoffs, in the second component of each payoff pair, are affected by the worker's type. Specifically, the firm earns $4 when the worker is hired as a C, regardless of her type, $10 when the worker is hired as an M and her productivity happens to be high but $0 otherwise. The firm's payoffs are, however, unaffected by whether the worker acquires education or not. In other words, education is *not productivity-enhancing*. This is, of course, a simplification from real-life education programs, so we can focus on the possibility that education becomes a signal to potential employers, facilitating information transmission from workers to firms. (We relax this assumption in Exercises 10.2 and 10.3, where we allow for education to increase the worker's productivity.)

10.3 BNE Prescribing Sequentially Irrational Behavior

What if we apply the BNE solution concept to find equilibrium behavior when players interact sequentially and operate under incomplete information? As the next example shows, the BNE may identify sequentially irrational behavior, that is, upon reaching a node or information set, a player acts in a way that does not maximize her payoff. In this sense, the BNE is analogous to the NE solution concept: we can apply it to both simultaneous- and sequential-move games, but when applying it to the latter we may identify "too many" equilibria, some of them prescribing sequentially irrational behavior. For this reason, in the rest of the chapter, we adapt the BNE so it requires sequentially rationality.

10.3 BNE Prescribing Sequentially Irrational Behavior

Example 10.2. Applying BNE to sequential-move games of incomplete information. Consider the game in Fig. 10.1 again. To find the BNE, we first need to represent its Bayesian normal form, following the same steps as in Chapter 8. We use this representation in Matrix 10.1, where each cell includes expected payoffs for each player. For instance, at $(O^B O^F, A)$, at the top left corner of the matrix, player 1 makes an offer to player 2 regardless of whether the investment is beneficial or not, and player 2 accepts the offer, yielding an expected payoff of $EU_1 = p3 + (1-p)4 = 4 - p$ for player 1 and $EU_2 = p4 + (1-p)0 = 4p$ for player 2. (You may practice with the expected payoffs in other cells to confirm you obtain the same result).

		Player 2	
		A	R
	$O^B O^F$	$4-p, \underline{4p}$	$-3+p, -3+5p$
	$O^B N^F$	$3p, \underline{4p}$	$-2p, 2p$
Player 1	$N^B O^F$	$\underline{4(1-p)}, \underline{0}$	$-3(1-p), -3(1-p)$
	$N^B N^F$	$0, \underline{0}$	$\underline{0}, \underline{0}$

Matrix 10.1. Bayesian normal-form representation of the game tree in Fig. 10.1.

Underlining expected payoffs, we find two BNEs in this game:

$$\left(O^B O^F, A\right) \text{ and } \left(N^B N^F, R\right).$$

Intuitively, the first BNE says that player 1 makes an offer regardless of whether the investment is beneficial for player 2 or not, and player 2 responds by accepting it. This occurs because, in expectation, accepting the offer yields a higher payoff than rejecting it, $4p > -3 + 5p$, which simplifies to $3 > p$; a condition that holds for all probabilities $p \in [0, 1]$.

The second BNE, however, indicates that player 1 never makes an offer (regardless of her type) and, yet, player 2 would respond by rejecting it if she ever receives one. This behavior is sequentially irrational: if player 2 receives an offer, at the only information set where she is called to move (see right-hand side of Fig. 10.1), she earns a higher expected payoff accepting than rejecting it, which holds *regardless* of the belief, μ, that she sustains about whether the investment is beneficial or not. Indeed, player 2's expected payoffs from accepting and rejecting are

$$EU_A = 4\mu + 0(1-\mu) = 4\mu, \text{ and}$$
$$EU_R = 2\mu + (-3)(1-\mu) = -3 + 5\mu$$

which satisfy $EU_A > EU_R$ since $4\mu > -3 + 5\mu$ simplifies to $3 > \mu$, which holds for all $\mu \in [0, 1]$. In other words, player 2 responds by accepting the offer regardless of her belief about the investment being beneficial, μ, which contradicts the second BNE, $(N^B N^F, R)$, where player 2 rejects the offer with certainty. As a consequence, we claim that the second BNE is sequentially irrational: once an offer is received, it is sequentially rational to accept it.

10.4 Perfect Bayesian Equilibrium—Definition

As shown in Sect. 10.3, the application of BNE can lead to a large number of equilibria in sequential-move games of incomplete information, some of them prescribing sequentially irrational outcomes. This problem is similar to that of applying the NE solution concept to sequential-move games of complete information. In that context, Chapter 6 developed a new solution concept, SPE, which requires every player to choose optimal actions at every node (or, more generally, at any information set) where they are called to move throughout the game tree. We will now impose a similar requirement, but in an incomplete information environment.

Simplifying our notation. As in our discussion about equilibrium strategies in BNE, let $s_i^*(\theta_i)$ denote player i's equilibrium strategy when her type is θ_i (which could be privately or publicly observed).[2] Similarly, let

$$s_{-i}^*(\theta_{-i}) \equiv \left(s_1^*(\theta_1), ..., s_{i-1}^*(\theta_{i-1}), s_{i+1}^*(\theta_{i+1}), ..., s_N^*(\theta_N) \right)$$

represent the strategy profile of player i's rivals, given their types, so an equilibrium strategy profile can be compactly expressed as $s^* \equiv \left(s_i^*(\theta_i), s_{-i}^*(\theta_{-i}) \right)$. In most applications, where we only consider two players (the first and second mover), this strategy profile simplifies to $s^* = \left(s_i^*(\theta_i), s_j^*(\theta_j) \right)$ where $j \neq i$.

We can also provide a compact representation of players' beliefs. First, let μ_j denote the list of player j's updated beliefs $\mu_j(\theta_i | s_i)$ for every type $\theta_i \in \Theta_i$ that player i may have and for every action $s_i \in S_i$ that she may take. Take, for instance, the simplest scenario where player i's types and actions are binary, $\Theta_i = \{H, L\}$ and $S_i = \{s_i^A, s_i^B\}$, i.e., her type is either high or low, and she can only choose between A and B. Player j's "system of beliefs" in this case has four components, as follows,

$$\mu_j = (\mu_j(L|s_i^A), \mu_j(H|s_i^A), \mu_j(L|s_i^B), \mu_j(H|s_i^B))$$

This vector can be alternatively represented by the first and third components alone because the second term is a function of the first term, $\mu_j(H|s_i^A) = 1 - \mu_j(L|s_i^A)$, and, similarly, the fourth term is function of the third, $\mu_j(H|s_i^B) = 1 - \mu_j(L|s_i^B)$.

---

[2] Technically, player i's strategy should be a function of her type, θ_i, and the history of play at the node she is called to move, h_i, so that $s_i(\theta_i, h_i)$. For compactness, we use $s_i(\theta_i)$ unless necessary to avoid misunderstandings.

10.4 Perfect Bayesian Equilibrium—Definition

More generally, player j's system of beliefs has $(\text{card}(\Theta_i) - 1) \times \text{card}(S_i)$ independent components, where "card" denotes the cardinality of the strategy set (i.e., the number of different strategies for that player). For example, if player i has three types, $\Theta_i = \{L, M, H\}$, and three possible actions, $S_i = \{s_i^A, s_i^B, s_i^C\}$, player j's system of beliefs has $2 \times 3 = 6$ components.[3] We can apply a similar definition to the system of beliefs by each of j's rivals, that is,

$$\mu_{-j} \equiv (\mu_1, \ldots, \mu_{j-1}, \mu_{j+1}, \ldots, \mu_N)$$

so a system of beliefs can be compactly expressed as $\mu \equiv (\mu_j, \mu_{-j})$ for all players.

We are now ready to use strategy profile $s^* \equiv (s_i^*(\theta_i), s_{-i}^*(\theta_{-i}))$ and system of beliefs μ to define a Perfect Bayesian Equilibrium.

Definition 10.1. **Perfect Bayesian Equilibrium (PBE).** A strategy profile s^* and a system of beliefs μ over all information sets is a Perfect Bayesian Equilibrium if:

(a) Every player i's strategies specify optimal actions at each information set where she is called to move, given the strategies of the other players, and given player i's system of beliefs μ_i; and
(b) Beliefs μ are consistent with Bayes' rule whenever possible.

Intuitively, condition (a) means that every player i chooses best responses to her rivals' strategies in an incomplete information environment, that is, given her beliefs about her rivals' types at that point of the game tree.

Condition (b) states that every player's beliefs must be consistent with Bayes' rule "whenever possible." To understand this requirement, first note that applying Bayes' rule is only possible along the equilibrium path, that is, at information sets that are reached in equilibrium. In contrast, when a player is called to move at an information set that is not reached in equilibrium, she cannot use Bayes' rule to update her beliefs of her rivals' types. If, she did, she would obtain an indeterminate result, $\mu_j(\theta_i|s_i) = \frac{0}{0}$. We refer to these beliefs as "off-the-equilibrium" beliefs and, graphically, recall that they are associated with information sets that are not reached in the strategy profile we examine.

Because we cannot use Bayes' rule to update off-the-equilibrium, we can specify any arbitrary value to them, i.e., $\mu_j(\theta_i|s_i) \in [0, 1]$. (Other solution concepts also impose consistency requirements on off-the-equilibrium beliefs, implying that we do not need the "whenever possible" clause. This is the case, for instance, of Sequential Equilibrium, which we examine in Chapter 11.)

We next illustrate how to use Bayes' rule to update beliefs in each of the strategy profiles described in the previous section. For simplicity, we consider the labor market signaling game of Fig. 10.2, as that allows for more concrete results.

---

[3] Technically, it has $3 \times 3 = 9$ components, but $\mu_j(H|s_i^k)$ is identified by $\mu_j(H|s_i^k) = 1 - \mu_j(L|s_i^k) - \mu_j(M|s_i^k)$ for every $k = \{A, B, C\}$, reducing the number of components to only 6.

Example 10.3. Applying Bayes' rule. In the labor market signaling game, the sender (job applicant) has one of two types describing her productivity level, $\Theta = \{L, H\}$, and chooses whether to acquire education, E, or not, NE. For compactness, we use E^H (NE^H) to represent that the high-productivity worker acquires (does not acquire) education; and a similar notation applies for the low-productivity worker, where E^L (NE^L, respectively).

1. *Separating strategy,* (E^H, NE^L). Consider that the job applicant acquires education only when her productivity is high. Applying Bayes' rule, we have that, upon observing that the job applicant acquires education, E, the employer's updated belief about the worker's productivity being H is

$$\mu(H|E) = \frac{\mu(H)\Pr(E|H)}{\Pr(E)} = \frac{\mu(H)\Pr(E|H)}{\mu(H)\Pr(E|H) + \mu(L)\Pr(E|L)}$$

$$= \frac{q \times 1}{(q \times 1) + ((1-q) \times 0)} = \frac{q}{q} = 1$$

since prior probabilities are $\mu(H) = q$ and $\mu(L) = 1 - q$ by assumption and, in this separating strategy profile, $\Pr(E|H) = 1$ and $\Pr(E|L) = 0$ because the H-type chooses E with certainty while the L-type never does. Intuitively, $\mu(H|E) = 1$ says that the employer, upon observing that the job applicant acquired education, becomes convinced of facing a H-type since only H-types choose E. In other words, E can only originate from the H-type, because it is the only player type who chooses education.[4] We can also evaluate the employer's posterior beliefs upon observing NE, that is,

$$\mu(H|NE) = \frac{\mu(H)\Pr(NE|H)}{\Pr(NE)} = \frac{\mu(H)\Pr(NE|H)}{\mu(H)\Pr(NE|H) + \mu(L)\Pr(NE|L)}$$

$$= \frac{q \times 0}{(q \times 0) + ((1-q) \times 1)} = \frac{0}{1-q} = 0$$

which means that NE can only originate from the L-type of sender, so observing NE helps the employer infer that she must face an L-type worker. In short, $\mu(H|NE) = 0$ and $\mu(H|E) = 1$.

2. *Separating strategy,* (NE^H, E^L). The opposite separating strategy profile, where only the low-productivity worker acquires education, yields, of course, the opposite results. Indeed, upon observing that the job applicant chooses NE, the employer's updated beliefs are

$$\mu(H|NE) = \frac{\mu(H)\Pr(NE|H)}{\Pr(NE)} = \frac{\mu(H)\Pr(NE|H)}{\mu(H)\Pr(NE|H) + \mu(L)\Pr(NE|L)}$$

---

[4] Needless to say, if $\mu(H|E) = 1$, then $\mu(L|E) = 0$ since $\mu(H|s_i) + \mu(L|s_i) = 1$ for every given strategy s_i.

10.4 Perfect Bayesian Equilibrium—Definition

$$= \frac{q \times 1}{(q \times 1) + ((1-q) \times 0)} = \frac{q}{q} = 1$$

meaning that *NE* cannot originate from the *L*-type of worker, stemming instead from the *H*-type worker, that is, $\mu(H|NE) = 1$ and $\mu(H|E) = 0$.

3. *Pooling strategy,* (E^H, E^L). In a pooling strategy profile in which the job applicant chooses *E* regardless of her type entails that the receiver's beliefs are unaffected after applying Bayes' rule because

$$\mu(H|E) = \frac{\mu(H)\Pr(E|H)}{\Pr(E)} = \frac{\mu(H)\Pr(E|H)}{\mu(H)\Pr(E|H) + \mu(L)\Pr(E|L)}$$

$$= \frac{q \times 1}{(q \times 1) + ((1-q) \times 1)} = \frac{q}{q + (1-q)} = q,$$

and since $\mu(H) = q$, we obtain that the employer's posterior and prior beliefs coincide, that is,

$$\mu(H|E) = \mu(H) = q, \text{ and}$$
$$\mu(L|E) = \mu(L) = 1 - q.$$

In other words, because all types of senders (job applicant) choose the same message (*E* in this strategy profile), the receiver (employer) cannot infer or extract any additional information she did not originally have. This is a useful property to remember when checking whether a pooling strategy profile can be sustained as a PBE. What if, instead, the employer observes the job applicant choosing *NE*? In this strategy profile, *NE* should not happen in equilibrium since every worker type is choosing *E* (what a surprise!). As explained above, we refer to this type of message as "off the equilibrium" to emphasize it should not occur in the strategy profile we analyze, (E^H, E^L). If, in this setting, the receiver (employer) tries to use Bayes' rule to update her beliefs, she finds an indetermination because

$$\mu(H|NE) = \frac{\mu(H)\Pr(NE|H)}{\Pr(NE)} = \frac{\mu(H)\Pr(NE|H)}{\mu(H)\Pr(NE|H) + \mu(L)\Pr(NE|L)}$$

$$= \frac{q \times 0}{(q \times 0) + ((1-q) \times 0)} = \frac{0}{0}.$$

In this case, off-the-equilibrium beliefs (since *NE* is an off-the-equilibrium strategy) can take any arbitrary number, that is, $\mu(H|NE) \in [0, 1]$. A similar argument applies to $\mu(L|NE)$ since, again, *NE* should have not been observed in this strategy profile.

4. *Pooling strategy,* (NE^H, NE^L). A similar reasoning applies to the pooling strategy profile where the job applicant does not acquire education, *NE*, regardless of her type, that is,

$$\mu(H|NE) = \frac{\mu(H)\Pr(NE|H)}{\Pr(NE)} = \frac{\mu(H)\Pr(NE|H)}{\mu(H)\Pr(NE|H) + \mu(L)\Pr(NE|L)}$$

$$= \frac{q \times 1}{(q \times 1) + ((1-q) \times 1)} = \frac{q}{q + (1-q)} = q,$$

meaning that posteriors and priors coincide, $\mu(H|NE) = \mu(H) = q$, which implies that $\mu(L|NE) = \mu(L) = 1 - q$. Following the same argument as in the pooling strategy profile (E^H, E^L), we can confirm that Bayes' rule does not help us pin down off-the-equilibrium beliefs in this case either since

$$\mu(H|E) = \frac{\mu(H)\Pr(E|H)}{\Pr(E)} = \frac{\mu(H)\Pr(E|H)}{\mu(H)\Pr(E|H) + \mu(L)\Pr(E|L)}$$

$$= \frac{q \times 0}{(q \times 0) + ((1-q) \times 0)} = \frac{0}{0}$$

and we say that off-the-equilibrium beliefs can take any arbitrary number, $\mu(H|E) \in [0, 1]$.

5. *Semi-separating strategy.* If the sender (job applicant) uses a semi-separating strategy such as E^H with probability 1 and E^L with probability $\sigma \in [0, 1]$, then player j's updated beliefs are

$$\mu(H|E) = \frac{\mu(H)\Pr(E|H)}{\Pr(E)} = \frac{\mu(H)\Pr(E|H)}{\mu(H)\Pr(E|H) + \mu(L)\Pr(E|L)}$$

$$= \frac{q \times 1}{(q \times 1) + ((1-q) \times \sigma)} = \frac{q}{q + (1-q)\sigma}$$

and $\mu(L|E) = 1 - \mu(H|E)$. Interestingly, when the L-type never plays E, $\sigma = 0$, the above expression simplifies to $\mu(H|E) = 1$, as in the separating strategy profile (E^H, NE^L). In contrast, when she plays E with certainty, $\sigma = 1$, the employer's posterior belief becomes $\mu(H|E) = q$, as in the pooling strategy profile (E^H, E^L). More generally, $\mu(H|E)$ decreases in the probability with which the L-type plays E, as captured by σ, starting at $\mu(H|E) = 1$ when $\sigma = 0$ and reaching $\mu(H|E) = q$ when $\sigma = 1$. Informally, the employer is more convinced of facing an H-type worker when the L-type rarely acquires education than when she often does.

10.5 A Tool to Find PBEs in Signaling Games

In this section, we take a practical approach to find PBEs in signaling games. First, we present a tool to check if a strategy profile for the sender, such as separating strategy (E^H, NE^L) or pooling strategy (E^H, E^L), can be supported as a PBE of the game. Sections 10.6 and 10.7, then, apply this tool to two game trees. For presentation purposes, we start with the game tree in Fig. 10.1, where the receiver only gets to play in one information set. Then, we continue with the game presented in Fig. 10.2, where she may be called to move in two information sets.

The following five-step tool can help us test if a specific strategy profile satisfies the definition of PBE presented in Sect. 10.4.

Tool 10.1. Finding PBEs in signaling games.

1. Specify a strategy profile for the sender, such as (E^H, NE^L) in the labor market signaling game.
2. Update the beliefs of the receiver using Bayes' rule, whenever possible.
3. Given the receiver's updated beliefs in Step 2, find her optimal response.
4. Given the receiver's response in Step 3, find the sender's optimal message.
5. If the sender's optimal message found in Step 4:
   a. Coincides with that initially postulated in Step 1, we have confirmed that the strategy profile listed in Step 1, along with the receiver's updated beliefs in Step 2 and response in Step 3, can be sustained as a PBE.
   b. Does not coincide with that initially postulated in Step 1, we can claim that the strategy profile in Step 1 cannot be supported as a PBE.

10.6 Finding PBEs in Games with one Information Set

Consider the game tree in Fig. 10.1 again. We will first check if separating strategy profile (O^B, N^F), where the sender only makes an offer when the investment is beneficial, can be supported as a PBE. We, then, extend a similar approach to test if the pooling strategy profile (O^B, O^F), where the sender makes an offer regardless of whether it is beneficial, can be sustained as a PBE. Testing whether the opposite separating strategy profile, (N^B, O^F) can be supported as a PBE, where the sender only makes an offer when the investment is not beneficial, is left as a practice for the reader.[5] A similar argument applies to the pooling strategy profile, (N^B, N^F), where the sender never makes an offer, which is also left to the reader.[6]

For presentation purposes, we refer to the sender as "she" and the receiver as "he."

---

[5] After trying to show that on your own, here is the answer: in separating strategy profile (N^B, O^F), player 2 updates its posterior belief to $\mu = 0$, responding with A. Given this response, player 1 prefers to deviate from N^B to O^B when the test is beneficial, implying that (N^B, O^F) cannot be sustained as a PBE.

[6] After trying to confirm that pooling strategy profile (N^B, N^F) cannot be supported as a PBE, here is the answer: player 2 cannot update its belief using Bayes' rule since the only information set is not reached in equilibrium, i.e., $\mu \in [0, 1]$. Therefore, player 2 responds with A if and only if $4\mu \geq 2\mu - 3(1 - \mu)$, which simplifies to $1 \geq \mu$, which holds for all values of admissible beliefs. Given this response, both types of player 1 prefer to deviate, implying that (N^B, N^F) cannot be sustained as a PBE.

10.6.1 Separating Strategy Profile (O^B, N^F)

1. *Specifying a strategy profile.* Applying the tool presented in Sect. 10.5, step 1 specifies a "candidate" of strategy profile for the sender that we seek to test as a PBE, (O^B, N^F). For easier reference, Fig. 10.3 highlights the tree branch corresponding to O^B, at the top right, and that corresponding to N^F, at the bottom left.
2. *Bayes' rule.* Step 2 is straightforward in this case since, as described in Example 10.3, $\mu = 1$, indicating that, if player 2 receives an offer in this separating strategy profile, he infers that the investment must be beneficial. Graphically, if he is called to move at the only information set of the game tree, he puts full probability weight on being at the top node of this information set. Informally, the receiver focuses his attention on the top right corner of the tree.
3. *Optimal response.* Given our result from Step 2, $\mu = 1$, the receiver responds by accepting the offer, A, since $4 > 2$. (Note that we only compare his payoffs at the top right corner of the tree, as he puts full probability weight on the top node.) To keep track of our results, we shade the branch corresponding to A, as we do in Fig. 10.3a. Note that A is shaded both at the top and bottom node since the receiver cannot condition his response on the sender's true type, which he does not observe.
4. *Optimal messages.* From our results in Step 3, we now need to identify the sender's optimal message, which needs to be separately done for each of the sender's types.
   a. When the investment is beneficial (at the top of the figure), choosing O^B, as prescribed by this strategy profile, yields 3, since the sender anticipates that the offer will be accepted by the receiver, as found in step 3. Graphically, we only need to follow the shaded branches. If, instead, the sender deviates

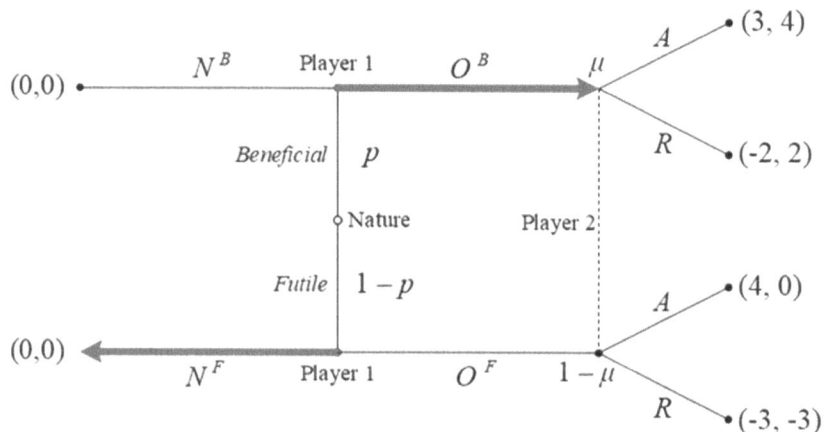

Fig. 10.3 Separating strategy profile (O^B, N^F)

10.6 Finding PBEs in Games with one Information Set

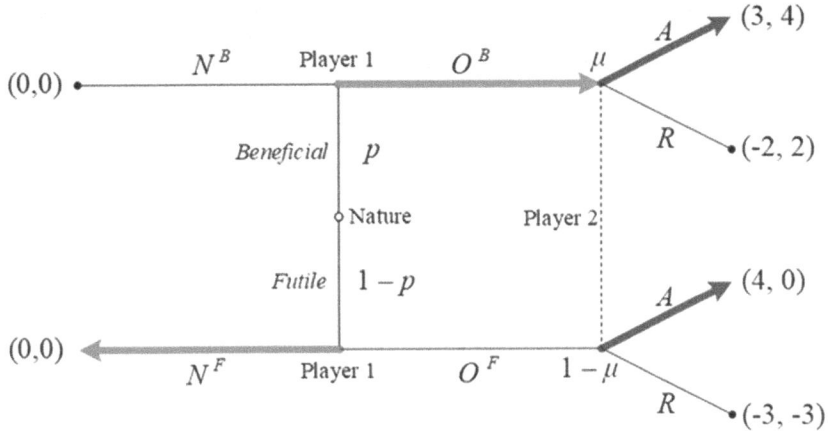

Fig. 10.3a Separating strategy profile (O^B, N^F)—Responses

toward N^B (top left in the figure), her payoff decreases to 0. Therefore, the sender chooses O^B when the investment is beneficial.

b. When the investment is futile (at the bottom of the figure), choosing N^F, as prescribed by this strategy profile, yields 0. Deviating toward O^F (bottom right hand of the figure) increases her payoff to 4, since she anticipates the offer is accepted.

5. *Summary.* From Step 4b, we found that at least one of the sender types has incentives to deviate from (O^B, N^F), namely, when the business opportunity is futile, N^F cannot be supported as optimal for the sender because of her incentives to deviate to O^F. In summary, the separating strategy profile (O^B, N^F) cannot be sustained as a PBE.

Had we found that all sender types behave as prescribed in Step 1—making an offer when the test is beneficial, O^B, but not making it otherwise, N^F—we would be able to claim that (O^B, N^F) can be sustained as a PBE.

10.6.2 Pooling Strategy Profile (O^B, O^F)

1. *Specifying a strategy profile.* We start specifying the strategy profile "candidate" that we test as a PBE, (O^B, O^F). As in our above analysis of the separating strategy profile, Fig. 10.4 shades the branches corresponding to O^B, at the top right, and O^F, at the bottom right, which will help our subsequent discussions.
2. *Bayes' rule.* In Step 2, we update the receiver's beliefs. As described in Sect. 10.4, posterior and prior beliefs coincide in this strategy profile, entailing that $\mu = p$. Intuitively, upon receiving an offer, the receiver cannot infer

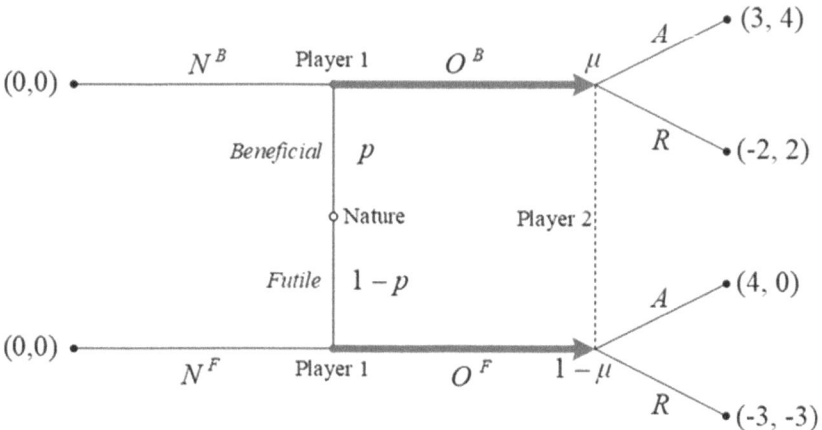

Fig. 10.4a Pooling strategy profile (O^B, O^F) —Responses

any information from this offer, as all sender types make offers, being left as uninformed as he was at the beginning of the game.

3. *Optimal response.* Given our result from Step 2, $\mu = p$, the receiver responds by accepting the offer, A, since his expected payoff satisfies

$$EU_A = 4p + 0(1-p) > 2p + (-3)(1-p) = EU_R,$$

which simplifies to

$$EU_A = 4p > -3 + 5p = EU_R$$

which ultimately reduces to $3 > p$, which holds for all $p \in [0, 1]$. In other words, player 2 responds by accepting the offer regardless of the probability that the investment is beneficial, p. To keep track of our results, we shade the branch corresponding to A, as we do in Fig. 10.4a.

4. *Optimal messages.* From our results in Step 3, we now identify the sender's optimal message, separately analyzing each sender's types.
    a. When the investment is beneficial (at the top of the figure), choosing O^B, as prescribed by this strategy profile, yields 3, whereas deviating to N^B (top left in the figure), decreases her payoff to 0. Therefore, the sender chooses O^B.
    b. When the investment is not beneficial (at the bottom of the figure), choosing O^F, as prescribed by this strategy profile, yields 4, while deviating to N^F would decrease her payoff to 0. Then, the sender chooses O^F.

5. *Summary.* From Step 4b, we found that all sender types prefer to behave as prescribed in Step 1, (O^B, O^F), implying that this pooling strategy profile can be supported as a PBE, with the receiver holding beliefs $\mu = p$ and responding accepting the offer.

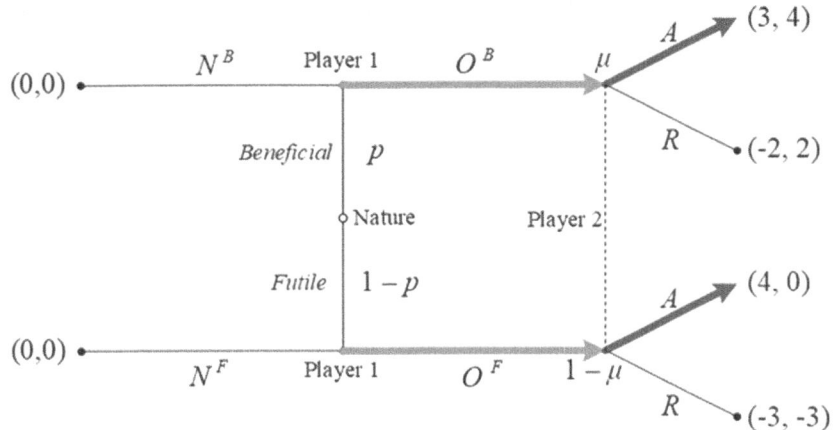

Fig. 10.4 Pooling strategy profile (O^B, O^F)

10.7 Finding PBEs in Games with Two Information Sets

We now apply the above 5-step tool to identify separating and pooling equilibria in Spence's labor market signaling game. While the procedure is very similar to that in the previous section, the presence of two information sets gives rise to an off-the-equilibrium information set in pooling strategy profiles and their associated off-the-equilibrium beliefs.

For simplicity, we consider that the probability of high-productivity workers is $q = 1/3$,[7] and start checking separating strategy profiles, such as (E^H, NE^L), where only the high-productivity worker acquires education. We then move on to pooling strategy profiles, such as (NE^H, NE^L), where no worker type acquires education.

10.7.1 Separating Strategy Profile (E^H, NE^L)

1. *Specifying a strategy profile.* We first specify the separating strategy profile that we seek to test as a PBE, (E^H, NE^L). As in our above analysis, Fig. 10.5 shades the branches corresponding to E^H, at the top right, and NE^L, at the bottom left, which will help our subsequent discussion.

---

[7] This assumption does not affect equilibrium behavior in the separating PBE. However, in pooling strategy profiles, this assumption decreases the worker's expected productivity and, thus, makes the firm more likely to respond by hiring her as a cashier. If, instead, $q > 1/2$, the worker's average type would be relatively high, inducing the firm to hire her as a manager, ultimately facilitating the emergence of pooling PBEs. (You may check this point, as a practice, at the end of Sect. 10.7.2.)

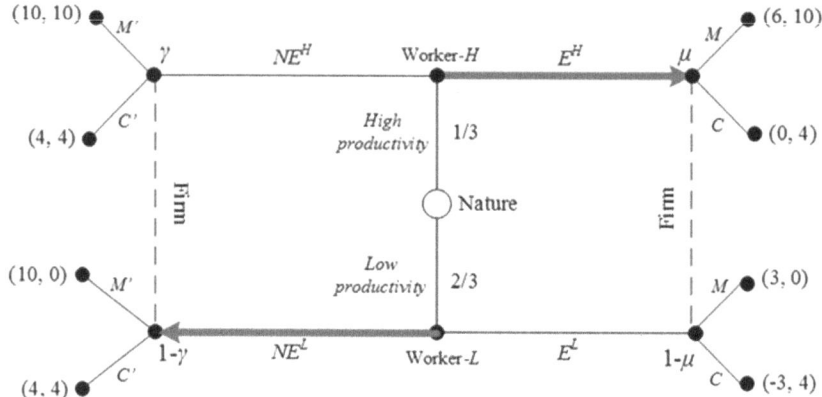

Fig. 10.5 Separating strategy profile (E^H, NE^L)

2. *Bayes' rule.* We can now update the firm's beliefs. As described in Sect. 10.4, upon observing education, beliefs in this strategy profile satisfy

$$\mu = \frac{\frac{1}{3}\alpha^H}{\frac{1}{3}\alpha^H + \frac{2}{3}\alpha^L} = \frac{\frac{1}{3}1}{\frac{1}{3}1 + \frac{2}{3}0} = 1$$

where α^H (α^L) denotes the probability that a high-productivity (low-productivity) worker acquires education. Intuitively, upon observing education, the firm believes it must only originate from a high-productivity worker, which corresponds to the top right node of the game tree. Upon observing no education, however, the firm's beliefs are

$$\gamma = \frac{\frac{1}{3}(1-\alpha^H)}{\frac{1}{3}(1-\alpha^H) + \frac{2}{3}(1-\alpha^L)} = \frac{\frac{1}{3}0}{\frac{1}{3}0 + \frac{2}{3}1} = 0$$

or, alternatively, the probability of being at the bottom left node, $1-\gamma$, is 100 percent. Hence, if the firm observes no education, it assigns full probability of facing a low-productivity worker.

3. *Optimal responses.* Given our results from Step 2, we now analyze the firm's responses upon observing each of the two possible messages (education or no education).

   a. Upon observing education, the firm responds by hiring the worker as a manager, M, since $10 > 4$ at the top right side of the game tree. (Recall that the firm believes, after observing education, that it deals with a high-productivity worker.)

   b. If, instead, the firm observes no education, it responds by hiring the worker as a cashier, C', because $4 > 0$ at the bottom left corner of the tree.

10.7 Finding PBEs in Games with Two Information Sets

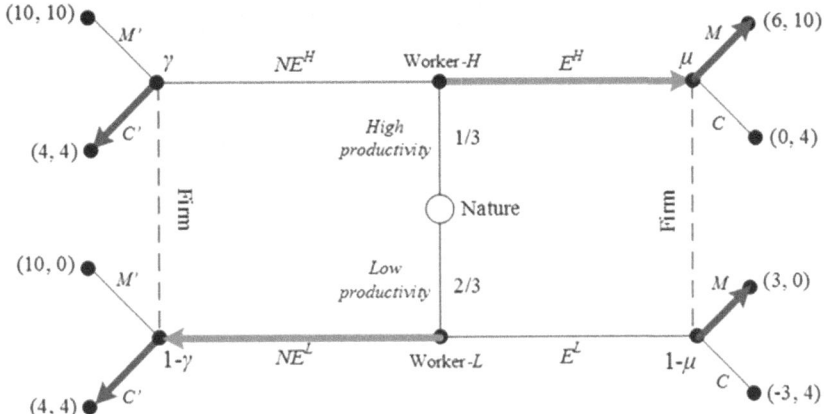

Fig. 10.5a Separating strategy profile E^H, NE^L—Responses

To keep track of our results, Fig. 10.5a shades the branches corresponding to M on the right side of the tree, and those corresponding to C' on the left side.

4. *Optimal messages.* From our results in Step 3, we now identify the worker's optimal message, separately analyzing each type.

   a. *High productivity.* At the top of the game tree, the high-productivity worker chooses E^H, moving rightward, instead of deviating to NE^H on the left side of the tree, since $6 > 4$. Intuitively, this worker type anticipates that education will be recognized as a signal of her high productivity, inducing the firm to respond by hiring her as a manager, as indicated by the shaded branches originating from the central node at the top of the figure and moving rightward. If this worker deviates to NE^H, she would save the education costs, but would be identified as a low-productivity type, hired as a cashier, and earning only 4. As a consequence, the cost of acquiring education (relatively low for this worker type) is offset by the wage gain that she experiences when she is hired as a manager.

   b. *Low productivity.* At the bottom of the game tree, the low-productivity worker chooses NE^L, moving leftward, which yields 4, instead of deviating to acquire education, on the right side of the tree, which would only yield 3. Intuitively, acquiring education helps her "fool" the firm into believing that she is a high-productivity worker and hiring her as a manager. Her wage gain, however, does not offset her cost of acquiring education, which is larger than that of the high-type worker. As a result, the low-productivity worker does not have incentives to mimic the high-productivity type acquiring education.

5. *Summary.* From Steps 4a–4b, we found that all sender types prefer to behave as prescribed in Step 1, (E^H, NE^L), implying that this separating strategy profile can be supported as a PBE, with the firm holding beliefs $\mu = 1$ and

$\gamma = 0$, and responding with (M, C'), i.e., hiring the worker as a manager upon observing education but as a cashier otherwise.

As a practice, you can show that the opposite separating strategy profile, (NE^H, E^L), where only the low-productivity worker acquires education, cannot be sustained as a PBE. The reason is straightforward: in such a setting, the low-productivity worker would invest in education and, despite this cost, she would be recognized as a low-productivity worker and hired as a cashier!

10.7.2 Pooling Strategy Profile (NE^H, NE^L)

1. *Specifying a strategy profile.* We start by specifying the pooling strategy profile (NE^H, NE^L) in which no worker type acquires education, as illustrated in Fig. 10.6.
2. *Bayes' rule.* Upon observing no education, the firm's beliefs are

$$\gamma = \frac{\frac{1}{3}(1-\alpha^H)}{\frac{1}{3}(1-\alpha^H) + \frac{2}{3}(1-\alpha^L)} = \frac{\frac{1}{3}1}{\frac{1}{3}1 + \frac{2}{3}1} = \frac{1}{3}$$

implying that posterior beliefs (γ) coincide with prior beliefs (1/3). In other words, observing that the worker did not acquire education provides no information about her type to the firm, since in this strategy profile all worker types do not acquire education, i.e., $1 - \alpha^H = 1 - \alpha^L = 1$. If, instead, the firm observes that the worker has education, which should not occur in this strategy

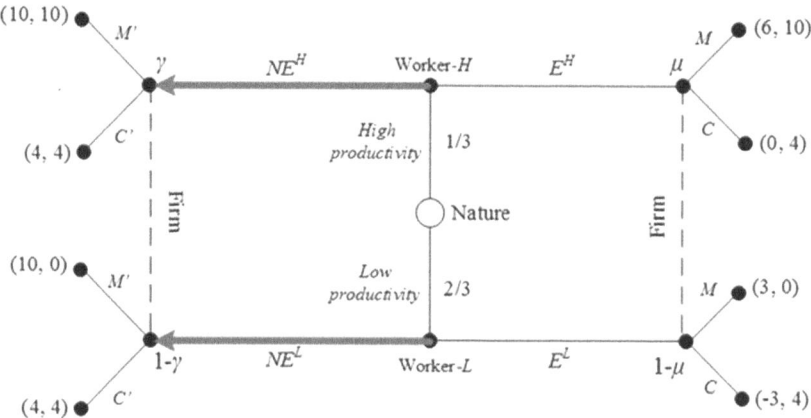

Fig. 10.6 Pooling strategy profile (NE^H, NE^L)

10.7 Finding PBEs in Games with Two Information Sets

profile, its updated beliefs are

$$\mu = \frac{\frac{1}{3}\alpha^H}{\frac{1}{3}\alpha^H + \frac{2}{3}\alpha^L} = \frac{\frac{1}{3}0}{\frac{1}{3}0 + \frac{2}{3}0} = \frac{0}{0}$$

thus being undefined. Graphically, the information set on the right side of the game tree (when the firm observes education) happens off-the-equilibrium path and the firm's beliefs in this information set are known as "off-the-equilibrium beliefs." These beliefs cannot be pinned down using Bayes' rule, as we only found an indetermination. For generality, these beliefs are left unrestricted, so that they can take any admissible value, i.e., $\mu \in [0, 1]$.

3. *Optimal responses.* Given our results from Step 2, we now analyze the firm's responses upon observing each of the two possible messages (education or no education).

   a. Upon observing an applicant with no education, the firm responds by hiring the worker as a manager, M', or as a cashier, C', based on its expected profit on the left side of the tree, as follows.[8]

   $$E\pi_{Firm}(M') = 10\frac{1}{3} + 0\frac{2}{3} = \frac{10}{3} \simeq 3.33, \text{ and}$$

   $$E\pi_{Firm}(C') = 4\frac{1}{3} + 4\frac{2}{3} = 4.$$

   Therefore, the firm responds by hiring the worker as a cashier, C', when she did not acquire education, since $4 > 3.33$.

   b. If, instead, the firm observes education, it responds by hiring the worker as a manager, M, if its expected profit at the right side of the game tree satisfies

   $$E\pi_{Firm}(M) = 10\mu + 0(1-\mu) = 10\mu, \text{ and}$$

   $$E\pi_{Firm}(C) = 4\mu + 4(1-\mu) = 4,$$

   and, comparing these expected profits, we obtain that $10\mu > 4$ holds if $\mu > \frac{2}{5}$. Then, the firm responds to an educated applicant hiring her as a manager, M, when its off-the-equilibrium beliefs satisfy $\mu > \frac{2}{5}$ (i.e., when it assigns a sufficiently high probability weight on facing a high-productivity worker), but responds by hiring the applicant as a cashier, C, otherwise. We then need to divide our following step, where we examine the worker's decisions, into two cases: (1) $\mu > \frac{2}{5}$, where the firm responds by hiring the worker as a manager, M, after observing education; and (2) $\mu \leq \frac{2}{5}$, where the firm responds by hiring her as a cashier, C, after observing education.

4. *Optimal messages.* From our results in Step 3, we now identify the worker's optimal message, separately analyzing cases (1)-(2).

---

[8] In this case, we cannot say that the firm focuses its attention on one of the nodes of the left information set, because its belief γ satisfies $\gamma = 1/3$, instead of being concentrated at 0 or 1.

Case 1: $\mu > \frac{2}{5}$. For easier reference, this case is illustrated in Fig. 10.6a, where the firm responds with C' after observing no education (left side of the tree) and with M after observing education (right side, given that $\mu > \frac{2}{5}$).

(a) *High productivity.* At the top of the game tree, if the high-productivity worker chooses NE^H, moving leftward, she earns 4, as she anticipates that the firm responds with C'. If, instead, the worker deviates to E^H, she earns 6, implying that she does not have incentives to behave as prescribed by this strategy profile.

At this point, we do not need to check if the low-productivity worker has incentives to choose NE^L, since we already found that a worker's type (high productivity) would deviate, implying that (NE^H, NE^L) cannot be sustained as a PBE when $\mu > \frac{2}{5}$. As a practice, however, we examine the low-productivity worker below.

(b) *Low productivity.* At the bottom of the game tree, if the low-productivity worker chooses NE^L, moving leftward, she earns 4, which exceeds her payoff from deviating to E^L, on the right side, 3. Intuitively, the low-productivity worker anticipates that she will be recognized as such when she does not acquire education, and hired as a cashier. Even if she could fool the firm into believing that it deals with a high-productivity worker when she acquires education, and be hired as a manager, the cost of investing in education is too high for this type of worker to undergo such a deviation.

Case 2: $\mu \leq \frac{2}{5}$. This case is illustrated in Fig. 10.6b, where the firm responds with C' after observing no education (left side of the tree) and with C after observing education (right side, given that $\mu \leq \frac{2}{5}$).

a. *High productivity.* At the top of the game tree, if the high-productivity worker chooses NE^H, moving leftward, she earns 4, as she anticipates that the firm responds with C'. If, instead, the worker deviates to E^H, she only earns 0, implying that she does not have incentives to deviate. Intuitively, the high-productivity worker is hired as a cashier regardless of whether she acquires education. It is, then, optimal for her to save on education as it does not affect the firm's response.
b. *Low productivity.* At the bottom of the game tree, if the low-productivity worker chooses NE^L, moving leftward, she earns 4, which exceeds her payoff from deviating to E^L, on the right side, -3. In this case, this type of worker faces similar incentives as the high-type above, as her educational level does not affect the firm's response, leading her to not invest in education.

5. *Summary.* When $\mu > \frac{2}{5}$ holds (Case 1), we found that one sender type (the high-productivity worker) prefers to deviate from the pooling strategy profile, implying that (NE^H, NE^L) cannot be supported as a PBE (see Step 4a). In contrast, when $\mu \leq \frac{2}{5}$ holds, both worker types have incentives to behave as

10.7 Finding PBEs in Games with Two Information Sets

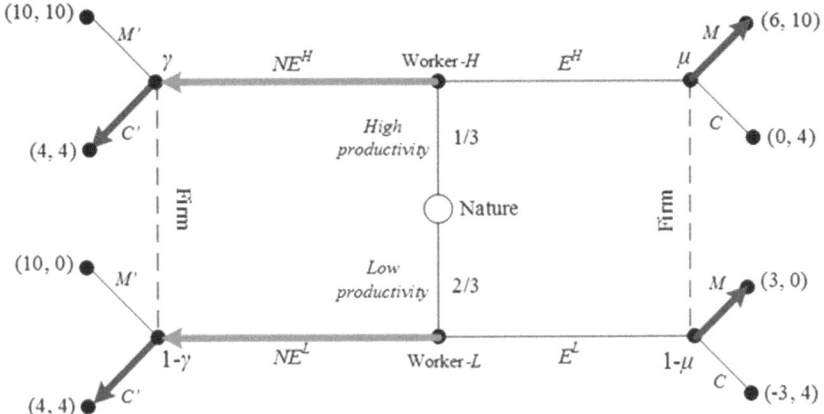

Fig. 10.6a Pooling strategy profile (NE^H, NE^L)—Responses (C', M)

prescribed by pooling strategy profile (NE^H, NE^L), implying that it is sustained as a PBE with the firm responding with (C, C') and with equilibrium beliefs $\gamma = 1/3$ and off-the-equilibrium beliefs $\mu \leq \frac{2}{5}$.

Extra practice. You may show that the opposite pooling strategy profile, ($E^H E^L$), where both worker types acquire education, *cannot* be sustained as a PBE regardless of the firm's off-the-equilibrium belief γ. To see this point, note first that, upon observing education, firm beliefs are $\mu = \frac{1}{3}$, but upon observing no education, its off-the-equilibrium beliefs are left unrestricted, $\gamma \in [0, 1]$. This means that, upon observing education, the firm responds with M, but upon observing no education, it responds with M' if and only if $\gamma > \frac{2}{5}$. When this condition on γ holds, both worker types prefer to deviate from education to no education, as the firm responds to both hiring the worker as a manager. When condition $\gamma > \frac{2}{5}$ does not hold, only the low-productivity worker deviates from education, seeing her payoff increase from 3 to 4. In summary, the pooling strategy profile ($E^H E^L$) cannot be sustained as a PBE under any off-the-equilibrium belief γ.

10.7.3 Insensible Off-the-Equilibrium Beliefs

The pooling strategy profile (NE^H, NE^L), where no worker type acquires education, requires that $\mu \leq \frac{2}{5}$. Intuitively, this means that, upon observing the off-the-equilibrium message of education, the firm assigns a relatively low probability weight on such a message originating from a high type. Informally, this is like saying that in a world where no worker type goes to college, if a firm ever observes a worker with a college degree (surprise!), this firm believes that it is likely facing a low-productivity worker.

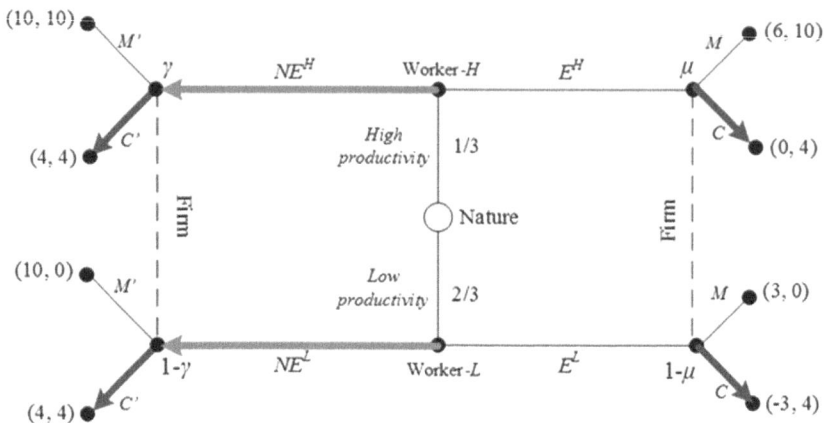

Fig. 10.6b Pooling strategy profile (NE^H, NE^L)—Responses (C', C)

You may feel that this off-the-equilibrium belief, $\mu \leq \frac{2}{5}$, is a bit insensible (fishy). To understand this point, recall that, from our discussion in Case 1 (Step 4b), the low-productivity worker would not like to deviate, from NE^L to E^L, even if it could fool the firm into believing that such a message stems from the high-productivity worker, and respond by hiring her as a manager. In other words, the cost of acquiring education for the low-productivity worker is so high that it prevents her from deviating from NE^L. In short, we say that, for the low type, deviating to acquire education, E^L, is "equilibrium dominated" because her equilibrium payoff with NE^L, 4, exceeds that from deviating to education, E^L, which holds *regardless* of how the firm responds to education (3 if hired as a manager, -3 if hired as a cashier). Chapter 11 formally defines equilibrium dominance and presents tools to eliminate pooling PBEs based on insensible off-the-equilibrium beliefs, such as (NE^H, NE^L) above, where $\mu \leq \frac{2}{5}$.

10.8 Evaluating PBE as a Solution Concept

Following our discussion in Sect. 1.6, we next evaluate PBE as a solution concept:

1. **Existence? Yes.** When we apply PBE to any game, we find that at least one equilibrium exists. Intuitively, this result is equivalent to the existence of SPE in sequential-move games of complete information, but extended to an incomplete information setting. For this finding to hold, however, we may need to allow

for mixed strategies; although all applications in this chapter produced a pure-strategy PBE. Section 10.9 considers, for completeness, a game where pure-strategy PBEs cannot be sustained, but a mixed strategy PBE (often known as "semi-separating" PBE) can be supported.

2. **Uniqueness? No.** This point is illustrated by the labor market signaling game in Sect. 10.7, where we found two PBEs: the separating strategy profile where only the high-type worker acquires education, (E^H, NE^L), and the pooling strategy profile where no types of worker does, (NE^H, NE^L). Other games, however, such as that in Sect. 10.6, have a unique PBE, (O^B, N^F), where player 1 only makes an investment offer when such investment is beneficial. Therefore, we cannot guarantee that the PBE solution concept provides a unique equilibrium prediction in *all* games, entailing that uniqueness does not hold.

3. **Robust to small payoff perturbations? Yes.** PBE yields the same equilibrium predictions if we were to change the payoff of one of the players by a small amount (e.g., 0.001 or, generally, *any* ε that approaches zero). This occurs because, if a strategy s_i is sequentially optimal for player i in our original game, it must still be optimal after we apply a small payoff perturbation.[9]

4. **Socially optimal? No.** As described in the labor market signaling game (Sect. 10.7), the presence of incomplete information gives rise to inefficiencies. In particular, in the separating PBE (E^H, NE^L), the high-type worker invests in a costly education (which does not improve her productivity) just to signal her type to the firm and, thus, be hired as a manager. In a complete information setting, instead, the firm would observe the worker's type, hiring the high (low) type as a manager (cashier). The worker would have no incentives to acquire education to convey her type to the firm. In that setting, their payoffs would be (10, 10) when the worker's productivity is high and (4, 4) when it is low. In contrast, in the separating PBE equilibrium payoffs are (6, 10) and (4, 4). As a consequence, if players behaved as under complete information, the high-type worker would improve her payoff from 6 to 10 (savings in education acquisition) while the payoffs of the low-type worker and the firm would remain unaffected.

   A similar argument applies to the pooling PBE (NE^H, NE^L), where equilibrium payoffs are (4, 4) regardless of the worker's type. In this context, the high-type worker and firm would improve their payoffs if they could behave as under complete information (increasing from 4 to 10 for both of them); while those of the low-type worker and the firm would remain unchanged.

---

[9] Technically, if the original game has payoff ties, making at least one of the players indifferent between two or more of her available strategies, we would find at least two PBEs. After applying a small payoff perturbation on the original game, we could have that the payoff ties are broken, implying that a subset of the PBEs in the original game can be supported. Alternatively, at least one of the PBEs in the original game can still be sustained while others cannot.

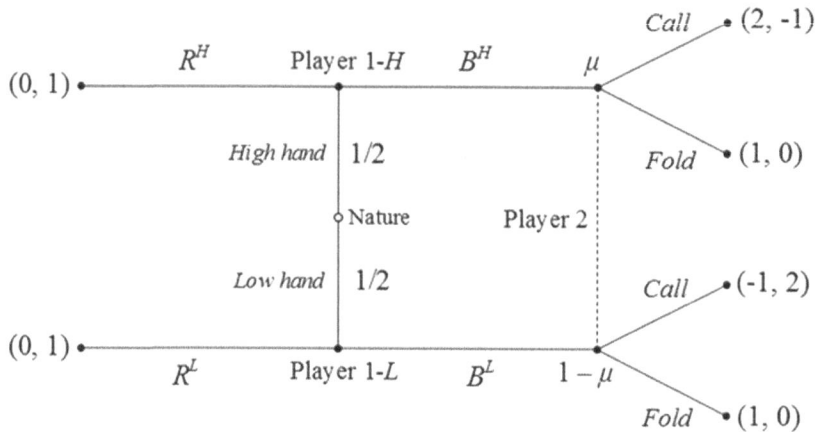

Fig. 10.7 Poker game

10.9 Semi-Separating PBE

What if, after checking for all the separating and pooling strategy profiles as candidates for PBEs (our "usual suspects"), we find that none of them can be supported as a PBE? Does it mean that the signaling game has no PBE? No, it just means that we need to allow at least one sender type to randomize her messages. This is analogous to our inability to find psNEs in simultaneous-move games of complete information, where we could, however, identify one msNE where at least one of the players randomizes.

Figure 10.7 depicts a simple poker game where, first, player 1 (sender) privately observes whether she has a high or low hand; then, player 1 chooses to bet (B) or resign (R). If she resigns the game is over and player 2 earns the initial pot. However, if she bets, player 2 (receiver) must respond by calling or folding without observing player 1's hand but knowing that high and low hands are equally likely.

No pure-strategy PBE. In this setting, none of the separating strategy profiles, (B^H, R^L) and (R^H, B^L), or the pooling strategy profiles, (B^H, B^L) or (R^H, R^L), can be sustained as a PBE. (You should confirm this point as a practice.)[10] Unlike in previous signaling games, this occurs because player 1, when holding a high hand, would like to induce player 2 to respond by calling it; but when holding a

---

[10] In (B^H, R^L), for instance, player 2's beliefs are $\mu = 1$, which induce this player to respond by folding since $0 > -1$. Anticipating player 2 folding, the high-hand player 1 prefers to deviate to R^H, implying that (B^H, R^L) cannot be sustained as a PBE. Similarly, in strategy profile (B^H, B^L), player 2's beliefs are $\mu = 1/2$, which induces player 2 to respond by calling. Anticipating this response, the low-hand player 1 deviates to R^L, implying that (B^H, B^L) cannot be supported as a PBE.

10.9 Semi-Separating PBE

low hand, she would prefer player 2 to respond by folding (informally, to not "call her bluff").[11]

In other words, each sender type would like the receiver (player 2) to *incorrectly infer* her type: (i) thinking that player 1's hand is low, calling it, when it is actually high (at the top right side of the tree); and (ii) thinking that her hand is high, folding, when it is in fact low (bottom right side). To prevent her type from being recognized, player 1 can create some "noise" in the signal, by betting with a positive probability. This randomization reduces player 2's ability to infer her type, as betting can now originate from the high- or low-hand players, with potentially different probabilities, having a similar effect as in msNEs, where a player mixed to keep her rivals guessing about her moves are.

Allowing for randomizations. Figure 10.7a reproduces 10.7 again, but highlighting the branch corresponding to betting for the high-hand player 1, B^H, who finds bet to be a strictly dominant strategy, i.e., B^H yields a payoff of 2, if player 2 responds by calling, or 1, if she responds by folding, both of them exceeding her payoff from resigning, R^H, where she earns 0 with certainty. Therefore, the high-hand player 1 plays B^H in pure strategies.[12] In contrast, the low-hand player 1 assigns probability p on B^L and $1 - p$ on R^L, where $p \in (0, 1)$.

Finally, we labeled player 2's probability of calling as q, where $q \in (0, 1)$, and that we seek to identify. Note that if, instead, we allowed for $q = 0$ ($q = 1$), and player 2 folded (called) in pure strategies, then player 1 would like to bet (resign) with certainty when her hand is low. As a consequence, $p = 1$ ($p = 0$), implying that a pooling (separating) could be supported as a PBE, which we know cannot hold. Therefore, player 2 must be randomizing with (non-degenerate) strategies, such that $q \in (0, 1)$.

We next follow our standard 5-step tool to identify PBEs.

1. *Specifying a strategy profile.* We first specify the semi-separating strategy profile that we seek to test as a PBE: player 1 chooses B^H with certainty, B^L with probability p, and player 2 responds by calling with probability q.
2. *Bayes' rule.* We can now update player 2's beliefs. Upon observing player 1 betting, beliefs satisfy

$$\mu = \frac{\frac{1}{2}1}{\frac{1}{2}1 + \frac{1}{2}p} = \frac{\frac{1}{2}}{\frac{1}{2} + \frac{1}{2}p} = \frac{1}{1+p}$$

---

[11] A scene from the movie *The Princess Bride* (1987) provides a similar setting where one of the characters, Wesley, succeeds in his bluffing to Prince Humperdinck. The latter does not know whether Wesley is still weak after being tortured or has regained his strength. Prince Humperdinck knows he would win a duel if Wesley is still weak, but would lose it otherwise. Wesley is actually weak, and says: "I might be bluffing. It's conceivable [...] that I'm only lying here because I lack the strength to stand. But, then again... perhaps I have the strength after all." For the complete scene, watch: https://www.youtube.com/watch?v=wUJccK4lV74.

[12] As shown in Chapter 6, a mixed strategy cannot assign a probability weight to strictly dominated strategies, such as R^H for the high-hand player 1.

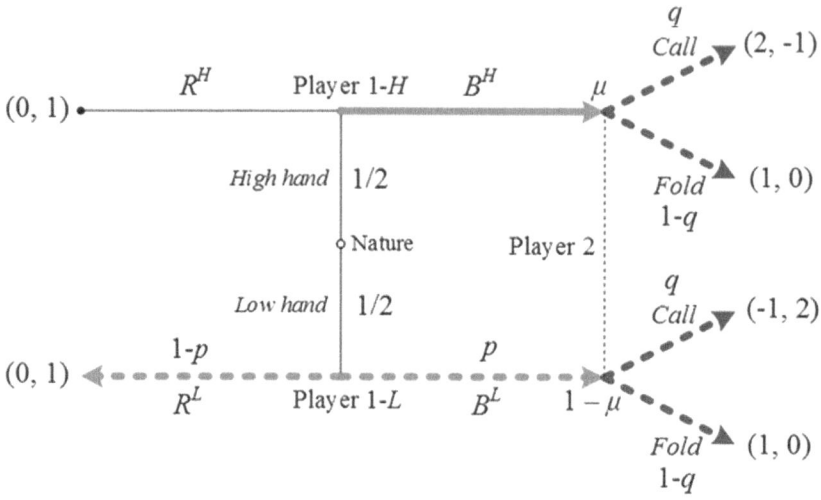

Fig. 10.7a Poker game—Semi-separating strategy profile

since player 1 bets with certainty when her hand is high, but bets with probability p otherwise.

3. *Optimal responses.* Given our results from Step 2, we now analyze player 2's response. Recall that player 2 must be mixing, $q \in (0, 1)$. Otherwise, as discussed above, the low-hand player 1 would have incentives to use pure strategies, which are not PBEs of this game. In addition, if player 2 is mixing, she must be indifferent between calling and folding, which entails that

$$\mu(-1) + (1 - \mu)2 = \mu 0 + (1 - \mu)0$$

where the left side represents player 2's expected utility of calling (which yields a payoff of -1 when player 1's hand is high, but 2 otherwise); while the right side denotes her expected utility from folding (which is 0 regardless of player 1's hand). Solving for μ, yields $\mu = \frac{2}{3}$. But, with which probability q does player 2 call? We know that, as in mixed strategy Nash equilibria (msNEs), player 2 must be mixing with a probability q that makes the low-hand player 1 indifferent between betting and resigning. This implies that $EU_1(B^L) = EU_1(R^L)$ or,

$$q(-1) + (1 - q)1 = 0$$

since the low-hand player 1 earns -1 when player 2 responds by calling but 1 when player 2 folds. Solving for q, we find that $q = \frac{1}{2}$, implying that player 2 calls with 50 percent probability.

4. *Optimal messages.* From our results in Step 3, we now identify player 1's optimal message. The high-hand player 1 finds betting, B^H, to be strictly dominant,

so we focus on the low-hand player, who calls with probability p. From Bayes' rule in Step 2, we know that

$$\frac{2}{3} = \frac{1}{1+p}$$

since $\mu = \frac{2}{3}$ from Step 3. Solving for p, we obtain that $p = \frac{1}{2}$, so the low-hand player 1 bets with 50 percent probability.

5. *Summary.* From the above steps, we found a PBE where the high-hand player 1 bets, B^H, with certainty; the low-hand player 1 bets, B^L, with probability $p = \frac{1}{2}$; and player 2 responds by calling with probability $q = \frac{1}{2}$ and sustains beliefs $\mu = \frac{2}{3}$.

10.10 Extensions

For presentation purposes, the labor market signaling game in previous sections considered only two types of senders (high- and low-productivity worker), who can choose between only two available messages (education or no education), and a receiver (firm) who only has two available responses (hire the applicant as a manager or as a cashier). In more general settings, we can allow for the signaling game to have more than two sender types, each choosing from more than two available messages, and a receiver who can respond with more than two actions. We next separately study each of these extensions, how it would affect the game tree representation of the signaling game and, more importantly, which steps in Tool 10.1 are affected when we seek to identify PBEs.

10.10.1 What if the Receiver has More than Two Available Responses?

In the labor market signaling game, the receiver (firm) has only two available responses, namely, hiring the worker as a manager or as a cashier. In more general settings, we could allow for the firm to have more than two available responses, such as hiring the worker as the company director, manager, or cashier; as depicted in Fig. 10.8. Graphically, the game tree in Fig. 10.2 only increases in the number of branches stemming from all nodes connected by each information set, from two to three branches originating from each node (director, manager, or cashier hire).[13]

---

[13] The cost of acquiring education for the high-type worker is still 4, as seen in the differences $12 - 8 = 4$ when hired as a director, $10 - 6 = 4$ when hired as manager, and $4 - 0 = 4$ when hired as a cashier. A similar argument applies to the low-type worker, who still experiences a cost of 7 to acquire education, as illustrated in the differences $12 - 5 = 7$ when hired as a director, $10 - 3 = 7$ as a manager, and $4 - 3 = 7$ as a cashier.

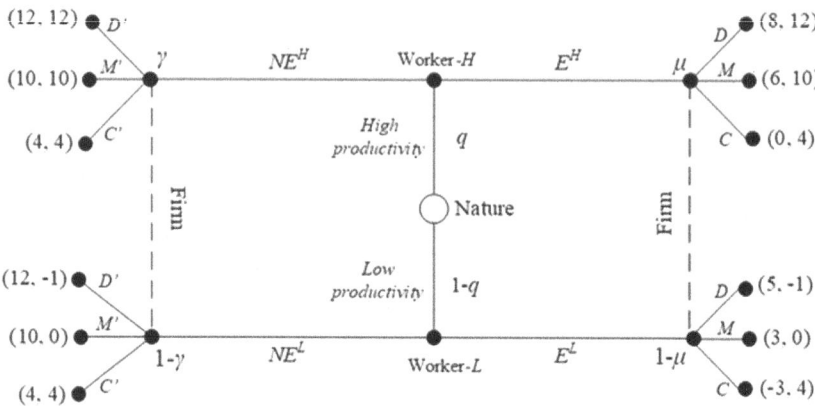

Fig. 10.8 Labor market signaling game with three responses

10.10.1.1 How is Tool 10.1 Affected by More Responses?

Our analysis in Sect. 10.7 would be mostly unaffected. In particular, the list of strategy profiles to test in Step 1 remains unchanged, (E^H, NE^L), (NE^H, E^L), (E^H, E^L), and (NE^H, NE^L). Similarly, the firm's updated beliefs in Step 2 are also unaffected, because the number of sender types and the number of available messages are unchanged, therefore, not modifying the firm's information sets.

However, the firm's optimal response (Step 3) is affected, as the firm would choose the response that yields the highest payoff, by comparing its three possible responses. This argument extends to other signaling games where we allow for $k \geq 2$ available responses, in which the firm would choose the one that yields the highest payoff. Exercise 10.9 asks you to test which strategy profiles can be sustained as PBEs.

10.10.2 What if the Sender has More than Two Available Messages?

Alternatively, the labor market signaling game could be extended by allowing the worker to choose between more than two available messages, such as acquiring an advanced graduate degree (A), an undergraduate degree (E), or no college education (NE).

Figure 10.9 illustrates this setting, showing that, relative to Fig. 10.2, each worker type (either high or low productivity) chooses now among three, instead of two, possible branches (A, E, or NE). Each branch, in turn, gives rise to a different information set: one after the firm observes A, another after observing E, and another after NE. In each information set, however, there are still two nodes, indicating that, upon observing a given message, the firm faces the same type of

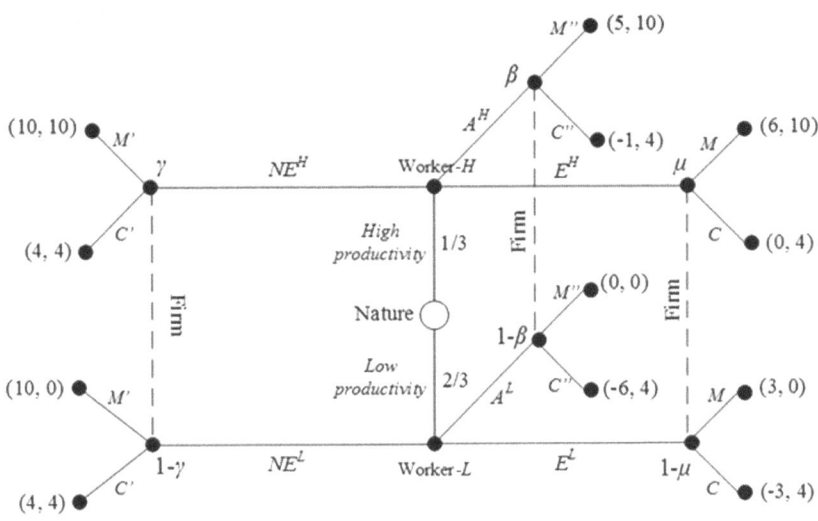

Fig. 10.9 Labor market signaling game with three messages

uncertainty as in Fig. 10.5, namely, not knowing whether the message originates from the high- or low-productivity worker.

While Fig. 10.9 is relatively similar to Fig. 10.2 (except for the branches where the worker chooses A), its tree representation may be difficult to interpret. For clarity, games with more than two messages are often represented with a game tree like that in Fig. 10.9a. The trees in Figs. 10.9 and 10.9a represent, of course the same games (same strategic setting, timing, information, and payoffs to each player). However, Fig. 10.9a can be understood as the "vertical" representation of the game, as nature chooses first, at the top of the tree, followed by each worker type choosing her education level, and by the firm responding to each possible message.

Regarding payoffs, note that the high-productivity worker incurs an additional cost of 1 when choosing A^H, relative to when she chooses E^H; whereas the low-productivity worker's additional cost is 3. The firm's payoffs are, of course, unaffected since education (either E or A) is not productivity-enhancing.

10.10.2.1 How is Tool 10.1 Affected by More Available Messages?

We now examine how the five-step tool to identify PBEs is affected when the sender has more available messages. We start at Step 1, where we consider a specific strategy profile as a candidate for PBE. In this setting, we have the following strategy profiles to check as PBEs: six separating, (A^H, E^L), (A^H, NE^L), (E^H, NE^L), (E^H, A^L), (NE^H, A^L), (NE^H, E^L), and three pooling, (A^H, A^L), (E^H, E^L), and (NE^H, NE^L).

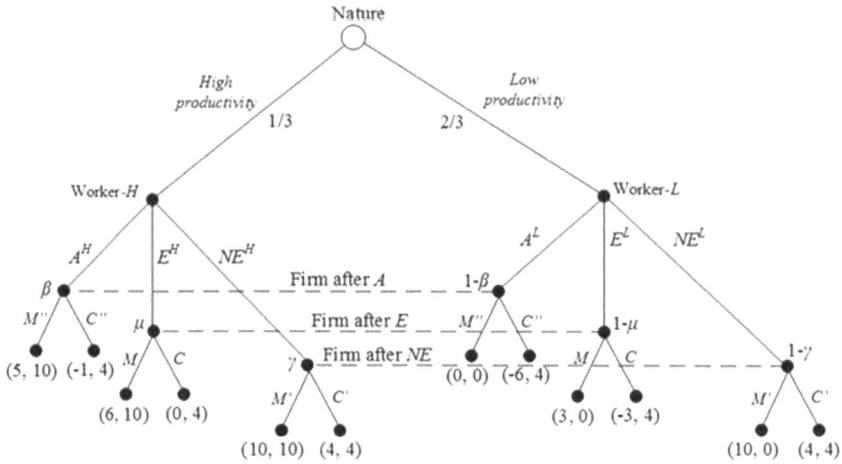

Fig. 10.9a Labor market signaling game with three messages—Alternative representation

More generally, if there are x sender types and each sender has y available messages, then there are a total of y^x different strategy profiles. In our example with $x = 2$ types and $y = 3$ messages, we face $3^2 = 9$ different profiles. For illustration purposes, we focus on strategy profile (A^H, E^L).

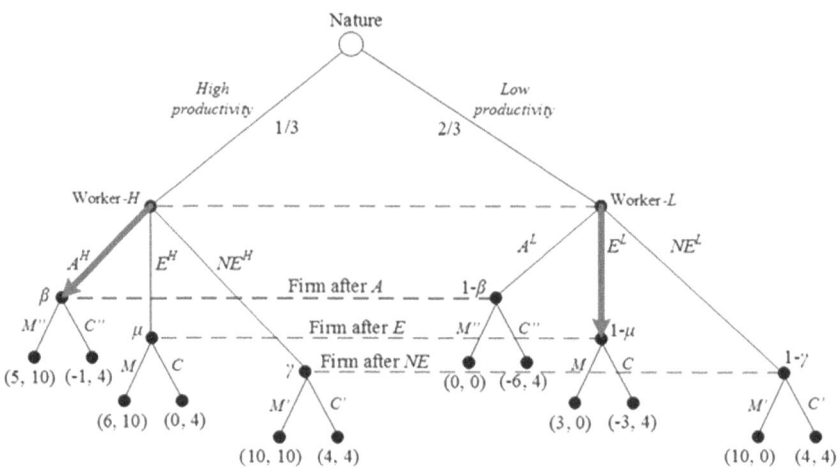

Fig. 10.9b Labor market signaling game with three messages—Strategy profile (A^H, E^L)

10.10 Extensions

1. *Specifying a strategy profile.* Consider strategy profile (A^H, E^L), depicted in Fig. 10.9b. Intuitively, both worker types acquire education, but the high type invests in an advanced graduate degree while the low type completes an undergraduate degree.
2. *Bayes' rule.* In this context, the firm's updated beliefs are

$$\mu(H|j) = \frac{\frac{1}{3}\alpha_j^H}{\frac{1}{3}\alpha_j^H + \frac{2}{3}\alpha_j^L}$$

where $j = \{A, E, NE\}$ denotes the message that the firm observes, and α_j^H (α_j^L) represents the probability that the high (low) type sends message j. In strategy profile (A^H, E^L), where $\alpha_A^H = \alpha_E^L = 1$ and all other α's are zero, this belief becomes $\mu(H|A) = \frac{\frac{1}{3}1}{\frac{1}{3}1 + \frac{2}{3}0} = 1$ after observing A, $\mu(H|E) = \frac{\frac{1}{3}0}{\frac{1}{3}0 + \frac{2}{3}1} = 0$ after observing E, and $\mu(H|NE) = \frac{\frac{1}{3}0}{\frac{1}{3}0 + \frac{2}{3}0} = \frac{0}{0}$ after NE.

In summary, upon observing A (E) the firm is convinced of facing a high (low) productivity worker, because only this worker type uses this message. Upon observing NE, which is off-the-equilibrium path, the firm cannot update its beliefs according to Bayes' rule, entailing that we must leave them unrestricted as $\mu(H|NE) \in [0, 1]$.

In most applications, however, researchers assume that, if the firm's beliefs are $\mu(H|E) = 0$ after E, they must also be $\mu(H|NE) = 0$ after NE. Intuitively, if a college degree signals that the worker's productivity is low, observing that the worker did not even complete college must provide a similar signal. From a practical approach, this assumption simplifies our analysis in the subsequent steps of Tool 10.1, and we consider it here.

3. *Optimal responses.* Given our results from Step 2, we now analyze the firm's responses upon observing each of the three possible messages.
   a. Upon observing A, the firm responds by hiring the worker as a manager, M'', since $10 > 4$ at the top center of the game tree. This argument is analogous to that in Sect. 10.7.1, where we analyzed separating strategy profile (E^H, NE^L), where the worker only had two available messages to send.[14]
   b. If the firm observes E, it responds by hiring the worker as a cashier, C, because $4 > 0$ at the bottom right corner of the tree.
   c. Finally, upon observing NE, the firm responds by hiring the worker as a cashier too, C', since we imposed that off-the-equilibrium beliefs satisfy $\mu(H|NE) = 0$, and $4 > 0$ at the far bottom right corner of the tree.

---

[14] The argument, actually, applies to all other separating strategy profiles listed above, (A^H, NE^L), (E^H, NE^L), (E^H, A^L), (NE^H, A^L), and (NE^H, E^L), because the firm can perfectly infer the worker's type upon observing equilibrium messages.

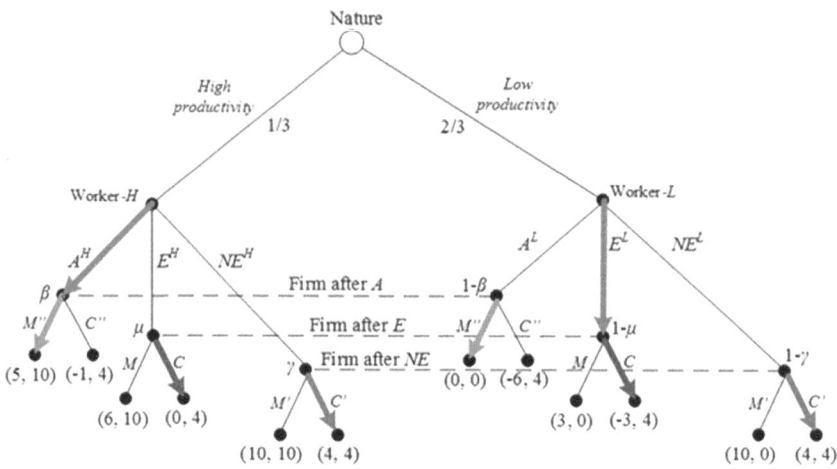

Fig. 10.9c Labor market signaling game with three messages—Optimal messages

To keep track of our results, Fig. 10.9c shades the branches corresponding to M'' in the left side of the tree, and those corresponding to C and C' in the right side.

4. *Optimal messages.* From our results in Step 3, we now identify the worker's optimal message, separately analyzing each type.
   a. *High productivity.* At the left of the game tree, if the high-productivity worker chooses A^H, moving upward, she earns 5, as she anticipates the firm responding with M''. If, instead, she deviates to E^H, she would earn 0 (as she is hired as a cashier, C). A similar argument applies if she does not acquire education, NE^H, where she is also hired as a cashier, C', and her payoff is 4. Therefore, she does not have incentives to deviate from A^H.
   b. *Low productivity.* At the right of the game tree, if the low-productivity worker chooses E^L, she earns -3. Then, she has incentives to deviate to NE^L, which yields 4. (This is her best deviation, as deviating to A^L only entails a payoff of 0.)
5. *Summary.* From Steps 4a–4b, we found that one sender type (the low-productivity worker) deviates from the strategy profile prescribed in Step 1, (A^H, E^L), implying that this profile cannot be supported as a PBE.

10.10.3 What if the Sender has More than Two Types?

Finally, Fig. 10.10 extends the signaling game of Fig. 10.2 to allow for the worker to have more than two types, such as a high, medium, or low productivity. Relative to the game tree in Fig. 10.2, a new branch stems from the initial node where

10.10 Extensions

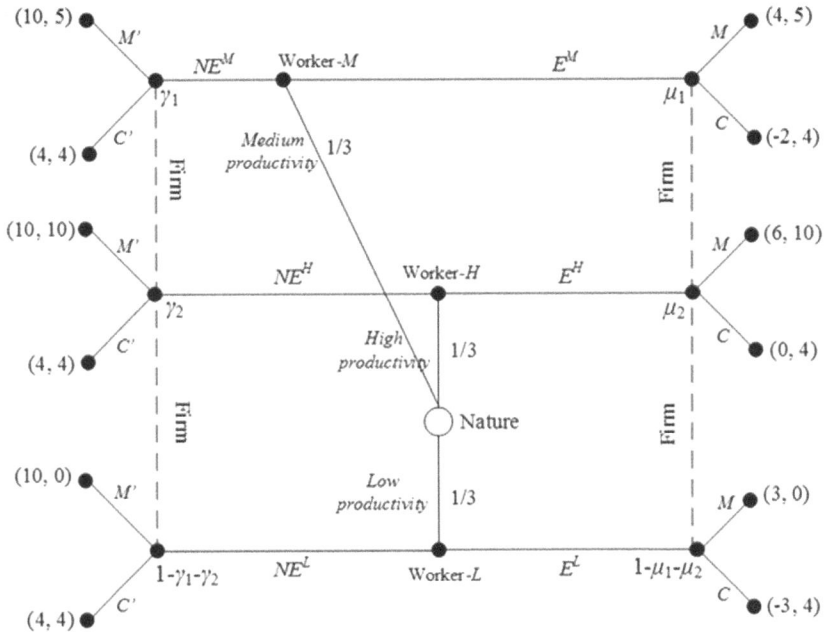

Fig. 10.10 Labor market signaling game with three worker types

nature draws the worker's type (at the top of the figure), corresponding to medium productivity.

For clarity, Fig. 10.10a depicts the "vertical" version of the game tree in Fig. 10.10, following the same approach as in Sect. 10.10.2. This representation helps illustrate that, after the initial move of nature, every type of worker privately observes her type and chooses whether to acquire education, E^k, or not, NE^k, where $k = \{H, M, L\}$. In this context, however, information sets connect three (instead of two) nodes. This means that, upon observing a worker with education, the firm does not know whether her productivity is high, medium, or low; as depicted by the dashed line with the label "Firm after E." A similar argument applies when the firm observes a worker with no education, in the information set with the label "Firm after NE" at the bottom of Fig. 10.10a.

Regarding players' payoffs, first note that the medium-productivity worker incurs a cost of 6 when acquiring education, which is in between that of the low-productivity worker (7) and that of the high-productivity worker (4). The firm's profits are always 4 when hiring the applicant as a cashier, regardless of her type (high, medium, or low) and regardless of her education. When hiring her as a manager, however, the firm's profit is 10 when the applicant's productivity is high, 5 when it is medium, and 0 otherwise.

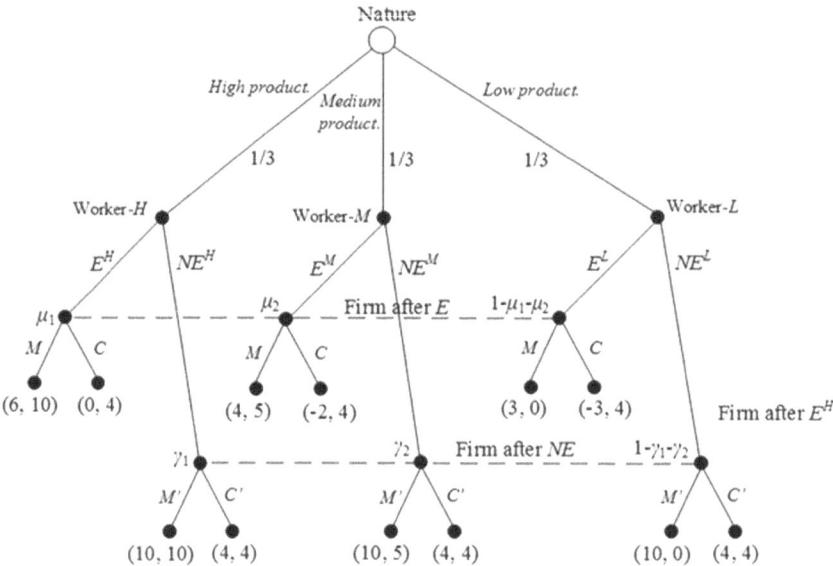

Fig. 10.10a Labor market signaling game with three worker types—Alternative representation

10.10.3.1 How is Tool 10.1 Affected by More than Two Types?

In this setting, we have different profiles to test:
(E^H, E^M, NE^L), (E^H, NE^M, E^L), (E^H, NE^M, NE^L),
(NE^H, E^M, E^L), (NE^H, E^M, NE^L), (NE^H, NE^M, E^L),
(E^H, E^M, E^L), and (NE^H, NE^M, NE^L).

While the last two strategy profiles are clearly pooling, because all worker types choose the same message (all acquire education or none do), the first six strategy profiles are neither fully separating (as two worker types pool into sending the same message) or fully pooling (as one worker type chooses a different message than the other two types). We may refer to them as "partially separating" profiles.

Following our discussion in Sect. 10.10.1, if there are $x = 3$ sender types and each sender has $y = 2$ available messages, then there are a total of $2^3 = 8$ different strategy profiles, as identified in our list above. For illustration purposes, we focus on (E^H, E^M, NE^L).

1. *Specifying a strategy profile.* Consider strategy profile (E^H, E^M, NE^L), as in Fig. 10.10b, where both the high- and medium-productivity worker acquire education while the low-productivity worker does not.
2. *Bayes' rule.* In this setting, the firm's updated beliefs are

$$\mu(H|E) = \frac{\frac{1}{3}\alpha^H}{\frac{1}{3}\alpha^H + \frac{1}{3}\alpha^M + \frac{1}{3}\alpha^L}$$

10.10 Extensions

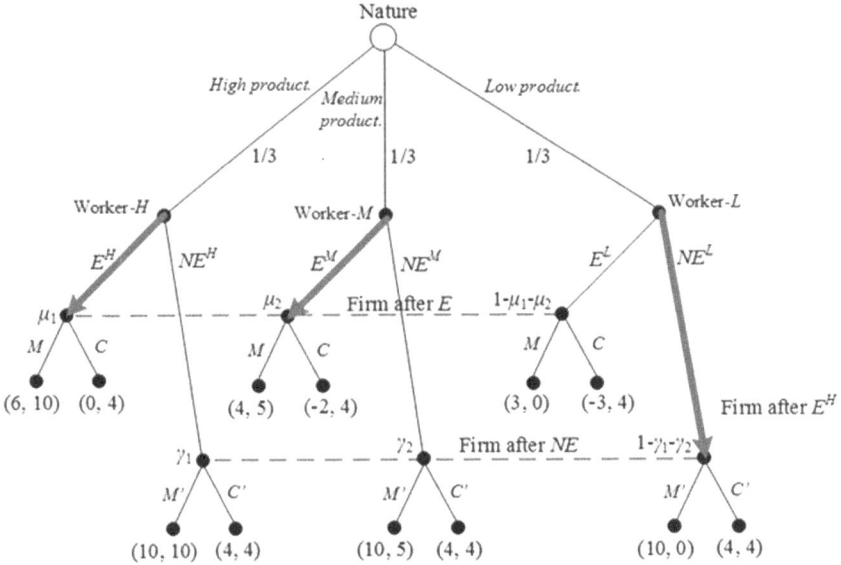

Fig. 10.10b Labor market signaling game with three worker types—Strategy profile (E^H, E^M, NE^L)

where Fig. 10.10b assumes, for simplicity, that all types are equally likely. In strategy profile (E^H, E^M, NE^L), we have that $\alpha^H = \alpha^M = 1$ but $\alpha^L = 0$, yielding an updated belief $\mu(H|E) = \frac{\frac{1}{3}}{\frac{1}{3}+\frac{1}{3}+0} = \frac{1}{2}$. Intuitively, the firm knows that education must originate from either the high or medium types and, since both are equally likely, the probability of facing a high-productivity worker is 50 percent. Similarly, $\mu(M|E) = \frac{\frac{1}{3}}{\frac{1}{3}+\frac{1}{3}+0} = \frac{1}{2}$, while $\mu(L|E) = \frac{0}{\frac{1}{3}+\frac{1}{3}+0} = 0$ since the low-productivity worker does not acquire education in this strategy profile. Upon not observing education, however, we obtain more concentrated beliefs, that is, $\mu(H|NE) = \mu(M|NE) = 0$ whereas $\mu(L|NE) = 1$, as the firm is convinced of facing a low-productivity worker.

3. *Optimal responses.* Given our results from Step 2, we now analyze the firm's responses upon observing each of the three possible messages.

   a. Upon observing E, the firm responds by hiring the worker as a manager, M, since

$$\underbrace{\frac{1}{2}10 + \frac{1}{2}5}_{=7.5} > 4$$

at the right side of the game tree. Unlike our argument in Sect. 10.7.1, where we analyzed separating strategy profile (E^H, NE^L), the firm now does not have concentrated beliefs upon observing E, as it can originate from the high or medium type. As a consequence, it must compute its expected profit from each of its decisions, rather than the certain profit.

b. Finally, upon observing NE, the firm responds by hiring the worker as a cashier too, C', since we found that beliefs satisfy $\mu(H|NE) = \mu(M|NE) = 0$.

To keep track of our results, Fig. 10.10c shades the branches corresponding to M in the middle section of the tree, and those corresponding to C' at the bottom of the tree.

4. *Optimal messages.* From our results in Step 3, we now identify the worker's optimal message, separately analyzing each type.
   a. *High productivity.* At the left of the game tree, the high-productivity worker chooses education, as her payoff from doing so, 6, exceeds that from not acquiring education, 4.
   b. *Medium productivity.* At the center of the tree, the medium-productivity worker is indifferent between choosing education, earning 4, or not acquiring education, earning 4.
   c. *Low productivity.* At the right of the tree, the low-productivity worker does not acquire education, yielding a payoff of 4, since acquiring it only entails a payoff of 3.
5. *Summary.* From Steps 4a-4b, we found that all sender types prefer to behave as prescribed in Step 1, (E^H, E^M, NE^L), implying that this separating strategy profile can be supported as a PBE, with the firm holding beliefs $\mu(H|E) = \mu(M|E) = \frac{1}{2}$ and $\mu(L|E) = 0$ upon observing education, and $\mu(H|NE) = \mu(M|NE) = 0$, $\mu(L|NE) = 1$ after no education, responding with (M, C'), i.e., hiring the worker as a manager upon observing education but as a cashier otherwise.

10.10.4 Other Extensions

The end-of-chapter exercises consider several other extensions of the standard labor market signaling game, seeking to evaluate how our main results are affected.

Productivity-enhancing education. Previous sections assumed, for simplicity, that education does not affect a worker's job productivity. In many real-life settings, however, education makes the worker more productive, and we explore this possibility in Exercises 10.2 and 10.3. Interestingly, we show that the pooling strategy profile where both worker types acquire education cannot yet be sustained: while educated workers may be more attractive for the firm, the worker's payoffs are unaffected, implying that the low-type worker still prefers to remain uneducated, and be hired as a cashier, than acquire education, and be hired as a manager.

More general cost differentials across types. Our above model assumed that the cost of acquiring education for the high (low) productivity worker was, specifically,

10.10 Extensions

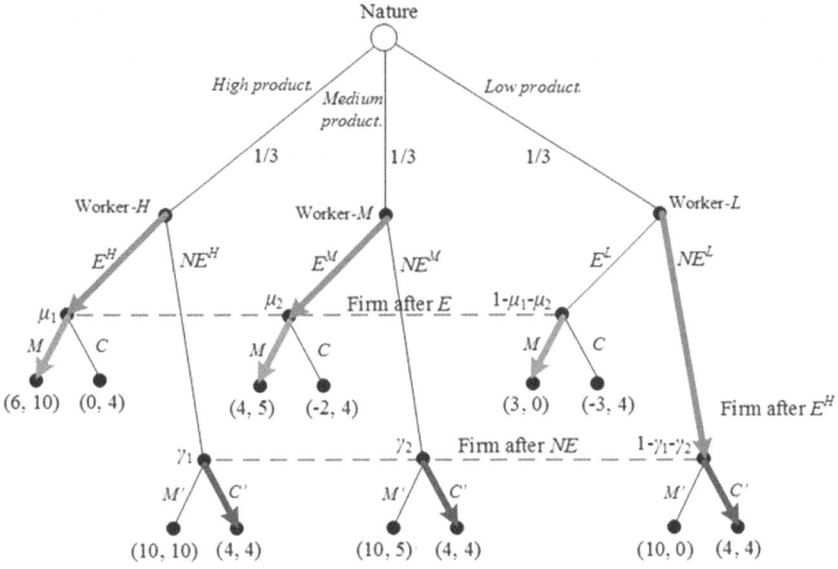

Fig. 10.10c Labor market signaling game with three worker types—Optimal responses

$c_H = 4$ ($c_L = 7$, respectively). Exercise 10.5 allows for more general costs: we assume that the low-productivity worker's cost is $c_L > 4$, so the difference $c = c_L - c_H > 0$ represents the "cost differential" that the low-productivity worker suffers relative to the high-productivity worker when acquiring education. That is, a higher cost differential may deter the low-productivity worker from mimicking the high-type's decision to acquire education, ultimately facilitating the emergence of separating PBEs.

More general profit differentials across types. The standard labor market signaling game considered that hiring a high (low) productivity worker in a managerial position provides the firm with a profit of $d_H = 10$ ($d_L = 0$, respectively); whereas hiring her as a cashier produces a profit of 4 regardless of the worker's type. In Exercise 10.6, we allow for a more general profit differential between worker types when hired as a manager, where we still assume that $d_H = 10$ but $d_L = 10 - d$, where $d = d_H - d_L$ and $0 \leq d \leq 10$. Intuitively, as d increases, the firm hiring the low-productivity worker as a manager entails a larger loss, which the firm seeks to avoid; while when $d = 0$, the firm earns the same profit hiring both worker types as a manager.

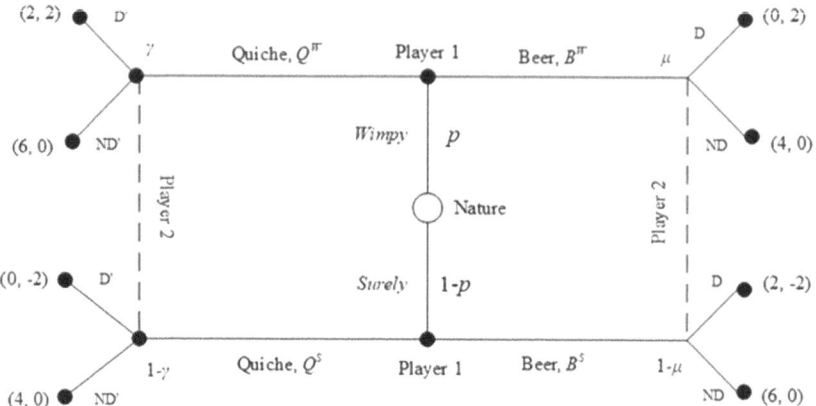

Fig. 10.11 Beer-Quiche game

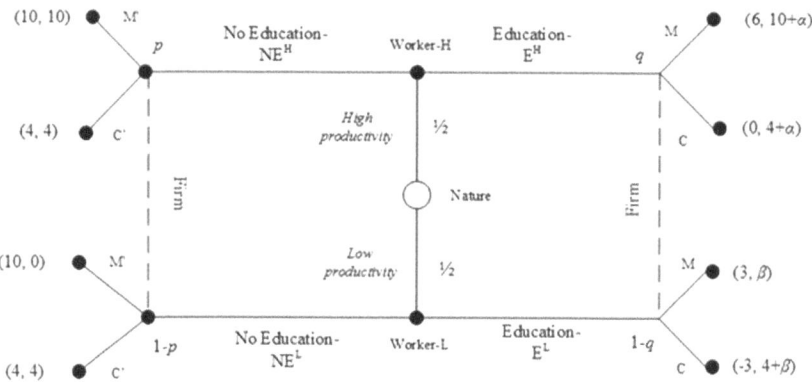

Fig. 10.12 - Labor market signaling when education is productivity enhancing

Exercises

10.1 **Beer-Quiche game.**[A] Consider the following Wild West scenario. A newcomer (player 1) enters a saloon. Player 1 observes his type (surely or wimpy) which determines whether he is a good or bad shooter. The sheriff in this remote town (player 2) does not observe the newcomer's type, but knows that it is surely with probability p and wimpy otherwise, where $p < \frac{1}{2}$. Player 1, then, chooses which breakfast to order at the bar: a quiche or a beer. Upon observing player 1's choice for breakfast, player 2 responds by challenging player 1 to a duel (which player 1 must accept) or not. Figure 10.11 identifies payoffs in each of the eight contingencies.

a. Can the separating strategy profile $B^S Q^W$, where only the surely type of player 1 orders a beer, be sustained as a PBE?
b. Can the separating strategy profile $Q^S B^W$, where only the wimpy type of player 1 orders a beer, be sustained as a PBE?
c. Can the pooling strategy profile $B^S B^W$, where both types of player 1 order a beer, be sustained as a PBE?
d. Can the pooling strategy profile $Q^S Q^W$, where both types of player 1 order a quiche, be sustained as a PBE?

10.2 **Labor market signaling when education is productivity enhancing—Separating equilibria.**[A] Consider the following labor market signaling game shown in Fig. 10.12. A worker privately observes whether he has High productivity or Low productivity with equal probability, and then decides whether to acquire some education that he will be able to use as a signal about his productivity level. The firm that is thinking of hiring him as a manager or a cashier without observing his productivity (only whether the worker acquired a college education or not). Additionally, there is no innative productivity differential between a low- and a high-type worker when they acquire no education but increases by α for the high-type worker and by β for the low-type worker.
   a. *Separating $NE^H E^L$*. Can you sustain a separating equilibrium where only the low-productivity worker acquires education?
   b. *Separating $E^H NE^L$*. Can you sustain a separating equilibrium where only the high-productivity worker acquires education?

10.3 **Labor market signaling when education is productivity enhancing—Pooling equilibria.**[B] Consider the setting in Exercise 10.1.
   a. *Pooling $E^H E^H$*. Can you sustain a pooling equilibrium where both worker types acquire education?
   b. *Pooling $NE^H NE^L$*. Can you sustain a pooling equilibrium where both worker types do not acquire education?

10.4 **Labor market signaling game with intrinsic value of education.**[B] Consider again the labor market signaling game of Sect. 10.7. Assume, however, that the high-productivity worker enjoys an intrinsic value from acquiring education (he loves to learn!) increases his utility from 6 to $6 + v_H$ when hired as a manager and, similarly, from 0 to v_H when hired as a cashier, where $v_H \geq 0$. The low-productivity worker also enjoys an intrinsic value from acquiring education, increasing his utility from 3 to $3 + v_L$ when hired as a manager and, similarly, from -3 to $-3 + v_L$ when hired as a cashier, where $v_L \geq 0$.
   a. Under which conditions of parameters v_H and v_L can you sustain a separating PBE where only the high-productivity worker acquires education?
   b. Under which conditions of parameters v_H and v_L can you support a separating PBE where only the low-productivity worker acquires education?

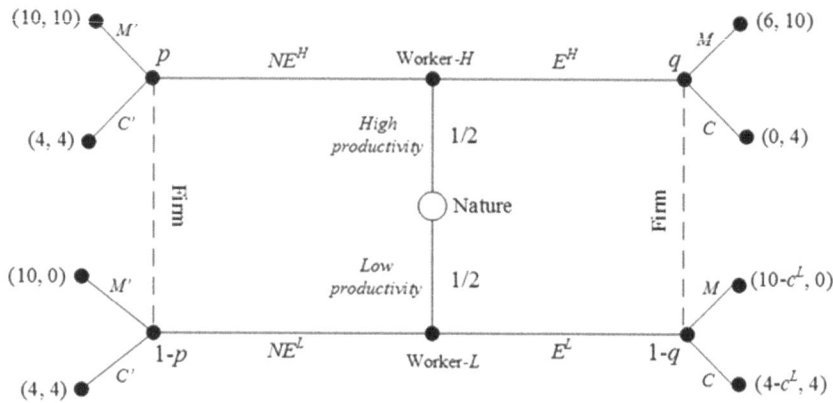

Fig. 10.13 - Labor market signaling game with cost differentials

c. Under which conditions of parameters v_H and v_L can you sustain a separating PBE where both worker types acquire education?

d. Under which conditions of parameters v_H and v_L can you support a separating PBE where no worker type acquires education?

5. Assume that both players assign the same intrinsic value to education, $v_H = v_L = v$. Characterize the set of separating and pooling PBEs.

6. Which strategy profiles can be evaluated in the special case where $v_H = v_L = v = 0$? How are the equilibrium results affected when v increases?

10.5 **Labor market signaling game with cost differentials.**[B] Consider an extension of the labor market signaling game, as shown in Fig. 10.13.

Like in the standard labor market game, consider that the high-type worker faces a cost of acquiring education of $c^H = 4$, but let us now assume that the cost of acquiring education for the low-type worker is $c^L > 4$, so the difference $c = c^L - c^H > 0$ represents the cost differential that the low-productivity worker suffers relative to the high-productivity worker at acquiring education.

a. *Separating NE'*. Can you sustain a separating equilibrium where the low-productivity worker acquires education, that is, NE'? For what values of c^L does the equilibrium exist?

b. *Separating EN'*. Can you sustain a separating equilibrium where the high-productivity worker acquires education, that is, EN'? For what values of c^L does the equilibrium exist?

c. *Pooling EE'*. Can you sustain a pooling equilibrium where both worker types acquire education, that is, EE'? For what values of c^L does the equilibrium exist?

d. *Pooling NN'*. Can you sustain a pooling equilibrium where no worker type acquires education, that is, NN'? For what values of c^L does the equilibrium exist?

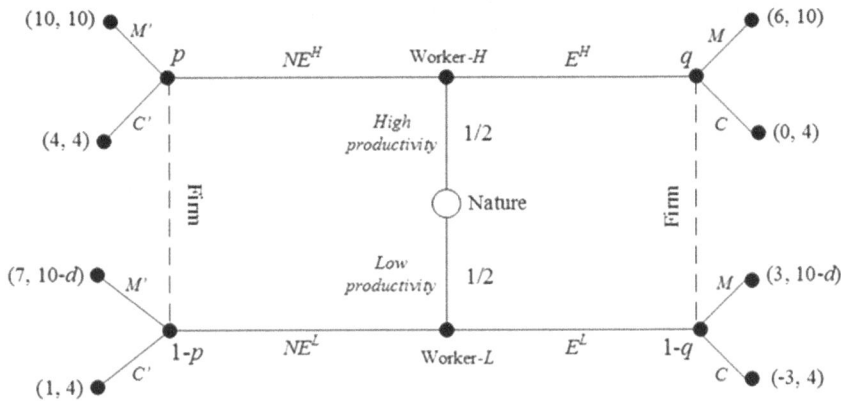

Fig. 10.14 - Labor market signaling game with profit differentials

10.6 **Labor market signaling game with profit differentials.**[B] Consider an extension of the labor market signaling game, as shown in Fig. 10.14.

Like in the standard labor market game, consider that hiring a high-type worker as manager provides a profit (d^H) of 10 for the firm whereas hiring a low-type worker as a manager provides a profit of $10 - d$ where $d = d^H - d^L > 0$.

a. *Separating NE'*. Can you sustain a separating equilibrium where the high- (low-) productivity worker does not (does) acquire education, that is, NE'? For what values of d does the equilibrium exist?

b. *Separating EN'*. Can you sustain a separating equilibrium where only the high-productivity worker acquires education, that is, EN'? For what values of d does the equilibrium exist?

c. *Pooling EE'*. Can you sustain a pooling equilibrium where both worker types acquire education, that is, EE'? For what values of d does the equilibrium exist?

d. *Pooling NN'*. Can you sustain a pooling equilibrium where both worker types do not acquire education, that is, NN'? For what values of d does the equilibrium exist?

10.7 **Brinkmanship game.**[B] Consider the sequential-move game between two players in a dispute, depicted in Fig. 10.15. In the first stage, player 1 privately observes whether it is wimpy or surely, and chooses whether to stand firm or cave to the other players' demands. Player 2 does not observe player 1's type, but knows that it is surely with probability $p \in [0, 1]$ and wimpy otherwise. If player 1 caves, the game is over. If player 1 stands firm, player 2 updates his belief about player 1' type, and responds by starting a fight or not.

a. Can the separating strategy profile $S^{SU}C^{WI}$, where only the surely type of player 1 stands firm, be sustained as a PBE?

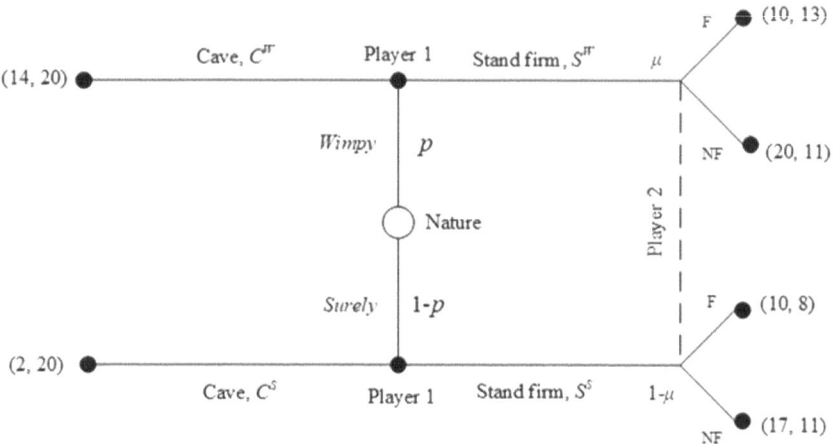

Fig. 10.15 - Brinkmanship game

b. Can the separating strategy profile $C^{SU} S^{WI}$, where only the wimpy type of player 1 stands firm, be sustained as a PBE?

c. Can the pooling strategy profile $S^{SU} S^{WI}$, where both types of player 1 stand firm, be sustained as a PBE?

d. Can the pooling strategy profile $C^{SU} C^{WI}$, where both types of player 1 cave, be sustained as a PBE?

e. Find a semi-separating PBE where the surely type of player 1 stands firm with certainty (playing pure strategies) but the wimpy type randomizes between standing firm and caving (playing mixed strategies). [Hint: For the wimpy player 1 to standing firm with probability q, player 2 must also respond by fighting with probability r.]

10.8 **Reputation without commitment, based on Kreps et al. (1982).**[C] Consider a sequential-move game of complete information depicted in Fig. 10.16a. In the first stage, player 1 chooses whether to continue the game or stop. If he stops, both players' payoffs are zero; while if he continues, it is player 2's turn to respond by continuing or stopping the game in the second stage. If player 2 stops the game, the game is over in the second stage, with payoffs of -4 for player 1 and zero for player 2. If player 2 continues, however, the game proceeds to its third stage, where player 1 gets to choose again whether to stop or continue. For simplicity, assume no payoff discounting.

a. *Complete information.* Find the SPE of this complete information game.

b. *Incomplete information, no separating PBEs.* In part (a) you must have found that subgame perfection yields an unfortunate result, where players stop their cooperation as soon as possible. Kreps et al. (1982) show that the introduction of a small element of incomplete information in this class of games can actually facilitate cooperation in settings where such

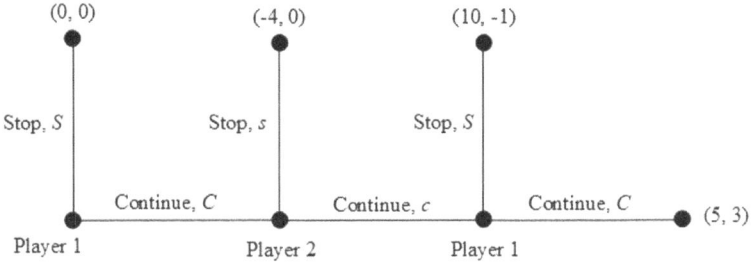

Fig. 10.16a - Reputation without commitment, Complete information

cooperation could not be sustained when players interact under complete information.[15]

In particular, consider that, at the beginning of the game, player 1 privately observes his type (selfish, with probability p, or cooperative, with probability $1-p$), where $p > \frac{5}{6}$. If selfish, he faces the same payoffs as in the above game tree. If cooperative, however, he finds continue to be a strictly dominant strategy, as shown in Fig. 10.16b.

Show that no separating strategy profile can be sustained as a PBE.

c. *Incomplete information, pooling PBEs.* Find under what values of p can no pooling strategy profile be sustained as a PBE.

d. *Incomplete information, Semi-separating PBE.* Find a semi-separating PBE where the cooperative type of player 1 continues, the selfish type randomizes his continuation decision with probability q, and player 2 responds by continuing with probability r.

e. What is the probability that both players choose to continue in the semi-separating PBE of part (d)? How is this probability affected by p? Interpret.

10.9 **Labor market signaling game with three responses.**[C] Consider the labor market signaling game presented in Sect. 10.10.1, where the firm can respond by hiring the worker in three different positions (director, D, manager, M, or cashier, C). For simplicity, assume that the prior probability of the worker's productivity being high, q, satisfies $q > \frac{5}{13}$.

a. Can you sustain a separating PBE where only the high-productivity worker acquires education, (E^H, NE^L)?

b. Can you support a separating PBE where only the low-productivity worker acquires education, (NE^H, E^L)?

c. Can you identify a pooling PBE where both worker types acquire education, (E^H, E^L)?

---

[15] This article is informally known as the "Gang of Four" given the four famous game theorists who co-authored it: David Kreps, Paul Milgrom, John Roberts, and Robert Wilson.

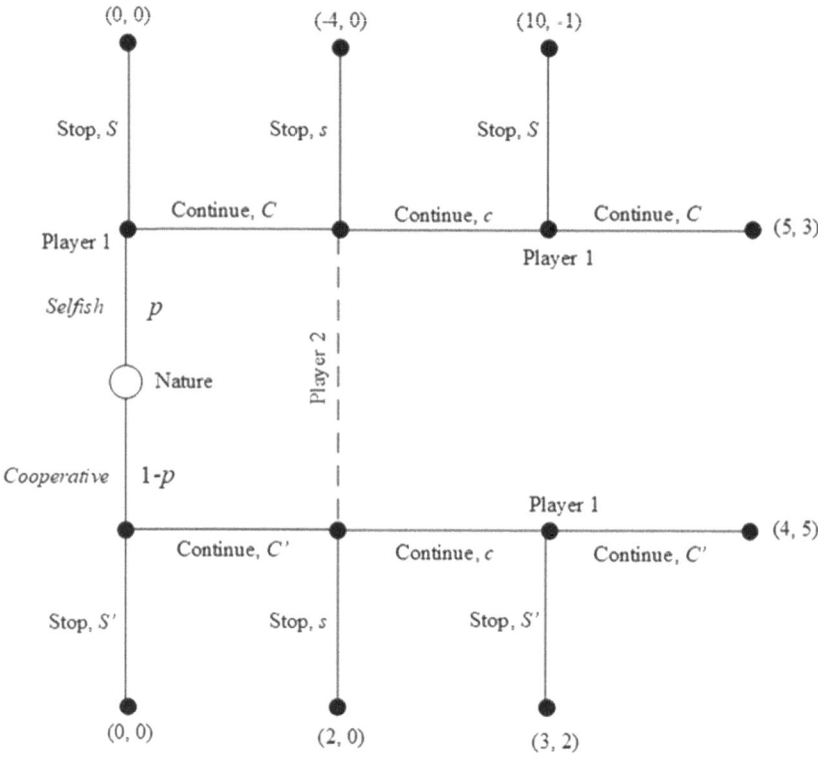

Fig. 10.16b - Reputation without commitment, Incomplete information

d. Can you support a pooling PBE where none of the worker types acquire education, (NE^H, NE^L)?

10.10 **Labor market signaling game with three messages.**[C] Consider the labor market signaling game presented in Sect. 10.10.2, where the worker can choose between three messages (advanced degree, A, undergraduate degree, E, or no education, NE).

   a. Can you sustain a separating PBE where the high-productivity (low-productivity) worker acquires an advanced degree (no education), (A^H, NE^L)?
   b. Can you support a pooling PBE where both worker types acquire an advanced degree, (A^H, A^L)?
   c. Can you support a pooling PBE where none of the worker types acquire education, (NE^H, NE^L)?

10.11 **Labor market signaling game with three worker types.**[C] Consider the labor market signaling game presented in Sect. 10.10.3, where the worker has three possible types (high, medium, or low productivity).

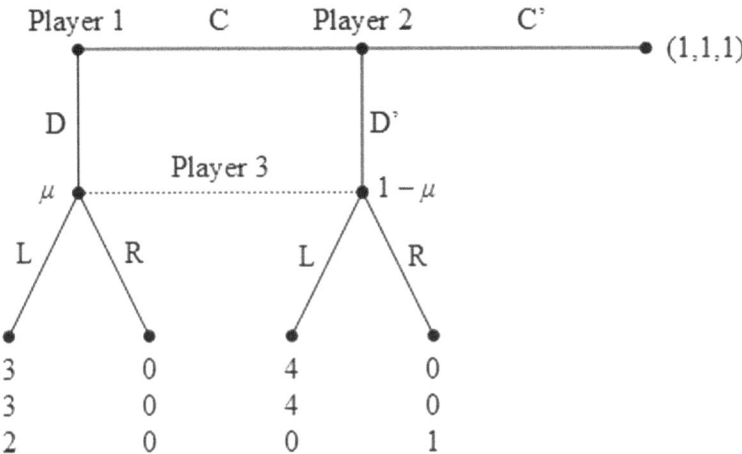

Fig. 10.17 Selten's horse

a. Can you sustain a partially separating PBE where only high-productivity worker acquires education, (E^H, NE^M, NE^L)?
b. Can you support a pooling PBE where all worker types acquire education, (E^H, E^M, E^L)?
c. Can you support a pooling PBE where none of the worker types acquire education, (NE^H, NE^M, NE^L)?

10.12 **Selten's horse.**[C] Consider the "Selten's Horse" game depicted in Fig. 10.17. Player 1 is the first mover in the game, choosing between C and D. If he chooses C, player 2 is called on to move between C' and D'. If player 2 selects C' the game is over. If player 1 chooses D or player 2 chooses D', then player 3 is called on to move without being informed whether player 1 chose D before him or whether it was player 2 who chose D'. Player 3 can choose between L and R, and then the game ends.

a. Define the strategy spaces for each player. Then find all pure-strategy Nash equilibria (psNE) of the game. [*Hint*: This is a three-player game, so you can consider that player 1 chooses rows, player 2 columns, and player 3 chooses matrices.]
b. Argue that one of the two psNEs you found in part (a) is not sequentially rational. A short verbal explanation suffices.
c. Show that there is only one Perfect Bayesian equilibrium and it coincides with one of the pure-strategy Nash equilibria you have identified in part (a).

10.13 **Players holding asymmetric beliefs.**[C] Consider three players interacting in the game tree depicted in Fig. 10.18, adapted from Fudenberg and Tirole (1991). Player 1 is the first mover, choosing A, L_1, or R_1. If he chooses A, the game is over, but if he chooses L_1 or R_1, player 2 responds with L_2 or R_2 without observing whether player 1 previously chose L_1 or R_1. Player 3

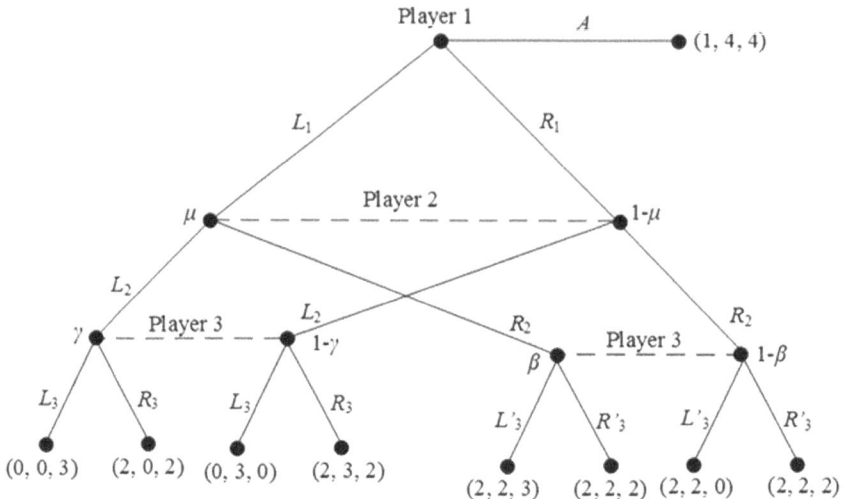

Fig. 10.18 - Players holding asymmetric beliefs

observes player 2's action, but does not observe whether player 1 chose L_1 or R_1. After player 2 chose L_2, player 3 responds with L_3 or R_3, whereas after player 2 selected R_2, player 3 responds with L'_3 or R'_3.

Consider the strategy profile $\{A, L_2, (L_3, L'_3)\}$, where player 1 chooses A, player 2 responds with L_2 (if his information set is reached, which should only happen off-the-equilibrium path), and player 3 responds with L_3 (L'_3) after observing player 2 choosing L_2 (R_2, respectively).

a. Under which conditions on off-the-equilibrium beliefs (μ, γ, and β) can $\{A, L_2, (L_3, L'_3)\}$ be sustained as a PBE?
b. Are the conditions you found in part (a) surprising?

10.14 **Checking if an SPE is also PBE.**[B] Consider the game tree below where player 1 is the first mover, choosing between X, Y, and Z. If she chooses Z, the game is over. If she chooses X or Y, player 2, without observing whether player 1 chose X or Y, responds with L or R. Importantly, players have no private information (no types), implying that this is a sequential-move game of complete information. Unlike in previous games studied in this chapter, player 1's actions do not convey or conceal her type to player 2. In this exercise, we seek to check which SPEs are also PBEs.

Exercises

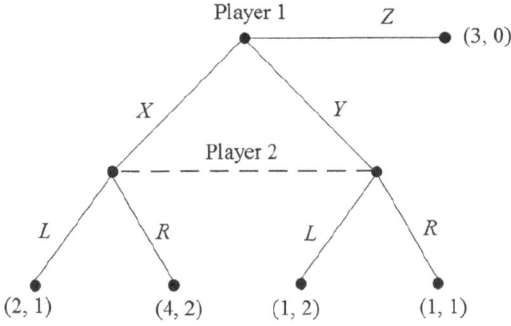

(a) Find the NEs in this game.
(b) Find the SPEs in this game.
(c) Check if the SPEs found in part (b) can be sustained as PBEs.

10.15 **Adding one more signal to the labor market signaling game, based on Dailey and Green** (2014). Consider the labor market signaling game from Sect. 10.7, but assume that the manager now observes two signals: whether the worker acquired education or not, and another signal (such as the worker's grades, or a recommendation letter from a previous employer). For compactness, let g denote the worker's grade, and let $R(g)$ denote the informativeness of grades. When $R(g) = 1$, grades offer no information about the worker's type; when $R(g) < 1$, the grade leads the firm to increase its updated belief of facing a high-productivity worker; and when $R(g) > 1$, the grade induces the firm to decrease update its belief about facing a high-productivity worker. As a hint, this occurs because the firm's updated belief upon observing education is now a function of $R(g)$, as follows

$$\mu(g) = \frac{\mu}{\mu + (1-\mu)R(g)}$$

and, similarly, upon observing no education, the firm's updated belief is $\gamma(g) = \frac{\gamma}{\gamma + (1-\gamma)R(g)}$.

a. Show that the separating strategy profile (E^H, NE^L), where only the high-productivity worker acquires education, can still be supported for all parameter values, and regardless of whether $R(g)$ is equal, higher, or lower than 1. Interpret.

b. Consider the pooling strategy profile where no worker type acquires education, (NE^H, NE^L). As shown in Sect. 10.7.2, this strategy profile can be sustained in the absence of grades if the firm's off-the-equilibrium belief (upon observing education) satisfies $\mu \leq \frac{2}{5}$. Assuming that $R(g) = 1/2$, find under which conditions can this pooling strategy profile be supported as a PBE. Does the presence of grades facilitate or hinder the emergence of this equilibrium? Interpret.

Reference

Dailey, B., and B. Green (2014) "Market Signaling with Grades," Journal of Economic Theory, 151, pp. 114–145.

Equilibrium Refinements 11

11.1 Introduction

Chapter 10 highlights that signaling games and the PBE solution concept are an excellent tool to explain a wide array of economic situations —from the role of education in the labor market, to pricing strategies seeking to deter firms' entry in an industry, to car warranties by new automakers operating in a market. However, we found that this class of games may yield a large number of PBEs and, more importantly, some of these equilibria can only be sustained if players hold insensible off-the-equilibrium beliefs. As an example, recall our discussion in subsection 10.7.3 where, in the context of the labor-market signaling game, we found that a PBE where no worker type acquires education can be supported if, upon observing the surprising event of an educated worker, the firm believes she must be of low productivity.

In this chapter, we present two commonly used tools to "refine" the set of PBEs to those satisfying a consistency requirement in their off-the-equilibrium beliefs. These tools are known as "equilibrium refinement" criteria, and are quite popular in the Industrial Organization literature, as well as other fields studying incomplete information environments.

We first describe the Cho and Kreps' (1987) " Intuitive Criterion,"using a non-technical six-step approach, which we apply to the labor-market signaling game of Chapter 10. We show that one of the pooling PBEs identified in that chapter violates this refinement criterion, thus confirming our suspicions about insensible off-the-equilibrium beliefs in subsection 10.7.3. The separating PBE, however, survives it, thus becoming the unique equilibrium prediction in this signaling game.

We then present the Banks and Sobel's (1987) "Divinity" Criterion (focusing on their D1-Criterion). For illustration purposes, we apply this refinement criterion to the labor-market signaling game, obtaining the same refinement result as with the Intuitive Criterion, that is, the pooling PBE violates the D1-Criterion but the separating PBE does not. The chapter ends with a discussion about games in which we can expect the Intuitive and D1 Criterion to produce different refinement results, namely, games with three or more types of senders but, for presentation purposes, we relegate a detailed analysis of these games to Chapter 12.

11.2 Intuitive Criterion

The Cho and Kreps' (1987) "Intuitive Criterion" helps us eliminate all PBEs that are sustained on insensible off-the-equilibrium beliefs, such as the pooling PBE (NE^H, NE^L) of section 10.7.2, reproduced in Fig. 11.1 for easier reference.

Essentially, the Intuitive Criterion seeks to answer a relatively basic question:

" If the receiver observes an off-the-equilibrium message,

such as education on the right side of Fig. 11.1,

which sender types could benefit from sending such an off-the-equilibrium message?"

If no sender types can benefit, then the equilibrium *survives* the Intuitive Criterion. In contrast, if some sender types can benefit, the receiver would then update her off-the-equilibrium beliefs and her response to this message. If this updated response induces some sender types to deviate from their equilibrium messages, we claim that the PBE we considered *violates* the Intuitive Criterion.

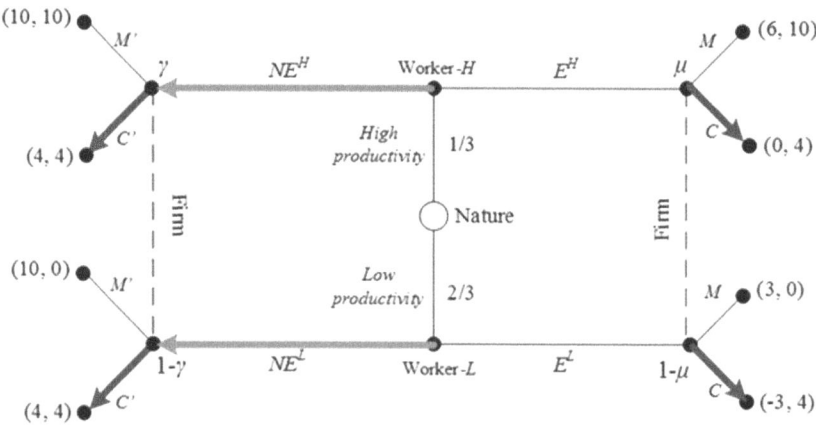

Fig. 11.1 Pooling strategy profile (NE^H, NE^L)—Responses (C', C)

11.2.1 A Six-Step Tool to Apply the Intuitive Criterion

The next tool presents the main steps of the Intuitive Criterion, using a non-technical approach. The appendix of Chapter 12 provides a more technical treatment, which applies to signaling games with discrete actions spaces (Chapters 10 and 11) and to those with continuous action spaces (Chapter 12).

Tool 11.1. Applying the Cho and Kreps' (1987) Intuitive Criterion.

Step 1. Consider a specific PBE.
   That is, consider, for instance the pooling PBE (NE^H, NE^L) found in section 10.7.2. Recall that (NE^H, NE^L) could be supported as a PBE if the firm responds with (C, C') and its beliefs are $\gamma = \frac{1}{3}$ upon observing no education (in the equilibrium path shaded in Fig. 11.1), and $\mu \leq \frac{2}{5}$ upon observing education (off-the-equilibrium path).

Step 2. Identify an off-the-equilibrium message for the sender.
   That is, find an action that no type of sender chooses in the PBE we are analyzing, e.g., education in the pooling PBE (NE^H, NE^L).

Step 3. Find which types of senders can obtain a higher utility level by deviating (i.e., when they send off-the-equilibrium messages) than by choosing their equilibrium message.
   In the pooling PBE (NE^H, NE^L), the high-productivity worker can benefit when she deviates from NE^H, where she earns a payoff of 4, to E^H, where she can earn a payoff of 6 if the firm responds hiring her as a manager. The low-productivity worker, however, cannot benefit when she deviates from NE^L, where she earns a payoff of 4, to E^L, where her highest payoff is 3 when the firm hires her as a manager.

Step 4. Restrict the off-the-equilibrium beliefs of the responder using the results of Step 3.
   Intuitively, if education is observed, the firm believes *it can only originate* from the high-productivity worker since the low-productivity worker cannot benefit from acquiring education. Therefore, the firm's beliefs upon observing education, μ, are restricted from $\mu \leq \frac{2}{5}$ to $\mu = 1$.

Step 5. Find optimal response given the restricted beliefs found in Step 4.
   In our ongoing example, upon observing education, the firm is convinced of dealing with a high-productivity worker, $\mu = 1$, and thus optimally responds hiring her as a manager, M (earning a profit of 10), which exceeds the profit of hiring the worker as a cashier (4).

Step 6. Given the optimal response found in Step 5, check if there is one or more sender types who can profitably deviate from her equilibrium message.
   (a) If there is, we say that the PBE we considered *violates* the Intuitive Criterion.
   (b) Otherwise, the PBE *survives* the Intuitive Criterion.

11.2.1.1 Pooling Equilibria May Violate the Intuitive Criterion

Consider the pooling PBE (NE^H, NE^L) of Fig. 11.1 (Step 1), where education is an off-the-equilibrium message because it is chosen by any worker type (Step 2). Moving on to Step 3, we note that only the high-productivity worker can profitably deviate towards education, as she could increase her payoff if the firm responds hiring her as a manager. In other words, the highest payoff she earns from deviating to education is 6, which exceeds her equilibrium payoff of 4.

In contrast, the low-productivity worker cannot profitably deviate, as the highest payoff she can earn from deviating is 3, which lies below her equilibrium payoff of 4. Informally, even if she fools the firm into believing that her productivity is high, and the firm hires her as a manager, her cost of acquiring education is too high, offsetting her wage increase. Formally, we say that the low-productivity worker finds education to be *equilibrium dominated*, since

$$u_L(NE^L, C') = 4 > 3 = \max_{a \geq 0} u_L(E^L, a)$$

where $a \in \{M, C\}$ denotes the firm's response. In contrast, the high-productivity worker does not find education to be equilibrium dominated

From Step 3, the firm can, in Step 4, restrict its off-the-equilibrium belief to $\mu = 1$. This means that, if the firm observes education, it must stem from the high-productivity worker. As a consequence, in Step 5, we find that the firm's optimal response to education, given $\mu = 1$, is to hire the worker as a manager, M, as its profit, 10, exceeds that of hiring her as a cashier, 4.

Finally, in Step 6, we obtain that one sender type (the high-productivity worker) has incentives to deviate from NE^H in the pooling PBE towards E^H, as she anticipates that the firm will respond hiring her as a manager (by Step 5). In conclusion, we can claim that this pooling PBE violates the Intuitive Criterion, implying that we can eliminate the pooling PBE (NE^H, NE^L) as being a solution to this game.

11.2.2 Separating Equilibria Survive the Intuitive Criterion

In the separating PBE (E^H, NE^L) of figure 10.5 there is no off-the-equilibrium message that we can identify in Step 2 of the Intuitive Criterion. In other words, all possible messages (education and no education) are being used by at least one worker type or, alternatively, no message is unused. In this setting, the separating PBE survives the Intuitive Criterion because, for a PBE to violate the Intuitive Criterion, we need to go through steps 1 to 6a, and in this case Step 2 does not identify off-the-equilibrium messages, entailing that we cannot move to step 3 and beyond. This argument applies to separating PBEs since all messages are used in equilibrium and, then, we cannot identify off-the-equilibrium messages. This point holds true for signaling games with the same number of sender types as messages (as in the standard labor-market signaling game of Chapter 10) and in games with more sender types than messages.

Table 11.1 Separating PBEs and the number of sender types

Separating PBEs	Survives Intuitive Criterion?
# sender types = # messages	√
# sender types > # messages	√
# sender types < # messages	Maybe

The argument, however, does not necessarily apply to "partially separating" PBEs, with fewer sender types than messages (e.g., two worker types and three education levels). Indeed, in a partially separating PBE leaving at least one of the messages is left unused, becoming an off-the-equilibrium message. In this setting, we can use off-the-equilibrium messages to check if the PBE survives the Intuitive Criterion, implying that the partially separating PBE may survive or violate this criterion. Table 11.1 summarizes our above discussion.

11.3 D1 Criterion

The D1 Criterion considers a similar approach as the Intuitive Criterion, as described in the six steps of tool 11.1, but differs in Step 3. In particular, for a given off-the-equilibrium message, the D1 Criterion seeks to answer the following question.

> If the receiver observes an off-the-equilibrium message,
>
> such as education on the right side of Fig. 11.1,
>
> for which sender type are most of the receiver's actions beneficial?

After that, we restrict the receiver's beliefs as in Step 4, and proceed through Steps 5 and 6 as in Tool 11.1. That is, the Intuitive Criterion seeks to identify which sender types could benefit from deviating towards the off-the-equilibrium message, which in some games could lead to a large number of sender types, thus not restricting the receiver's beliefs in Step 4 enough to be able to rule out PBEs.

The D1-Criterion poses a similar question, but in a sense it "goes deeper" by finding not only which sender types have incentives to deviate but which sender type is "the most likely to deviate" as we describe next.

1. First, we measure the number of the receiver's responses (e.g., salary offers from the firm) that would weakly improve the sender's payoff, relative to her equilibrium payoff, repeating this process for each sender type.
2. Second, we compare which sender type has the largest set of responses that would make her better off than in equilibrium.

11.3.1 Applying the D1 Criterion—An Example

For further illustration, we next test whether the (NE^H, NE^L) pooling PBE survives the D1-Criterion.

Step 1. Consider a specific PBE, such as the pooling PBE (NE^H, NE^L) we found in section 10.7.2.
Step 2. Identify an off-the-equilibrium message for the sender. In (NE^H, NE^L), the only off-the-equilibrium message is that of education, on the right-side information set.
Step 3. Find which sender type is *more likely to benefit* from the deviation of Step 2. (Recall that this is the main difference relative to the Intuitive Criterion.) To identify this type of sender, we go through the following steps:
   (a) When the high-productivity worker acquires education, she only has one response from the firm that can improve her equilibrium payoff, M, where her payoff increases from 4 to 6. When the low-productivity worker acquires education, however, she has no responses from the firm that could improve her equilibrium payoff of 4.
   (b) Comparing the number of payoff-improving responses for the high- and low-productivity workers, we find that the former has more than the latter (1 vs. 0) implying that the high type is more likely to deviate.
Step 4. (From this point, all steps coincide with those in the Intuitive Criterion.) Restrict the off-the-equilibrium beliefs of the responder using your results from Step 3.

   Intuitively, if education is observed, the firm believes it must originate from the high-productivity worker, restricting its off-the-equilibrium beliefs from $\mu \leq \frac{2}{5}$ to $\mu = 1$.
Step 5. Find optimal response given the restricted beliefs found in Step 4. Upon observing education, the firm is convinced of facing a high-productivity worker. If it responds hiring her as a manager, M, it earns a profit of 10, which exceeds the profit of hiring her as a cashier, 4.
Step 6. Given the optimal response found in Step 5, check if there is one or more sender types who can profitably deviate from her equilibrium message.
   (a) If there is, we say that the PBE we considered *violates* the D1-Criterion.
   (b) Otherwise, the PBE *survives* the D1-Criterion.

In our ongoing example, after Step 5, we see that the high-productivity (low-productivity) deviates (does not deviate) as she anticipates the firm will respond to education hiring her as a manager, M. As a consequence, we can conclude that pooling PBE (NE^H, NE^L) violates the D1-Criterion.

11.3.2 Discrete and Continuous Responses

The application of these two steps slightly differs depending on whether the receiver has a discrete or continuous strategy space. For presentation purposes, we separately discuss each case next.

Discrete responses. In a setting with discrete responses, such as when the firm hires the applicant as a CEO, manager, or cashier, the first step would just count how many of the firm responses can improve the high-productivity worker's equilibrium payoff, how many can improve the low-productivity worker's, and similarly for each worker's type. In the second step, we would compare which worker type has more payoff-improving responses, identifying her as the one who is " the most likely to deviate."

Continuous responses. In a context with continuous responses, such as when the firm responds with a salary $w > 0$ to the worker's observed education, these two steps are slightly different. In particular, the first step would measure the set of firm responses that improve the worker's productivity (i.e., a wage interval). The second step would compare the size of these wage intervals across worker types. In this context, the worker with the longest "utility-improving wage interval" is the one who is the "most likely to deviate." (We return to signaling games with continuous action spaces in Chapter 12, and apply the Intuitive and D1 Criteria in sections 12.4 and 12.5.)

11.3.3 Comparing Intuitive and Divinity Criteria

We found the same refinement result as when applying the Intuitive Criterion; but that is not generally the case. Both refinement criteria yield the same results in signaling games with only two sender types, explaining why the Intuitive Criterion is used in this class of games, as it is often easier to apply and provides the same refinement power. When the sender has three or more possible types, however, the D1-Criterion gives us weakly more refinement. Formally, this means that, if strategy profile $s = (s_1, s_2, ..., s_N)$ can be sustained as a PBE,

$$s \text{ survives the D1-Criterion} \underset{\not\Leftarrow}{\implies} s \text{ survives the Intuitive Criterion}$$

We return to this point in our analysis of signaling games with continuous action spaces in Chapter 12 (section 12.7) but, intuitively, their differences arise from Step 3. In the Intuitive Criterion, Step 3 only helps us to eliminate those sender types for which deviations are equilibrium dominated (i.e., they are better off behaving as prescribed by the PBE). In the D1-Criterion, in contrast, Step 3 identifies which sender type is the most likely to benefit from the deviation, implying that we focus our attention on a specific type of sender (or types) found in Step 3 of the Intuitive Criterion.

11.3.4 Other Refinement Criteria

The literature offers other refinement criteria in signaling games which you may consider when analyzing PBEs where the Intuitive or D1 Criterion have no bite, or where a large number of PBEs survive these two criteria. Examples include the "divinity" and "universal divinity" criteria, after Banks and Sobel (1987). Divinity is a weakening of the D1-Criterion, as it requires that the receiver's posterior belief after observing an off-the-equilibrium message cannot increase the likelihood ratio of the sender being type θ relative to θ'.

The D1-Criterion, in contrast, sets a stronger requirement, as it puts zero probability weight on this type of sender. Universal divinity, instead, requires that a sender type θ is eliminated, making updated beliefs independent on priors. These refinement criteria are, however, beyond the scope of this book and the advanced reader can refer to Banks and Sobel (1987) for more details. For experiments testing if subject behave as prescribed by the Intuitive and D1 Criteria, among other refinements, see Brandts and Holt (1992) and Banks et al. (1994).

11.4 Sequential Equilibrium

Another refinement criterion that helps us eliminate PBEs that are sustained on insensible off-the-equilibrium beliefs is the "sequential equilibrium" solution concept that we define next. While refining PBEs, it yields similar results as the Intuitive or D1 Criteria (i.e., same set of surviving PBEs) and, importantly, it is not as straightforward to apply as these two criteria, where we only need to check for sender types who could benefit from deviating to an off-the-equilibrium message. For this reason, most of our analysis in subsequent chapters, where we often need to deploy some refinement tools to trim the set of PBEs, uses the Intuitive or D1 Criteria for simplicity.

For the following definition, consider a behavioral strategy profile $b = (b_i, b_{-i})$. Recall from Chapter 6 (appendix), that player i's behavioral strategy b_i prescribes which action she selects when she is called to move at information set h_i, allowing her to choose one action with certainty (e.g., one branch) among those available at information set h_i or, instead, to randomize over two or more actions (e.g., branches) available to her at h_i.

Definition 11.1. **Sequential equilibrium (SE).** Behavioral strategy profile $b = (b_i, b_{-i})$ and beliefs μ over all information sets are sustained as a sequential equilibrium if:

1. *Sequential rationality.* Behavioral strategy b_i specifies optimal actions for every player i at each information set where she is called to move, given the strategies of the other players, b_{-i}, and given the system of beliefs μ; and
2. *Belief consistency.* There is a sequence $\{(b^k, \mu^k)\}_{k=1}^{\infty}$ such that, for all k:

11.4 Sequential Equilibrium

(a) b^k is a totally mixed behavioral strategy, assigning a positive probability weight to all available actions at each information set h_i where player i gets to play.
(b) μ^k is consistent with Bayes' rule given b^k.
(c) Sequence $\{(b^k, \mu^k)\}_{k=1}^{\infty}$ converges to (b, μ).

While sequential rationality (point 1) is analogous to the requirement on the PBE definition, belief consistency (point 2) differs because it uses totally mixed behavioral strategies. Graphically, point 2a implies that every information set on the game tree is reached with positive probability, which did not necessarily happen in PBEs where some information sets may not be reached in equilibrium (see, for instance, pooling PBE in Fig. 11.1). By reaching all information sets, Bayes' rule is no longer undefined, and helps us update beliefs given players' behavior (point 2b). Finally, point 2c states that the perturbations in the sequence $\{(b^k, \mu^k)\}_{k=1}^{\infty}$ become smaller as k increases, ultimately converging to (b, μ).

Intuitively, both PBE and SE require players to choose sequentially rational actions at every information set they are called to move. However, PBE imposes no restrictions on off-the-equilibrium beliefs. The SE solution concept, instead, requires that off-the-equilibrium beliefs are consistent with Bayes' rule when players deviate from their equilibrium strategies with a positive probability (they suffer from a perturbation, or "tremble"). Therefore, SE helps us identify which PBEs are sustained on consistent off-the-equilibrium beliefs (refining the set of PBEs). However, SE exhibits no refining power on separating PBEs since all information sets are reached in equilibrium or, in other words, there are no off-the-equilibrium beliefs.

This is an important point to remember: every separating PBE is also a SE, but not all pooling PBEs are necessarily SEs as some may be sustained on insensible off-the-equilibrium beliefs. In summary, SE coincides or is a subset of all PBEs, that is,

$$(b, \mu) \text{ is a } SE \Rightarrow (b, \mu) \text{ is a } PBE.$$
$$\not\Leftarrow$$

The next section illustrates how to find SEs in the labor-market signaling game.

11.4.1 Finding Sequential Equilibria

Consider a variation of the labor market signaling game where the probability of the worker's productivity being high is now 2/3, as depicted in Fig. 11.2. In this setting, three strategy profiles can be sustained as PBEs (exercise 11.19 asks you to show these results as a practice):

1. The separating PBE $(E^H, NE^L; M, C')$, where only the high-productivity worker acquires education, sustained with equilibrium beliefs $\mu = 1$ ($\gamma = 0$) after observing education (no education, respectively).

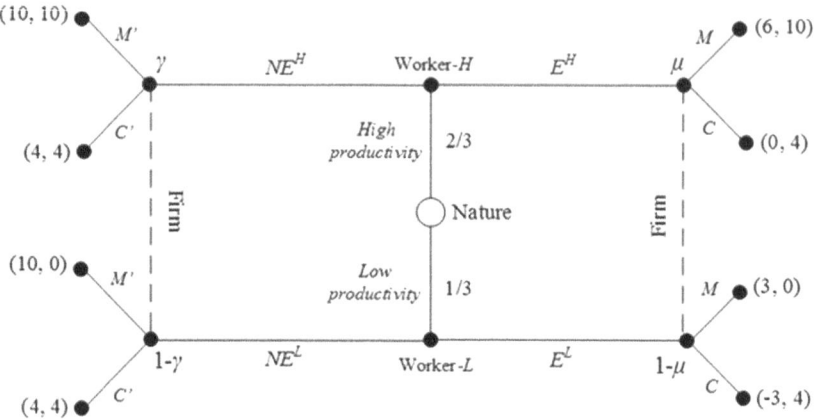

Fig. 11.2 Modified labor market signaling game

2. The pooling PBE $(NE^H, NE^L; M, M')$, where no type of worker acquires education, the firm holds off-the-equilibrium beliefs $\mu > 2/5$ and responds with M upon observing education.
3. The pooling PBE $(NE^H, NE^L; C, M')$, where no type of worker acquires education, the firm holds off-the-equilibrium beliefs $\mu \leq 2/5$ and responds with C upon observing education.

We next study if each of these strategy profiles can be sustained as a SE.

11.4.2 Separating PBEs that Are Also SEs

From our above discussion, every separating PBE is a SE. For completeness, we next confirm this result in the separating strategy profile $(E^H, NE^L; M, C')$, depicted in Fig. 11.3, which is supported as a PBE with beliefs $\mu = 1$ and $\gamma = 0$.

First, consider the following totally mixed behavioral strategy

$$b^k = \left(\underbrace{\left(1 - \varepsilon^k, \varepsilon^k\right)}_{\text{Worker-}H}, \underbrace{\left(\varepsilon^k, 1 - \varepsilon^k\right)}_{\text{Worker-}L}, \underbrace{\left(1 - 2\varepsilon^k, 2\varepsilon^k\right)}_{\text{Firm after educ.}}, \underbrace{\left(2\varepsilon^k, 1 - 2\varepsilon^k\right)}_{\text{Firm after no educ.}} \right)$$

where $\varepsilon^k > 0$ denotes a small perturbation. The first component of the high (low) type worker's behavioral strategy denotes the probability that she chooses E^H (E^L), and the first component in the third (fourth) parenthesis represents the probability that the firm responds with M (M') upon observing education (no education, respectively). This totally mixed behavioral strategy converges to $(1, 0; 1, 0)$ when

11.4 Sequential Equilibrium

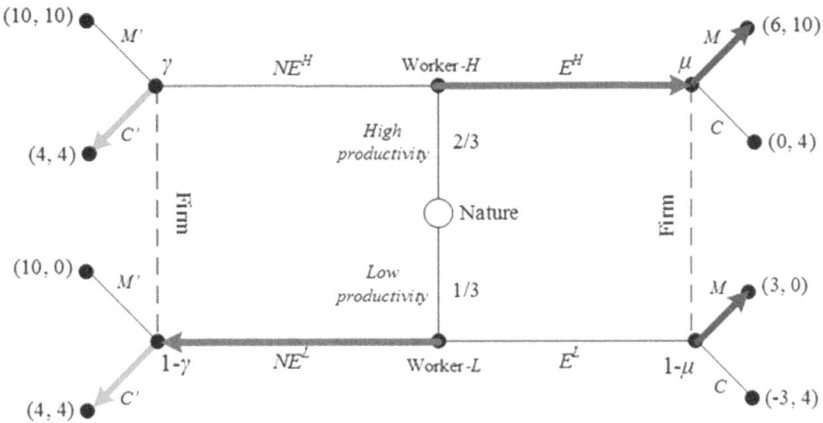

Fig. 11.3 Separating strategy profile $(E^H, NE^L; M, C')$

$k \to +\infty$, yielding outcome $(E^H, NE^L; M, C')$. In addition, the belief system (μ^k, γ^k) is consistent with Bayes rule if

$$\mu^k = \frac{\frac{2}{3}(1-\varepsilon^k)}{\frac{2}{3}(1-\varepsilon^k) + \frac{1}{3}\varepsilon^k} = \frac{2(1-\varepsilon^k)}{2-\varepsilon^k} \quad \text{upon observing education, and}$$

$$\gamma^k = \frac{\frac{2}{3}\varepsilon^k}{\frac{2}{3}\varepsilon^k + \frac{1}{3}(1-\varepsilon^k)} = \frac{2\varepsilon^k}{1+\varepsilon^k} \quad \text{upon observing no education.}$$

For illustration purposes, Fig. 11.4 plots these beliefs as a function of k, showing that they converge to $\lim_{k \to +\infty} \mu^k = 1$ and $\lim_{k \to +\infty} \gamma^k = 0$. Therefore, updated beliefs coincide with those in the separating PBE, $\mu = 1$ and $\gamma = 0$, as required.

Finally, given these beliefs, the firm's responses, (M, C'), satisfy sequential rationality, as required by SE, since its profits are $\pi(M) = 10 > 4 = \pi(C)$ after observing education and $\pi(C') = 4 > 0 = \pi(M')$ after no education. Therefore, the PBE $(E^H, NE^L; M, C')$ sustained with beliefs $\mu = 1$ and $\gamma = 0$ is also a SE of this game.

11.4.3 A Pooling PBE that Is Not a SE

Let us now consider pooling PBE $(NE^H, NE^L; M, M')$, depicted in Fig. 11.5, which is based on insensible off-the-equilibrium beliefs. Indeed, $\mu \leq 2/5$ means that, upon observing the off-the-equilibrium message of education, the firm assigns a relatively low weight to such a message originating from the high-productivity worker. Condition $\mu \leq 2/5$ even allows for $\mu = 0$, meaning that, upon observing an educated worker, the firm is convinced to face a low-productivity worker! We next demonstrate that this PBE is not a SE.

Fig. 11.4 Updated beliefs μ^k and γ^k as a function of k

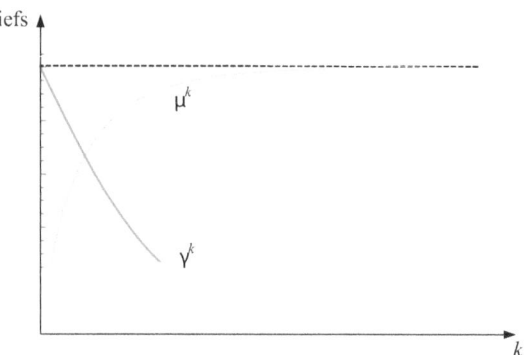

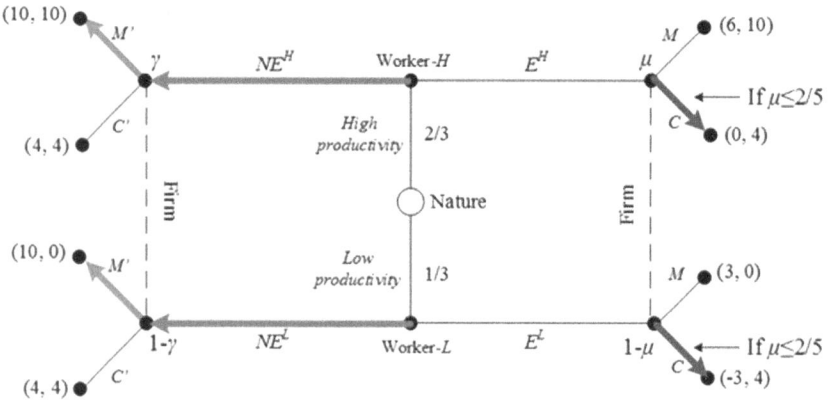

Fig. 11.5 Pooling strategy profile $(NE^H, NE^L; M, M')$

To show this result, recall that when players choose a sequence of totally mixed behavioral strategies $\{b^k\}_{k=1}^{\infty}$, the firm's right-hand information set is reached with positive probability (i.e., E^H and E^L are played with positive probabilities). Because this information set is reached, we no longer face off-the-equilibrium information sets, and the firm's beliefs can be updated by Bayes' rule. In particular, if both types of workers choose education with the same small perturbation, ε^k, Bayes' rule entails that

$$\mu^k = \frac{\frac{2}{3}\varepsilon^k}{\frac{2}{3}\varepsilon^k + \frac{1}{3}\varepsilon^k} = \frac{2}{3}$$

implying that posteriors and priors coincide. Indeed, when perturbation ε^k is symmetric across players, the two nodes of the right-hand information set are reached with the same (small) probability, as in pooling strategy profiles. Given this updated belief, the firm responds with M after observing education since its

11.4 Sequential Equilibrium

expected profits satisfy $E\pi(M) = \frac{20}{3} > 4 = E\pi(C)$. Therefore, responding with C cannot be part of a SE. We next provide a formal proof, describing the sequence of totally mixed behavioral strategies.

Consider the following totally mixed behavioral strategy

$$b^k = \left(\underbrace{\left(\varepsilon^k, 1-\varepsilon^k\right)}_{\text{Worker-}H}, \underbrace{\left(\varepsilon^k, 1-\varepsilon^k\right)}_{\text{Worker-}L}, \underbrace{\left(2\varepsilon^k, 1-2\varepsilon^k\right)}_{\text{Firm after educ.}}, \underbrace{\left(1-2\varepsilon^k, 2\varepsilon^k\right)}_{\text{Firm after no educ.}} \right)$$

which converges to $(0, 0; 0, 1)$ when $k \to +\infty$, yielding strategy profile $(NE^H, NE^L; C, M')$. In addition, the belief system (μ^k, γ^k) is consistent with Bayes rule if

$$\mu^k = \frac{\frac{2}{3}\varepsilon^k}{\frac{2}{3}\varepsilon^k + \frac{1}{3}\varepsilon^k} = \frac{2}{3} \quad \text{upon observing education, and}$$

$$\gamma^k = \frac{\frac{2}{3}(1-\varepsilon^k)}{\frac{2}{3}(1-\varepsilon^k) + \frac{1}{3}(1-\varepsilon^k)} = \frac{2}{3} \quad \text{upon observing no education.}$$

which converge to $\lim_{k \to +\infty} \mu^k = \gamma^k = \frac{2}{3}$.

Finally, given these beliefs, the firm's responses, (C, M'), *violate* sequential rationality, since $E\pi(M) = \frac{20}{3} > 4 = E\pi(C)$ after observing education and, similarly, $E\pi(M') = \frac{20}{3} > 4 = E\pi(C')$ after no education. In conclusion, the PBE $(NE^H, NE^L; C, M')$, supported with $\mu \leq 2/5$, is not a SE.

11.4.4 A Pooling PBE that Is Also A SE

Consider now the pooling PBE $(NE^H, NE^L; M, M')$, depicted in Fig. 11.6. As shown in section 10.7.2, this PBE is supported if the firm, upon observing the off-the-equilibrium message of education, holds beliefs $\mu > 2/5$. In this case, off-the-equilibrium beliefs indicate that, when the firm observes education, it assigns relatively high probability on it originating from the high-productivity worker; which seem relatively sensible. We next confirm that this PBE is also a SE.

To see this point, consider the totally mixed behavioral strategy

$$b^k = \left(\underbrace{\left(\varepsilon^k, 1-\varepsilon^k\right)}_{\text{Worker-}H}, \underbrace{\left(\varepsilon^k, 1-\varepsilon^k\right)}_{\text{Worker-}L}, \underbrace{\left(1-2\varepsilon^k, 2\varepsilon^k\right)}_{\text{Firm after educ.}}, \underbrace{\left(1-2\varepsilon^k, 2\varepsilon^k\right)}_{\text{Firm after no educ.}} \right)$$

which converges to $(0, 0; 1, 1)$ when $k \to +\infty$, yielding outcome $(NE^H, NE^L; M, M')$. In addition, the belief system (μ^k, γ^k) is consistent with

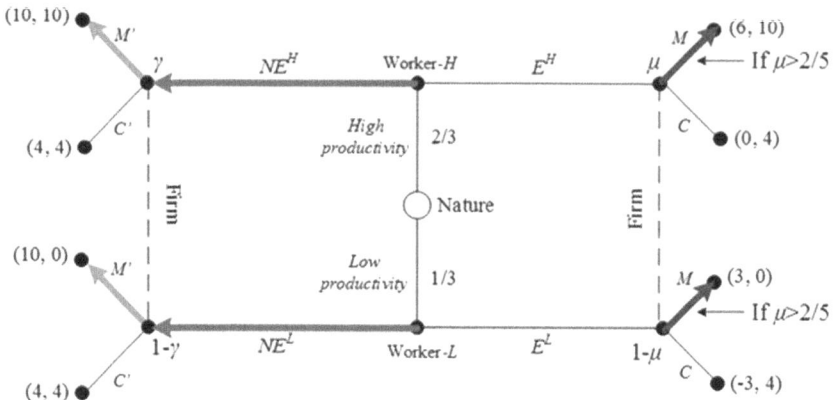

Fig. 11.6 Pooling strategy profile $(NE^H, NE^L; M, M')$

Bayes rule, and produces the same results as b^k in Sect. 11.4.1, that is, $\mu^k = \frac{2}{3}$ upon observing education, and $\gamma^k = \frac{2}{3}$ upon observing no education; converging to $\lim_{k \to +\infty} \mu^k = \gamma^k = \frac{2}{3}$.

Finally, given these beliefs, the firm's responses, (M, M'), *satisfy* sequential rationality, since $E\pi(M) = \frac{20}{3} > 4 = E\pi(C)$ after observing education and, similarly, $E\pi(M') = \frac{20}{3} > 4 = E\pi(M')$ after no education. Consequently, the PBE $(NE^H, NE^L; M, M')$, supported with $\mu > 2/5$, is also a SE.

Exercises

11.1 **Applying the Intuitive Criterion to the Beer-Quiche game.**[A] Consider the Beer-Quiche game of exercise 10.1. In that exercise, we found two pooling PBEs: (i) one where both types of player 1 order beer for breakfast, (B^S, B^W), which can be sustained if player 2's off-the-equilibrium belief upon observing quiche, γ, satisfies $\gamma > \frac{1}{2}$, leading player 2 to respond with (ND, D'); and (ii) another where both types of player 1 order quiche for breakfast, (Q^S, Q^W), which can be sustained if player 2's off-the-equilibrium belief upon observing beer, μ, satisfies $\mu > \frac{1}{2}$, leading player 2 to respond with (D, ND'). We now test if these equilibria survive the Intuitive Criterion.
   (a) Does the pooling PBE (B^S, B^W) survive the Intuitive Criterion?
   (b) Does the pooling PBE (Q^S, Q^W) survive the Intuitive Criterion?

11.2 **Applying the Intuitive Criterion to the labor market signaling game when education is productivity enhancing.**[B] Consider the labor market signaling game in exercises 10.2 and 10.3.

(a) In exercise 10.2, we found only one separating PBE, where the high-productivity worker acquires education, (E^H, NE^L). Show that this PBE survives the Intuitive Criterion.

(b) In exercise 10.3, we found only one pooling PBEs, where none of the worker types acquire education, (NE^H, NE^L), which can be supported regardless of the firm's off-the-equilibrium belief upon observing education, $q \in [0, 1]$. Check if this PBE survives the Intuitive Criterion.

11.3 **Applying the Intuitive Criterion to the labor market signaling game with intrinsic value of education.**[B] Consider the labor market signaling game in exercise 10.4, assuming for simplicity that both worker types assign the same intrinsic value to education, $v_H = v_L = v \geq 0$. As shown in exercise 10.4, part (e), two pooling strategy profiles can be sustained as PBEs: (i) one where both worker types acquire education, (E^H, E^L), if $v \geq 7$ and $p > \frac{2}{5}$, or $v \geq 1$ and $p \leq \frac{2}{5}$; and (ii) one where none of the worker types acquires education, (NE^H, NE^L) if $v \leq 4$ and $q > \frac{2}{5}$, or $v \leq 10$ and $q \leq \frac{2}{5}$.

(a) Does the pooling PBE (E^H, E^L) survive the Intuitive Criterion?

(b) Does the pooling PBE (NE^H, NE^L) survive the Intuitive Criterion?

11.4 **Applying the Intuitive Criterion to the labor market signaling game with cost differentials.**[B] Consider the labor market signaling game of exercise 10.5. In this game, we found two pooling PBEs: (i) one where both worker types acquire education, (E, E'), which can be sustained if the firm's off-the-equilibrium belief upon observing no education, p, satisfies $p \leq \frac{2}{5}$ and $c^L \leq 6$; and (ii) one where none of the worker types acquires education, (NE, NE'), which can be sustained regardless of the firm's off-the-equilibrium belief upon observing education, $q \in [0, 1]$.

(a) Does the pooling PBE (E, E') survive the Intuitive Criterion?

(b) Does the pooling PBE (NE, NE') survive the Intuitive Criterion?

11.5 **Applying the Intuitive Criterion to the labor market signaling game with profit differentials.**[B] Consider the labor market signaling game in exercise 10.6. In that exercise, we found two pooling PBEs: (i) one where both worker types acquire education, (E, E'), when the profit differential between worker types is relatively low, $d < 12$, and the firm's off-the-equilibrium belief upon observing no education, p, satisfies $p \leq 1 - \frac{6}{d}$; and (ii) one where no worker type acquires education, (NE, NE'), when the profit differential is low, $d < 12$, regardless of the firm's off-the-equilibrium belief upon observing education, $q \in [0, 1]$, and when the profit differential is high, $d \geq 12$, and $q \leq 1 - \frac{6}{d}$.

(a) Does the pooling PBE (E, E') survive the Intuitive Criterion?

(b) Does the pooling PBE (NE, NE') survive the Intuitive Criterion?

11.6 **Applying the Intuitive Criterion to the Brinkmanship game.**[A] Consider the Brinkmanship game from exercise 10.7. In that setting, we found: (i) a pooling PBE, (S^{SU}, S^{WI}), where player 2's beliefs $p < \frac{3}{5}$; and (ii) a semi-separating PBE where the wimpy type of player 1 stands firm with probability $q = \frac{3(1-p)}{2p}$, where p denotes the prior probability of player 1 being wimpy, the surely type of player 1 chooses to stand firm in pure strategies, player 2's updated beliefs are $\mu = \frac{3}{5}$ and responds fighting with probability $r = \frac{3}{5}$.

(a) Does the pooling PBE (S^{SU}, S^{WI}) survive the Intuitive Criterion?
(b) Does the semi-separating PBE described above survive the Intuitive Criterion?

11.7 **Applying the Intuitive Criterion to the labor market signaling game with three responses.**[B] Consider the labor market signaling game of exercise 10.9, where the worker still has two types (high or low productivity), two messages (education or no education), but the firm has three potential responses (hiring the worker as a director, manager, or cashier). In that exercise, we found: (i) a pooling PBE where both worker types acquire education, (E^H, E^L), which can be sustained if the firm's off-the-equilibrium beliefs upon observing no education, γ, satisfy $\gamma \leq \frac{5}{13}$; and (ii) a pooling PBE where none of the worker types acquires education, (NE^H, NE^L), which can be supported regardless of the firm's off-the-equilibrium beliefs, $\mu \in [0, 1]$.

(a) Does the pooling PBE (E^H, E^L) survive the Intuitive Criterion?
(b) Does the pooling PBE (NE^H, NE^L) survive the Intuitive Criterion?

11.8 **Applying the Divinity Criterion to the labor market signaling game with three types.**[B] Consider the labor market signaling game of exercise 11.7.

(a) Does the pooling PBE (E^H, E^L) survive the D1 Criterion?
(b) Does the pooling PBE (NE^H, NE^L) survive the D1 Criterion?

11.9 **Applying the Intuitive Criterion to the labor market signaling game with three messages.**[C] Consider the labor market signaling game of exercise 10.10, where the worker still has two types (high or low productivity), but three messages (advanced degree, education, or no education), and the firm still has two potential responses (hiring the worker as a manager or cashier). In that exercise, we found: (i) a separating PBE where the high (low) productivity worker acquires an advanced degree (no education, respectively), (A^H, NE^L), which can be sustained if the firm's off-the-equilibrium beliefs upon observing education, μ, satisfy $\mu < \frac{2}{5}$; and (ii) a pooling PBE where none of the worker types acquires education, (NE^H, NE^L), which can be supported if the firm's off-the-equilibrium beliefs upon observing an advanced degree, β, or education, μ, satisfy $\beta, \mu < \frac{2}{5}$.

(a) Does the separating PBE (A^H, NE^L) survive the Intuitive Criterion?
(b) Does the pooling PBE (NE^H, NE^L) survive the Intuitive Criterion?
(c) Does any of the above pooling PBEs survive the D1 criterion?

11.10 **Applying the Intuitive Criterion to the labor market signaling game with three worker types.**[C] Consider the labor market signaling game of exercise 10.10, where the worker has three possible types (high, medium, or low productivity), chooses whether or not to acquire education, and the firm responds hiring him as a manager or cashier. In that exercise, we found: (i)

Fig. 11.7 PBEs when all players are uninformed-I

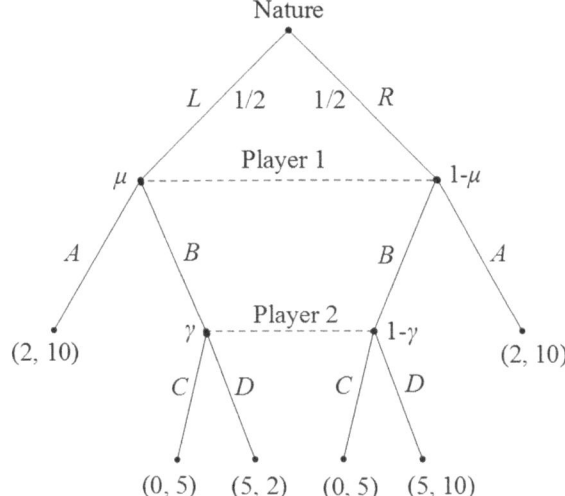

a pooling PBE where all worker types acquire education, (E^H, E^M, E^L), which is supported when the firm's off-the-equilibrium beliefs, γ_1 and γ_2, satisfy $2\gamma_1 + \gamma_2 < \frac{4}{5}$; and (ii) a pooling PBE where no worker type acquires education, (NE^H, NE^M, NE^L), which is sustained regardless of the firm's off-the-equilibrium beliefs.

(a) Does the pooling PBE (E^H, E^M, E^L) survive the Intuitive Criterion?
(b) Does the pooling PBE (NE^H, NE^M, NE^L) survive the Intuitive Criterion?

11.11 **Applying the Divinity Criterion to the labor market signaling game with three worker types.**[B] Consider the labor market signaling game of exercise 10.11. As shown in exercise 11.10, the two pooling PBEs in this game survive the Intuitive Criterion.
(a) Does the pooling PBE (E^H, E^M, E^L) survive the D1 Criterion?
(b) Does the pooling PBE (NE^H, NE^M, NE^L) survive the D1 Criterion?

11.12 **PBEs when all players are uninformed-I.**[C] Consider the sequential-move game in Fig. 11.7. First, nature selects L or R, both equally likely (probability is 1/2). Second, player 1 responds choosing A or B but, unlike in standard signaling games, he cannot observe the move of nature; that is, player 1 does not observe whether L or R where selected, which explains his information set. If player 1 responds with A, the game is over. If he chooses B, then player 2 responds with C or D, not observing the move of nature, so he faces an information set.
(a) Find the Bayesian Nash Equilibria (BNEs) of this game. For simplicity, consider only pure strategies.
(b) Argue that the BNEs found in part (a) are also Subgame Perfect Nash Equilibria (SPNEs).

(c) Check if all BNEs found in part (a) are also Perfect Bayesian Equilibria (PBEs). Interpret your results.
(d) Argue that all PBEs identified in part (c) survive Cho and Kreps' Intuitive Criterion. Argue that one of the PBEs is based on insensible off-the-equilibrium beliefs.

11.13 **PBEs when all players are uninformed-II.**^{C} Consider the sequential-move game in exercise 11.12, but assume that player 1 observes the move of nature.
(a) Find the set of PBEs in this game.
(b) Do all of them survive Cho and Kreps' Intuitive Criterion?
(c) Compare your results in part (b) with those in part (d) of the previous exercise.

11.14 **PBEs when three players are uninformed.**^{C} Consider the game tree in Fig. 11.8, based on Gonzalez-Diaz and Melendez-Jimenez (2014), where player 3 has type I with probability p and type II with probability $1 - p$, where $p \in [0, 1]$. Once the type of player 3 is realized, but without observing it, player 1 chooses whether to continue the game (C) or to stop (S). If player 1 chooses S, the game is over; but if he chooses C, player 2 responds with U or D. Player 2 observes player 1's action (C), but does not observe player 3's type. Finally, player 3 is called to move if his type is I, responding with u or d, but without observing player 2's action.
(a) Show that strategy profile (C, U, u) can be supported as a PBE of the game.

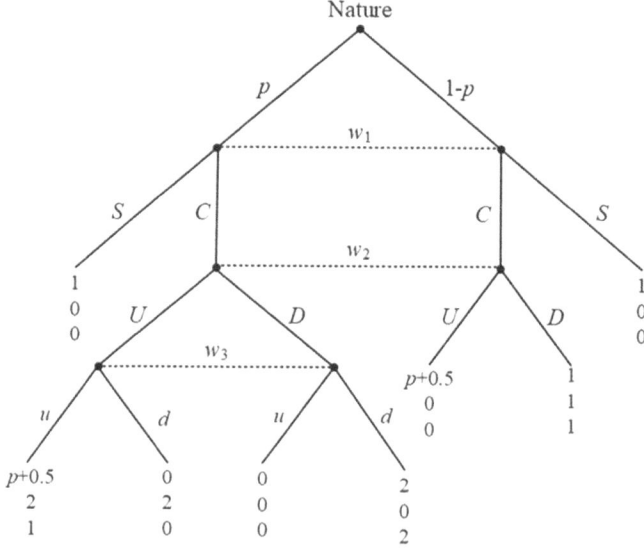

Fig. 11.8 PBEs when three players are uninformed

(b) Show that strategy profile (S, U, d) cannot be supported as a PBE of the game.

(c) Show that strategy profile (S, D, u) cannot be supported as a PBE of the game.

(d) Which of the PBEs found in parts (a)-(c) survive the Intuitive Criterion?

(e) Which of the PBEs found in parts (a)-(c) are sequential equilibria (SE)?

11.15 **Applying the Intuitive Criterion in the Selten's horse.**[C] Consider exercise 10.12, about the so-called Selten's horse. In that exercise, we showed that $\{C, C', R\}$ is the unique PBE of this game, which can be sustained when player 3's belief, μ, satisfies $\mu \leq \frac{1}{3}$. Check if this PBE survives the Intuitive Criterion.

11.16 **Finding sequential equilibria in the Selten's horse.**[C] Consider exercise 10.12, about the Selten's horse, where we found that $\{C, C', R\}$ is the unique PBE, which can be supported when player 3's off-the-equilibrium belief, μ, satisfies $\mu \leq \frac{1}{3}$. Show that this PBE is also a sequential equilibrium.

11.17 **PBEs with players holding asymmetric off-the-equilibrium beliefs.**[C] Consider exercise 10.13, where we showed that strategy profile $\{A, L_2, (L_3, L'_3)\}$ can only be supported as a PBE when off-the-equilibrium beliefs satisfy $\mu \leq \frac{1}{3}$ and $\gamma \geq \frac{2}{3}$, for all $\beta \in [0, 1]$. Show that this PBE is not a sequential equilibrium.

11.18 **Checking if PBEs are also SEs.**[B] Consider the game tree in exercise 10.14, which we reproduce below for easier reference. In that exercise, we found two PBEs: (i) (X, R) if player 2's beliefs are $\mu = 1$; and (ii) (Z, L) if $\mu \leq \frac{1}{2}$. In this exercise, we seek to show that these PBEs can also be supported as SEs.

(a) Show that (X, R) with $\mu = 1$ can be supported as a SE.

(b) Show that (Z, L) with $\mu \leq \frac{1}{2}$ can be supported as a SE.

11.19 **Finding PBEs in figure** 11.2.[A] Sect. 11.4.1 describes a modified version of the labor-market signaling game, depicted in Fig. 11.2, and mentions that only three strategy profiles can be sustained as PBEs. In this exercise, we show that result.

(a) Show that the separating strategy profile (E^H, NE^L), where only the high-productivity worker acquires education, can be sustained as a PBE.

(b) Show that the opposite separating strategy profile, (NE^H, E^L), where only the low-productivity worker acquires education, cannot be supported as a PBE.

(c) Show that the pooling strategy profile (NE^H, NE^L), where no type of worker acquires education, cannot be sustained as a PBE if the firm holds off the equilibrium beliefs $\mu \geq \frac{2}{5}$.

(d) Show that the pooling strategy profile (NE^H, NE^L), where no type of worker acquires education, can be supported as a PBE if the firm holds off-the-equilibrium beliefs $\mu < 2/5$.

(e) Compare your results in parts (c) and (d). Interpret.

11.20 **PBEs sustainable as SEs but only with extreme beliefs.**[B] Consider the game tree below where player 1 is the first mover, choosing between X and Y. If she chooses X, player 2 responds with A or B. If player 2 chooses A, the game is over. If she chooses B, player 3 responds, choosing between L and R. In this exercise, we seek to illustrate settings where player 3's beliefs in the SE are concentrated at zero or one.

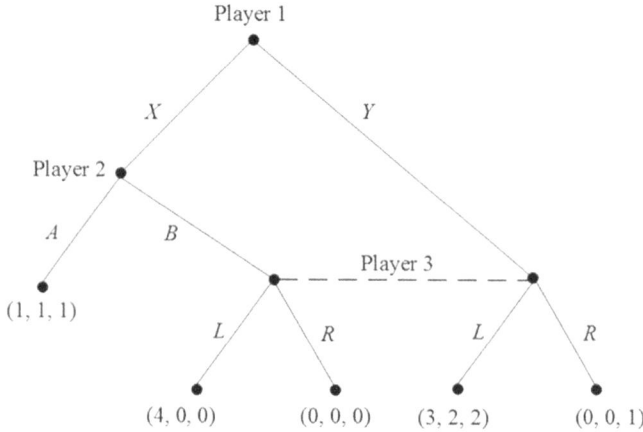

(a) Find the NEs in this game.
(b) Find the SPEs in this game.
(c) Check if the SPEs found in part (b) can be sustained as SEs.

References

Banks, J. S., and J. Sobel (1987) "Equilibrium selection in signaling games," Econometrica, 55, pp. 647–661.

Banks, J. S., F. C. F. Camerer and D. Porter (1994) "Experimental Tests of Nash Re-nements in Signaling Games," Games and Economic Behavior, 6, pp. 1–31.

Brandts, J. and C. A. Holt (1992) "An Experimental Test of Equilibrium Dominance in Signaling Games," American Economic Review, 82, pp. 1350–1365.

Cho, I.-K., and D. M. Kreps (1987) "Signaling Games and Stable Equilibria," Quarterly Journal of Economics, 102(2), pp. 179–221.

Gonzalez-Diaz, J. and M. A. Melendez-Jimenez (2014) "On the Notion of Perfect Bayesian Equilibrium," Top, 22, pp. 128–143.

12 Signaling Games with Continuous Messages

12.1 Introduction

Chapters 10 and 11 consider, for simplicity, players choosing from discrete action spaces (e.g., two messages from the worker and two responses from the firm). More general settings, however, may allow for the sender to choose her message from a continuous action space (e.g., years of education, $e \geq 0$) and, similarly, the receiver to respond with a continuous action (e.g., a wage $w \geq 0$).

This chapter extends Spence's labor market signaling game to a setting with continuous messages and responses. First, it presents equilibrium behavior under complete information, then we find the separating and pooling PBEs of the incomplete information game, and apply the refinement criteria studied in Chapter 11 (the Intuitive and D1 Criterion) to discard those PBEs sustained on insensible off-the-equilibrium beliefs. This presentation also helps us compare equilibrium results under complete and incomplete information to answer a more fundamental question: are workers better off when firms observe their productivity (as under complete information) or when they need to use education to signal their type to the employers (as under incomplete information)? Or, more informally, which worker types benefit when the firm becomes better informed, and which worker types would prefer the firm to "remain in the dark"?

We then further extend our model to a setting with three worker types, identifying the set of separating PBEs, and then refining it with the help of the Intuitive and D1 Criterion. Because of the presence of three types, the Intuitive Criterion has no bite, not letting us eliminate any separating PBE, but the D1 Criterion helps us discard all but one separating PBE, that where the sender invests the least in education to signal her type (i.e., the most efficient separating PBE, also known as the "Riley" outcome).

12.2 Utility Functions with Continuous Actions

Consider a worker with utility function

$$u(e, w | \theta_K) = w - c(e, \theta_K)$$

where $w \geq 0$ represents the salary that she receives from the firm, while $c(e, \theta_K)$ denotes her cost of acquiring e units (e.g., years) of education. As in previous chapters, the worker's productivity is denoted by θ_K where $K = \{H, L\}$. For simplicity, function $c(e, \theta_K)$ often satisfies the following four assumptions:

1. *Zero cost of no education.* When the worker acquires no education, she incurs no cost, that is, $c(0, \theta_K) = 0$.
2. *Cost is increasing and convex.* Cost function $c(e, \theta_K)$ is strictly increasing and convex in education, i.e., $c_e > 0$ and $c_{ee} > 0$, indicating that education is costly and that additional years of education become progressively more costly.
3. *Cost is decreasing in productivity.* Cost function $c(e, \theta_K)$ is decreasing in the worker's productivity, θ_K, i.e., $c(e, \theta_H) < c(e, \theta_L)$, implying that a given education e (e.g., a university degree) is easier to complete for the high-productivity than for the low-productivity worker.
4. *Marginal cost is decreasing in productivity (single-crossing property).* A similar property applies to the marginal cost of education, $c_e(e, \theta_K)$, which is also lower for the high-productivity than the low-productivity worker, that is, $c_e(e, \theta_H) < c_e(e, \theta_L)$. Our discussion of Fig. 12.1 justifies why $c_e(e, \theta_H) < c_e(e, \theta_L)$ is often referred to as the "single-crossing property."

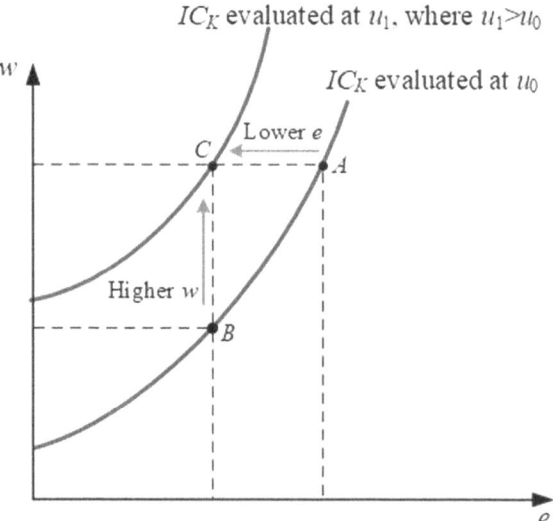

Fig. 12.1 Indifference curves for a representative worker

Example 12.1 Cost of acquiring education. As a practice, consider a cost function such as

$$c(e, \theta_K) = \frac{Ae^2}{\theta_K},$$

where $A > 0$ is a constant. We next check that $c(e, \theta_K)$ satisfies the above four assumptions. Graphically, we can depict the worker's indifference curve in the (e, w)–quadrant by, first, solving for w, which yields $w = u + c(e, \theta_K)$. (If $c(e, \theta_K) = \frac{Ae^2}{\theta_K}$, this indifference curve becomes $w = u + \frac{Ae^2}{\theta_K}$.) Since the cost of education is strictly increasing and convex in e, indifference curves are also monotonically increasing and convex in e; as depicted in Fig. 12.1.

Intuitively, a higher education level must be accompanied by an increase in her salary for the worker's utility to remain unchanged. Additionally, indifference curves to the northwest are associated with a higher utility level since, for a given education level e, the worker receives a higher wage (movement from point B to C in Fig. 12.1) or, for a given wage w, the worker acquires less education (movement from A to C in the figure).

In addition, since $\theta_H > \theta_L$, the indifference curve of the high-productivity worker is flatter than that of the low-productivity worker. Together with the property that indifference curves are increasing and convex for both worker types, we obtain that these curves only cross each other at one point in the (w, e)-quadrant. As an illustration, if $A = \theta_L = 1$ and $\theta_H = 2$, the indifference curve of the high-productivity worker is $u_H - \frac{e^2}{2}$, while that of the low-productivity worker is $u_L - e^2$, entailing that they cross once at $u_H - \frac{e^2}{2} = u_L - e^2$, or $\frac{e^2}{2} = u_H - u_L$, which yields $e = 4(u_H - u_L)^{1/2}$. □

Remark: When messages are discrete, the fourth assumption can be expressed as "cost increments," by saying that the cost the worker experiences when increasing her education from e to e', where $e' > e$, as captured by $c(e', \theta_K) - c(e, \theta_K)$, is smaller for the high- than the low-productivity worker, that is,

$$c(e', \theta_H) - c(e, \theta_H) < c(e', \theta_L) - c(e, \theta_L).$$

12.3 Complete Information

We assume that the labor market is perfectly competitive. In a complete information setting where the firm observes the worker's type, this assumption implies that every firm pays a salary equal to the worker's productivity $w = \theta_K$. If a firm pays, instead, a lower salary, $w < \theta_K$, other firms could offer a slightly higher salary w' where $w < w' \leq \theta_K$, which attracts the worker and yields a weakly positive profit. If, however, a firm pays a higher salary $w > \theta_K$, it would attract the worker but make a negative profit.

As a benchmark, we consider a complete information context where the firm observes the worker's productivity, θ_K. In this setting, the only subgame perfect equilibrium has the worker acquiring zero education regardless of her productivity, $e_H = e_L = 0$. The firm responds with salary $w(\theta_H) = \theta_H$ when the worker's productivity is high and $w(\theta_L) = \theta_L$ when her productivity is low. Intuitively, the worker cannot use education as a signal about her type since the firm observes her productivity.

The next section shows that, under incomplete information, education can become an informative (although costly) signal. Assume that the worker privately observes her type θ_K before choosing her education level e. The firm observes education e but does not know the worker's type θ_K. However, it assigns a prior probability $p \in [0, 1]$ to the worker's productivity being high and $1 - p$ to her productivity being low. This probability distribution is common knowledge among players.

12.4 Separating PBE

We next check if a separating strategy profile where each type of worker acquires a different education level can be sustained as a PBE. We follow the same four steps as in the discrete version of the game in Chapter 10.

First step, strategy profile. In a separating strategy profile, the high-productivity worker chooses education level e_H while the low-productivity worker chooses e_L, where $e_L \neq e_H$.

Second step, updating beliefs. Upon observing education level e_H, the firm assigns full probability to facing a high-productivity worker, that is, $\mu(\theta_H|e_H) = 1$. In contrast, after observing education level e_L, the firm assigns no probability to facing the high-productivity worker, i.e., $\mu(\theta_H|e_L) = 0$, as it is convinced of dealing with a low-productivity worker. If, instead, the firm observes the worker acquiring an off-the-equilibrium education level, i.e., an education e different than e_H and e_L, it cannot update its beliefs using Bayes' rule, leaving them unrestricted, that is, $\mu(\theta_H|e) \in [0, 1]$ for all $e \neq e_H \neq e_L$. (We impose some conditions on off-the-equilibrium beliefs below, but at this point we leave these beliefs unconstrained.)

Third step, optimal responses. Given the above beliefs, we must now find the firm's optimal responses. Upon observing education e_H, the firm pays a salary $w(e_H) = \theta_H$ since it is convinced of facing a high-productivity worker. Similarly, upon observing education e_L, the firm pays a salary $w(e_L) = \theta_L$ since it puts full probability at dealing with a low type. Intuitively, salaries coincide with the worker's productivity, thus being the same as under complete information. Yet, education levels in the separating PBE do not coincide with those under complete information, as we show below.

Upon observing off-the-equilibrium education $e \neq e_H \neq e_L$, the firm beliefs are $\mu(\theta_H|e) \in [0, 1]$, as discussed in the second step above. Therefore, the firm

12.4 Separating PBE

Fig. 12.2 Separating PBE—Two wage schedules

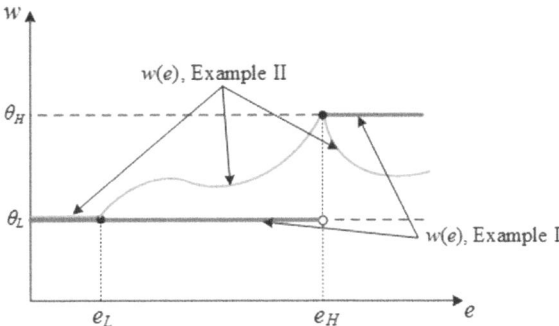

pays a salary between $w(e) = \theta_L$, which occurs when its beliefs are $\mu(\theta_H|e) = 0$, and $w(e) = \theta_H$, which happens when $\mu(\theta_H|e) = 1$, that is, $w(e) \in [\theta_L, \theta_H]$. As illustrated in Fig. 12.2, this wage schedule means that $w(e_L) = \theta_L$ at education level e_L and $w(e_H) = \theta_H$ at e_H (see dotted lines stemming from the horizontal axis). However, for all other education levels $e \neq e_H \neq e_L$, the wage can lie weakly above θ_L and below θ_H (but, of course, bounded between θ_L and θ_H). The wage schedule denoted as "Example I" in Fig. 12.2 is relatively straightforward since it determines a wage $w(e) = \theta_L$ for all education levels below e_H, but a wage $w(e) = \theta_H$ otherwise. Example II in this figure, however, pays a wage $w(e) = \theta_L$ to low education levels, $e \leq e_L$, a wage $w(e) = \theta_H$ to high education levels, $e \geq e_H$, and a non-linear schedule to intermediate education levels.

At first glance, one could expect wages to be increasing in the worker's education, so $w(e)$ increases in e, as Example I in Fig. 12.2 which is weakly increasing. However, our above discussion about the firm's optimal responses does not preclude the possibility of a decreasing portion in the wage schedule $w(e)$, as illustrated in Example II of Fig. 12.2 and in the only wage schedule of Fig. 12.3.

Fourth step, optimal messages. Given the above wage schedule from the firm, we must identify under which conditions the low-productivity worker has incentives to choose education e_L while the high-productivity worker chooses e_H.

Low-productivity worker. Starting with the low-type, let us find when e_L satisfies $e_L = 0$ (the lowest possible education). Figure 12.3 depicts the indifference curve of the low-productivity worker, IC_L, that passes through point $(e, w) = (0, \theta_L)$. At education level $e_L = 0$, the firm is convinced to deal with a low-productivity worker, paying a wage $w(e_L) = \theta_L$. Figure 12.3 also depicts a wage schedule $w(e)$ which guarantees that this type of worker has no incentives to deviate to education level $e_L = 0$. For instance, at point A, the worker acquires education level e_1, receiving a salary above θ_L, but the indifference curve passing through point A, IC_L^A, lies to the southeast of IC_L, thus associated to a lower utility level. Intuitively, the worker finds that the additional cost of education she incurs when deviating from $e_L = 0$ to $e_1 > 0$ offsets the extra salary she receives. A similar argument applies for any other education levels $e > e_L$, since mapping

Fig. 12.3 Separating PBE—Low-productivity worker

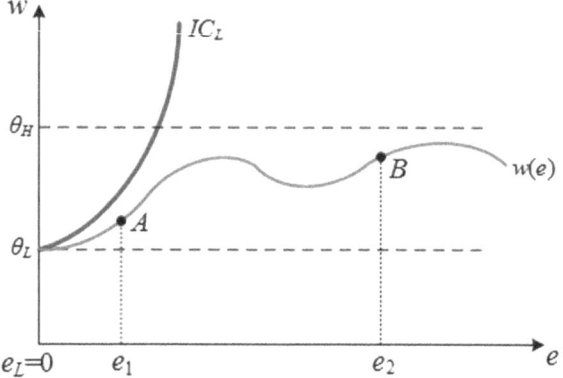

them into the firm's wage schedule $w(e)$ we obtain (e, w) pairs lying to the southeast of IC_L. More generally, for the low-productivity worker to stick to $e_L = 0$, we need that the firm's wage schedule lies below IC_L.[1]

High-productivity worker. The high-productivity worker chooses the education level prescribed in this separating strategy profile, e_H, rather than deviating to the low type's education, $e_L = 0$, if

$$\theta_H - c(e_H, \theta_H) \geq \theta_L - c(0, \theta_H)$$

since she can anticipate that acquiring e_H identifies her as a high-productivity worker, yielding a salary $w(e_H) = \theta_H$ while acquiring $e_L = 0$ identifies her as a low-productivity worker, where she receives $w(e_L) = \theta_L$ (see third step). Since cost $c(0, \theta_H) = 0$ by assumption, we can rearrange the above inequality as

$$\theta_H - \theta_L \geq c(e_H, \theta_H).$$

Intuitively, the high-productivity worker chooses education level e_H when the wage differential she enjoys, relative to her salary when deviating to $e_L = 0$, as captured by $\theta_H - \theta_L$, compensates her for the additional cost of education, $c(e_H, \theta_H) - c(0, \theta_H) = c(e_H, \theta_H)$. We can write a similar condition to express her incentives to not deviate toward off-the-equilibrium education levels $e \neq e_H \neq e_L$, as follows

$$\theta_H - c(e_H, \theta_H) \geq w(e) - c(e, \theta_H).$$

---

[1] To see this point, depict a different wage schedule $w(e)$ in Fig. 12.3, starting at $(e, w) = (0, \theta_L)$ but lying above IC_L for at least some (e, w) pairs, i.e., the wage schedule could cross through IC_L once or more than once. The low-productivity worker will now have incentives to deviate from education level $e_L = 0$ to a positive education e' since the additional salary she earns compensates her for the additional cost of education.

12.4 Separating PBE

Figure 12.4 depicts a wage schedule $w(e)$ that provides the high-productivity worker with incentives to choose education level e_H, where she reaches indifference curve IC_H, rather than deviating toward any other education $e \neq e_H$. For instance, at a lower education e_1, her salary is represented by the height of point A. An indifference curve crossing through this point yields a lower utility level than that at IC_H since it lies to the southeast of IC_H. Therefore, the high-productivity worker does not have incentives to deviate from e_H to e_1. A similar argument applies to all other education levels $e < e_H$, including $e_L = 0$; and to all education levels above e_H, such as e_2, where her salary is represented by the height of point B.

Other wage schedules also induce the high-productivity worker to choose education e_H, such as that of Fig. 12.5a. Intuitively, this wage schedule indicates that the firm pays the lowest salary $w(e) = \theta_L$ upon observing $e < e_H$, but pays the highest salary otherwise. As a practice, the figure also depicts education levels e_1 and e_2, confirming that the worker does not have incentives to deviate from e_H.

Interestingly, different wage schedules help us support different education levels e_H for the high-productivity worker, such as that in Fig. 12.5b, where the firm only pays the high salary θ_H when e_H is extremely high, that is, $e_H = e_4$. In this case,

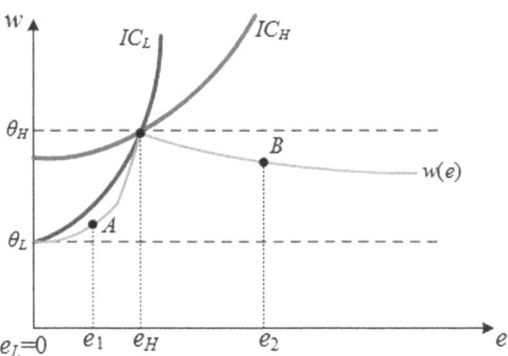

Fig. 12.4 Separating PBE—Low- and high-productivity worker

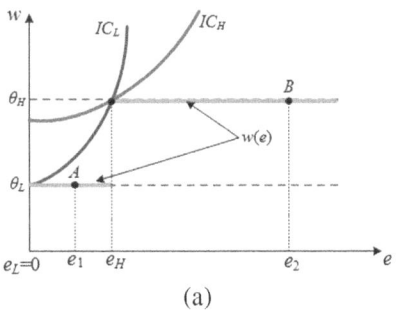

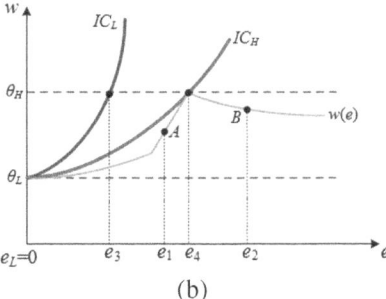

Fig. 12.5 Separating PBE: a Case I, b Case II

the high-productivity worker, despite receiving salary $w(e_4) = \theta_H$, is indifferent between acquiring education e_4 or deviating to $e_L = 0$. Specifically, e_4 solves

$$\theta_H - \theta_L = c(e_4, \theta_H).$$

The least-costly separating PBE, e_3, cannot be supported with the wage schedule depicted in Fig. 12.5b. It could be sustained with wage schedules such as those in Figs. 12.4 or 12.5a. Overall, a wage schedule may help us sustain one equilibrium but not another one.

Example 12.2 Identifying the highest education that achieves separation. Following with our setting in Example 12.1, consider that $\theta_H = 2$, $\theta_L = 1$, and that the cost of education function is given by $c(e, \theta_K) = \frac{e^2}{\theta_K}$. Solving for e_4 in the above equation, $\theta_H - \theta_L = \frac{e_4^2}{\theta_H}$, to obtain $e_4 = \sqrt{2} \simeq 1.41$. □

More generally, we can claim that, in separating PBEs of the labor market signaling game, while the low-productivity worker chooses $e_L = 0$, the high-productivity worker selects an education level e_H in the range $[e_3, e_4]$, where $e_H = e_3$ represents the "least-costly separating PBE," since the high type conveys her type to the firm acquiring the lowest education level, whereas $e_H = e_4$ represents the "most-costly separating PBE." (Generally, we refer to least-costly separating equilibria when the sender can convey her type spending as little as possible in costly signals.) Importantly, in all education levels $e_H \in [e_3, e_4]$, the low-productivity worker has no incentives to choose e_H (i.e., to mimic the high type) as that would yield a lower utility level than that at IC_L.

The least-costly separating education level $e_H = e_3$ solves

$$\theta_H - c(e_3, \theta_L) = \theta_L - c(0, \theta_L).$$

Intuitively, the low-productivity worker is indifferent between her equilibrium strategy $e_L = 0$, receiving a wage of θ_L, and deviating to education level e_3 which provides her with wage θ_H. Since $c(0, \theta_L) = 0$ by assumption, the above equation simplifies to

$$\theta_H - \theta_L = c(e_3, \theta_L).$$

Example 12.3 Finding the lowest education that achieves separation. In the setting of Example 12.1, education level e_3 solves $\theta_H - \theta_L = \frac{e_3^2}{\theta_L}$. Since $\theta_H = 2$ and $\theta_L = 1$, we obtain that $e_3 = 1$. Therefore, the low-productivity worker chooses an education level $e_L = 0$ in equilibrium, while the high-productivity worker selects e_H in the interval $[1, \sqrt{2}]$. □

Fig. 12.6 Applying the Intuitive Criterion to the separating PBEs

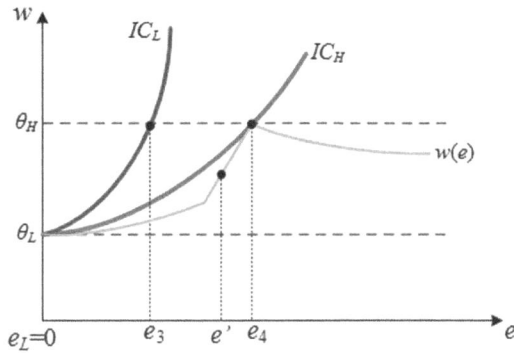

12.4.1 Separating PBE—Applying the Intuitive Criterion

The most-costly separating PBE where the high-productivity worker chooses $e_H = e_4$ can only be sustained if the firm, upon observing any off-the-equilibrium education level in the interval $e_H \in [e_3, e_4)$, strictly below e_4, believes that it cannot be facing a high-productivity worker, i.e., $\mu(\theta_H|e_H) < 1$, and thus pays her a wage strictly lower than θ_H. But are these off-the-equilibrium beliefs sensible? No, which we can show using the Cho and Kreps' (1987) Intuitive Criterion, following our six-step approach from Chapter 11.

Step 1, identify a PBE to test. Consider a specific PBE, such as the most-costly separating PBE where $(e_L, e_H) = (0, e_4)$.

Step 2, off-the-equilibrium message. Identify an off-the-equilibrium message for the worker, such as $e' \in [e_3, e_4)$, as depicted in Fig. 12.6.[2]

Step 3, types who profitably deviate. Find which types of workers can benefit by deviating to e'. The low-productivity worker cannot benefit since, even if the firm responds by paying her the highest salary, $w(e') = \theta_H$, the cost of effort is too large for this type of worker. Formally, $\theta_L - 0 \geq \theta_H - c(e', \theta_L)$, or $c(e', \theta_L) \geq \theta_H - \theta_L$. The high-productivity worker, however, can benefit from choosing e'. Intuitively, if the firm keeps paying her the highest salary, $w(e') = \theta_H$, but she incurs less education costs when acquiring e' rather than e_4, she will certainly deviate to e'. Formally, $\theta_H - c(e', \theta_H) \geq \theta_H - c(e_4, \theta_H)$ simplifies to $c(e_4, \theta_H) \geq c(e', \theta_H)$, which holds given that $e_4 > e'$. Overall, this means that education level e' can only originate from the high-productivity worker.

Step 4, restricting off-the-equilibrium beliefs. We can now restrict the firm's off-the-equilibrium beliefs. If education level e' is observed, it can only originate from the high-productivity worker, i.e., $\mu(\theta_H|e') = 1$.

---

[2] This is an off-the-equilibrium education level since e' does not coincide with the education level of any type of worker in the above separating PBE, where $e_L = 0$ and $e_H = e_4$, that is, $e' \neq 0 \neq e_4$.

Step 5, updated responses. Let us find the optimal response given the restricted belief $\mu(\theta_H|e') = 1$. As the firm is convinced of dealing with a high-productivity worker, it optimally responds by paying $w(e') = \theta_H$.

Step 6, conclusion. Given the optimal response found in Step 5, we can see that the high-productivity worker has incentives to deviate from her equilibrium strategy of e_4 to education e'. Therefore, the most-costly separating PBE $(e_L, e_H) = (0, e_4)$ violates the Intuitive Criterion.

A similar argument applies to any other separating PBE where $e_H \in (e_3, e_4]$, since only the high-productivity worker has incentives to deviate from a lower education level $e' < e_H$. Her educational choice at the least-costly separating PBE, $e_H = e_3$, however, survives the Intuitive Criterion. To see this, note that both types of worker can benefit from deviating to education levels strictly below e_3, implying that the firm cannot restrict its beliefs upon observing off-the-equilibrium education e' satisfying $e' < e_3$, keeping its beliefs unaltered relative to those in the separating PBE. Therefore, only one separating PBE survives the Intuitive Criterion, namely, the least-costly separating PBE where $(e_L, e_H) = (0, e_3)$.

12.4.2 Separating PBE—Applying the D1 Criterion

From Sect. 11.3, we know that, in signaling games with only two sender types, the Intuitive and D1 Criterion generally have the same refinement power. In other words, the same set of PBEs survives both refinement criteria. As a practice, show that the separating PBE in Fig. 12.7, where $e_L = 0$ and $e_H = e_2$, violates the Intuitive Criterion. We next show that it also violates the D1 Criterion.

Step 1, identify a PBE to test. Consider a specific PBE, such as the separating PBE where $(e_L, e_H) = (0, e_2)$.

Step 2, off-the-equilibrium message. Identify an off-the-equilibrium message for the worker, such as e', where $e' < e_1$, as depicted in Fig. 12.7.

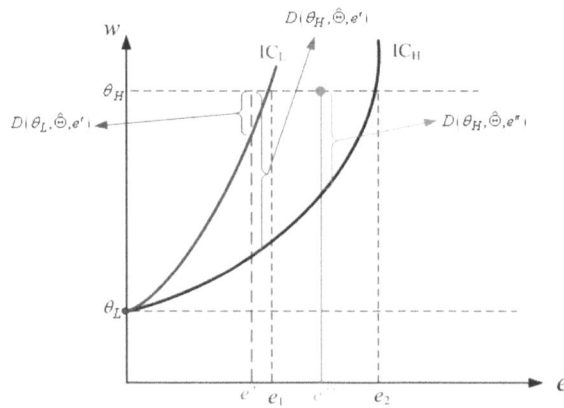

Fig. 12.7 Applying the D1 Criterion to the separating PBEs

12.4 Separating PBE

Step 3, type most likely to deviate. Find which worker type is most likely to benefit from deviating to e', that is, who faces a wider range of wage offers (firm responses) that would help her improve her equilibrium utility. Figure 12.7 identifies the set of wages that help the low-productivity worker be better off after deviating to e', which we denote as $D\left(\theta_L, \widehat{\Theta}, e'\right)$. Intuitively, wages such as $w = \theta_H$ would be associated with an indifference curve to the northwest of IC_L, thus yielding a higher utility than in equilibrium. Similarly, the set of wages for which the high-productivity worker's utility improves when she deviates to e', relative to the separating PBE, is labeled as $D\left(\theta_H, \widehat{\Theta}, e'\right)$.

Comparing $D\left(\theta_L, \widehat{\Theta}, e'\right)$ and $D\left(\theta_H, \widehat{\Theta}, e'\right)$, we see that $D\left(\theta_L, \widehat{\Theta}, e'\right)$ is a subset of $D\left(\theta_H, \widehat{\Theta}, e'\right)$, indicating that, after deviating to e', there are more wage offers that improve the equilibrium payoff of the high-productivity worker than those improving the low-productivity worker's. We can, then, conclude that the high-productivity worker is more likely to send message e', which ultimately helps the firm update its off-the-equilibrium beliefs: upon observing e', $\mu\left(\theta_H | e'\right) = 1$.

For completeness, the figure also includes deviations to off-the-equilibrium message e'', where $e'' \in (e_1, e_2)$. The low-productivity worker would never benefit from deviating to e'', entailing that the set of wage offers improving her equilibrium utility, $D\left(\theta_L, \widehat{\Theta}, e''\right)$, is in this case nil. In contrast, sending e'' might be profitable for the high-productivity worker, as depicted in the figure. Therefore, upon observing e'', the firm's updated beliefs are also $\mu\left(\theta_H | e''\right) = 1$.

Step 4, restricting off-the-equilibrium beliefs. We can now restrict the firm's off-the-equilibrium beliefs. From Step 3, we found that, any deviation from the separating PBE $(e_L, e_H) = (0, e_2)$, where $e' < e_2$, is more likely to originate from the high- than the low-productivity worker, so the firm's beliefs become $\mu\left(\theta_H | e'\right) = 1$.[3]

Step 5, updated responses. Because the firm is convinced of dealing with a high-productivity worker, it optimally responds by offering a salary $w(e') = \theta_H$.

Step 6, conclusion. Given the optimal response found in Step 5, the high-productivity worker deviate from her equilibrium strategy of e_2 to e'. Therefore, separating PBE $(e_L, e_H) = (0, e_2)$ violates the D1 Criterion.

Like in the Intuitive Criterion, one can show that all separating PBEs in this game can be eliminated using the D1 Criterion, except for the efficient (Riley) outcome, where the low-productivity worker acquires no education, $e_L = 0$, and the high-productivity worker acquires the lowest education level that helps convey her type to the firm, $e_H = e_1$.

---

[3] Deviations to e', where $e' > e_2$, would not be profitable for any worker type, even if the firm responds by offering the most generous salary, $w = \theta_H$, implying that the firm does not update its off-the-equilibrium beliefs in this context.

Fig. 12.8 Pooling PBE—Example of wage schedule

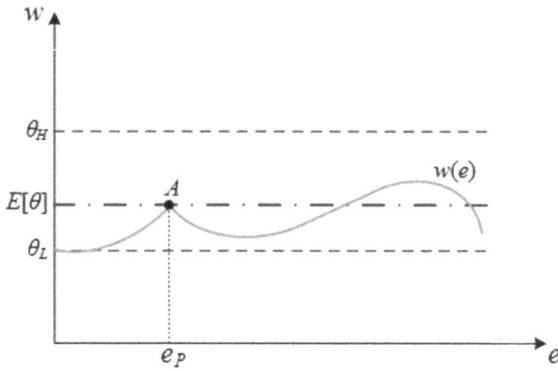

12.5 Pooling PBE

We next check if a pooling strategy profile where both types of workers acquire the same education level can be sustained as a PBE. We follow the same four steps as in the discrete version of the game.

First step, strategy profile. In a pooling strategy profile, both the high- and low-productivity workers choose education level e_P, where the subscript P denotes pooling equilibrium.

Second step, updating beliefs. Upon observing education level e_P, firm posterior beliefs coincide with its prior, i.e., $\mu(\theta_H|e_P) = p$. Intuitively, the observation of education level e^P provides the firm with no additional information about the worker's productivity since all worker types choose to acquire the same education. Upon observing the off-the-equilibrium education level $e \neq e_P$, the firm cannot update its beliefs using Bayes' rule, leaving them unrestricted, that is, $\mu(\theta_H|e) \in [0, 1]$. (As with the separating PBEs, we will impose some conditions on off-the-equilibrium beliefs below.)

Third step, optimal responses. Given the above beliefs, upon observing the pooling education level e^P, the firm optimally responds with salary $w(e^P) = p\theta_H + (1-p)\theta_L$, which captures the worker's *expected* productivity, $E[\theta] \equiv p\theta_H + (1-p)\theta_L$. After observing any off-the-equilibrium education $e \neq e_P$, the firm responds with a salary $w(e) \in [\theta_L, \theta_H]$ since its beliefs in this case are left unrestricted, $\mu(\theta_H|e) \in [0, 1]$; as described in the second step. Figure 12.8 depicts, as an example, a wage schedule that satisfies the above two properties: at the pooling education level e^P, the salary is $w(e^P) = E[\theta]$, and for all other education levels $e \neq e^P$ the firm pays salaries bounded between θ_L and θ_H.[4]

---

[4] Other permissible wage schedules could be, for instance, $w(e) = \theta_L$ for all $e < e_P$, $w(e) = E[\theta]$ at exactly $e = e^P$, but $w(e) = \theta_H$ otherwise; or $w(e) = \theta_L$ for all $e \neq e_P$, but $w(e) = E[\theta]$ at exactly $e = e^P$.

12.5 Pooling PBE

Fourth step, optimal messages. Given the above wage schedule from the firm, we next identify under which conditions both types of workers choose the same education level e^P.

Low-productivity worker. Let us start with the low-productivity worker. Figure 12.9 depicts an indifference curve of the low-productivity worker, IC_L, that originates at $(0, \theta_L)$ on the vertical axis, where the worker acquires no education and receives the lowest salary θ_L, and passes through point A, which represents $(e, w) = (e^P, E[\theta])$. This type of worker is, then, indifferent between identifying herself as a low-productivity worker (i.e., acquiring no education and receiving salary θ_L) and acquiring the pooling education level e^P. Formally, this means that the IC_L curve in Fig. 12.9 must satisfy

$$\theta_L - c(0, \theta_L) = E[\theta] - c(e^P, \theta_L)$$

or, given that $c(0, \theta_L) = 0$ by definition,

$$c(e^P, \theta_L) = E[\theta] - \theta_L,$$

where $E[\theta] > \theta_L$ since $p > 0$. For this to be the case, we need the wage schedule $w(e)$ to lie weakly below IC_L since that implies that a deviation toward any education level $e \neq e_P$ produces an overall utility lower than that at e^P for the low-productivity worker.

Example 12.4 Finding a pooling education level. Consider the parametric example of Example 12.1, where $\theta_H = 2$, $\theta_L = 1$, and the cost of education is $c(e, \theta_K) = \frac{e^2}{\theta_K}$. If we assume $p = 1/3$, we obtain an expected productivity $E[\theta] = \frac{1}{3}2 + \frac{2}{3}1 = \frac{4}{3}$.

Fig. 12.9 Pooling PBE—Low-productivity worker

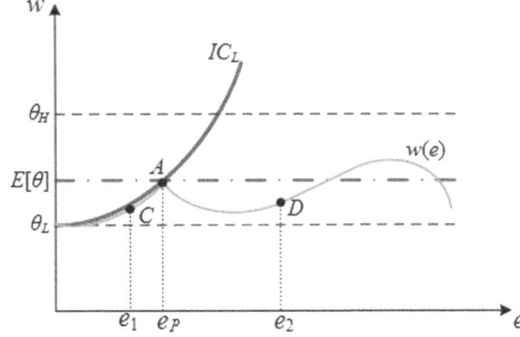

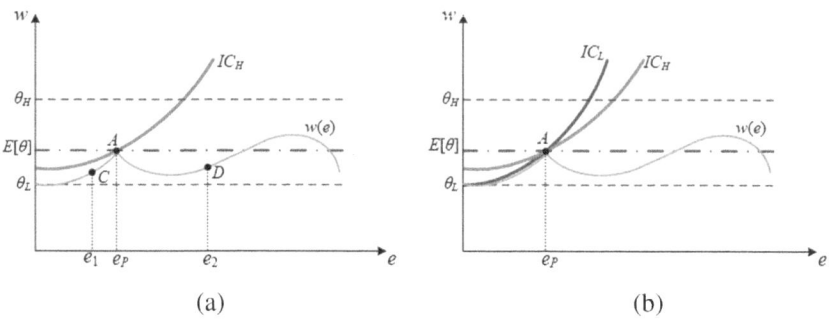

Fig. 12.10 Pooling PBE: **a** High-productivity worker, **b** Both worker types

The above equation, $c(e^P, \theta_L) = E[\theta] - \theta_L$, becomes $(e^P)^2 = \frac{4}{3} - 1$, which yields a pooling education level of $e^P = 0.58$.[5] $\square$

High-productivity worker. We now turn to the high-productivity worker, who must have incentives to choose the pooling education level e^P rather than deviating toward a different education $e \neq e_P$. Figure 12.10a illustrates a wage schedule $w(e)$ that does not provide incentives to deviate from e^P to this type of worker. Deviations to, for instance, education level e_1 yields a lower salary than $w(e^P) = E[\theta]$ and incurring a higher education cost, as depicted at point C of the figure. A similar argument applies in deviations to e_2 (point D in the figure), where she earns a salary $w < E[\theta]$ and incurs a higher education cost than at the pooling education level e_P since $e_2 > e_P$.[6]

Figure 12.10b superimposes Fig. 12.9, from our analysis of the low-productivity worker, and 12.10a, from our discussion of the high-productivity worker. Note that the wage schedule $w(e)$ must lie weakly below both indifference curves IC_L and IC_H, passing through point A for both types of workers, to have incentives to choose e^P rather than deviating. The pooling education level lies at the point where IC_L and IC_H cross each other.

---

[5] Technically, solving for e^P in equation $(e^P)^2 = \frac{4}{3} - 1$ we find two roots: $e^P = \frac{1}{\sqrt{3}}$ and $e^P = -\frac{1}{\sqrt{3}}$. However, since education must be positive or zero, we only report the first root, $e^P = \frac{1}{\sqrt{3}} \simeq 0.58$.

[6] Deviations to education levels much higher than e_2, are not worth it for the high-productivity worker either. Indeed, even if she earns a higher salary than $w(e^P) = E[\theta]$ (see portion of the wage schedule $w(e)$ on the right side of the figure), the worker needs to incur such a large education cost that, overall, her utility level would be lower than when choosing the pooling education level e_P.

Fig. 12.11 Pooling PBE—Other equilibria

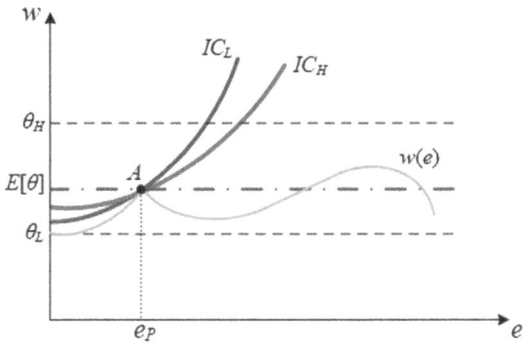

12.5.1 Other Pooling PBEs

Are there any other pooling PBEs? Yes, if IC_L originates strictly above θ_L, the crossing point of IC_L and IC_H happens closer to the origin, as depicted in Fig. 12.11.

A similar argument applies if we keep increasing the origin of IC_L until we reach a crossing point at $e^P = 0$. In this case, both types of workers acquire zero education and they receive a salary equal to their expected productivity $E[\theta]$; producing the same result as in an incomplete information game where workers cannot acquire education to signal their types.

Therefore, we can summarize all pooling PBEs as (e, w)-pairs where both workers choose $e^P \in [0, e_A]$, and education e_A solves

$$\theta_L - c(0, \theta_L) = E[\theta] - c(e^P, \theta_L)$$

or, given that $c(0, \theta_L) = 0$ by definition,

$$c(e^P, \theta_L) = E[\theta] - \theta_L.$$

This equation just identifies the education level where the IC_L starting at θ_L crosses horizontal line $E[\theta]$ in Fig. 12.11. Intuitively, at this education level the low-productivity worker is indifferent between the pooling education e^P, earning a salary $w(e^P) = E[\theta]$, and a zero education level receiving a salary of θ_L. In all these pooling PBEs, the firm responds with a salary $w(e^P) = E[\theta]$ and a wage schedule $w(e)$ that lies below both IC_L and IC_H for all $e \neq e_P$.[7] Like in our above discussion of separating PBEs, our results indicate that we have a range of pooling PBEs: from the least-costly pooling PBE where both types of workers acquire zero education, $e^P = 0$, to the most-costly pooling PBE where both

---

[7] Other wage schedules satisfy this property, such as $w(e) = \theta_L$ for all $e < e^P$ and $w(e) = E[\theta]$ otherwise.

acquire the highest education $e^P = e_A$. (Education levels above e_A are, of course, unsustainable since the low-productivity worker could have incentives to deviate to a zero education.)

12.5.2 Pooling PBE—Applying the Intuitive Criterion

Our above discussion correctly identified under which conditions we can sustain pooling PBEs in this game, but do they survive the Intuitive Criterion? The answer is no, but before we formally show this, let us build some intuition by considering the above off-the-equilibrium beliefs for any education level $e \neq e_P$. Intuitively, the condition that $w(e)$ lies below both IC_L and IC_H for all $e \neq e_P$ means that, upon observing deviations from e^P, even to relatively high education levels, the firm infers that such deviation is not likely originating from the high-productivity worker, and thus pays a relatively low wage; as depicted by the height of $w(e)$ in the right-hand side of Fig. 12.12. This off-the-equilibrium beliefs are, of course, not sensible, since one could argue that deviations toward high education levels are more likely to stem from the high-type worker, as we show next following our six-steps approach to the Intuitive Criterion.

Step 1, identify a PBE to test. Consider a specific PBE, such as the most-costly pooling PBE where $e^P = e_A$.

Step 2, off-the-equilibrium message. Identify an off-the-equilibrium education, such as $e' > e_A$.

Step 3, types who profitably deviate. Find which types of workers can benefit by deviating to e'. The low-productivity worker cannot benefit since, even if the firm responds by paying her the highest salary, $w(e') = \theta_H$, the cost of effort is too large for this type of worker to offset the cost from additional education. Formally, $E[\theta] - c(e_A, \theta_L) \geq \theta_H - c(e', \theta_L)$, or

$$c(e', \theta_L) - c(e_A, \theta_L) \geq \theta_H - E[\theta].$$

Intuitively, the additional cost that the worker must incur offsets the wage increase she experiences. This is depicted in Fig. 8.32, where deviating toward education e', even if responded with the highest salary $w(e') = \theta_H$ at point B, yields a lower utility for the low-type worker than that she obtains in equilibrium. Graphically, the indifference curve crossing through point B lies to the southeast of IC_L.

The high-productivity worker, however, can benefit from choosing e', that is, $E[\theta] - c(e_A, \theta_H) \leq \theta_H - c(e', \theta_H)$, or

$$c(e', \theta_H) - c(e_A, \theta_H) \leq \theta_H - E[\theta].$$

Intuitively, if the firm responds to education level e' by paying her the highest salary $w(e') = \theta_H$, her wage increase would offset the additional education cost.

Fig. 12.12 Applying the Intuitive Criterion to Pooling PBEs

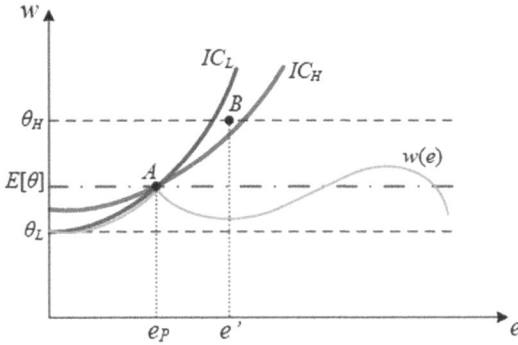

At this point, we can combine the above two inequalities to obtain

$$c(e', \theta_H) - c(e_A, \theta_H) \leq \theta_H - E[\theta] \leq c(e', \theta_L) - c(e_A, \theta_L)$$

which simplifies to

$$c(e', \theta_H) - c(e_A, \theta_H) \leq c(e', \theta_L) - c(e_A, \theta_L).$$

This condition holds from our initial assumptions: the marginal cost of education (increasing e from e_A to e') is larger for the low- than the high-productivity worker. Overall, this means that education level e' can only originate from the high-productivity worker.

Step 4, restricting off-the-equilibrium beliefs. We can now restrict the firm's off-the-equilibrium beliefs as follows: If education level e' is observed, it can only originate from the high-productivity worker, i.e., $\mu(\theta_H | e') = 1$.

Step 5, updated responses. Let us find the optimal response given the restricted belief $\mu(\theta_H | e') = 1$. As the firm is convinced of dealing with a high-productivity worker, it optimally responds by paying $w(e') = \theta_H$.

Step 6, conclusion. Given the optimal response found in Step 5, we can see that the high-productivity worker has incentives to deviate from her equilibrium strategy of e_A to e'. Therefore, the most-costly pooling PBE $e^P = e_A$ violates the Intuitive Criterion.

A similar argument applies to all other pooling PBEs where $e^P < e_A$, including the least-costly pooling PBE where $e^P = 0$. Hence, no pooling PBE in the labor market signaling game survives the Intuitive Criterion, implying that the only PBE of the labor market signaling game, among those surviving the Intuitive Criterion, is the least-costly separating PBE where $(e_L, e_H) = (0, e_3)$. These are great news in terms of information transmission from the privately informed worker to the uninformed firm: education is used to convey the worker's type, not to conceal it, and it is done in the most efficient manner.

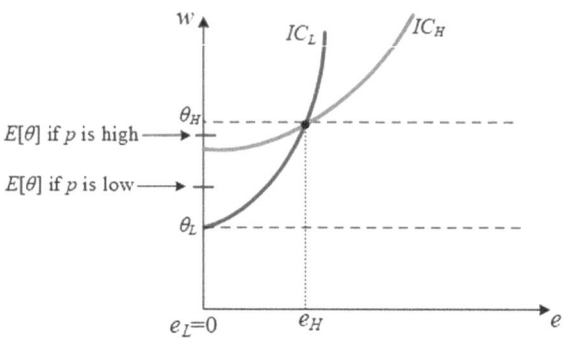

Fig. 12.13 Utility comparison across information settings

12.6 Can Signaling Be Welfare Improving?

From our above results, a natural question is whether the worker is better off when she uses education to signal her type (in the least-costly separating PBE found above) than when such a signal is not available. When the worker cannot use education as a signal, the equilibrium outcome is a BNE, where the firm (being "in the dark" about the worker's type) pays a salary equal to her expected productivity, $w = E[\theta]$; and the worker, anticipating this wage, acquires no education regardless of her type. Figure 12.13 compares the indifference curve that each type of worker reaches in these two information settings. The figure also includes the wage $E[\theta] = p\theta_H + (1-p)\theta_H$ evaluated at two values of p, where $E[\theta]$ is increasing in p.[8]

Low-productivity worker. The low-type worker is unambiguously worse off with signaling, where she acquires zero education but receives the lowest salary θ_L, than without signaling, where she still acquires no education but earns a higher salary $E[\theta]$. Graphically, the indifference curve passing through point $(0, E[\theta])$, on the vertical axis, reaches a higher utility level than IC_L does. This result holds regardless of the specific probability that the worker is of high productivity, p, which graphically means regardless of where $E[\theta]$ lies in the vertical axis.

High-productivity worker. In contrast, the high-type worker is better off with signaling, where she reaches IC_H, than without signaling, where she acquires no education and earns a salary $E[\theta]$, only if $E[\theta]$ is sufficiently low, which occurs when p is relatively low. Intuitively, this type of worker is better off acquiring education to earn the highest wage θ_H, despite its cost, than not investing in education and receiving $E[\theta]$ when this wage is sufficiently low, which occurs when the firm believes that the high-productivity worker is relatively unlikely. If, instead, probability p (as then $E[\theta]$) is sufficiently high, the high-type worker is better off in the setting where education cannot be used as a signal to firms.

---

[8] You can confirm that differentiating $E[\theta]$ with respect to p yields $\theta_H - \theta_L$, which is positive since $\theta_H > \theta_L$ by assumption.

12.7 What If the Sender Has Three Types?

Example 12.5 Firms left in the dark? Consider again the setting in Examples 12.1–12.3, where $\theta_H = 2$ and $\theta_L = 1$. Recall that, in the least-costly separating PBE, the high-productivity worker chooses education $e_3 = 1$. Therefore, her utility in equilibrium is

$$u_H = w - \frac{e^2}{\theta_H} = \theta_H - \frac{1^2}{\theta_H} = 2 - \frac{1}{2} = \frac{3}{2}.$$

However, when signaling is not available, she earns a salary

$$E[\theta] = p2 + (1-p)1 = 1 + p,$$

yielding a utility of $1 + p$ since in this setting she acquires zero education. Therefore, this type of worker is better off in the environment where education cannot be used as signal to firms if $1 + p > \frac{3}{2}$ or, solving for p, when $p > 1/2$; as described in our above discussion. Intuitively, the frequency of high types is so high that the high-productivity worker benefits from firms being "in the dark" about her type. In short, her salary is relatively high and her education cost is nil.□

12.7 What If the Sender Has Three Types?

Consider an extension of the previous signaling game, where the sender can now have three equally likely types, θ_L, θ_M, and θ_H, where $\theta_L < \theta_M < \theta_H$. We first identify the set of separating PBEs; then show that applying the Intuitive Criterion in this setting would have no bite, meaning that no separating PBE violates the Intuitive Criterion; and, finally, show that the D1 Criterion helps us delete some separating PBEs in this signaling game with three sender types. Indeed, only the least-costly separating PBE (Riley outcome) survives the application of the D1 Criterion.

12.7.1 Separating PBEs

For compactness, we describe here the set of separating PBEs. (As a practice, you can go through the 6-step approach to confirm that you can reproduce the results.) The set of all separating PBEs satisfies

$$(e_L, e_M, e_H) = \left(0, e_M^*, e_H^*\right)$$

where $e_M^* \in \left[e_1^M, e_2^M\right]$ and $e_H^* \in \left[e_1^H, e_2^H\right]$. Wages satisfy $w(e_K) = \theta_i$ for every $i = \{L, M, H\}$ after observing equilibrium education levels $(0, e_M^*, e_H^*)$, and $w(e)$ after observing off-the-equilibrium education levels $e \neq e_i$, where $w(e)$ lies below the indifference curve of all worker types, as depicted in Fig. 12.14.

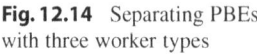

Fig. 12.14 Separating PBEs with three worker types

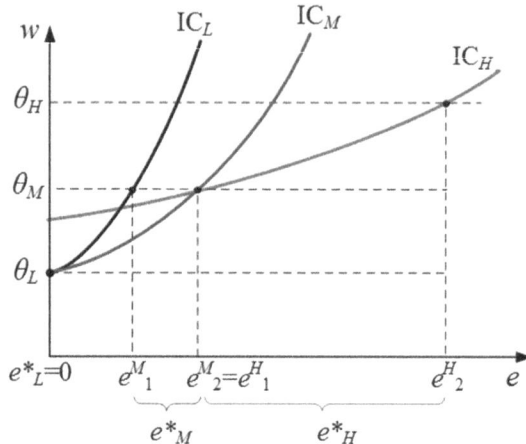

While Fig. 12.14 summarizes all these separating PBEs (see intervals of e_M^* and e_H^* on the horizontal axis), the indifference curves correspond to a specific separating PBE, namely, $(0, e_2^M, e_2^H)$, where both the medium- and high-type worker choose the most-costly education level in their corresponding intervals. Importantly, the high-type worker has no incentives to deviate to the medium-type worker's education level, e_2^M, which would yield a wage of $w = \theta_M$. This education and wage pair, (e_2^M, θ_M), however, would provide her with the same utility as her equilibrium education, e_2^H, as depicted by IC_H in the figure.[9]

12.7.2 Separating PBE—Applying the Intuitive Criterion

Step 1, identify a PBE to test. Consider a specific PBE, such as the separating equilibrium described above, $(e_L, e_M, e_H) = \left(0, e_M^*, e_H^*\right)$.

Step 2, off-the-equilibrium message. Identify an off-the-equilibrium education, e, where $e \in \left(\hat{e}, e_H^*\right)$, as depicted in Fig. 12.15.

Step 3, types who profitably deviate. Find which types of workers can benefit by deviating to e:

- *Low type.* The low-productivity worker cannot benefit since, even if the firm responds by paying her the highest salary, $w(e) = \theta_H$, the cost of effort is too

---

[9] Recall that, as in the setting with two worker types, other wage schedules are admissible in equilibrium as long as they lie below IC_L for all $e < e_1^M$, and below both IC_L and IC_H for all $e_1^M < e < e_2^M$.

Fig. 12.15 Applying the Intuitive Criterion with three worker types

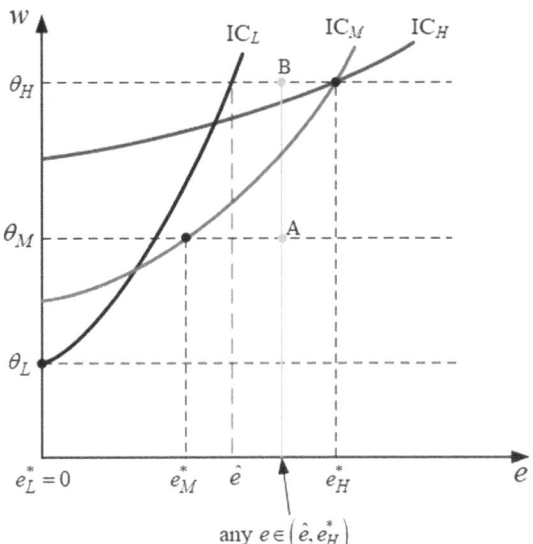

large for this type of worker. Formally, her utility in equilibrium is

$$u_L^* = \theta_L - c(0, \theta_L) = \theta_L$$

given that $c(0, \theta_L) = 0$, whereas the highest utility that she can earn by deviating to e is $\theta_H - c(e, \theta_L)$. Therefore, $\theta_H - c(e, \theta_L) < \theta_L$ holds if

$$c(e, \theta_L) > \theta_H - \theta_L,$$

indicating that the cost of acquiring e years of education for the low-productivity worker exceeds her highest salary gain, $\theta_H - \theta_L$. Graphically, her equilibrium utility level, u_L^*, is represented by the indifference curve IC_L, and her utility from deviating to e, $\theta_H - c(e, \theta_L)$, would correspond to a downward shift of the indifference curve IC_L that passes through point B (when the worker is paid the high-productivity wage θ_H). So, the low-productivity worker does not have incentives to deviate to $e \in (\hat{e}, e_H)$.

- **Medium type.** In contrast, medium-productivity workers could send message $e \in (\hat{e}, e_H)$ because

$$\theta_M - c(e_M^*, \theta_M) < \theta_H - c(e, \theta_M),$$

or alternatively, $c(e, \theta_M) - c(e_M^*, \theta_M) < \theta_H - \theta_M$, reflecting that the cost of acquiring $e - e_M^*$ additional years of education is offset by her salary increase when the firm offers her a high-productivity wage $w(e) = \theta_H$. Graphically, $\theta_H - c(e, \theta_M)$ is represented by the indifference curve of the medium-productivity worker, passing through point B, which lies to the northwest of her indifference curve in equilibrium, IC_M, implying that this worker type can benefit from deviating to e.

- *High type.* A similar argument applies to the high-productivity worker, who can benefit by deviating to e since

$$\theta_H - c(e_H^*, \theta_H) < \theta_H - c(e, \theta_H),$$

simplifies to $c(e_H^*, \theta_H) > c(e, \theta_H)$, which holds by assumption. Intuitively, by deviating toward e, she does not modify her salary (if the firm maintains a wage of $w(e) = \theta_H$), but saves on education costs. In Fig. 12.15, $\theta_H - c(e, \theta_H)$ is illustrated by her indifference curve passing through point B, which lies to the northwest of her indifference curve in equilibrium, IC_H, thus yielding a higher utility level.

Step 4, restricting off-the-equilibrium beliefs. We can now restrict the firm's off-the-equilibrium beliefs as follows: If education level $e \in (\hat{e}, e_H)$ is observed, it can only originate from the high- or medium-productivity worker, but not from the low type. Technically, we say that deviations toward $e \in (\hat{e}, e_H)$ are *not* equilibrium dominated for the high- and medium-productivity workers, but is equilibrium dominated for the low-productivity worker. Therefore, upon observing $e \in (\hat{e}, e_H)$, the firm concentrates its beliefs on these two worker types:

$$\Theta^{**}(e) = \{\theta_M, \theta_H\} \quad \text{for all } e \in (\hat{e}, e_H).$$

Step 5, updating responses. Given $\Theta^{**}(e) = \{\theta_M, \theta_H\}$ for all $e \in (\hat{e}, e_H)$, we need to find which types of worker may have incentives to deviate to $e \in (\hat{e}, e_H)$ anticipating that firm's best response to this education level will be a wage offer somewhere in between $w(e) = \theta_M$ and $w(e) = \theta_H$.

- First, for the medium-productivity worker, the lowest wage offer she can receive is $w(e) = \theta_M$, implying that her indifference curve will pass through point A, being *below* (to the southeast) of her indifference curve in equilibrium, IC_M. In summary, this worker type's equilibrium payoff is higher than the lowest payoff she can obtain from deviating toward e. Alternatively, even after the firm has restricted its beliefs so that deviations to e must originate from the high and medium type, the medium-productivity worker cannot guarantee that her equilibrium utility will be strictly improved by deviating to e.
- We can extend a similar argument to the high-productivity worker. The lowest wage that she can earn after deviating to e is $w(e) = \theta_M$, yielding an indifference curve that passes through point A. This indifference curve is also *below* her indifference curve in equilibrium, IC_H, providing her with a lower utility level. As a consequence, she cannot guarantee that her equilibrium utility will improve by deviating to e.

12.7 What If the Sender Has Three Types?

Step 6, conclusion. Given the optimal responses found in Step 5, we see that no worker type has incentives to deviate from her equilibrium strategy to e. In conclusion, the separating PBE specified in Fig. 12.15 survives the Intuitive Criterion. Thus, the application of the Intuitive Criterion does not necessarily eliminate separating PBE with more than two types of senders.

12.7.3 Separating PBE—Applying the D1 Criterion

We now show that the separating PBE examined in the previous section violates the D1 Criterion. For compactness, we do not reproduce all steps here, as they all coincide, except for Step 3. Recall that, in this step, we identify the worker type who most probably deviated to message e'. As illustrated in Fig. 12.16, when deviating to e' the worker with the largest set of wage offers that improve her equilibrium utility is the medium-productivity worker, i.e., $D(\theta_M, \widehat{\Theta}, e')$ is longer than that of the other worker types, $D(\theta_L, \widehat{\Theta}, e')$ and $D(\theta_H, \widehat{\Theta}, e')$.

Given our results in Step 3, we can move to Step 4, and claim that, upon observing e', the firm believes that it must originate from the medium-productivity worker alone, $\Theta^{**}(e') = \{\theta_M\}$.

As a consequence, in Step 5, we identify that the firm's best response. Given its restricted beliefs (the worker's productivity must be medium), the firm offers a wage $w(e) = \theta_M$ after observing e'.

Finally, in Step 6, we obtain that the medium-productivity worker's deviation payoff is

$$w(e') - c(e', \theta_M) = \theta_M - c(e', \theta_M)$$

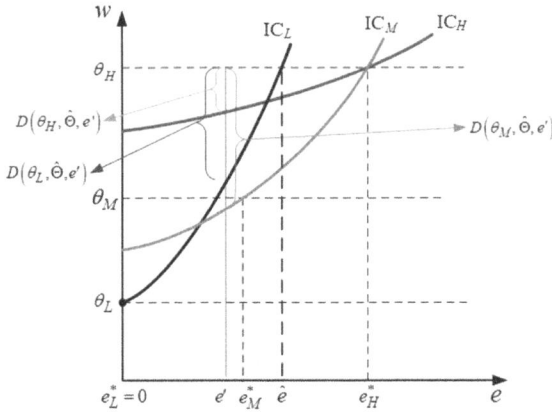

Fig. 12.16 Applying the D1 Criterion with three worker types

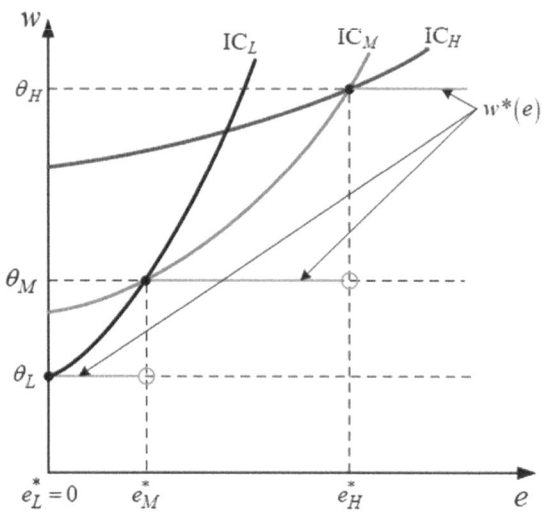

Fig. 12.17 Unique separating PBE surviving the D1 Criterion with three types

while her equilibrium payoff is $u_M^* = \theta_M - c(e_M, \theta_M)$. And given that $e' < e_M$ and education is costly, we have that $c(e', \theta_M) < c(e_M, \theta_M)$; which implies

$$\theta_M - c(e', \theta_M) > \theta_M - c(e_M, \theta_M).$$

Hence, we have found that the medium-productivity worker has incentives to deviate toward e, entailing that the separating PBE $(e_L, e_M, e_H) = (0, e_M^*, e_H^*)$ *violates* the D1 Criterion. In contrast, the low and high-productivity workers do not find this deviation profitable: since the firm pays $w(e') = \theta_M$, they would be reaching an indifference curve to the southeast of IC_L and IC_H, respectively.

Repeating this process for all off-the-equilibrium messages, we can delete all separating PBEs, except for the efficient (Riley) equilibrium outcome described in Fig. 12.17.

Appendix: Equilibrium Refinements

We now present a more formal definition of the Intuitive and D1 Criterion, following Fudenberg and Tirole (1991, pp. 452-453).

Intuitive Criterion

The first step focuses on those types of senders who can improve their equilibrium utility level by deviating. Formally, for any off-the-equilibrium message m, we construct a subset of types $\Theta^{**}(m) \subset \Theta$ for which sending m is not equilibrium dominated. That is,

$$\Theta^{**}(m) = \left\{\theta \in \Theta | u_i^*(\theta) \leq \max_{a \in A^*(\Theta, m)} u_i(m, a, \theta)\right\} \quad (12.1)$$

Intuitively, expression (12.1) states that, from all types in Θ, we restrict our attention to those types who *could* improve their equilibrium utility level, $u_i^*(\theta)$. Note the emphasis on "could" since $\max_{a \in A^*(\Theta, m)} u_i(m, a, \theta)$ represents the highest payoff that a θ-type can achieve by deviating to the off-the-equilibrium message[10] m. In short, we can interpret $\Theta^{**}(m)$ as the subset of senders who *could improve* their equilibrium utility by deviating from m^* to m.

Given $\Theta^{**}(m)$, we now check if the following inequality holds for some type $\theta \in \Theta^{**}(m)$,

$$\min_{a \in A^*(\Theta^{**}(m), m)} u_i(m, a, \theta) > u_i^*(\theta) \quad (12.2)$$

Intuitively, this inequality states that, once beliefs are restricted to $\Theta^{**}(m)$, there is at least one sender type who prefers to deviate to a message m, as this deviation provides her with a strictly higher utility level than her equilibrium message m^*, *regardless* of the response of the receiver, i.e., even if the receiver responds in the least beneficial way for the sender, as captured by $\min_{a \in A^*(\Theta^{**}(m), m)} u_i(m, a, \theta)$.

In settings where the first step of the Intuitive Criterion helps us restrict off-the-equilibrium beliefs to a single type, as if the receiver was "convinced" of facing a specific type of sender, yielding a specific best response. As a consequence, the application of the second step is really unnecessary, as that sender type has incentives to deviate from her equilibrium message. However, in contexts where the first step leaves us with two or more sender types in $\Theta^{**}(m)$, off-the-equilibrium beliefs are not restricted to a single sender type, and the application of the second step is still necessary. (As an illustration, see our discussion in Sect. 12.7.2.)

Formally, an equilibrium strategy profile (m^*, a^*) *violates* the Intuitive Criterion if there is a type of agent θ and an action m such that condition (12.2) is satisfied. Otherwise, we say that the equilibrium strategy profile *survives* the Intuitive Criterion.

D1 Criterion

The D1 Criterion, instead, considers that, among all potential deviators, the receiver restricts her beliefs to only those types of senders who *most likely* send

---

[10] Note that the maximization problem is with respect to the receiver's response, a, among the set of best responses to message, m, given the set of all possible senders, Θ, and the receiver's off-the-equilibrium beliefs.

the off-the-equilibrium message. In particular, we now identify the sender for whom most of the responder's actions (in a setting with discrete actions) or the sender with the largest set of responses (in a context with continuous actions) provide a payoff strictly higher than her equilibrium payoff. Formally, for any off-the-equilibrium message m, let us define

$$D\left(\theta, \widehat{\Theta}, m\right) := \bigcup_{\mu : \mu\left(\widehat{\Theta}|m\right)=1} \{a \in MBR(\mu, m) \mid u_i^*(\theta) < u_i(m, a, \theta)\} \quad (12.3)$$

as the set of mixed best responses[11] (MBR) of the receiver for which the θ-type of sender is *strictly better-off* deviating toward message m than sending her equilibrium message m^*. Note that $\mu\left(\widehat{\Theta} \mid m\right) = 1$ in expression (12.3) represents that the receiver believes that message m only comes from types in the subset $\widehat{\Theta} \in \Theta$. Let us also define

$$D^\circ\left(\theta, \widehat{\Theta}, m\right) := \bigcup_{\mu : \mu\left(\widehat{\Theta}|m\right)=1} \{a \in MBR(\mu, m) \mid u_i^*(\theta) = u_i(m, a, \theta)\} \quad (12.4)$$

as the set of MBR of the receiver that make the θ-type *indifferent* between deviating toward message m and sending her equilibrium message m^*. Let us next describe the first step of the D1 Criterion.

First, we say that θ-type can be *deleted* if there is another θ'-type such that, when the off-the-equilibrium message m is observed

$$\left[D\left(\theta, \widehat{\Theta}, m\right) \cup D^\circ\left(\theta, \widehat{\Theta}, m\right)\right] \subset D\left(\theta', \widehat{\Theta}, m\right) \quad (12.5)$$

That is, for a given message m, the set of receiver's actions which make the θ'-type of sender strictly better off (relative to the equilibrium), $D\left(\theta', \widehat{\Theta}, m\right)$, is larger than the set of actions making the θ-type of sender strictly better off, $D\left(\theta, \widehat{\Theta}, m\right)$, or indifferent, $D^\circ\left(\theta, \widehat{\Theta}, m\right)$. Intuitively, after receiving message m, there are more best responses of the receiver that improve the θ'-type's equilibrium payoff than there are for the θ-type. As a consequence, the θ'-type is the most likely sender type to deviate from her to off-the-equilibrium message m. We continue this comparison for all types of senders, deleting those for which there is another type of sender who is more likely to deviate toward m. Finally, the set of types that cannot be deleted after using this procedure is denoted by $\Theta^{**}(m)$.

---

[11] The set of mixed best responses (MBR) of the receiver to a given message m from the sender includes both the actions that the receiver chooses using pure strategies, and those involving mixed strategies.

Once the set of types has been restricted to $\Theta^{**}(m)$, the Intuitive and the D1 Criterion proceed similarly, seeking to find at least a sender type $\theta \in \Theta^{**}(m)$ with incentives to deviate even if the receiver responds in the least beneficial way for the sender, that is,

$$\min_{a \in A^*(\Theta^{**}(m), m)} u_i(m, a, \theta) > u_i^*(\theta) \tag{12.6}$$

which coincides with expression (12.2) in our application of the Intuitive Criterion.

Exercises

12.1. **Labor market signaling game—Role of the cost function.**[B] Consider the setting in Chapter 12, where we found a separating PBE where the low-productivity worker chooses an education level $e_L = 0$, while the high-productivity worker chooses the least-costly separating education e_H that solves $\theta_H - \theta_L = c(e, \theta_H)$. Assume that $c(e, \theta_i) = \frac{e^\alpha}{\theta_i}$ for every worker type $i = \{H, L\}$, where $\alpha \geq 1$.
   a. Find the least-costly separating education e_H in this context. How is e_H affected by parameter α? Can the high-productivity worker convey his type to the firm when the cost of acquiring education is linear (i.e., $\alpha = 1$)?
   b. How is e_H affected by the productivity differential $\theta_H - \theta_L$? Interpret.
   c. Find the high-productivity worker's utility in equilibrium. Compare it against her utility when signaling is unavailable (as in Example 12.5), to determine for which values of probability p this worker is better off than in the separating equilibrium.

12.2. **Labor market signaling with productivity-enhancing education.**[B] Consider the setting in Chapter 12, but assume that education improves the worker's productivity. In particular, a worker with type θ_i who acquires e years of education, has productivity $\theta_i(1 + \alpha e)$, where $\alpha \geq 0$ denotes the productivity enhancing effect of education. When $\alpha = 0$, the worker's productivity is θ_i, thus being unaffected by e, as in Chapter 12. When $\alpha > 0$, however, education improves the worker's productivity. For simplicity, let us assume that $\alpha = 1$.
   a. *Complete information.* As a benchmark, find the education level that the high- and low-type workers choose when the firm observes his type, θ_H or θ_L, respectively.
   b. *Separating PBEs.* Assume for the remainder of the exercise that the firm cannot observe the worker's type. Find the set of separating PBEs in this game, where each worker type chooses a different education level.
   c. *Pooling PBEs.* Find the set of pooling PBEs where both worker types choose the same education level.

12.3. **Labor market signaling with productivity-enhancing education—Parametric example.**[A] Consider the setting in Example 12.2, where worker types are $\theta_H = 2$ and $\theta_L = 1$, his cost function is $c(e, \theta_K) = \frac{e^2}{\theta_K}$, and assume that the probability of a high-productivity worker is $p = \frac{1}{3}$.
   a. Evaluate the complete information setting with and without productivity-enhancing education. Interpret your results.
   b. Evaluate the separating PBE from Exercise 12.2 and compare your results against those when education is not productivity enhancing.
   c. Evaluate the pooling PBE from Exercise 12.2 and compare your results against those when education is not productivity enhancing.

12.3. **Applying the Intuitive Criterion to the labor market signaling model with productivity-enhancing education.**[B] Consider the labor market signaling model in Exercise 12.2.
   a. Which of the separating PBEs identified in part (b) survive the Intuitive Criterion?
   b. Compare your results against those in a labor market signaling model where education is not productivity enhancing.
   c. Which pooling PBEs found in part (c) survive the Intuitive Criterion?

12.4. **Labor market signaling game with continuous responses.**[B] Consider a labor market signaling game where, first, a worker privately observes his productivity (high, θ_H, or low, θ_L) and chooses between two messages (education, e, or no education, ne). The firm does not observe the worker's type, but knows that it is high with probability p and low otherwise. The firm, then, responds with a continuous wage offer, which we denote as $w_e \geq 0$ after observing education, and $w_{ne} \geq 0$ and after observing no education. As in previous exercises, let $c(e, \theta_i) > 0$ denote worker i's cost of acquiring education, where $i = \{H, L\}$, where $c(e, \theta_L) > c(e, \theta_H)$, and $c(0, \theta_i) = 0$.
   a. *Complete information.* As a benchmark, assume that the firm can observe the worker's productivity. Find the subgame perfect equilibrium (SPE) of the game.
   b. *Separating PBEs.* Check if the separating strategy profile where only the high-productivity worker acquires education, (e^H, ne^L), can be supported as a PBE. Can (ne^H, e^L) be sustained as a PBE?
   c. *Pooling PBEs.* Check if the pooling strategy profile where both worker types acquire education, (e^H, e^L), can be supported as a PBE. Can the opposite pooling strategy profile, (ne^H, ne^L), be sustained as a PBE?
   d. *Parametric example.* Assume that $\theta_H = 2$, $\theta_L = 1$, $c(e, \theta_i) = \frac{e^2}{\theta_i}$ where $i = \{H, L\}$, and $p = \frac{1}{3}$. Evaluate your equilibrium results in parts (b) and (c).
   e. *Intuitive Criterion.* Which of the PBEs found in parts (b) and (c) survive the Intuitive Criterion?
   f. *D1 Criterion.* Which of the PBEs found in parts (b) and (c) survive the D1 Criterion?

12.5. **Countersignaling, based on Feltovich et al.** (2002).[C] Consider the labor market signaling game where the worker has three possible types (high, medium, or low) from Sect. 10.10.3. After privately observing its type, the worker chooses whether to acquire education. Upon observing the worker's decision (E or NE), but without observing the worker's type, the firm responds with a wage offer to the worker, $w \geq 0$. For generality, assume that their productivities are θ_H, θ_M, and θ_L, where $\theta_H > \theta_M > \theta_L$, which are all equally likely.

In this exercise, we consider a "twist" of the above setting where, first, the worker privately observes his type (high, medium, or low); second, the firm receives a recommendation letter from a previous employer describing whether the worker was good or bad (G or B); third, the worker (without observing the content of the letter) chooses whether to acquire education (E or NE); and fourth, the firm observes the worker's education decision and the content of the letter and responds with a wage offer w. If the worker's type is high (low), the recommendation letter is good (bad) with certainty; but if the worker's type is medium, the recommendation letter is good with probability p, where $1 > p > 0$.

a. Under which conditions can you sustain a PBE where the low and high types choose NE, while the medium type chooses E? This is often known as a "countersignaling" equilibrium. Interpret.

b. Show that the semi-pooling equilibrium, where only the low type chooses NE while the medium and high choose E, cannot be sustained.

c. For the remainder of the exercise, assume that $\theta_H = \frac{3}{2}$, $\theta_M = 1$, $\theta_L = \frac{1}{2}$, $c(e, \theta_H) = \frac{1}{8}$, $c(e, \theta_M) = \frac{1}{4}$, and $c(e, \theta_L) = \frac{1}{2}$. Find under which conditions on probability p the PBE found in part (a) can be sustained.

d. Does the PBE found in part (a) survive the Intuitive Criterion?

e. Does the PBE found in part (a) survive the D1 Criterion?

12.6. **Finding separating PBEs in limit pricing.**[C] Consider a market with inverse demand function $p(Q) = 1 - Q$, where $Q = q_1 + q_2$ denotes aggregate output. Let us analyze an entry game with an incumbent monopolist (Firm 1) and an entrant (Firm 2) who analyzes whether or not to join the market. The incumbent's marginal costs are either high H or low L, i.e., $c_1^H = \frac{1}{2} > c_1^L = \frac{1}{3}$, while it is common knowledge that the entrant's marginal costs are high, i.e., $c_2 = \frac{1}{2}$. To make the entry decision interesting, assume that when the incumbent's costs are low, entry is unprofitable; whereas when the incumbent's costs are high, entry is profitable. (Otherwise, the entrant would enter regardless of the incumbent's cost, or stay out regardless of the incumbent's cost.) For simplicity, assume no discounting of future payoffs throughout all the exercises.

a. *Complete information.* Let us first examine the case in which the entrant and the incumbent are informed about each others' marginal costs. Consider a two-stage game where, in the first stage, the incumbent has monopoly power and selects an output level, q. In the second stage, a potential entrant decides whether or not to enter. If entry occurs, agents compete as Cournot duopolists, simultaneously and independently selecting production levels, x_1 and x_2. If entry does not occur, the incumbent maintains its monopoly power in both periods (producing q in the first period and x in the second period). Find the subgame perfect equilibrium (SPNE) of this complete information game.

b. *Incomplete information.* In this section we investigate the case where the incumbent is privately informed about its marginal costs, while the entrant only observes the incumbent's first-period output which the entrant uses as a signal to infer the incumbent's cost. The time structure of this signaling game is as follows:

1. Nature decides the realization of the incumbent's marginal costs, either high or low, with probabilities $p \in (0, 1)$ and $1 - p$, respectively. The incumbent privately observes this realization but the entrant does not.
2. The incumbent chooses its first-period output level, q.
3. Observing the incumbent's output decision, the entrant forms beliefs about the incumbent's initial marginal costs. Let $\mu(c_1^H|q)$ denote the entrant's posterior belief about the initial costs being high after observing a particular first-period output from the incumbent q.
4. Given the above beliefs, the entrant decides whether or not to enter the industry.
5. If entry does not occur, the incumbent maintains its monopoly power; whereas if entry occurs, both agents compete as Cournot duopolists and the entrant observes the incumbent's type.

c. Write down the incentive compatibility conditions that must hold for a separating Perfect Bayesian Equilibrium (PBE) to be sustained. Then find the set of separating PBEs.

d. Which separating PBEs of those you found in part (b) survive the Cho and Kreps' Intuitive Criterion?

12.7. **Finding pooling PBEs in limit pricing.**[C] Consider the limit pricing setting in Exercise 12.6.

a. Find pooling PBEs where both types of firms choose the same output level, q.

b. Does any pooling PBE survive the Cho and Kreps' Intuitive Criterion? Interpret.

c. Does any pooling PBE survive the D1 Criterion? Interpret.

References

Cho, I.-K., and D. M. Kreps (1987) "Signaling Games and Stable Equilibria," Quarterly Journal of Economics, 102(2), pp. 179–221.

Feltovich, N., R. Harbaugh, and T. To (2002) "Too Cool for School? Signaling and Countersignaling," RAND Journal of Economics, 33(4), pp. 630–649.

Fudenberg, D., and J. Tirole (1991) *Game Theory*, The MIT Press, Cambridge, Massachussets.

Cheap Talk Games 13

13.1 Introduction

In this chapter, we explore a class of signaling games where the sender incurs no costs in sending different messages. Unlike in the previous chapter, where messages were costly, such as acquiring years of education, we now consider *costless messages*, often known as "cheap talk." Examples include a lobbyist (sender) informing a member of Congress (receiver) about the situation of an industry she works for, a medical doctor (sender) describing a patient (receiver) a diagnosis and the medical tests she should take, or an investment banker (sender) recommending a client (receiver) which stocks to purchase.

While in Chapters 10–12 we allowed for the sender's utility function to be $u(m, a, \theta)$, thus being a function of her message m, her type θ, and the receiver's response a, now the sender's utility function is unaffected by m, implying that

$$u(m, a, \theta) = u(m', a, \theta)$$

for every two messages $m \neq m'$. This property in cheap talk games helps us present the sender's utility more compactly as $u(a, \theta)$, thus not being a function of m. Our main goal is to identify if separating PBEs can still be supported where information flows from the privately informed player (sender) to the uninformed player (receiver).

At first glance, this sounds as a formidable task: in previous chapters, we often found that to sustain separating PBEs, we need the cost of sending messages to differ across sender types (as captured by the single-crossing property). If all sender types now face zero costs from sending messages, how can we support separation? As we show, separating PBEs can still emerge if the sender and receiver are sufficiently similar in their preferences. Otherwise, only pooling PBEs arise, where all sender types send the same message, and information is concealed from the receiver.

13.2 Cheap Talk with Discrete Messages and Responses

Consider the signaling game in Fig. 13.1, where nature determines the sender's type, either high or low. Privately observing her type, the sender sends a message, either m_1 or m_2, which the receiver observes (without observing the sender's type) and responds with a, b, or c after m_1, and similarly with a', b', or c' after m_2. Importantly, the game represents a cheap talk setting because the sender's payoff is unaffected by her message (i.e., coincides with m_1 and m_2). This property is graphically represented by the fact that payoffs on the left side of the tree (when the sender chooses m_1) coincide with those on the right side (when she chooses m_2). Her payoff, nevertheless, depends on the receiver's response (either a, b, or c) and nature's type. In the example of a lobbyist (sender) and a politician (receiver), the lobbyist is unaffected by her messages, as they are costless (e.g., a few seconds of speech), but is affected by the politician's response (which policy is implemented in Congress) and by the true state of nature (whether the industry she represents is in a good or bad condition).

A similar argument applies to the receiver's payoffs, which are unaffected by the sender's message. Intuitively, the receiver only cares about the sender's message because of its informational content about the true state of nature (which she cannot observe), such as a politician listening to a lobbyist saying that an industry, or a natural park, is in terrible or terrific condition.

This game is, of course, an extreme simplification of cheap talk (or strategic information transmission) between two players, but helps us highlight the main incentives and results. In Sect. 13.3, we extend the model to a setting where types are not discrete, but continuous, and where both the sender's messages and receiver's responses are continuous. For presentation purposes, we next separately

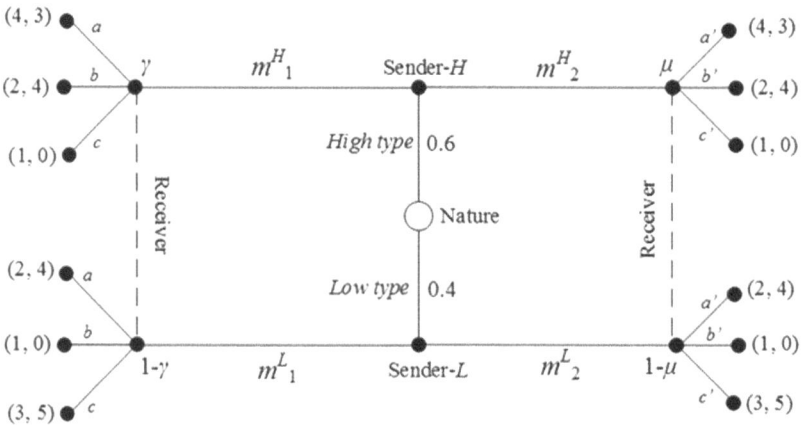

Fig. 13.1 A cheap talk game with two sender types and two messages

13.2 Cheap Talk with Discrete Messages and Responses

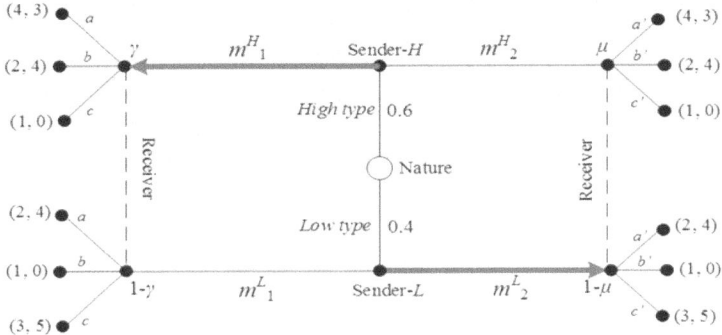

Fig. 13.2 Separating strategy profile (m_1^H, m_2^L) in the cheap talk game

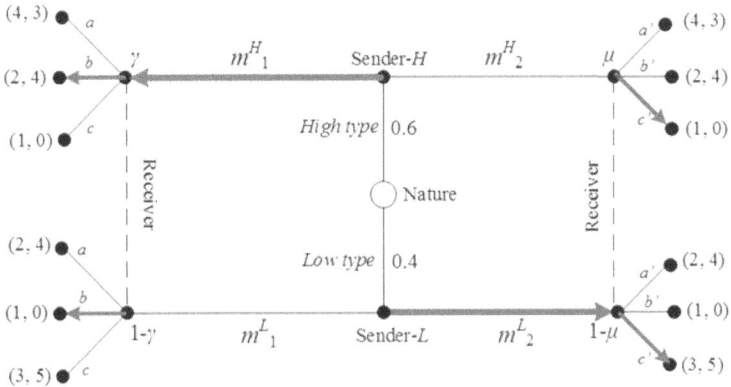

Fig. 13.2a Separating strategy profile (m_1^H, m_2^L) in the cheap talk game—Optimal responses

examine under which conditions a separating or a pooling PBE can be supported, following the 5-step approach introduced in Sect. 10.5.

13.2.1 Separating PBE

1. *Specifying a strategy profile.* We first specify a "candidate" of strategy profile that we seek to test as a PBE, such as (m_1^H, m_2^L). For easier reference, Fig. 13.2 highlights the branches corresponding to m_1^H, at the top left, and that corresponding to m_2^L, at the bottom right.
2. *Bayes' rule.* Upon observing m_1 (m_2), the receiver believes that this message must originate from the high (low) type, entailing that $\mu(H|m_1) = 1$ at the top left side of the figure ($\mu(H|m_2) = 0$ at the bottom right side of the figure).

3. *Optimal response.* Given our result from Step 2, the receiver responds as follows:
   (a) Upon observing m_1, since $\mu(H|m_1) = 1$, she responds with b since it yields a payoff of 4, which exceeds her payoff from a, 3, and from c, 0.
   (b) Upon observing m_2, since $\mu(H|m_2) = 0$, she responds with c' because it generates a payoff of 5, which exceeds her payoff from a', 4, and from b', 0.

   To keep track of our results, Fig. 13.2a shades the branches corresponding to b, upon observing m_1 in the left side of the tree, and those corresponding to c', upon observing m_2 in its right side.

4. *Optimal messages.* From our results in Step 3, we now identify the sender's optimal messages.
   (a) *High type.* If she sends a message of m_1^H, as prescribed in this strategy profile, she earns a payoff of 2 (since m_1 is subsequently responded with b), which exceeds her payoff from deviating toward m_2^H, 1, where the receiver responds with c'.
   (b) *Low type.* If she sends a message of m_2^L, as prescribed in this strategy profile, she obtains a payoff of 3 (as m_2 is responded with c'), which exceeds her payoff from deviating toward m_2^H, 1, where the receiver responds with b.

5. *Summary.* From Step 4, we found that no sender types have incentives to deviate from (m_1^H, m_2^L), implying that this separating strategy profile can be supported as a PBE.

As a practice, Exercise 13.1 (part a) asks you to show that the opposite separating strategy profile, (m_2, m_1'), can also be sustained as a PBE in this cheap talk game.

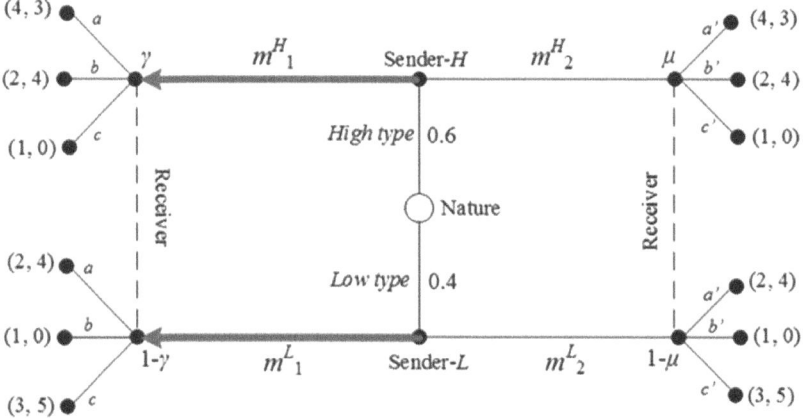

Fig. 13.3 Pooling strategy profile (m_1^H, m_1^L) in the cheap talk game

13.2 Cheap Talk with Discrete Messages and Responses

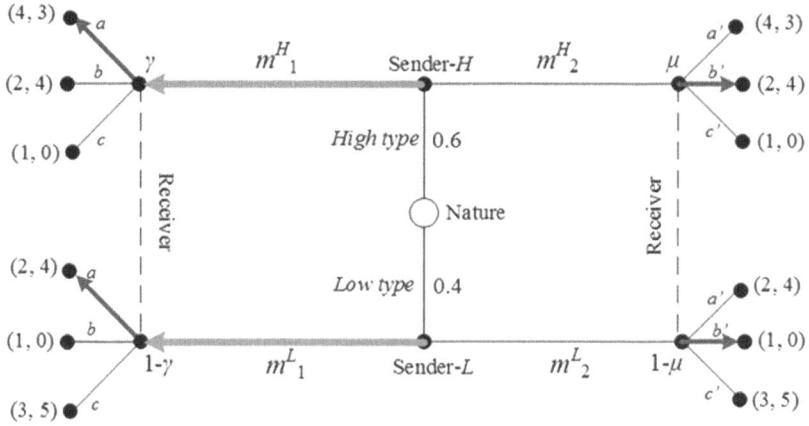

Fig. 13.3a Pooling strategy profile (m_1^H, m_1^L) in the cheap talk game—Optimal responses

13.2.2 Pooling PBEs

1. *Specifying a strategy profile.* We now test if the pooling strategy profile (m_1^H, m_1^L), where both sender types move leftward in the tree, can be sustained as a PBE. Figure 13.3 depicts this strategy profile by highlighting branches m_1^H and m_1^L.
2. *Bayes' rule.* Upon observing m_1, in equilibrium, the receiver's posterior beliefs coincide with her priors, that is, $\mu(H|m_1) = 0.6$; a property we encountered in other pooling strategy profiles. Upon observing the off-the-equilibrium message m_2 (on the right side of the tree), however, Bayes' rule does not help the receiver update her belief, which remains generic at $\mu(H|m_2) \in [0, 1]$.
3. *Optimal response.* Given our result from Step 2, the receiver responds as follows:
   (a) Upon observing m_1, since $\mu(H|m_1) = 0.6$, the receiver's expected utilities from responding with a, b, or c are

   $$EU_R(a) = (0.6 \times 3) + (0.4 \times 4) = 3.4,$$
   $$EU_R(b) = (0.6 \times 4) + (0.4 \times 0) = 2.4, \text{ and}$$
   $$EU_R(c) = (0.6 \times 0) + (0.4 \times 5) = 2.0,$$

   implying that the receiver responds with a.
   (b) Upon observing m_2, since $\mu(H|m_2)$ is unrestricted, we define $\mu(H|m_2) = \mu$ for compactness, and obtain the following expected utilities from responding with a', b', or c',

   $$EU_R(a') = (\mu \times 3) + ((1 - \mu) \times 4) = 4 - \mu,$$
   $$EU_R(b') = (\mu \times 4) + ((1 - \mu) \times 0) = 4\mu, \text{ and}$$

$$EU_R(c') = (\mu \times 0) + ((1-\mu) \times 5) = 5(1-\mu).$$

Comparing $EU_R(a')$ and $EU_R(b')$, we find that $4\mu > 4-\mu$ holds if and only if $\mu > \frac{4}{5}$. Similarly, comparing $EU_R(a')$ and $EU_R(c')$, we obtain that $4-\mu > 5(1-\mu)$ if and only if $\mu > \frac{1}{4}$. And comparing $EU_R(b')$ and $EU_R(c')$, we find that $4\mu > 5(1-\mu)$ holds if and only if $\mu > \frac{5}{9}$. Therefore, we identified three regions: (i) if $\mu \leq \frac{1}{4}$, $5(1-\mu)$ is the highest expected utility, inducing the receiver to respond with c'; (ii) if $\frac{1}{4} < \mu \leq \frac{4}{5}$, $4-\mu$ is the highest expected utility, and the receiver responds with a'; and (iii) if $\mu > \frac{4}{5}$, 4μ is the highest expected utility, and the receiver responds with b'. For simplicity, we next consider only the last case, where the receiver responds with b' off-the-equilibrium path.[1]

To keep track of our results, Fig. 13.3a shades the branches corresponding to a, upon observing m_1 in equilibrium (on the left side of the tree), and those corresponding to b' upon observing m_2 off-the-equilibrium (on the right side).

4. *Optimal messages.* From our results in Step 3, we now find the sender's optimal messages.

   (a) *High type.* If she sends a message of m_1^H, as prescribed in this strategy profile, she earns a payoff of 4 (since m_1 is subsequently responded with a), which exceeds her payoff from deviating toward m_2^H, 2, where the receiver responds with b'.

   (b) *Low type.* If she sends a message of m_1^L, as prescribed in this strategy profile, she obtains a payoff of 2 (as m_1 is responded with a), which exceeds her payoff from deviating toward m_2^L, 1, where the receiver responds with b'.

5. *Summary.* Since no sender type has incentives to deviate, the pooling strategy profile (m_1^H, m_1^L) can be supported as a PBE with responses (a, b') and equilibrium beliefs $\mu(H|m_1) = 0.6$ and off-the-equilibrium beliefs $\mu(H|m_2) > \frac{4}{5}$.

Exercise 13.1 (part b) asks you to find for which conditions the opposite pooling strategy profile, (m_2^H, m_2^L), where both sender types move rightward, can be sustained as a PBE.

---

[1] As a practice, you may reproduce Steps 4–5 below in case (i), showing that the pooling strategy profile (m_1^H, m_1^L) cannot be supported because the low-type sender has incentives to deviate to m_2^L. In case (ii), this pooling strategy profile can be sustained as a PBE, since none of the sender types has strict incentives to deviate.

13.3 Cheap Talk with Discrete Messages But Continuous Responses

Let us extend the cheap talk model in Sect. 13.1 by allowing for continuous responses from the receiver (politician). Figure 13.4 depicts the game tree, where the arcs next to the terminal nodes indicate that the last mover chooses her response (policy choice, p, for the politician) from a continuum of available policies, that is, $p \in [0, 1]$.

As in the previous section, the lobbyist privately observes the state of nature, θ_H or θ_L, both equally likely. The lobbyist, then, chooses a message to send to the politician which, for simplicity, is still binary (either θ_H or θ_L). Upon observing this message, the politician responds with a policy $p > 0$.

Quadratic loss functions. The payoffs of the government (politician) follow quadratic loss functions, as in Crawford and Sobel (1982),

$$U_G(p, \theta) = -(p - \theta)^2$$

which becomes zero when the government responds with a policy that coincides with the true state of the world, that is, $p = \theta$; but is negative otherwise (both when $p < \theta$ and when $p > \theta$). Graphically, $U_G(p, \theta)$ has an inverted-U shape, lying in the negative quadrant for all $p \neq \theta$, but has a height of zero at exactly $p = \theta$; as depicted in Fig. 13.5.

Similarly, the lobbyist's utility is given by a quadratic loss function

$$U_L(p, \theta) = -(p - (\theta + \delta))^2$$

which becomes zero when the policy coincides with the lobbyist's ideal, $p = \theta + \delta$, but is negative otherwise. Intuitively, parameter $\delta > 0$ represents the lobbyist's bias. When $\delta = 0$ the utility functions of both lobbyist and politician coincide, and we can say that their preferences are aligned; but otherwise the lobbyist's ideal policy, $p = \theta + \delta$, exceeds the politician's, $p = \theta$. This bias will play an important role in our subsequent analysis.

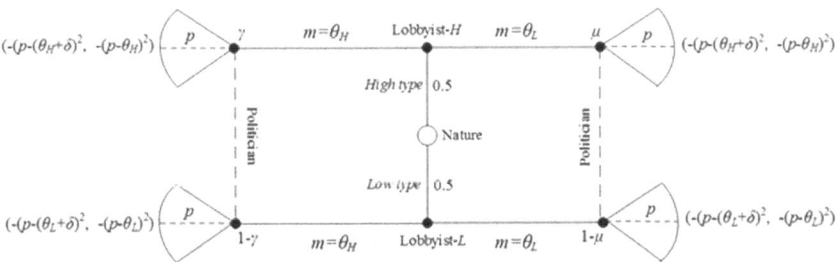

Fig. 13.4 Cheap talk with two messages but continuous responses

Fig. 13.5 Quadratic loss function for each player

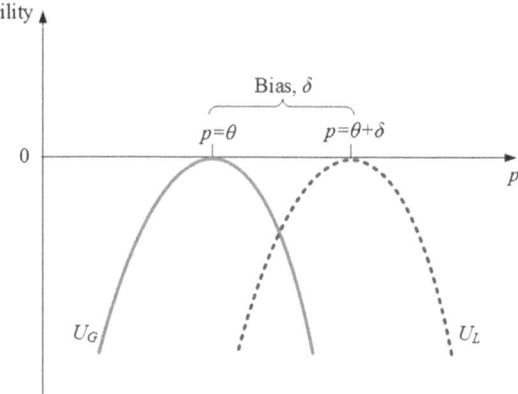

13.3.1 Separating PBE

1. *Specifying a strategy profile.* We first specify a "candidate" of strategy profile that we seek to test as a PBE, (θ_H, θ_L), where the θ_H-type lobbyist chooses θ_H, on the top left side of Fig. 13.6; whereas the θ_L-type lobbyist selects θ_L, on the bottom right side of the figure.
2. *Bayes' rule.* Upon observing θ_H, the politician believes that this message must originate from a high-type lobbyist, entailing that $\mu(\theta_H|\theta_H) = 1$, at the top left side of the figure, which implies that $\mu(\theta_L|\theta_H) = 0$. Similarly, upon observing θ_L, the politician beliefs are $\mu(\theta_L|\theta_L) = 1$, at the bottom right side of the figure, entailing that $\mu(\theta_H|\theta_L) = 0$.
3. *Optimal responses.* Given our result from Step 2, the politician responds as follows:
   (a) Upon observing θ_H, she responds with policy $p = \theta_H$, as this policy minimizes her quadratic loss, yielding a payoff of zero.[2]
   (b) Similarly, upon observing θ_L, she responds with policy $p = \theta_L$, since this policy minimizes her quadratic loss, also yielding a payoff of zero.
4. *Optimal messages.* From our results in Step 3, we now identify the lobbyist's optimal messages.
   (a) *High type.* If she sends a message of θ_H, as prescribed in this strategy profile, she anticipates that the politician will respond with policy $p = \theta_H$, yielding a payoff of $-(\theta_H - (\theta_H + \delta))^2 = -\delta^2$. If, instead, this lobbyist

---

[2] Formally, one can first consider the politician's utility function, $U_G(p, \theta_H) = -(p - \theta_H)^2$, evaluated at θ_H given the politician's updated beliefs from Step 2, and then differentiate it with respect to policy p, which yields $-2(p - \theta_H) = 0$ that holds if and only if $p = \theta_H$.

13.3 Cheap Talk with Discrete Messages But Continuous Responses

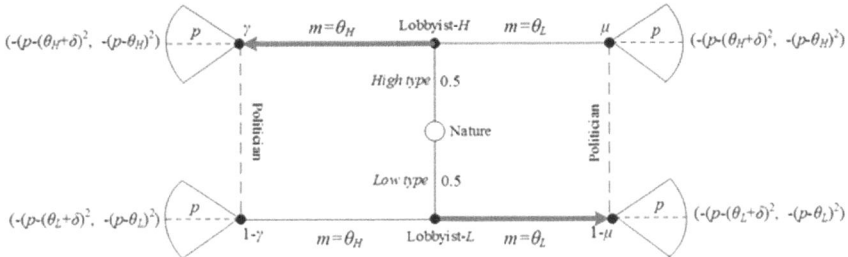

Fig. 13.6 Cheap talk with two messages and continuous responses—Separating profile (θ_H, θ_L)

deviates toward message θ_L, the politician believes this message, responding with policy $p = \theta_L$, with an associated payoff of $-(\theta_L - (\theta_H + \delta))^2 = -(\theta_L - \theta_H - \delta)^2$ for the lobbyist. Therefore, she does not have incentives to deviate if

$$-\delta^2 > -(\theta_L - \theta_H - \delta)^2$$

which holds since $\theta_H > \theta_L$, implying that the high-type lobbyist has no incentive to deviate from message θ_H. This result is quite intuitive because of her bias, the lobbyist prefers a policy above the state of nature. Then, sending message θ_H, she induces a policy of at least $p = \theta_H$, but deviating to message θ_L she would induce a lower policy, $p = \theta_L$, which is further away from her ideal, $p = \theta_H + \delta$; indicating that she does not have incentives to send θ_L.

(b) *Low type.* If she sends a message of θ_L, as prescribed in this strategy profile, the politician responds with $p = \theta_L$, yielding a payoff $-(\theta_L - (\theta_L + \delta))^2 = -\delta^2$ for the lobbyist. If, instead, she deviates to message θ_H (e.g., an overestimation of the true state of the lobbyist's industry), the politician responds with $p = \theta_H$, and the lobbyist earns $-(\theta_H - (\theta_L + \delta))^2$. Therefore, the low-type lobbyist does not have incentives to misrepresent the state of nature if

$$-\delta^2 \geq -(\theta_H - (\theta_L + \delta))^2$$

simplifying, and solving for δ, yields

$$\delta \leq \frac{\theta_H - \theta_L}{2}.$$

Therefore, the low-type lobbyist truthfully reports the state of nature if her bias, δ, is sufficiently small or, alternatively, when her preferences and the politician's are sufficiently aligned.

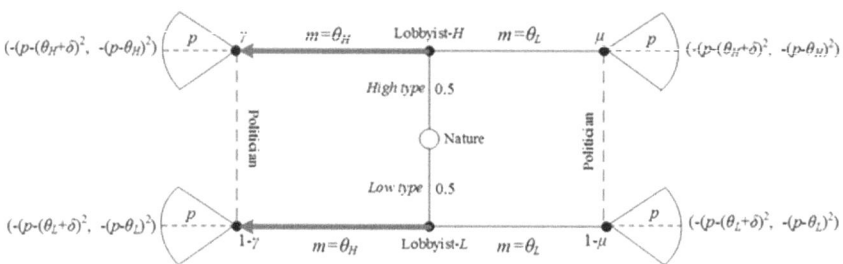

Fig. 13.7 Cheap talk with two messages and continuous responses—Pooling profile (θ_H, θ_H)

5. *Summary.* From Step 4, we found that no sender type has incentives to deviate from (θ_H, θ_L), implying that this separating strategy profile can be supported as a PBE if $\delta \leq \frac{\theta_H - \theta_L}{2}$. In this PBE, the politician, upon observing message θ_i, holds beliefs $\mu(\theta_i|\theta_i) = 1$ and $\mu(\theta_j|\theta_i) = 0$ where $j \neq i$, and responds with policy $p = \theta_i$.

13.3.2 Pooling PBEs

1. *Specifying a strategy profile.* We now test if the pooling strategy profile (θ_H, θ_H), where both lobbyists send message θ_H, can be sustained as a PBE. Figure 13.7 depicts this strategy profile.
2. *Bayes' rule.* Upon observing message θ_H, in equilibrium, the politician cannot infer any information from this message, and her posterior beliefs coincide with her priors, that is, $\mu(\theta_H|\theta_H) = 1/2$. Upon observing message θ_L, however, which occurs off-the-equilibrium path, beliefs cannot be updated using Bayes' rule, and we leave them unrestricted, $\mu(\theta_L|\theta_L) = \mu \in [0, 1]$.
3. *Optimal responses.* Given our result from Step 2, the politician responds as follows:
  (a) Upon observing θ_H, the politician chooses the policy p that solves

$$\max_{p \geq 0} \underbrace{-\frac{1}{2}(p - \theta_H)^2}_{\text{if } \theta=\theta_H} - \underbrace{\frac{1}{2}(p - \theta_L)^2}_{\text{if } \theta=\theta_L}$$

since, from Step 2, the politician's beliefs are $\mu(\theta_H|\theta_H) = 1/2$ and $\mu(\theta_L|\theta_H) = 1/2$. Differentiating with respect to p, we obtain $-(p - \theta_H) - (p - \theta_L) = 0$, and solving for p, we find that her optimal policy response becomes

$$p = \frac{\theta_H + \theta_L}{2},$$

13.3 Cheap Talk with Discrete Messages But Continuous Responses

i.e., the expected state of nature. Intuitively, when the politician believes that both states of nature are equally likely, she implements a policy that coincides with the expected state of nature.

(b) Upon observing θ_L, which occurs off-the-equilibrium path, the politician solves

$$\max_{p \geq 0} \underbrace{-\mu(p-\theta_H)^2}_{\text{if } \theta=\theta_H} \underbrace{-(1-\mu)(p-\theta_L)^2}_{\text{if } \theta=\theta_L}$$

where $\mu(\theta_H|\theta_L) = \mu$ and $\mu(\theta_L|\theta_L) = 1-\mu$. Differentiating with respect to p, we obtain

$$-2\mu(p-\theta_H) - 2(1-\mu)(p-\theta_L) = 0,$$

and, after solving for p, we find that the politician's optimal response is

$$p = \mu\theta_H + (1-\mu)\theta_L$$

which can also be interpreted as the expected state of nature, given the politician's off-the-equilibrium beliefs μ and $1-\mu$ on the high and low state occurring, respectively.

4. *Optimal messages.* From our results in Step 3, we now identify the lobbyist's optimal messages.

   (a) *High type.* From point 4a in the separating PBE (see Sect. 13.3.1), the θ_H-type lobbyists send message θ_H, and this result holds for all parameter values (i.e., for all δ). Recall that this is due to the lobbyist's upward bias.

   (b) *Low type.* From point 4a in the separating PBE (Sect. 13.3.1), we know that the θ_L-type lobbyists send message θ_H, thus misreporting the true state of nature, if

   $$\delta^2 < (\theta_H - (\theta_L + \delta))^2$$

   which simplifying, and solving for δ, yields

   $$\delta > \frac{\theta_H - \theta_L}{2}.$$

   Therefore, the low-type lobbyist misreports the state of nature, θ_L, if her bias, δ, is sufficiently large or, in other words, if her preferences and the politicians are sufficiently misaligned.

5. *Summary.* From Step 4, we found that no sender type has incentives to deviate from the pooling strategy profile (θ_H, θ_H), sustaining it as a PBE, if $\delta > \frac{\theta_H - \theta_L}{2}$. In this PBE, the politician, upon observing message θ_H, in equilibrium, holds beliefs $\mu(\theta_H|\theta_H) = 1/2$, responding with $p = \frac{\theta_H + \theta_L}{2}$; and upon observing θ_L off-the-equilibrium path, her beliefs are unrestricted, $\mu(\theta_L|\theta_L) = \mu$, responding with policy $p = \mu\theta_H + (1-\mu)\theta_L$.

As a practice, Exercise 13.2 asks you to examine under which conditions can the opposite pooling strategy profile, (θ_L, θ_L), be supported as a PBE.

13.4 Cheap Talk with Continuous Messages and Responses

Let us now extend the above cheap talk model to allow for continuous messages and responses. While the setting in Sect. 13.3 considered continuous responses by the politician, it restricted the lobbyist's messages to only two (binary), as she chooses either message θ_H or θ_L. In addition, we now allow for the state of nature, θ, to be continuous, in particular, we assume that $\theta \sim U[0, 1]$. Otherwise, it would look unnatural to still consider only two states of nature (θ_H or θ_L) but a continuum of potential messages, $m \geq 0$, since we seek to identify under which setting information transmission can be sustained in equilibrium. "Information transmission" refers here to the sender truthfully reporting the state of nature, θ, to the receiver, with message $m = \theta$ for all values of θ.

Our discussion, based on Crawford and Sobel (1982), helps us confirm one of the results we found in the previous section, namely, that separating strategy profiles can emerge in equilibrium if the lobbyist and politician's preferences are relatively aligned (low δ). This more general setting, however, allows us a new result that the quality of information transmission, understood as the number of different messages that the lobbyist sends, also depends on the players' preference alignment. Intuitively, as their preferences become more aligned, the lobbyist has incentives to emit a wider array of distinct messages, like different words in a language, ultimately improving the information that the politician receives.

13.4.1 Separating PBE

1. *Specifying a strategy profile.* We first specify a "candidate" of strategy profile that we seek to test as a PBE, as that depicted in Fig. 13.8, where the lobbyist sends message m_1 when θ lies in the first interval, that is, $\theta \in [\theta_0, \theta_1]$; sends message m_2 when θ lies in the second interval, that is, $\theta \in [\theta_1, \theta_2)$; and similarly for the next intervals until reaching the last interval N.
2. *Bayes' rule.* Upon observing message m_k, the politician believes that the state of nature must be in interval k, that is, $\theta \in [\theta_{k-1}, \theta_k]$, such that

$$\mu(\text{Interval } k | m_k) = 1 \text{ and}$$
$$\mu(\text{Interval } j | m_k) = 0 \text{ for all } j \neq k.$$

3. *Optimal responses.* Given our result from Step 2, the politician responds as follows:
   - Upon observing message m_k, she believes to be in interval k, thus responding with policy $p = \frac{\theta_{k-1} + \theta_k}{2}$, which minimizes the quadratic loss in that interval.

13.4 Cheap Talk with Continuous Messages and Responses

To formally prove this result, note that, upon observing message m_k, the politician solves

$$\max_{p \geq 0} \; - E\left[(p - \theta)^2\right]$$

such that $\theta \in [\theta_{k-1}, \theta_k]$

Since the politician believes that $\theta \in [\theta_{k-1}, \theta_k]$, the expected value of θ is $\frac{\theta_{k-1}+\theta_k}{2}$, which simplifies the above problem to

$$\max_{p \geq 0} \; -\left(p - \frac{\theta_{k-1}+\theta_k}{2}\right)^2$$

Differentiating with respect to policy p, yields

$$-2\left(p - \frac{\theta_{k-1}+\theta_k}{2}\right)(-1) = 2\left(p - \frac{\theta_{k-1}+\theta_k}{2}\right) = 0$$

which entails an optimal policy of

$$p = \frac{\theta_{k-1}+\theta_k}{2}.$$

In other words, after receiving message m_k, the politician (receiver) responds with a policy that coincides with the expected state of the nature in this interval, $p = \frac{\theta_{k-1}+\theta_k}{2}$.

4. *Optimal messages.* From our results in Step 3, we now find the lobbyist's optimal messages.

  - When θ_k is the true state of the world, the lobbyist sends a message in the kth interval, as prescribed by this strategy profile, inducing the politician to respond with policy $p = \frac{\theta_{k-1}+\theta_k}{2}$. We now check that the sender does not have incentives to deviate, separately showing that she does not want to overreport or underreport. Without loss of generality, it suffices to show that the kth type sender sends neither m_{k-1} or m_{k+1} (in the intervals immediately below and above interval k, respectively).[3]

  - *No incentives to overreport.* First, we check that the kth sender has no incentives to overreport by sending message m_{k+1}, which occurs if

$$\underbrace{-\left(\frac{\theta_{k-1}+\theta_k}{2} - (\theta_k + \delta)\right)^2}_{\text{Utility from sending message } m_k} \geq \underbrace{-\left(\frac{\theta_{k+1}+\theta_k}{2} - (\theta_k + \delta)\right)^2}_{\text{Utility from sending message } m_{k+1}}$$

---

[3] Formally, if sending message m_{k-1} and m_{k+1} is dominated by message m_k, then sending messages m_{k-j} or m_{k+j}, where $j \geq 2$ thus indicating intervals further away from interval k, would also be dominated.

Since $\frac{\theta_{k-1}+\theta_k}{2} < \frac{\theta_{k+1}+\theta_k}{2}$, the left-hand side of this inequality is positive while that of the right-hand side is negative, yielding

$$\frac{\theta_{k-1}+\theta_k}{2} - (\theta_k + \delta) \geq (\theta_k + \delta) - \frac{\theta_{k+1}+\theta_k}{2}$$

Rearranging the above expression, we obtain

$$\theta_{k-1} + \theta_k - 2(\theta_k + \delta) \geq 2(\theta_k + \delta) - \theta_{k+1} - \theta_k$$

and further simplifying, yields

$$\theta_{k+1} \geq 2\theta_k - \theta_{k-1} + 4\delta.$$

As an illustration, let us check the initial condition, for which the sender of type θ_1 has no incentives to overreport by sending message θ_2.

$$-\left(\frac{\theta_0+\theta_1}{2} - (\theta_1 + \delta)\right)^2 \geq -\left(\frac{\theta_2+\theta_1}{2} - (\theta_1 + \delta)\right)^2.$$

After rearranging, we find

$$\theta_0 + \theta_1 - 2(\theta_1 + \delta) \geq 2(\theta_1 + \delta) - \theta_2 - \theta_1$$

and further simplifying, yields

$$\theta_2 \geq 2\theta_1 - \theta_0 + 4\delta.$$

- *No incentives to underreport.* Second, we check that the $(k+1)^{th}$ sender has no incentives to underreport by sending m_k, which entails

$$-\left(\frac{\theta_{k+1}+\theta_k}{2} - (\theta_{k+1} + \delta)\right)^2 \geq -\left(\frac{\theta_k+\theta_{k-1}}{2} - (\theta_{k+1} + \delta)\right)^2$$

Following the same approach as in the case of "no incentives to overreport," we now have that $\frac{\theta_{k+1}+\theta_k}{2} > \frac{\theta_k+\theta_{k-1}}{2}$, implying that the left-hand side of the above inequality is negative whereas its right-hand side is positive. Rearranging the above expression, we obtain

$$\theta_{k+1} + \delta - \frac{\theta_{k+1}+\theta_k}{2} \geq \frac{\theta_k+\theta_{k-1}}{2} - (\theta_{k+1} + \delta)$$

Simplifying,

$$\theta_{k+1} \geq \frac{1}{3}(2\theta_k + \theta_{k-1} - 4\delta)$$

13.4 Cheap Talk with Continuous Messages and Responses

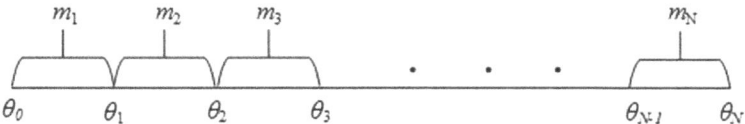

Fig. 13.8 Partially informative strategy profile

Since $\theta_{k+1} > \theta_k$ by construction, the kth sender has no incentives to underreport.[4] Intuitively, the lobbyist finds it unprofitable to report a lower type to the politician, for all values of the bias parameter δ. Therefore, in general, the condition for the kth sender to send the appropriate message is

$$\theta_{k+1} \geq 2\theta_k - \theta_{k-1} + 4\delta.$$

Remark: At this point of our analysis, we have found under which conditions the lobbyist does not have incentives to under or overreport, meaning that her messages about the interval where θ lies are truthful, and the separating strategy profile described in Step 1 can be supported as a PBE. There are, nonetheless, some details about this strategy profile that we have not characterized yet, in particular: (i) the number of partitions that can be sustained in equilibrium, N; (ii) how this number of partitions is affected by the preference divergence parameter, δ; and (iii) the length of each of these partitions (intervals), as they are not necessarily equally long. We analyze each of them in the following subsections.

13.4.2 Equilibrium Number of Partitions

To answer these questions, recall that, since $\theta \sim U[0, 1]$, the first interval starts at $\theta_0 = 0$, and the last interval finishes at $\theta_N = 1$ (see Fig. 13.8). For easier reference, we denote the length of the first interval as $d \equiv \theta_1 - \theta_0$, where $d \geq 0$. We can now rearrange the incentive compatibility condition describing the lobbyist's no incentives to overreport, $\theta_{k+1} \geq 2\theta_k - \theta_{k-1} + 4\delta$, as follows

$$\theta_{k+1} - \theta_k \geq (\theta_k - \theta_{k-1}) + 4\delta$$

Intuitively, this inequality says that each interval must be at least 4δ longer than its predecessor. If this condition binds, we obtain that the second interval length is

$$\theta_2 - \theta_1 = \underbrace{(\theta_1 - \theta_0)}_{d} + 4\delta = d + 4\delta$$

---

[4] This means that condition $\theta_{k+1} > \theta_k > \frac{2}{3}\theta_k + \frac{1}{3}\theta_{k-1} = \frac{1}{3}(2\theta_k + \theta_{k-1}) > \frac{1}{3}(2\theta_k + \theta_{k-1} - 4\delta)$ holds for all values of the bias parameter δ. As a consequence, the incentive compatibility constraint for no underreporting becomes slack.

while the length of the third interval is

$$\theta_3 - \theta_2 = \underbrace{(\theta_2 - \theta_1)}_{d+4\delta} + 4\delta = d + (2 \times 4\delta)$$

and similarly for subsequent intervals. By recursion, the length of the k-th interval is, then,

$$\begin{aligned}
\theta_k - \theta_{k-1} &= (\theta_{k-1} - \theta_{k-2}) + 4\delta \\
&= (\theta_{k-2} - \theta_{k-3}) + (2 \times 4\delta) \\
&= (\theta_{k-3} - \theta_{k-4}) + (3 \times 4\delta) \\
&= \ldots \\
&= (\theta_1 - \theta_0) + [(k-1) \times 4\delta] \\
&= d + 4(k-1)\delta
\end{aligned}$$

As an illustration, the length of the final interval, where $k = N$, is

$$\theta_N - \theta_{N-1} = d + 4(N-1)\delta$$

We can now express the length of the unit interval, $\theta_N - \theta_0 = 1$, as the sum of N partitions, as follows

$$\theta_N - \theta_0 = \underbrace{(\theta_N - \theta_{N-1})}_{d+4(N-1)\delta} + \underbrace{(\theta_{N-1} - \theta_{N-2})}_{d+4(N-2)\delta} + \ldots + \underbrace{(\theta_2 - \theta_1)}_{d+4\delta} + \underbrace{(\theta_1 - \theta_0)}_{d}$$

$$= Nd + 4\delta[(N-1) + (N-2) + \ldots + 1]$$

where the term in square brackets can be simplified as follows

$$\underbrace{[(N-1) + (N-2) + \ldots + 1]}_{N-1 \text{ terms}} = N(N-1) - [1 + 2 + \ldots + (N-1)]$$

$$= N(N-1) - \frac{N(N-1)}{2}$$

$$= \frac{N(N-1)}{2}$$

since $1 + 2 + \cdots + (N-1) = \frac{(N-1)N}{2}$. Therefore, the above expression of $\theta_N - \theta_0$ further simplifies to

$$\begin{aligned}
\theta_N - \theta_0 &= Nd + 4\delta \frac{N(N-1)}{2} \\
&= Nd + 2\delta N(N-1)
\end{aligned}$$

13.4 Cheap Talk with Continuous Messages and Responses

And since the left-hand side is $\theta_N - \theta_0 = 1$ (θ lies in the unit interval), we can write the above equation as $1 = Nd + 2\delta N(N-1)$, and solve for the bias parameter, δ, to obtain

$$\underline{\delta}(d) = \frac{1 - Nd}{2N(N-1)}.$$

Cutoff $\underline{\delta}(d)$ decreases in the length of the first interval, d. Intuitively, as the first interval becomes wider (higher d), a given number of partitions N becomes more difficult to be sustained as a PBE.[5] This result suggests that when the length of the first interval is nil, $d = 0$, we obtain that the maximal value of cutoff $\underline{\delta}(d)$ becomes

$$\underline{\delta}(0) = \frac{1}{2N(N-1)}.$$

Therefore, more partitions (higher N) can only be supported as a PBE if the bias parameter δ becomes smaller. That is, more informative PBEs can be sustained when the preferences of lobbyist and politician are more similar (lower δ). Solving for N, we can also find the maximum number of partitions, $N(\delta)$, as a function of δ, as follows

$$N(N-1) \leq \frac{1}{2\delta}$$

which we can rearrange as

$$N^2 - N - \frac{1}{2\delta} \leq 0$$

Factorizing the above inequality, yields

$$\left(N - \frac{1 + \sqrt{1 + \frac{2}{\delta}}}{2}\right)\left(N - \frac{1 - \sqrt{1 + \frac{2}{\delta}}}{2}\right) \leq 0$$

Since $N \geq 1$ (that is, there must be at least one partition), we can rule out the negative root such that the relevant result is

$$N \leq \frac{1 + \sqrt{1 + \frac{2}{\delta}}}{2}$$

---

[5] Note that this finding cannot be interpreted as a standard comparative statics result since we have not found yet the length d that arises in equilibrium. We do that below.

Furthermore, since N is a positive integer, we have that

$$N \leq \overline{N}(\delta) \equiv \left\lfloor \frac{1 + \sqrt{1 + \frac{2}{\delta}}}{2} \right\rfloor$$

where the $\lfloor \cdot \rfloor$ sign rounds to the next integer from below, e.g., $\lfloor 3.7 \rfloor = 3$. Cutoff $\overline{N}(\delta)$ represents the maximum number of partitions that can be supported for a given value of δ. As the bias parameter, δ, increases, fraction $\frac{2}{\delta}$ becomes smaller, ultimately decreasing cutoff $\overline{N}(\delta)$. Intuitively, as the lobbyist and the politician become more divergent in their preferences, the lobbyist has more incentives to overreport her type, so her messages become less informative (fewer partitions of the unit interval in Fig. 13.8).

Figure 13.9 depicts cutoff $\overline{N}(\delta)$ and illustrates that the PBE yields a smaller number of partitions as the bias parameter δ increases, thus supporting less information transmission from the (informed) lobbyist to the (uninformed) politician. Alternatively, solving for δ in cutoff $\overline{N}(\delta)$, we find that

$$\delta \leq \frac{1}{2N(N-1)},$$

which also indicates that, as we seek a larger number of partitions (higher N) in equilibrium, the preference divergence parameter must be lower.

Example 13.1 Equilibrium number of partitions. If we seek to support $N = 2$ partitions in equilibrium, we need $\delta \leq \frac{1}{2 \times 2(2-1)} = \frac{1}{4}$. To sustain $N = 3$ partitions, however, we require $\delta \leq \frac{1}{2 \times 3(3-1)} = \frac{1}{12}$, which imposes a more restrictive condition on players' preference alignment. □

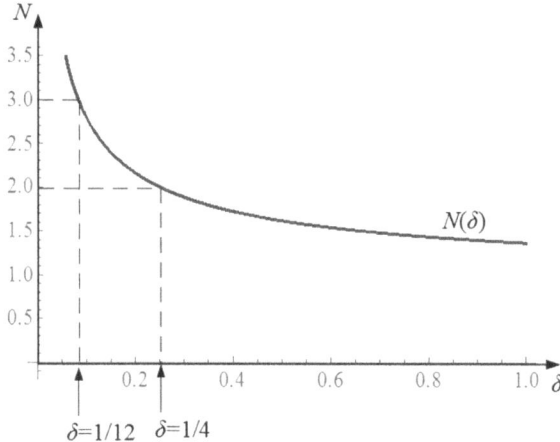

Fig. 13.9 Cutoff $\overline{N}(\delta)$ as a function of the bias parameter δ

13.4.3 Interval Lengths in Equilibrium

Let us now find the equilibrium length of the first interval, d^*, which will provide us with the length of all subsequent intervals. Consider again the lobbyist's incentive compatibility condition for not overreporting,

$$\theta_{k+1} - 2\theta_k + \theta_{k-1} = 4\delta$$

This expression can be understood as a second-order linear difference equation,[6] and rearrange it as $\theta_{k+1} = 2\theta_k - \theta_{k-1} + 4\delta$. Since the unit interval starts at $\theta_0 = 0$, we can write the above equation for $k = 1$ as follows

$$\theta_2 = 2\theta_1 - \theta_0 + 4\delta = 2\theta_1 + 4\delta$$

and, similarly, for $k = 2$ this equation becomes

$$\begin{aligned}\theta_3 &= 2\theta_2 - \theta_1 + 4\delta \\ &= 2\underbrace{(2\theta_1 + 4\delta)}_{\theta_2} - \theta_1 + 4\delta \\ &= 3\theta_1 + 12\delta\end{aligned}$$

which helps us express it more generally, for any value of k, as follows

$$\theta_k = k\theta_1 + 2k(k-1)\delta.$$

Therefore, evaluating this expression at $k = N$, we obtain $\theta_N = Nd + 2N(N-1)\delta$, which helps us write $\theta_N - \theta_0 = Nd + 2N(N-1)\delta$ since $\theta_0 = 0$. In addition, because $\theta_N - \theta_0 = 1$, we can express $1 = Nd + 2N(N-1)\delta$ and, solving for d, we find that the length of the first interval is

$$d^* = \frac{1}{N} - 2(N-1)\delta$$

which also satisfies $d^* = \theta_1 - \theta_0 = \theta_1$ given that $\theta_0 = 0$. Interestingly, this length decreases in the number of partitions in that equilibrium, N, since $\frac{\partial d^*}{\partial N} = -2\delta - \frac{1}{N^2} \leq 0$, which means that the first interval shrinks to "make room" for subsequent partitions to its right side.

---

[6] Recall that linear second-order difference equations generally take the form $x_{t+2} + ax_{t+1} + bx_t = c_t$, where a and b are constants and c_t is a number which can vary with t. In our setting, $x_{t+2} = \theta_{k+1}$, $a = -2$, $x_{t+1} = \theta_k$, $b = 1$, $x_t = \theta_{k-1}$, and $c_t = 4\delta$ for all t. For an introduction to linear difference equations, see for instance, Simon and Blume (1994, chapter 23).

Example 13.2 First interval decreasing in N. If $\delta = \frac{1}{20}$ and $N = 2$, which is compatible with condition $N \leq \overline{N}(\delta)$ found in the previous section, we obtain that the length of the first interval is

$$d^* = \frac{1}{2} - 2(2-1)\frac{1}{20} = \frac{2}{5}.$$

However, when N increases to $N = 3$, this length shrinks to $d^* = \frac{1}{3} - 2(3-1)\frac{1}{20} = \frac{2}{15}$. $\square$

We can now use the above results to find the length of the kth interval, $\theta_k - \theta_{k-1}$. First, substituting $\theta_1 = \frac{1}{N} - 2(N-1)\delta$ into expression $\theta_k = k\theta_1 + 2k(k-1)\delta$, we obtain

$$\theta_k = k\underbrace{\left(\frac{1}{N} - 2(N-1)\delta\right)}_{\theta_1} + 2k(k-1)\delta$$

$$= \frac{k}{N} - 2k(N-k)\delta$$

which implies that the length of the kth interval is

$$\theta_k - \theta_{k-1} = \underbrace{\left(\frac{k}{N} - 2k(N-k)\delta\right)}_{\theta_k} - \underbrace{\left(\frac{k-1}{N} - 2(k-1)(N-k+1)\delta\right)}_{\theta_{k-1}}$$

$$= \frac{1}{N} - 2(N+1-2k)\delta.$$

As a remark, we can confirm that the length of the first interval coincides with the expression found above, d^*. Indeed, evaluating $\theta_k - \theta_{k-1}$ at the first interval (i.e., $k = 1$), this expression simplifies to

$$\theta_1 - \theta_0 = \frac{1}{N} - 2(N+1-2)\delta$$

$$= \frac{1}{N} - 2(N-1)\delta = d^*$$

which coincides with the result of d^* found above.

Example 13.3 Length of each interval in equilibrium. Following with Example 13.2, in a context where $\delta = \frac{1}{20}$ and $N = 2$, the length of the first interval is $d^* = \frac{2}{5}$. The length of the second interval is

$$\theta_2 - \theta_1 = \frac{1}{2} - 2[2+1-(2\times 2)]\frac{1}{20} = \frac{3}{5}$$

which, together, account for the total length of the unit interval, i.e., $\frac{2}{5} + \frac{3}{5} = 1$. Similarly, in a setting with $N = 3$ partitions, it is easy to show that the first interval's length is $d^* = \frac{2}{15}$, as shown in Example 13.2, that of the second interval is

$$\theta_2 - \theta_1 = \frac{1}{3} - 2[3 + 1 - (2 \times 2)]\frac{1}{20} = \frac{1}{3},$$

and that of the third interval is

$$\theta_3 - \theta_2 = \frac{1}{3} - 2[3 + 1 - (2 \times 3)]\frac{1}{20} = \frac{8}{15},$$

with their sum satisfying $\frac{2}{15} + \frac{1}{3} + \frac{8}{15} = 1$, spanning the unit interval. □

13.5   Extensions

The cheap talk model in Sect. 13.4, based on Crawford and Sobel (1982), made some simplifying assumptions that have been relaxed by the literature. We next present some of these extensions.

Allowing for "conversations." Krishna and Morgan (2004) consider how the above results would be affected if the politician could also send messages to the lobbyist. Allowing the uninformed politician to communicate with the informed lobbyist does not seem, at first glance, as facilitating information transmission (more partitions in equilibrium), but the authors identify conditions under which this "conversation" between the players yields more partitions than otherwise.

Cheap talk vs. Delegation. Dessein (2002) poses whether allowing the politician to delegate the decision of policy p to the lobbyist can improve information transmission. Delegation entails a trade-off: on one hand, it shifts the policy decision to the informed player (the lobbyist), which should be positive; but, on the other hand, it allows the lobbyist to choose a policy according to her bias, δ, which reduces the politician's utility. As expected, when δ is relatively small, the first (positive) effect dominates the second (negative) effect, implying that the politician is better-off delegating. Otherwise, the lobbyist's bias is too large, and the politician keeps the choice of p, as in the standard Crawford and Sobel (1982) model. Exercise 13.6 helps you show this result in a tractable context.

Open vs. Closed rules. An application of this finding is the comparison of the so-called "open rule" and "closed rule" in the House of Representatives. Specifically, when a specialized committee (expert) sends a bill to the floor of the House (politician), the latter can freely amend the bill when operating under open rule (as the politician in the standard cheap talk model) but faces limited ability to amend the bill under closed rule (as the politician in Dessein's article). Gilligan and Krehbiel (1987) show that the closed rule, despite "tying the hands" of the voting body on the floor of the House, can yield more information transmission in equilibrium than the open rule. Exercise 13.8 studies this setting.

Several lobbyists. Another natural question is whether information transmission is facilitated when the politician receives messages from more than one lobbyist (sender). A related question is, then, whether the composition of a cabinet matters: should all lobbyists exhibit similar biases (e.g., $\delta_1, \delta_2 > 0$) or opposite biases (e.g., $\delta_1 > 0 > \delta_2$)? Krishna and Morgan (2001) study these questions allowing, first, for lobbyists to send their messages simultaneously. In this context, the authors show that every lobbyist conveys the true state to the politician, $m_i = \theta$ for every $i = \{1, 2\}$, entailing full information; which is robust to lobbyists exhibiting similar or opposite biases. Then, Krishna and Morgan (2001) consider that messages are, instead, sent sequentially by each lobbyist, showing that, when experts have similar (opposite) biases, the politician is better-off ignoring the lobbyist with the smaller bias (considering the messages sent by both lobbyists, respectively).

The literature considers several other extensions and their effects on information transmission. Examples include allowing for multiple receivers (politicians), as in Farrell and Gibbons (1989); repeated cheap talk, as in Ottaviani and Sorensen (2006); allowing for the expert's private information to be multidimensional, such as Chakraborty and Harbaugh (2007); the presence of a mediator (uninformed about θ, receiving messages from the lobbyist, and reporting any information she likes to the politician), as in Ganguly and Ray (2005); or noise in the lobbyist messages before the politician receives them, as in Blume et al. (2007). Another interesting line of the literature proposes equilibrium refinement criteria in cheap talk environments, such as Farrell (1993), Chen et al. (2008), and de Groot Ruiz et al. (2015). For related literature, see Farrell and Rabin (1996) and Krishna and Morgan (2008), and for a survey of experiments testing the theoretical predictions in controlled environments see Crawford (1998) and Blume et al. (2007).

Exercises

13.1. **Two types of senders and two responses.**[A] Consider the cheap talk game with two sender types and two messages depicted in Fig. 13.1.
   (a) Show that separating strategy profile (m_2^H, m_1^L) can be sustained as a PBE.
   (b) Show that pooling strategy profile (m_2^H, m_2^L), where both sender types move rightward, can be supported as a PBE.

13.2. **Two types of sender and continuous responses.**[B] Consider the cheap talk game with two sender types and a continuum of responses from Sect. 13.3. Under which conditions can the pooling strategy profile (θ_L, θ_L) be sustained as a PBE?

13.3. **A binary version of Crawford and Sobel.**[A] Consider the cheap talk model in Sect. 13.4, but assume that $\theta = \theta_H$ with probability p and $\theta = \theta_L$ with probability $1 - p$. The sender (lobbyist) privately observes the realization of θ (whether it is θ_H or θ_L) and sends a binary message, either H or L, to the politician. The politician observes this message, updates her beliefs about θ, and responds with a binary response as well (h or l upon observing H, and h' or l' upon observing L).

(a) Describe the payoffs of each player in each terminal node.

(b) Under which condition on the preference divergence parameter, δ, can you sustain a separating PBE where the sender chooses message H (L) when his type is θ_H (θ_L, respectively)?

(c) If θ_H increases, is the above separating PBE more likely to hold? Interpret.

13.4. **Ex-ante expected utilities in cheap talk games.**[B] Consider the Crawford-Sobel cheap talk game with a continuum of types. In Sect 13.4, we discussed the maximal number of partitions that can be sustained as a PBE of the game.

(a) Find the ex-ante expected utility that the sender obtains in equilibrium. (By "ex-ante" we mean before observing the realization of parameter θ.)

(b) Find the *ex-ante* expected utility that the receiver obtains in equilibrium.

(c) How are the above *ex-ante* expected utilities affected by an increase in the preference divergence parameter δ? And how are they affected by an increase in the number of partitions?

13.5. **Cheap talk vs. Commitment, based on Melumad and Shibano (1991).**[B] Consider the Crawford and Sobel (1982) cheap talk game, where an expert privately observes the realization of parameter $\theta \sim U[0, 1]$, sends a message $m \in [0, 1]$ to the uninformed politician who updates his beliefs about θ, and responds with a policy $p \in [0, 1]$. The politician's utility function is given by $u_P(p, \theta) = -(p - \theta)^2$, while the expert's utility function is $u_E(p, \theta) = -(p - (\theta + \delta))^2$, where parameter $\delta > 0$ represents the divergence between the preferences of the politician and the expert (also known as the expert's bias parameter).

Consider now an alternative communication setting, where the expert (sender) commits to truthtelling, that is, his message coincides with the state of nature θ, $m(\theta) = \theta$ for all realizations of θ. In this exercise, we seek to understand under which contexts the expert prefers to "tie his hands" by ex-ante committing to this truthtelling strategy before he observes the realization of parameter θ, and in which settings he prefers to keep playing the standard cheap talk game. For simplicity, assume that the most informative equilibrium emerges from the cheap talk game.

(a) Find the ex-ante expected utilities of expert and politician in the most informative equilibrium of the cheap talk game, labeling them u_E^{CT} and u_P^{CT} where the superscript CT denotes cheap talk.

(b) Find the ex-ante expected utilities of expert and politician if the expert commits to the truthtelling strategy before he observes the realization of parameter θ, labeling them u_E^{Commit} and u_P^{Commit}.

(c) Compare the expert's ex-ante expected utility from committing to the truthtelling strategy, as found in part (b), against that from playing the most informative cheap talk game, as found in part (a). Under which parameter conditions the expert prefer to "tie his hands" by ex-ante committing to this truthtelling strategy?

13.6. **Cheap talk vs. Delegation, based on Dessein (2002).**[B] Consider the Crawford and Sobel (1982) cheap talk game, where an expert privately observes the realization of parameter $\theta \sim U[0, 1]$, sends a message $m \in [0, 1]$ to the uninformed politician who updates his beliefs about θ, and responds with a policy $p \in [0, 1]$. The politician's utility function is given by $u_P(p, \theta) = -(p-\theta)^2$, while the expert's utility function is $u_E(p, \theta) = -(p-(\theta+\delta))^2$, where parameter $\delta > 0$ represents the divergence between the preferences of the politician and the expert (also known as the expert's bias parameter).

Consider now an alternative communication setting, where the politician delegates the decision of policy p to the expert. Intuitively, the politician seeks to minimize the information loss from the cheap talk setting, at the cost of having the expert implement a policy that considers his own bias. We next examine which policy the expert implements in under delegation, and in which cases the politician is better-off delegating the decision to the expert rather than operating in the standard cheap talk context.

(a) *Delegation.* Which policy the expert chooses under delegation? What are the expected utilities for expert and politician?

(b) *Do-it-yourself.* Consider now that the receiver (politician) ignores the sender's messages, which is often known as if the politician took a "do-it-yourself" approach. Find the policy he responds with, and his expected utility in this setting. Under which value of the preference divergence parameter δ the politician prefers the "do-it-yourself" approach over the delegation approach studied in part (a) of the exercise?

(c) From the previous exercises, we found that expected utilities in the Crawford and Sobel (1982) cheap talk game are $u_P^{CT} = -\frac{1}{12N^2} - \frac{\delta^2(N^2-1)}{3}$ for the politician, where the superscript CT indicates "cheap talk"; and $u_E^{CT} = u_P^{CT} - \delta^2$ for the expert. In addition, recall that N denotes the number of partitions in the $[0, 1]$ interval, or different messages that the expert sends; where this N-partition equilibrium can be sustained if the preference divergence parameter is sufficiently small, that is, $\delta \leq \frac{1}{2N(N-1)}$. Show that the expert prefers delegation rather than sending messages to the politician in the cheap talk game, and that this result holds under all parameter conditions. Then show that the politician prefers delegation only if the expert's bias, as captured by δ, is sufficiently small.

13.7. **Signaling when the sender observes imprecise signals.**[C] Consider the following signaling model between a lobbyist and a politician. For simplicity, assume that the state of the world is discrete, either $\theta = 1$ or $\theta = 0$ with prior probability $p \in (0, 1)$ and $1 - p$, respectively. The lobbyist privately observes an informative but noisy signal s, which also takes two discrete values $s \in \{0, 1\}$. The precision of the signal is given by the conditional probability

$$\text{prob}(s = k | \theta = k) = q,$$

where $k = \{0, 1\}$, and $q > \frac{1}{2}$. In words, the probability that the signal s coincides with the true state of the world θ is q (precise signal), while the probability of an imprecise signal where $s \neq \theta$ is $1 - q$. The time structure of the game is as follows:
(1) Nature chooses θ according to the prior p.
(2) Expert observes signal s and reports a message $m \in \{0, 1\}$
(3) Decision maker observes m and responds with $x \in \{0, 1\}$
(4) θ is observed and payoffs are realized.
The payoff function for the politician is $u(x, \theta) = \left(\theta - \frac{1}{2}\right)x$, while that of the lobbyist is $v(m, \theta) = 1$ if $\theta = m$ but $v(m, \theta) = 0$ otherwise. Therefore, the lobbyist's payoff is 1 when the message she sends coincides with the true realization of the state of the world, but becomes zero otherwise. Importantly, her payoff is unaffected by the signal, which she only uses to infer the actual realization of parameter θ.

Can a strategy profile where the lobbyist reports her signal truthfully be sustained as a PBE?

13.8. **Open or closed rule in committees, based on Gilligan and Krehbiel (1987).**[C] Consider the setting in the Crawfold and Sobel's cheap talk game, but assume that the state of the world, θ, is either $\theta > 0$ or $\theta < 0$, both of them equally likely. In this exercise, we compare closed and open rules in committees. Under closed rule, the politician must either follow the recommendation from the sender (a lobbyist or, more generally, a committee) or keep the status quo (choosing the policy without receiving any information from the sender). In contrast, under open rule, the politician can follow the recommendation from the sender or responds with any other policy. We analyze each setting separately, and then compare them to identify in which settings closed rule, despite allowing for less policy flexibility of the receiver (politician), can facilitate information transmission from the sender (lobbyist or committee).
(a) *Status quo.* If the lobbyist cannot send a message to the politician, what policy would the politician choose?
(b) *Open rule.* Under which conditions can you sustain a separating strategy profile (m, m'), where the lobbyist sends message m upon observing that the state of the world is $\theta > 0$, but message $m' \neq m$ when $\theta < 0$?
(c) *Closed rule.* Under which conditions can you sustain a separating strategy profile (m, m'), where message $m = \theta + \delta$ with $\theta > 0$ and message $m' = \theta + \delta$ with $\theta < 0$? For simplicity, separately consider the case of $\delta \leq \theta$, and then that where $\delta > \theta$.

References

Blume, A., O. Board, and K. Kawamura (2007) "Noisy Talk," Theoretical Economics, 2, pp. 395–440.
Chakraborty, A. and R. Harbaugh (2007) "Comparative Cheap Talk," Journal of Economic Theory, 132, pp. 70–94.
Chen, Y., N. Kartik, and J. Sobel (2008) "Selecting Cheap-talk Equilibria," Econometrica, 76, pp. 117–136.
Crawford, V. (1998) "A Survey of Experiments on Communication via Cheap Talk," Journal of Economic Theory, 78(2), pp. 286–298.
Crawford, V. P., and J. Sobel (1982) "Strategic Information Transmission," Econometrica, 50(6), pp. 1431–1451.
de Groot Ruiz, A., T. Offerman, and S. Onderstal (2015) "Equilibrium Selection in Experimental Cheap Talk Games," Games and Economic Behavior, 91, pp. 14–25.
Dessein, W. (2002) "Authority and Communication in Organizations," *Review of Economic Studies*, 69, pp. 811–838.
Farrell, J. (1993) "Meaning and Credibility in Cheap-talk Games," Games and Economics Behavior, 5, pp. 514–531.
Farrell, J., and R. Gibbons (1989) "Cheap Talk with Two Audiences," American Economic Review, 79, pp. 1214–1223.
Farrell, J., and M. Rabin (1996) "Cheap Talk," The Journal of Economic Perspectives, 10(3), pp. 103–118.
Ganguly, C., and I. Ray (2005) "On Mediated Equilibria of Cheap-talk Games," working paper, University of Birmingham.
Gilligan, T., and K. Krehbiel (1987) "Collective Decision-making and Standing Committees: An Informational Rationale for Restrictive Amendment Procedures," Journal of Law, Economics, and Organization, 3, pp. 287–335.
Krishna, V., and J. Morgan (2001) "A Model of Expertise," Quarterly Journal of Economics, 116, pp. 747–775.
Krishna, V., and J. Morgan (2004) "The Art of Conversation: Eliciting Information from Experts Through Multi-stage Communication," Journal of Economic Theory, 117, pp. 147–179.
Krishna, V., and J. Morgan (2008) "Cheap Talk," in: *The New Palgrave Dictionary of Economics*, eds. Steven N. Durlauf and Lawrence E. Blume, Second Edition, Palgrave Macmillan.
Melumad, N. D., and T. Shibano (1991) " Communication in Settings with no Transfers," Rand Journal of Economics, 22, pp. 173–198.
Ottaviani, M., and P. T. Sorensen (2006) " Reputational Cheap Talk," RAND Journal of Economics, 37(1), pp. 155–175.
Simon, C., and L. Blume (1994) *Mathematics for Economists*, Worth Publishers.

Mathematical Appendix

This mathematical appendix provides a brief summary of basic concepts used in the book, such as sets, sequences, functions, limits and continuity, differentiation, and integration. The reader is recommended to refer to books such as Klein (2001) and Wainwright and Chiang (2004) for an undergraduate introduction to these topics, and to books such as Simon and Blume (1994) and Corbae et al. (2009) for a graduate-level presentation.

A.1 Sets

We understand a set S as a list of objects, and use $x \in S$ to denote that element x belongs to set S, meaning that x is one of the elements in set S. Examples include strategy sets where a player can choose between high or low prices, h or l, so that $S = \{h, l\}$. Common set operators are:

1. **Intersection.** The intersection of two sets A and B, denoted as $A \cap B$, which lists the elements that belong to both sets A and B.
2. **Union.** The union of two sets A and B, denoted as $A \cup B$, which represents the elements that belong to either set A or B or both.
3. **Complement.** The complement of set B relative to A, denoted as $A \setminus B$, which lists the elements in A that do not belong to B.
4. **Cartesian product.** The Cartesian product of two sets, $A \times B$, is the set of all combinations of pairs (a, b) where the first component originates from set A, $a \in A$, and the second component comes from set B, $b \in B$. This is common when considering the strategy profiles in a N-player game, where we take the Cartesian product of all players' strategy spaces, $S = S_1 \times S_2 \times ... \times S_N$, where an element in S is a specific strategy profile $s = (s_1, s_2, ..., s_N)$.

A.2 Sequences

A sequence $\{x_n\}$, where $n > 0$ is an integer, is an ordered list, and each of its elements represents a term in the sequence. For instance, if $x_n = \frac{1}{n}$, the sequence starts $\frac{1}{1} = 1$ when $n = 1$, decreases to $\frac{1}{2}$ when $n = 2$, and further decreases to $\frac{1}{3}$ when $n = 3$, and so on, decreasing in n, and converging to zero when $n \to +\infty$. In contrast, if $x_n = \frac{n}{1+2n}$ the sequence starts at $\frac{1}{3}$ when $n = 1$, but increases to $\frac{2}{5} = 0.4$ when $n = 2$, keeps increasing to $\frac{3}{7} \simeq 0.429$ when $n = 3$, and further increases, converging to $\frac{1}{2}$ when $n \to +\infty$.

Convergent sequence. We say that a sequence $\{x_n\}$ converges to a real number r if, for every $\varepsilon > 0$, there is a positive integer N such that the distance $|x_n - r|$ satisfies

$$|x_n - r| < \varepsilon \text{ for all } n \geq N,$$

and compactly denote it at $x_n \to r$.

Intuitively, a sequence converges to a real number r if x_n is close to r beyond position N. Since this must hold "for every $\varepsilon > 0$," we need a sufficiently high integer N. A common trick is to express N as a function of ε, such as $N = \frac{1}{\varepsilon}$ or $N = \frac{1}{2\varepsilon}$. In our above example, where $x_n = \frac{1}{n}$, if we define $N = \frac{1}{\varepsilon}$, the distance $|x_n - r|$ simplifies to $\left|\frac{1}{n} - 0\right| = \frac{1}{n}$ when $r = 0$. This distance is smaller than ε for all $n \geq N = \frac{1}{\varepsilon}$. Specifically, at the Nth term, $n = N$, this distance is exactly equal to ε since

$$|x_N - r| = \left|\frac{1}{N} - 0\right| = \left|\frac{1}{\frac{1}{\varepsilon}} - 0\right| = \varepsilon,$$

but becomes smaller than ε for all $n > N$.

A.3 Functions

A function $f(x)$ describes the relationship (a mapping) between an input and its output. Real-valued functions are common in economics, usually defined as $f : \mathbb{R}^N \to \mathbb{R}^N$. Examples include the linear function

$$f(x) = a + bx,$$

where $a, b \in \mathbb{R}$ are constants and $b \neq 0$; or non-linear functions

$$f(x) = ax^2 + d \text{ or } f(x) = ax^b + cx + d,$$

where $a, b, c, d \in \mathbb{R}$ are constants and $a \neq 0$.

In the linear function, parameter a represents the vertical intercept where the function originates, whereas b denotes the slope of the function (which can be positive if the function increases in x or negative if the function decreases in x). Since slope b is constant, the graphical representation of the linear function is a straight line. In addition, setting the linear function equal to zero, $a + bx = 0$, and solving for x, helps us obtain the function's horizontal intercept, $x = -\frac{a}{b}$, which graphically represents the point where the linear function crosses the horizontal axis.[1]

When the set of possible inputs of the function is restricted to $A \subset \mathbb{R}^N$ (its domain) and the set of possible outputs to $B \subset \mathbb{R}^N$ (its range or image), we express the function as $f : A \to B$. For example, for function $f(x) = x^2$, its domain is the set of all real numbers (positive and negative), $A = \mathbb{R}$; while its image is the set of positive real numbers, $B = \mathbb{R}_+$. Another example of functions commonly used in the book are "payoff functions," which map the strategy profile being played in a game, $s \in S$, where $s = (s_1, s_2, ..., s_N)$, into a payoff for that player (positive or negative), $u_i : S \to \mathbb{R}$.

We next list some basic properties of functions:

1. **One-to-one.** A function $f : A \to B$ is one-to-one (or injective) if, for every two points in its domain, $x \neq y$, their images differ, $f(x) \neq f(y)$.

   Function $f(x) = x^2$ where $x \in \mathbb{R}$, for instance, is *not* one-to-one, since we can find two points, such as $x = 3$ and $y = -3$, both yielding the same image, $f(x) = f(y) = 9$. However, if we restrict the domain of the function to the positive reals, $x \in \mathbb{R}_+$, the function is one-to-one. Functions that are monotonically increasing or decreasing, such as $f(x) = a + bx$, are one-to-one.

2. **Onto.** A function $f : A \to B$ is onto (or surjective) if, for every image $b \in B$, there is an element in the domain $a \in A$ that yields b, i.e., $f(a) = b$.

   The linear function $f(x) = a + bx$ is onto, while function $f(x) = x^2$ is not since we can find a real number y, where $y < 0$, that is not mapped from $x \in \mathbb{R}$.

3. **Bijection.** A function $f : A \to B$ is a bijection if it is one-to-one and onto. The linear function $f(x) = a + bx$ is, of course, bijective.

   An interesting feature of bijective functions is that they are invertible, so starting from function $f(x)$, we can find the input that generated such output, and denote such inverse as $f^{-1}(x)$. Graphically, one moves from the images of the function on the vertical axis to the domain that generated such images on the horizontal axis. In the case of linear function $f(x) = a + bx$ where $b \neq 0$, for instance, its inverse is $f^{-1}(x) = \frac{x}{b} - \frac{a}{b}$.

---

[1] In the non-linear function, parameter d can still be interpreted as the function's vertical intercept, but parameters a, b, or c do not denote the slope of the function. As we discuss below, the slope of $f(x) = ax^b + cx + d$ is obtained by differentiating this function with respect to x, obtaining $f'(x) = abx^{b-1} + c$.

A.4 Limits and Continuity

Consider function $f : A \to B$, where $A, B \subset \mathbb{R}$. We say that function $f(x)$ has a limit $p \in B$ at point $c \in A$ if, for every $\varepsilon > 0$ such that $|f(x) - p| < \varepsilon$, there exists a real number $\delta > 0$ such that x and c are δ-close, meaning $|x - c| < \delta$. For compactness, we often write

$$\lim_{x \to c} f(x) = p,$$

which intuitively means that, as x becomes sufficiently close to c, denoted as $x \to c$, the function approaches p. Basic properties include the following:

1. $\lim_{x \to c} af(x) = a \lim_{x \to c} f(x)$ when a is a constant.
2. $\lim_{x \to c} [f(x) + g(x)] = \lim_{x \to c} f(x) + \lim_{x \to c} g(x)$, where f and g are real-valued functions.
3. $\lim_{x \to c} [f(x) g(x)] = \lim_{x \to c} f(x) \lim_{x \to c} g(x)$.
4. $\lim_{x \to c} \frac{f(x)}{g(x)} = \frac{\lim_{x \to c} f(x)}{\lim_{x \to c} g(x)}$ if the limit in the denominator is non-zero, $\lim_{x \to c} g(x) \neq 0$.

A function is continuous if, informally, does not have breaks or "jumps." Function $f(x)$ is continuous at point $x = a$ if its limit at this point satisfies $\lim_{x \to a} f(x) = f(a)$. If this property holds for all points in the function's domain (i.e., for all $x \in A$), we say that the function is continuous.

Otherwise, the function is discontinuous, which is formally defined as that there is a point $x = a$ for which the limit of $f(x)$ "from the left of a" and "from the right of a" does not coincide, that is, $\lim_{x \to a^-} f(x) \neq \lim_{x \to a^+} f(x)$. A function like $f(x) = \frac{1}{x-2}$ is discontinuous at $x = 2$, or more generally $f(x) = \frac{1}{x-a}$ which is discontinuous at $a > 0$. Indeed, $\lim_{x \to a^-} f(x) = -\infty$ whereas $\lim_{x \to a^+} f(x) = +\infty$. Similarly, function $f(x) = a^{\frac{x}{1-x^2}}$, where $a \neq 1$, is discontinuous at $x = -1$ and $x = 1$.

A.5 Quadratic Equation

In this section, we seek to solve a quadratic equation of the form $f(x) = ax^2 + bx + c$, where a, b, c are constants (positive or negative). The roots that solve this equation are

$$x_1 = \frac{-b + \sqrt{b^2 - 4ac}}{2a} \quad \text{and} \quad x_2 = \frac{-b - \sqrt{b^2 - 4ac}}{2a}$$

which are real numbers if the discriminant is non-negative, that is, $\Delta \equiv b^2 - 4ac \geq 0$. (If the discriminant is strictly negative, $\Delta < 0$, no roots are real numbers.)

Mathematical Appendix

To show that x_1 and x_2 are roots of the quadratic equation $f(x) = ax^2 + bx + c$, we first set it equal to zero, and divide both sides by a, yielding

$$x^2 + \frac{bx}{a} + \frac{c}{a} = 0.$$

Adding $\frac{b^2}{4a_2}$ on both sides and rearranging, we find that

$$x^2 + \frac{bx}{a} + \frac{b^2}{4a^2} = \frac{b^2}{4a^2} - \frac{c}{a}$$

that can be factorized into

$$\left(x + \frac{b}{2a}\right)^2 = \frac{b^2 - 4ac}{4a^2}$$

Taking the square root on both sides, we obtain

$$x + \frac{b}{2a} = \frac{\sqrt{b^2 - 4ac}}{2a} \quad \text{or} \quad x + \frac{b}{2a} = -\frac{\sqrt{b^2 - 4ac}}{2a}$$

so that solving for x, we find

$$x = \frac{-b + \sqrt{b^2 - 4ac}}{2a} \quad \text{or} \quad x = \frac{-b - \sqrt{b^2 - 4ac}}{2a}$$

as required, which is more compactly expressed as $x = \frac{-b \pm \sqrt{b^2 - 4ac}}{2a}$.

Examples. *Positive discriminant.* When the quadratic equation is $f(x) = x^2 + 9x + 16$, the discriminant, Δ, is unambiguously positive since

$$\Delta = 9^2 - (4 \times 1 \times 16) = 81 - 64 = 17 > 0,$$

so that real roots exist, and are given by

$$x = \frac{-9 \pm \sqrt{9^2 - (4 \times 1 \times 16)}}{2 \times 1}$$

which we simplify to

$$x = \frac{-9 + \sqrt{17}}{2} \quad \text{or} \quad x = \frac{-9 - \sqrt{17}}{2}$$

Negative discriminant. When the quadratic equation is $f(x) = 3x^2 - 4x + 2$, we find that the discriminant is

$$\Delta = (-4)^2 - 4 \times 3 \times 2 = 16 - 24 = -8 < 0,$$

so that no real roots exist. Graphically, $f(x)$ lies above the horizontal axis for all values of $x \in \mathbb{R}$. The opposite argument applies for the quadratic equation $f(x) = -2x^2 + 5x - 4$, whose discriminant is $\Delta = 5^2 - [4 \times (-2) \times (-4)] = 25 - 32 = -7 < 0$, so that no real roots exist. In this case, $f(x)$ graphically lies below the horizontal axis for all values of $x \in \mathbb{R}$.

Zero discriminant. Finally, consider the quadratic equation $f(x) = 4x^2 - 28x + 49$. In this case, we find that the discriminant is zero, $\Delta = (-28)^2 - (4 \times 4 \times 49) = 784 - 784 = 0$, so that real root exists. To find them, a common trick is to factorize $f(x) = 0$ into $(2x - 7)^2 = 0$ to yield a unique root of x on the real line, that is, $x = \frac{7}{2}$.

In summary, we find the following cases, according to the discriminant's sign, $\Delta \equiv b^2 - 4ac$:

1. If $\Delta > 0$, there are two distinct real roots.
2. If $\Delta = 0$, there is only one real root.
3. If $\Delta < 0$ there are no real roots since the quadratic function $f(x) = ax^2 + bx + c$ lies above (below) the horizontal axis if $a > 0$ ($a < 0$, respectively).

A.6 Cramer's Rule

In the book, we encounter systems of k linear equations with k variables. For instance, $k = 2$ variables (x_1 and x_2) and two linear equations $f(x_1, x_2)$ and $g(x_1, x_2)$. We often use the so-called "substitution method," where we solve for one of the variables in one of the equations, and then plug the result into the remaining equation. However, some systems of equations are solved using Cramer's rule, which are more convenient than the substitution method. We present Cramer's rule below where, for simplicity, we consider $k = 2$ variables and two linear equations:

$$ax_1 + bx_2 + c = 0$$
$$dx_1 + ex_2 + f = 0$$

where $ae - bd \neq 0$. Rearranging the above system of equations in matrix form, we have

$$\begin{bmatrix} a & b \\ d & e \end{bmatrix} \begin{bmatrix} x_1 \\ x_2 \end{bmatrix} = -\begin{bmatrix} c \\ f \end{bmatrix}$$

Applying the Cramer's rule, we find

$$x_1 = -\frac{\begin{bmatrix} c & b \\ f & e \end{bmatrix}}{\begin{bmatrix} a & b \\ d & e \end{bmatrix}} = -\frac{ce - bf}{ae - bd} = \frac{bf - ce}{ae - bd}$$

$$x_2 = -\frac{\begin{bmatrix} a & c \\ d & f \end{bmatrix}}{\begin{bmatrix} a & b \\ d & e \end{bmatrix}} = -\frac{af - cd}{ae - bd} = \frac{cd - af}{ae - bd}$$

where to solve for x_1 (x_2), we replace the first (second) column of the matrix in the numerator with the vector on the right-hand side, divided by its determinant, $ae - bd$.

As an example, consider the following two equations.

$$x_1 + 3x_2 + 2 = 0$$
$$2x_1 + x_2 + 1 = 0$$

Applying the Cramer's rule, we find

$$x_1 = -\frac{\begin{bmatrix} 2 & 3 \\ 1 & 1 \end{bmatrix}}{\begin{bmatrix} 1 & 3 \\ 2 & 1 \end{bmatrix}} = -\frac{2 - 3}{1 - 6} = -\frac{1}{5}$$

$$x_2 = -\frac{\begin{bmatrix} 1 & 2 \\ 2 & 1 \end{bmatrix}}{\begin{bmatrix} 1 & 3 \\ 2 & 1 \end{bmatrix}} = -\frac{1 - 4}{1 - 6} = -\frac{3}{5}$$

A.7 Differentiation

We are often interested in measuring how quickly a function increases or decreases, graphically understood as the slope of the function at a given point. In particular, starting at point x with image $f(x)$, we consider a given increase in x, denoted as Δx, whose output is $f(x + \Delta x)$. The derivative of function f is, then, defined as the "rise over run," as follows

$$f'(x) = \lim_{\Delta x \to 0} \frac{f(x + \Delta x) - f(x)}{\Delta x}.$$

When this limit exists, we say that the function is differentiable. When $f'(x) > 0$, we say the function "increases in x," when $f'(x) < 0$ we say the function "decreases in x," and when $f'(x) = 0$, we say the function is "constant in x." We next find derivatives for common functions:

1. If $f(x) = a$, then $f'(x) = 0$, implying that the derivative of a constant function is zero.
2. If $f(x) = a + bx$, then $f'(x) = b$, so the derivative of a linear function is its slope, b, and it is unaffected by the function's vertical intercept, a.
3. If $f(x) = ax^2 + d$, then $f'(x) = 2ax$, meaning that the slope of the function is $f'(0) = 0$ at the origin, but increases in x at a rate of $2a$.
4. If $f(x) = x^a$, then $f'(x) = ax^{a-1}$.
5. If $f(x) = a + b\ln(x)$, then $f'(x) = \frac{b}{x}$.
6. If $f(x) = a^x$ where $a > 0$, then $f'(x) = a^x \ln(a)$.
7. If $f(x) = e^x$, then $f'(x) = e^x$.

Some basic properties about differentiation used in the book include the following:

1. **Chain rule.** Consider two real-valued functions f and g. The composite function $f(g(x))$ has derivative $f'(g(x)) = f'(g(x))g'(x)$. For instance, if $f(g(x)) = g(x)^n$, then $f'(g(x)) = n[g(x)]^{n-1}g'(x)$.
2. **Basic operations.**
   - *Multiplication by a constant.* If $f(x)$ is a real-valued function, then the product $g(x) = af(x)$ has derivative $g'(x) = af'(x)$.
   - If f and g are two real-valued functions:
     - Their sum $f(x) + g(x)$, has derivative $f'(x) + g'(x)$.
     - Their difference $f(x) - g(x)$ has derivative $f'(x) - g'(x)$.
     - *Product rule.* Their product $f(x)g(x)$ has derivative $f'(x)g(x) + f(x)g'(x)$.
     - *Quotient rule.* Their quotient $\frac{f(x)}{g(x)}$ has derivative $\frac{f'(x)g(x) - f(x)g'(x)}{[g(x)]^2}$ where $g'(x) \neq 0$.
3. **L'Hopital's rule.** Consider two real-valued functions f and g such that their limit at point c satisfies $\lim_{x \to c} \frac{f(x)}{g(x)} = \frac{0}{0}$ or $\frac{\pm\infty}{\pm\infty}$. Then, $\lim_{x \to c} \frac{f(x)}{g(x)} = \lim_{x \to c} \frac{f'(x)}{g'(x)}$ where $g'(x) \neq 0$.

A.7.1 First- and Second-order Conditions

The use of derivatives is pervasive in economics problems to identify a maximum or minimum. Maxima and minima are generally known as "extremum" points. If the function is differentiable, a necessary condition for an extremum at point x^* is that its derivative of the function at that point is zero, $f'(x^*) = 0$. This condition is often referred to as the first-order necessary condition, or first-order condition for short.

If, in addition, the function is concave at point x^*, which entails that its second-order derivative is negative at that point, $f''(x^*) < 0$, then we say that x^* is a maximum. If, instead, the function is convex at that point, $f''(x^*) > 0$, we say that x^* is a minimum. These conditions are known as second-order conditions, since we use the second-order derivative, and help us identify local maxima or minima, where "local" means that there may be other maxima or minima.

However, if the function is concave (convex) over all its domain, i.e., $f''(x) < 0$ for all x ($f''(x) > 0$ for all x), we can then guarantee that point x^*, which satisfies the first-order condition $f'(x^*) = 0$, is a global maximum (minimum, respectively). Finally, if $f''(x) = 0$, then the function is indeterminate, meaning that an inflection point is neither maximum nor minimum, such as the function $f(x) = x^3$, whose first-order derivative is $f'(x) = 3x^2$, implying that $f'(x) = 0$ at $x = 0$.

A.7.2 Inverse Function Theorem

The inverse function theorem says that if a function $f(x)$ is a continuously differentiable function with non-zero derivative at point $x = a$, $f'(a) \neq 0$, then the derivative of the inverse function at $b = f(a)$, which we denote as $(f^{-1})'(b)$, is the reciprocal of the derivative of function f at point $x = a$, that is,

$$(f^{-1})'(b) = \frac{1}{f'(a)} = \frac{1}{f'(f^{-1}(b))}$$

since $f^{-1}(b) = a$. In a linear function $f(x) = a + bx$, we can see that $f'(x) = b$ and its inverse is $(f^{-1})' = \frac{1}{b}$. The opposite route yields the same result: the inverse of the linear function is $f^{-1}(x) = \frac{x}{b} - \frac{a}{b}$, and its derivative is $\frac{1}{b}$. For another example, consider function $f(x) = \ln x$, where $x > 0$, so that its inverse is $f^{-1}(x) = e^x$, and the derivative of the inverse is $f^{-1\prime}(x) = e^x$, so that

$$\frac{1}{f'(f^{-1}(x))} = \frac{1}{f'(e^x)} = \frac{1}{\frac{1}{e^x}} = e^x$$

confirming the inverse function theorem.

In our analysis of equilibrium bids in the auctions (chapter 9), this theorem says that if bidder i's bidding function $b_i(v_i)$ is continuous and differentiable at her valuation for the object, v_i, then the derivative of the inverse of this function is

$$\frac{\partial b_i^{-1}(x_i)}{\partial x_i} = \frac{1}{b_i'(v_i)} = \frac{1}{b'\left(b_i^{-1}(x_i)\right)}$$

since the inverse of the bidding function is bidder i's valuation, $b_i^{-1}(x_i) = v_i$.

A.8 Integration

Consider a real-valued function $f : \mathbb{R} \to \mathbb{R}$. The integral of f is $F(x) = \int f(x)dx$, implying that the derivative of F is f, that is $F'(x) = f(x)$. Graphically, the integral measures the area under function $f(x)$ for all points in f's domain.

Similarly, a definite integral focuses on an interval of this domain, such as all $x \in [x_0, x_1]$, so that $F(x) = \int_{x_0}^{x_1} f(x)dx$, and it is graphically understood as the area under function $f(x)$ between x_0 and x_1. This is often presented with the so-called fundamental theorem of calculus, as follows $\int_{x_0}^{x_1} f(x)dx = F(x_1) - F(x_0)$. For instance, if the function is linear, $f(x) = a + bx$, its integral is $F(x) = ax + \frac{bx^2}{2}$, and its definite integral between $x_0 = 0$ and $x_1 = y$ is

$$\int_0^y (a + bx)dx = ax + \frac{bx^2}{2}\Big|_0^y = ay + \frac{by^2}{2}.$$

We next list some common integrals (in all cases, C denotes the constant of integration):

1. If $f(x) = a$, then $\int a\,dx = ax + C$.
2. If $f(x) = x$, then $\int x\,dx = \frac{x^2}{2} + C$.
3. If $f(x) = x^a$ where $a > 0$, then $\int x^a dx = \frac{x^{a+1}}{a+1} + C$.
4. If $f(x) = \frac{1}{x}$, then $\int \frac{1}{x}dx = \ln(x) + C$.
5. If $f(x) = e^x$, then $\int e^x dx = e^x + C$.
6. If $f(x) = a^x$ where $a > 0$, then $\int a^x dx = \frac{a^x}{\ln(a)} + C$.
7. If $f(x) = \ln(x)$, then $\int \ln(x)dx = x\ln(x) - x + C$.

Some basic properties about integration:

1. *Multiplication by a constant.* If $f(x)$ is a real-valued function, then the product $af(x)$ has integral $a \int f(x)dx$
2. If f and g are two real-valued functions:
   - Their sum $f(x) + g(x)$ has integral $\int f(x)dx + \int g(x)dx$.
   - Their difference $f(x) - g(x)$ has integral $\int f(x)dx - \int g(x)dx$.
   - *Integration by parts.* From our previous discussion, we know that the derivative of the product of two functions, $fg = f(x)g(x)$, is $(fg)' = f'g + fg'$. Rearranging the above results, we obtain $fg' = (fg)' - f'g$. Integrating both sides of this equation, yields

$$\int (fg')dx = fg - \int (f'g)dx.$$

For instance, if we seek to integrate the product xe^x, we can apply integration by parts, defining $f \equiv x$ and $g \equiv e^x$, which implies that $df = f' = 1$

and $dg = g' = e^x$. Inserting these elements in the above equation, yields

$$\int xe^x dx = xe^x - \int (e^x) dx$$
$$= xe^x - e^x + C.$$

References

Aragonès, E., A. Postlewaite, and T. Palfrey (2007) "Political Reputations and Campaign Promises," Journal of the European Economic Association, 5(4), pp. 846–884

Aumann, R. (1974) "Subjectivity and Correlation in Randomized Strategies," Journal of Mathematical Economics, 1(1), pp. 67–96.

Azar, O. (2015) "A Linear City Model with Asymmetric Consumer Distribution," PLoS ONE, 10(6).

Banks, J. S., and J. Sobel (1987) "Equilibrium selection in signaling games," Econometrica, 55, pp. 647–661.

Banks, J. S., F. C. F. Camerer, and D. Porter (1994) "Experimental Tests of Nash Refinements in Signaling Games," Games and Economic Behavior, 6, pp. 1–31.

Bertrand, J. (1883) "Théorie Mathématique de la Richesse Sociale," Journal des Savants, pp. 499–508.

Blume, A., O. Board, and K. Kawamura (2007) "Noisy Talk," Theoretical Economics, 2, pp. 395–440.

Blume, A., E. K. Lai, and W. Lim (2020) "Strategic Information Transmission: A Survey of Experiments and Theoretical Foundations." in: *Handbook of Experimental Game Theory*, eds. C. M. Capra, R. T. A. Croson, M. L. Rigdon, and T. S. Rosenblat, Edward Elgar Publishers.

Bolton, G., and A. Ockenfels (2000) "ERC: A Theory of Equity, Reciprocity, and Competition," American Economic Review, 90(1), pp. 166–193.

Border, K. C. (1985) *Fixed Point Theorems with Applications to Economics and Game Theory*, Cambridge University Press, Cambridge.

Brandts, J., and C. A. Holt (1992) "An Experimental Test of Equilibrium Dominance in Signaling Games," American Economic Review, 82, pp. 1350–1365.

Camerer, C. F. (2003) *Behavioral Game Theory: Experiments in Strategic Interaction* (The Roundtable Series in Behavioral Economics), Princeton University Press.

Camerer, C. F., T. Ho, and J-K Chong (2004) "A Cognitive Hierarchy Model of Games," The Quarterly Journal of Economics, 119(3), pp. 861–898.

Chakraborty, A. and R. Harbaugh (2007) " Comparative Cheap Talk," Journal of Economic Theory, 132, pp. 70–94.

Chen, Y., N. Kartik, and J. Sobel (2008) " Selecting Cheap-talk Equilibria," Econometrica, 76, pp. 117–136.

Chisik, R. A., and R. J. Lemke (2006) "When Winning is the Only Thing: Pure Strategy Nash Equilibria in a Three-candidate Spatial Voting Model," Social Choice and Welfare, 26, pp. 209–215.

Cho, I.-K., and D. M. Kreps (1987) "Signaling Games and Stable Equilibria," Quarterly Journal of Economics, 102(2), pp. 179–221.

Congleton, R. (2003) "The Median Voter Model," in: *The Encyclopedia of Public Choice*, eds. C. K. Rowley and F. Schneider, Kluwer Academic Press.

Corbae, D., M. D. Stinchombe, and J. Zeeman (2009) *An Introduction to Mathematical Analysis for Economic Theory and Econometrics*, Princeton University Press, Princeton.

© The Editor(s) (if applicable) and The Author(s), under exclusive license to Springer Nature Switzerland AG 2023
A. Espinola-Arredondo and F. Muñoz-Garcia, *Game Theory*,
https://doi.org/10.1007/978-3-031-37574-3

Cournot, A. A. (1838) *Cournot, Recherches sur les Principes Mathematiques de la Theorie des Richesses*, Hachette, Paris. Translated to English as: *Researches into the Mathematical Principles of the Theory of Wealth* (1927), by N. T. Bacon, translator, Macmillan.

Cox, G. W. (1987) "Electoral Equilibrium Under Alternative Voting Institutions," American Journal of Political Science, 31, pp. 82–108

Crawford, V. (1998) "A Survey of Experiments on Communication via Cheap Talk," Journal of Economic Theory, 78(2), pp. 286–298.

Crawford, V. P., and J. Sobel (1982) "Strategic Information Transmission," Econometrica, 50(6), pp. 1431–1451.

Crawford, V. P., M. A. Costa-Gomes, and N. Irriberri (2013) "Structural Models of Nonequilibrium Strategic Thinking: Theory, Evidence, and Applications," Journal of Economic Literature, 51(1), pp. 5–62.

Dailey, B., and B. Green (2014) "Market Signaling with Grades," Journal of Economic Theory, 151, pp. 114–145.

Dasgupta, P., and E. Maskin (1986) "The Existence of Equilibrium in Discontinuous Economic Games, II: Applications," Review of Economic Studies, 46, pp. 27–41.

de Groot Ruiz, A., T. Offerman, and S. Onderstal (2015) "Equilibrium Selection in Experimental Cheap Talk Games," Games and Economic Behavior, 91, pp. 14–25.

Dessein, W. (2002) "Authority and Communication in Organizations," Review of Economic Studies, 69, pp. 811–838.

Dixit, A. K., S. Skeath, and D. H. Reiley Jr. (2015) *Games of Strategy*, Fourth edition, W.W. Norton & Company.

Downs, A. (1957) "An Economic Theory of Political Action in a Democracy," Journal of Political Economy, 65(2), pp. 135–150.

Dunaway, E., and F. Munoz-Garcia (2020) "Campaign Limits and Policy Convergence with Asymmetric Agents," Public Choice, 184, pp. 429–461.

Dutta, P. K., and W. Vergote (2022) *Strategies and Games, Theory and Practice*, Second edition, MIT Press.

Farrell, J. (1993) "Meaning and Credibility in Cheap-talk Games," Games and Economics Behavior, 5, pp. 514–531.

Farrell, J., and R. Gibbons (1989) "Cheap Talk with Two Audiences," American Economic Review, 79, pp. 1214–1223.

Farrell, J., and M. Rabin (1996) "Cheap Talk," The Journal of Economic Perspectives, 10(3), pp. 103–118.

Fehr, E., and K. M. Schmidt (1999) "A Theory of Fairness, Competition, and Cooperation," The Quarterly Journal of Economics, 114(3), pp. 817–868.

Feltovich, N., R. Harbaugh, and T. To (2002) " Too Cool for School? Signaling and Countersignaling," RAND Journal of Economics, 33(4), pp. 630–649.

Fudenberg, D., and J. Tirole (1991) *Game Theory*, The MIT Press, Cambridge, Massachussets.

Ganguly, C., and I. Ray (2005) "On Mediated Equilibria of Cheap-talk Games," working paper, University of Birmingham.

Gibbons, R. (1992) *Game Theory for Applied Economists*, Princeton University Press.

Gilligan, T., and K. Krehbiel (1987) "Collective Decision-making and Standing Committees: An Informational Rationale for Restrictive Amendment Procedures," Journal of Law, Economics, and Organization, 3, pp. 287–335.

Glicksberg, I. L. (1952) "A Further Generalization of the Kakutani Fixed Point Theorem with Application to Nash Equilibrium Points," Proceedings of the National Academy of Sciences, 38, pp. 170–174.

Gonzalez-Diaz, J., and M. A. Melendez-Jimenez (2014) "On the Notion of Perfect Bayesian Equilibrium," Top, 22, pp. 128–143.

Grossman, G. M., and E. Helpman (2001) *Special Interest Politics*, The MIT Press, Cambridge, Massachussets.

Harrington, J. E. (2006) "How Do Cartels Operate?," Foundations and Trends in Microeconomics, 2(1), pp. 1–105.

References

Harrington, J. E. (2014) *Games, Strategies, and Decision Making*, Worth Publishers.

Harsanyi, J. C. (1967) "Games with Incomplete Information Played by "Bayesian" Players, Part I–III. Part I. The Basic Model." Management Science, 14, pp. 159–182.

Harsanyi, J. C. (1973) "Oddness of the Number of Equilibrium Points: A New Proof," International Journal of Game Theory, 2, pp. 235–250.

Hotelling, H. (1929) "Stability in Competition," The Economic Journal, 39, pp. 41–57.

Kandori, M. (2002) "Introduction to Repeated Games with Private Monitoring," Journal of Economic Theory, 102(1), pp. 1–15.

Klein, M. (2001) *Mathematical Methods for Economics*, Second edition, Pearson Publishing.

Kreps, D., P. Milgrom, J. Roberts, and R. Wilson (1982) "Rational Cooperation in the Finitely-Repeated Prisoners' Dilemma," Journal of Economic Theory, 27(2), pp. 245–252.

Krishna, V., and J. Morgan (2001) "A Model of Expertise," Quarterly Journal of Economics, 116, pp. 747–775.

Krishna, V., and J. Morgan (2004) "The Art of Conversation: Eliciting Information from Experts Through Multi-stage Communication," Journal of Economic Theory, 117, pp. 147–179.

Krishna, V., and J. Morgan (2008) "Cheap Talk," in: *The New Palgrave Dictionary of Economics*, eds. Steven N. Durlauf and Lawrence E. Blume, Second Edition, Palgrave Macmillan.

Kuhn, H. W. (1953) "Extensive Games and the Problem of Information" , in: *Contributions to the Theory of Games II*, eds. H. W. Kuhn and A. W. Tucker, Annals of Mathematics Studies, 28, , pp. 193–216, Princeton University Press.

Levenstein, M. C., and V. Y. Suslow (2006) "What Determines Cartel Success?," Journal of Economic Literature, 44(1), pp. 43–95.

Luce, R. D. (1959) *Individual Choice Behavior*, John Wiley, New York.

Luce, R. D., and H. Raiffa (1957) *Games and Decisions: Introduction and Critical Survey*, John Wiley & Sons, Inc., New York.

Mailath, G., and L. Samuelson (2006) *Repeated Games and Reputations: Long-run Relationships*, Oxford University Press, Oxford.

Melumad, N. D., and T. Shibano (1991) " Communication in Settings with no Transfers," Rand Journal of Economics, 22, pp. 173–198.

Myerson, R. (1978) "Refinements of the Nash Equilibrium Concept," International Journal of Game Theory, 7, pp. 73–80.

Nagel, R. (1995) "Unraveling in Guessing Games: An Experimental Study," American Economic Review, 85(5), pp. 1313–1326.

Nash, J. F. (1950) "Non-Cooperative Games," Annals of Mathematics, 54(2), pp. 286–295.

Osborne, M. and A. Rubinstein (1994) *A Course in Game Theory*, The MIT Press.

Osborne, M. (2003) *An Introduction to Game Theory*, Oxford University Press.

Ottaviani, M., and P. T. Sorensen (2006) " Reputational Cheap Talk," RAND Journal of Economics, 37(1), pp. 155–175.

Pearce, D. (1984) "Rationalizable Strategic Behavior and the Problem of Perfection," Econometrica, 52, pp. 1029–1050.

Rasmusen, E. (2006) *Games and Information: An Introduction to Game Theory*, Blackwell Publishing.

Schelling, T. C. (1960) *The Strategy of Conflict*, First edition, Harvard University Press, Cambridge.

Selten, R. (1975) "Reexamination of the Perfectness Concept for Equilibrium Points in Extensive Games," International Journal of Game Theory, 4, pp. 301–324.

Simon, C., and L. Blume (1994) *Mathematics for Economists*, Worth Publishers.

Smith, S. B. (2015) *Chance, Strategy, and Choice. An Introduction to the Mathematics of Games and Elections*, Cambridge University Press, Cambridge.

Spence, M. (1973) "Job Market Signaling," The Quarterly Journal of Economics, 87(3), pp. 355–374.

Stahl, D. O., and P. W. Wilson (1995) "On Players' Models of Other Players: Theory and Experimental Evidence," Games and Economic Behavior, 10(1), pp. 218–254.

Tadelis, S. (2013) *Game Theory, An Introduction*, Princeton University Press.

Tirole, J. (1988) *The Theory of Industrial Organization*, The MIT Press, Cambridge.

Von Neumann, J. (1928) "Zur Theorie der Gesellschaftsspiele," Mathematische Annalen 100, pp. 295–320.

Wainwright, K., and A. Chiang (2004) *Fundamental Methods of Mathematical Economics*, Fourth Edition, McGraw-Hill Education.

Watson, J. (2013) *Strategy: An Introduction to Game Theory*, Third edition, Worth Publishers.

Wilson, R. (1971) "Computing Equilibria in N-person Games," SIAM Journal of Applied Mathematics, 21, pp. 80–87.

Wittman, D. (1983) "Candidate Motivation: A Synthesis of Alternative Theories," The American Political Science Review, 77(1), pp. 142–157.

Zermelo, E. (1913) "On an Application of Set Theory to the Theory of the Game of Chess," in: "Zermelo and the Early History of Game Theory," Games and Economic Behavior, 34, pp. 123–137, eds. U. Schwalbe and P. Walker.

Index

A

Actions, 1, 9, 186, 221, 222, 236, 260, 263, 309, 310, 316, 337, 361, 366, 367, 404
Allocation mechanism, 282
Alternating-offer bargaining game with infinite periods, 181, 193
Altruistic players, 191
An eye for an eye (in repeated games), 232
Anticoordination game, 26, 27, 30, 40, 60, 63, 67, 68, 109, 112, 241
Assignment rule, 282, 284, 287, 292
Asymmetric beliefs, 355
Asymmetric games, 28
Auction types, 301

B

Backward induction, 160, 161, 165, 168, 169, 173, 175, 177, 180, 188, 189, 205, 207, 209, 210, 231
Bargaining, 175, 177
Bargaining under incomplete information, 272
Battle of market shares game, 189, 191
Battle of the sexes game, 23–26, 29, 30, 40, 60–63, 65, 68, 116, 141
Bayes' rule, 132, 312, 317, 318, 321–323, 326, 328, 329, 335, 341, 344, 367, 370, 382, 390, 411, 413, 416, 418
Bayesian Nash Equilibrium (BNE), 11, 303, 309, 310, 314–316, 396
Bayesian-normal form, 254, 275
Beer-Quiche game, 347, 348
Behavioral strategies, 184, 366
Beliefs, 52–54, 57, 151, 310–313, 316, 317, 323, 325, 331, 332, 338, 359, 360, 363, 366, 367, 369–371, 382, 388, 389, 400, 403
Belief updating, 382, 390

Best response, 47, 49, 50, 52–54, 58, 59, 61, 62, 64, 77–79, 82, 83, 85, 87, 92, 106, 109, 113, 115, 129, 133, 134, 139, 157, 160, 207, 209, 210, 254, 257–259, 263, 268, 404
Best responses, graphical representation, 113
Bid, 275, 281, 282, 285
Bid shading, 287, 290, 300, 305
Binary, 6, 316, 415, 420, 430
Brinkmanship game, 351, 373
Budget constrained bidders, 306
Burning money game, 191

C

Centipede game, 193
Chain-store paradox, 250
Cheap talk games, 409, 433
Cheap talk with continuous messages and responses, 420
Cheap talk with discrete messages and responses, 410
Cheap talk with discrete messages but continuous responses, 415
Chicken game, 69, 105, 116, 129, 133
Closed rule, 429, 433
Collusion, 204, 224, 227, 229, 238
Collusion with cost shocks, 247
Collusion with imperfect monitoring, 245–247
Common pool resource, 101, 102, 196, 253
Common value auctions, 300, 301, 307
Comparative statics, 179, 197, 306, 307, 425
Concave, 51, 255, 299, 305
Concealing information, 310, 409
Conditional probability, 279, 433
Constant-sum games, 123
Continuous, 138, 255, 260, 361, 410, 420
Continuous responses, 365, 415, 420
Continuous strategy space, 3, 6, 50, 52, 108, 141, 151, 365

Contracts when effort is observable, 200
Convex, 139, 140, 197, 218, 219, 381
Conveying information, 310
Cooperative outcome, 212, 214, 217, 224, 231, 233, 234, 237, 248, 250
Coordination game, 23–25, 27, 30, 60, 61, 68
Correlated equilibrium, 129–133
Correspondence, 95, 139, 185
Cost differentials, 346, 350, 373
Cost externalities, 98, 102
Cost-reducing investment, 199, 200
Countersignaling, 407
Cramer's rule, 440

D
D1 Criterion, 360, 363, 366, 379, 388, 389, 397, 401–406
Delegation, 429, 432
Deletion order, 36
Differentiation, 77, 87, 99
Discontinuous, 67
Discount factor, 177, 180, 183, 184, 211, 214–216, 221, 223, 227–230, 235, 236, 238, 242, 246–248
Discrete responses, 365
Discrete strategy space, 48, 67, 77, 151
Disutility, 192, 200
Dominant strategy, 39, 40, 60, 61, 204, 248, 284, 286, 287, 335, 353
Dominated strategy, 13, 14, 16, 18, 19, 23–25, 27, 30, 40, 60, 116, 134, 146
Downward deviation, 285

E
Efficiency (in auctions), 292
Electoral competition game, 96, 97
Endogenous location game, 198
Entry deterring, 201
Entry fees, 304
Envious players, 192
Equilibrium behavior, 8, 10, 121, 151, 152, 160, 161, 164, 167–169, 182, 191, 199, 207, 221, 257, 314, 325, 379
Equilibrium bid, 281, 282, 286, 288, 292, 293, 301
Equilibrium dominance, 13, 46, 332
Equilibrium existence, 168
Equilibrium number of partitions (in cheap talk games), 423, 426
Equilibrium prediction, 11, 12, 19, 24, 27, 32, 33, 36, 38, 42, 60, 61, 63, 116, 121, 128, 267, 359

Equilibrium refinements, 402
Equilibrium robustness, 34
Equilibrium selection, 68
Equilibrium uniqueness, 168
Ex-ante and ex-post stability, 259
Ex-ante symmetry, 259
Expected revenue (in auctions), 281
Exponential distribution, 304, 307
Externalities, 98, 102

F
Feasible and individually rational payoffs (FIR), 218
Feasible payoffs (FR), 218
Finitely repeated game, 203
First-mover, 175, 310
First-mover advantage, 171, 194
First-order condition, 289, 290, 302, 306
First-price auctions, 282, 287
Fixed-point theorem, 138–140
Folk theorem, 217, 221, 223, 224, 229, 232
Follower, 7, 9, 169–171, 173, 175, 194, 433
Free riding, 77
Function, 52, 57, 77–79, 82, 84, 85, 92, 95, 113, 114, 139, 141, 169, 171, 173, 230, 255, 263, 268, 281, 287, 291–293, 297, 299, 316, 409, 415

G
Game of chicken, 26, 27, 29, 30, 63, 64, 70, 142, 187, 241, 275
Game trees, 1, 5, 7, 9, 151, 152, 158, 161, 168, 257, 309, 320
Gift exchange game, 192
Grim-trigger strategy (GTS), 203, 204, 212–214, 220–222, 225, 228, 229, 231, 232, 234–237

H
Hawk-Dove game with incomplete information, 276
Heterogeneous goods, 77, 82, 87, 194
Homogeneous goods, 77, 84, 98
Hotelling game, 197, 198

I
Imperfect information, 9, 257
Imperfect monitoring (in repeated games), 204, 236
Impure public good, 101

Incomplete information, 11, 253, 254, 256, 257, 264, 267, 268, 276, 277, 281, 309, 310, 312, 314, 316, 317, 332, 352, 379, 393, 408
Incredible beliefs, vii
Individually rational payoffs (IR), 218, 220
Infinitely repeated game, 203, 229
Information set, 9, 10, 151, 155, 156, 158, 161, 165, 184, 204, 260, 271, 310, 313, 314, 316, 317, 320, 321, 325, 329, 337, 338, 343, 366, 367, 370
Information transmission, 314, 395, 410, 420, 426, 429, 430, 433
Insensible beliefs, 331, 359, 366, 369, 379
Integration, 289, 290, 294
Integration by parts, 289, 290, 294
Interval lengths (in cheap talk games), 424, 427
Intrinsic value of education, 349, 373
Intuitive Criterion, 359–365, 372, 379, 387–389, 394, 395, 397, 401–403
Iterated Deletion of Strictly Dominated Strategies (IDSDS), 13, 15, 16, 18, 21, 23, 24, 33, 38–40, 44, 47, 52, 53, 55, 60, 63, 65
Iterated Deletion of Weakly Dominated Strategies (IDWDS), 36, 38, 39

J
Job applications game, 143
Joint profits, 195, 199, 200, 224, 229
Jury voting game, 278

L
Leader, 7, 169–171, 173, 175
Lemons problem, 272
Limit, 81, 135
Limit pricing, 407
Lobbying, 409, 410, 415, 420, 423, 425–427, 429, 430, 433
Lottery auction, 282, 299

M
Market entry game with incomplete information, 277
Matching pennies game, 65, 66, 124, 267, 276
Matrices, 1, 5, 6, 11
Maximum, 140, 425, 426
Mergers, 98
Minimal discount factor, 214–217, 227–230, 235, 236

Minimum, 72, 249
Mixed strategy, 107, 108, 134, 136, 184–186
Mixed strategy Nash equilibrium (msNE), 105
More information may hurt, 278
More than two available messages, 337, 338
More than two available responses, 310, 337
More than two types, 310, 342, 401
Multiple equilibria, 19

N
Nash equilibrium, 11, 33, 98
Nash existence theorem, 140
Never a best response (NBR), 52, 72
No bite, 11, 30, 33, 59, 61, 62, 72, 366, 379, 397
N players, 4, 43

O
Off-the-equilibrium beliefs, 317, 319, 320, 325, 329, 331, 332, 359, 360, 366, 367, 369, 371, 379, 382, 387, 394, 403
Open rule, 429, 433
Output, 4, 12, 15, 47, 50, 52, 57, 77–79

P
Pareto coordination game, 25, 26, 142, 186, 188, 240
Pareto dominance, 68, 69
Pareto optimal, 12
Pareto optimality, 22
Partitions (in cheap talk games), 423, 426, 429
Payment rule, 283, 284, 287
Payoffs, 1, 5–7, 10–12, 14, 15, 19, 22–24, 29–31, 34, 49, 59, 61, 64, 66, 67, 69, 106, 117, 118, 124, 129, 136, 160, 167, 179, 180, 183, 203, 204, 208, 209, 217, 220, 230, 262, 284, 309, 312–315, 322, 333, 339, 343, 346, 348, 352, 410, 415
Perfect Bayesian Equilibrium (PBE), 11
Permanent punishments, 243, 244
Players, 1, 3, 4, 6, 12, 16, 24, 27, 50, 62, 66, 79, 92, 107, 108, 129, 136, 253–255, 257, 259, 260, 281, 309, 313, 335, 367, 379
Player's type, 253, 254, 312
Politician, 410, 415, 420, 425, 426, 429, 430
Pooling strategy profile, 319–321, 323–325, 328, 331, 333, 334, 346, 370, 371, 373, 390, 414

Posterior beliefs, 311, 328, 390, 413, 418
Preference alignment, 420, 426
Preplay offer, 188
Price, 77, 82, 85, 141, 156, 175
Price competition (Bertrand model), 77, 84
Prior beliefs, 311, 319, 328
Prisoner's Dilemma game, 20, 22, 29, 39, 40, 60, 65, 69, 116, 146, 203, 204, 214, 218, 221, 223, 230, 232
Private recommendations, 131
Probability distribution over types, 254, 311
Productivity-enhancing education, 346
Profit function, 2, 51, 79, 89, 225
Proper equilibrium, 64, 136, 137
Proper subgame, 205
Public good game, 90, 248
Public good game with incomplete information, 274
Public recommendations, 131

Q

Quadratic equation, 415
Quadratic loss function, 415, 416
Quantity competition (Cournot model), 50, 77, 78, 83, 98, 170

R

Randomization, 14, 33, 67, 107, 116, 127, 184, 335
Randomizations in IDSDS, 30
Rationalizability, 53–56
Repeated games, 203, 212, 221, 231
Repeated public good game, 248
Revenue equivalence principle, 282, 299
Risk averse bidders, 305
Risk dominance, 64, 69
Robustness to payoff changes, 11
Root, 7, 161

S

Second-mover, 6, 162, 310, 316
Second-mover advantage, 188
Second-order condition, 51
Second-price auction, 281, 283, 284, 304
Security strategies (max-min), 121, 124, 126, 128
Selten's horse, 355
Semi-separating strategy profile, 335
Separating strategy profile, 310, 318, 321–323, 325, 327, 328, 333, 334, 341, 368, 382, 420, 423

Sequence, 134, 136, 211, 233, 367, 371
Sequential anticoordination game, 187
Sequential equilibrium (SE), 317, 366
Sequentially irrational behavior, 309, 310, 314
Sequential-move games of incomplete information, 310, 312, 316
Sequential Pareto coordination game, 186
Sequential Prisoner's Dilemma game, 186
Sequential public good game, 172
Sequential quantity competition, 169
Sequential quantity competition (Stackelberg model), 169
Set, 13, 52, 65, 77, 111, 133, 156
Short and nasty punishments (in repeated games), 235, 240
Signaling games, 310, 320, 321, 334, 338, 359, 361, 362, 365, 366, 375, 388, 409
Socially optimal, 34, 41, 61, 94, 248
Split-or-Steal game, 74
Stag Hunt game, 25, 26, 62, 64, 145
Stick-and-carrot strategy, 231
Stop-Go game, 143
Strategic complements, 85, 88
Strategic substitutes, 52, 78, 91
Strategies, 47–49, 53, 60, 62, 63, 86, 96, 108, 109, 112, 117
Strategy set, 3–5, 14, 16, 28, 52, 53, 55, 107, 108, 140, 260, 261, 317
Strategy space, 4–6, 8, 48, 53, 141, 186
Strictly competitive games, 121, 122, 124, 126
Strictly dominant equilibrium (SDE), 14, 39, 73
Strictly dominated strategies, 14, 16, 20, 21, 25, 30, 32, 40
Subgame, 158
Subgame Perfect Equilibrium (SPE), 11, 151, 249, 272, 309
Survive the Intuitive Criterion, 362, 372
Symmetric games, 28

T

Temporary punishments, 214, 243
Terminal nodes, 7, 160, 189, 204, 231, 415
Three players, 194, 355
Tiebreaking rules (in auctions), 303
Tit-for-tat, 250
Tree branches, 322
Tree rules, 151, 152
Trembling-hand Perfect Equilibrium (THPE), 133–135
Twice-repeated game, 210, 231, 241, 248

Two-period alternating offers bargaining game, 177
Two players, 30, 48, 261, 272, 351, 410

U
Ultimatum bargaining game, 175, 176, 178, 180
Uncooperative outcome, 210, 211, 217, 221
Underlining best responses, 61, 64, 160, 167, 207, 209, 210, 230, 262
Uniform distribution, 275, 300, 306
Upward deviation, 286

V
Valuation, 281, 283, 284, 290–292, 297
Violate the Intuitive Criterion, 362

W
War of attrition, 250
Weakly dominated strategies, 34–36
Winner's curse, 282, 300, 301, 303

Z
Zero-sum games, 123

SPRINGER NATURE

GPSR Compliance

The European Union's (EU) General Product Safety Regulation (GPSR) is a set of rules that requires consumer products to be safe and our obligations to ensure this.

If you have any concerns about our products, you can contact us on ProductSafety@springernature.com

In case Publisher is established outside the EU, the EU authorized representative is:

Springer Nature Customer Service Center GmbH
Europaplatz 3
69115 Heidelberg, Germany